Genomics, Personalized Medicine and Oral Disease

Stephen T. Sonis
Editor

Genomics, Personalized Medicine and Oral Disease

Springer

Editor
Stephen T. Sonis
Brigham and Women's Hospital
Boston, Massachusetts
USA

and

Biomodels, LLC
Watertown, Massachusetts
USA

ISBN 978-3-319-17941-4 ISBN 978-3-319-17942-1 (eBook)
DOI 10.1007/978-3-319-17942-1

Library of Congress Control Number: 2015947129

Springer Cham Heidelberg New York Dordrecht London

Printed on acid-free paper

Springer International Publishing AG Switzerland is part of Springer Science+Business Media (www.springer.com)

This book is dedicated to the memory of Dr. Marco Ramoni, my friend, teacher and collaborator.

Preface

Clinical applications of genomics and personalized medicine have transitioned from being on a theoretical wish list to becoming a transformational driver of medical practice. In the mere decade since the completion of the Human Genome Project, commercially available genetic tests now predict the behavior of certain breast cancers, help establish effective doses of Coumadin, determine the toxic potential of certain cancer drugs, or identify patients at risk for periodontitis. Many more clinical applications of genomics are in the pipeline which will have impact as diagnostics, risk predictors, or treatment determinants. Furthermore, gene-based therapy is maturing.

The mouth and its related structures represent a unique part of the human body. It is the only site in which two hard tissues (teeth and bone), different types of epithelium, and glandular tissue dynamically interact in an environment consisting of a myriad of microorganisms that is constantly bathed in a heterogeneous salivary fluid comprised of immunoglobulins, enzymes, and buffering agents. The opportunities for genes to influence the behavior of cells, saliva composition, and microorganisms are remarkable. Furthermore, the heterogeneity of its composition predisposes the mouth to a wide range of infectious, neoplastic and autoimmune diseases which range broadly in their frequency, severity and impact. And the mucosa and bone are frequent targets of toxicities of a range of therapeutic modalities. Genes govern the risk, course or response of almost every one of these conditions, whether their etiology is natural or iatrogenic.

The objective of this book is to catalyze the application of genomics to the diagnosis and treatment of oral diseases by comprehensively presenting focused discussions on the current state of knowledge. The first section of the book provides basic information about genetics, genomics, and personalized medicine and the informatical methods available to apply and organize genetic data so that it has clinical relevance. Recognizing the genetic robustness of the oral cavity, the introductory section also includes chapters on the oral microbiome and host genomics and response to infectious agents. The next two sections contain chapters which describe the genomics of specific oral diseases and conditions, including the genetic basis for mechanism and risk of treatment toxicities associated with cancer therapy and bisphosphonates. Four chapters focus on gene-based therapies and the pharmacogenomics applied to oral disease. The book concludes with a provocative summary which describes a comprehensive vision of the melding of genomics to personalized medicine and the potential actionable outcomes that will likely affect clinical practice in the upcoming years.

Despite the biological complexities of many oral diseases, their heterogeneous etiology, and the opportunity for the genome to impact their risk, course and response to therapy, there is no comprehensive (or even incremental) discussion of the topic among the many fine texts on genomics and personalized medicine. It is my hope that this book will fill that void.

Stephen T. Sonis, DMD, DMSc

Contents

Contributors

Amit Acharya Institute for Oral and Systemic Health, Marshfield Clinic Research Foundation, Marshfield, WI, USA

A. Razzaque Ahmed Center for Blistering Diseases, Boston, MA, USA

Priscila Maria de Andrade Medicine Department, Federal University of São Carlos (UFSCar), São Carlos, Brazil

Bruce J. Baum Gene Transfer Section, National Institute of Dental and Craniofacial Research, National Institutes of Health, Bethesda, MD, USA

Ricardo Carneiro Borra Genetic and Evolution Department, Federal University of São Carlos (UFSCar), São Carlos, Brazil

Nicola Luigi Bragazzi Department of Experimental Medicine (DIMES), Nanobiotechnology and Biophysics Laboratories (NBL), University of Genoa, Genoa, Italy

Murray H. Brilliant Center for Human Genetics, Marshfield Clinic Research Foundation, Marshfield, WI, USA

J. Robert Chang OsteoInvent Inc., West Roxbury, MA, USA

Gordon W. Duff University of Sheffield Medical School, Sheffield, UK

Leif W. Ellisen Massachusetts General Hospital Cancer Center and Harvard Medical School, Boston, MA, USA

Michael W. Epperly Department of Radiation Oncology, University of Pittsburgh Medical Center, Pittsburgh, PA, USA

Xiang Gao Department of Pharmaceutical Science, University of Pittsburgh, Pittsburgh, PA, USA

Geoffrey S. Ginsburg Center for Genomic Medicine, Duke Translational Medicine Institute, Durham, NC, USA

Ingrid Glurich Institute for Oral and Systemic Health, Marshfield Clinic Research Foundation, Marshfield, WI, USA

Joel S Greenberger Department of Radiation Oncology, University of Pittsburgh Medical Center, Pittsburgh, PA, USA

Jan Ching Chun Hu University of Michigan School of Dentistry, Ann Arbor, MI, USA

Kenneth Huttner Edimer Pharmaceuticals, Cambridge, MA, USA

Valerian Kagan Department of Environmental and Occupational Health, School of Public Health, University of Pittsburgh, Pittsburgh, PA, USA

Kenneth S. Kornman Interleukin Genetics, Waltham, MA, USA

Winston Patrick Kuo IES Diagnostics, Cambridge, MA, USA

Harvard Catalyst, Laboratory for Innovative Translational Technologies, Harvard Medical School, Boston, MA, USA

Department of Developmental Biology, Harvard School of Dental Medicine, Boston, MA, USA

Elizabeth J. Leslie Center for Craniofacial and Dental Genetics, Department of Oral Biology, School of Dental Medicine, University of Pittsburgh, Pittsburgh, PA, USA

Christopher J. Lessard Arthritis and Clinical Immunology Research Program, Oklahoma Medical Research Foundation, Oklahoma City, OK, USA

Department of Pathology, University of Oklahoma Health Sciences Center, Oklahoma City, OK, USA

Chia-Cheng Li Division of Genetics, Department of Medicine, Brigham and Women's Hospital and Harvard Medical School, Boston, MA, USA

Department of Oral Medicine, Infection and Immunity, Harvard School of Dental Medicine, Boston, MA, USA

Song Li Department of Pharmaceutical Science, University of Pittsburgh, Pittsburgh, PA, USA

Zhe Li Division of Genetics, Department of Medicine, Brigham and Women's Hospital and Harvard Medical School, Boston, MA, USA

Mary L. Marazita Center for Craniofacial and Dental Genetics, Department of Oral Biology, School of Dental Medicine, Department of Human Genetics, Graduate School of Public Health, Clinical and Translational Science Institute, and Department of Psychiatry, School of Medicine, University of Pittsburgh, Pittsburgh, PA, USA

Department of Human Genetics, Graduate School of Public Health, University of Pittsburgh, Pittsburgh, PA, USA

Center for Craniofacial and Dental Genetics, Department of Oral Biology, School of Dental Medicine, University of Pittsburgh, Pittsburgh, PA, USA

Department of Psychiatry and Clinical and Translational Science Institute, School of Medicine: University of Pittsburgh, Pittsburgh, PA, USA

Raman Mehrzad Center for Blistering Diseases, Boston, MA, USA

Reshma S Menon Department of Oral Medicine, Infection and Immunity, Harvard School of Dental Medicine, Boston, MA, USA

Manuela Moraru Immunogenetics & Histocompatibility, Instituto de Investigación Sanitaria Puerta de Hierro, Majadahonda, Spain

Claudio Nicolini Department of Experimental Medicine (DIMES), Nano-biotechnology and Biophysics Laboratories (NBL), University of Genoa, Genoa, Italy

Nanoworld Institute Fondazione EL.B.A Nicolini (FEN), Bergamo, Italy

Peter J. Polverini University of Michigan School of Dentistry, Ann Arbor, MI, USA

Enkhtsetseg Purev National Heart, Lung, and Blood Institute, National Institutes of Health, Bethesda, MD, USA

Astrid Rasmussen Arthritis and Clinical Immunology Research Program, Oklahoma Medical Research Foundation, Oklahoma City, OK, USA

Srinivas Vinod Saladi Massachusetts General Hospital Cancer Center and Harvard Medical School, Boston, MA, USA

Arne S. Schäfer Research Center ImmunoSciences (RCIS) CC03, Center of Dento-Maxillo-Facial Medicine Dept. of Periodontology and Synoptic Dentistry, Charité - University Medicine Berlin, Berlin, Deutschland

John R. Shaffer Department of Human Genetics, Graduate School of Public Health, University of Pittsburgh, Pittsburgh, PA, USA

Shawn Shetty Center for Blistering Diseases, Boston, MA, USA

Sanjay Shukla Center for Human Genetics, Marshfield Clinic Research Foundation, Marshfield, WI, USA

Kathy L. Sivils Arthritis and Clinical Immunology Research Program, Oklahoma Medical Research Foundation, Oklahoma City, OK, USA

Stephen T. Sonis Brigham and Women's Hospital, Boston, MA, USA

Biomodels, LLC, Watertown, MA, USA

Harvard School of Dental Medicine, Boston, MA, USA

Carlos Vilches Immunogenetics & Histocompatibility, Instituto de Investigación Sanitaria Puerta de Hierro, Majadahonda, Spain

Peter Wipf Department of Chemistry and Center for Chemical Methodologies and Library Development, University of Pittsburgh, Pittsburgh, PA, USA

Sook-Bin Woo Department of Oral Medicine, Infection and Immunity, Harvard School of Dental Medicine, Boston, MA, USA

Division of Oral Medicine and Dentistry, Brigham and Women's Hospital, Boston, MA, USA

Athanasios Zavras Department of Pediatric Dentistry, Boston University, Boston, USA

Henry M. Goldman School of Dental Medicine, Boston, MA, USA

Fundamentals of Genetics and Genomics

Stephen T. Sonis

Introduction

When we think about genetics, we typically think of patterns of inheritance that affect us and our environment. Will our kids have blue eyes or brown? Is there a risk of a particular disease in our family? Can I eat a gluten-dense pizza with impunity? Rarely do most of us give much thought to the biological processes that control the variables that impact phenotypes. But as more and more has been learned about biology, and especially human molecular biology, it has become clear that almost every physiologic function and risk of pathology, whether organic or behavioral is, at least in part, genetically controlled. Genetics studies the individual genes, while genomics is more dynamic in that it looks at the interaction between genes and genes and the environment.

Historically, the diagnosis and treatment of diseases has been based on the belief that if we effectively address the normal distribution of disease risk, diagnosis and response to treatment, we're effectively addressing the proper clinical problem. But is this true? Probably not. What if you developed a drug that was incredibly effective for a deadly disease, but only for those individuals who had a specific gene to metabolize the agent? And what if that gene was only present in 15 % of the population? If you designed a classic clinical trial in which you tested your drug against a placebo and only 15 % of the study population responded, the test might be deemed a failure.

Gregor Mendel, that famous Austrian Monk with the peas, published his Laws of Inheritance at around the time of the American Civil War. But it wasn't until 1902 that an English physician, Archibald Garrod, made a connection between genetic traits and disease risk when he noted familial patterns of an obscure condition called alkaptonuria. And while DNA was described in 1869, it wasn't until the early 1950's that its role in mediating heredity and its structure were noted. Since then

S. T. Sonis (✉)
Brigham and Women's Hospital, 75 Francis Street, 01778 Boston, MA, USA
e-mail: ssonis@partners.org; sonis@biomodels.com

S. T. Sonis (ed.), *Genomics, Personalized Medicine and Oral Disease*,
DOI 10.1007/978-3-319-17942-1_1

major advances in cell and molecular biology, genetics and genomics have established, not only the biological importance of the genome in affecting disease risk, but also have provided major opportunities for the translation of genomic information into clinically meaningful and actionable information. Genes have now been associated with cancer and heart disease risk and also with how patients respond to certain drugs, both therapeutically and in terms of toxicity or adverse reactions.

Recognizing the clinical potential of genomics, in 1990 the Office of Health and Environmental Research of the U.S. Department of Energy set about establishing the Human Genome Project. As described in a monograph on the topic by Palladino [15], the HGP had eight objectives of which the first four were probably the most directly relevant to clinical genomics:

- Create genetic and physical maps of human chromosomes.
- Identify the entire set of genes in the DNA of human cells.
- Determine the nucleotide sequence of DNA base pairs that comprise the human genome.
- And analyze genetics variations among humans, including the identification of single nucleotide polymorphisms (SNPs).

The objective of this book is to take a look at the most current information around genetics as it relates to oral diseases and to understand how all of this sophisticated science can be used in a way that ultimately is translatable to patients.

The Fundamentals: Chromosomes, DNA and Genes (Figure 1)

The genetic epicenter of the cell is its nucleus. In humans (not all animals have the same number of chromosomes) the nucleus contains 23 pairs of chromosomes of which 22 pairs are similar looking and called autosomes. The remaining pair are the sex-determining X and Y chromosomes (Fig. 2). Chromosome numbering reflects size—one is the largest. If the number of chromosomes is abnormal because of a consequence of faulty division, the result is an anomaly, often reflected as a defect at birth. Probably the most common example is Down syndrome in which there are 3, rather than a pair of chromosome 21.

The most significant structural component of chromosomes is DNA (deoxyribonucleic acid). Each chromosome contains a coiled strand of DNA which is wound around an alkaline protein core of histones. The unit of DNA and histone forms a fiber which is termed chromatin. One complete copy of a chromosome pair is designated as the chromatid and is joined to the other copy by the centromere.

If there is one molecule that is ubiquitously associated with genetics, it would have to be DNA. The story behind the discovery of DNA, the realization of its role in genetics and its structure and mechanism of action is among one of the most compelling in the history of science and was comprehensively reviewed by Petter Portin [16].

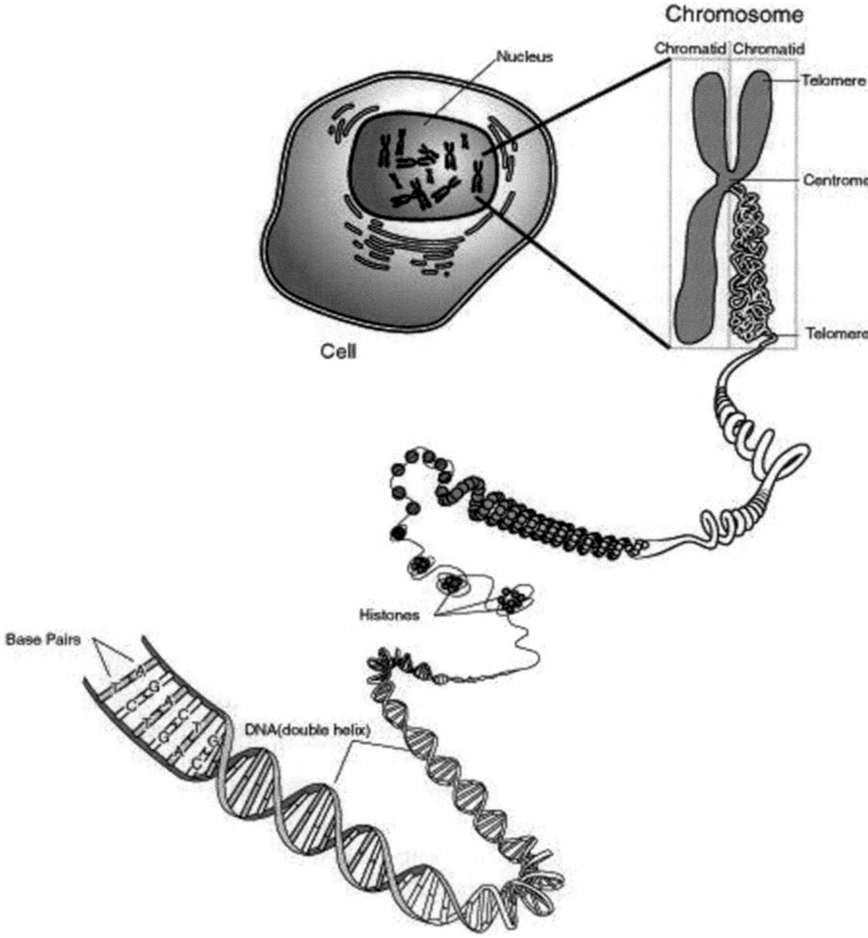

Fig. 1 The basics. The nucleus contains 23 pairs of chromosomes which contain strands of DNA wrapped around a histone core. DNA is composed of opposing strands (a double helix) joined together by base pairs (adenine [A], guanine [G], cytosine [C], and thymine [T]. Base pairs always join in a specific way: A-T or G-C. Each base is joined to a sugar phosphate backbone which together (base, deoxyribose sugar, and a phosphate group) define a nucleotide. Courtesy: National Human Genome Research Institute

Although it could be said that the DNA story culminated with the Nobel Prize winning description of its structure by Watson and Crick in 1953 [21], at least half a dozen other events were critical to its understanding [17]. At about the same time that Mendel was working out his Laws of Inheritance, Freidrich Miescher (1869) identified DNA from the nuclei of human white blood cells obtained from pus which he called nuclein. Shortly thereafter, the botanist Edward Zacharias made the link between nucleic acids and chromosomes. A critical discovery which localized genes in nuclear chromosomes was made in the early 1900's by Boveri and Sutton.

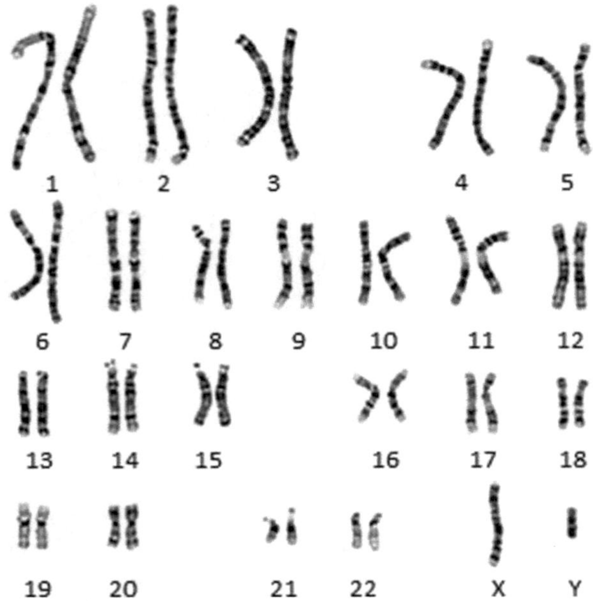

Fig. 2 Karyotype of human chromosomes. Chromosomes contain genes within their DNA. The size of chromosomes varies—1 being the largest and 22 the smallest. Likewise, the number of genes on each chromosome also varies. Genes occur along parts of the DNA and there's plenty of DNA that is not associated with protein-coding genes. In the past the non-gene DNA was given the misleading title of "junk DNA"

The completion of the chain linking chromosomes, genes and DNA occurred while World War II was raging. In 1944, in studies using pneumococcus, Oswald Avery, Collin MacLeod and Maclyn McCarty working at the Rockefeller Institute concluded that DNA was the carrier code and responsible for hereditary characteristics. Shortly thereafter, Edwin Chargaff successfully established the proportions of DNA's nucleic acids (Chargaff's rule) in which the amounts of adenine and thymine were equal to each other as was the case with guanine and cytosine.

In 1953, Watson and Crick described the double-helical structure of DNA to which we refer today (Fig. 3). Two strands DNA are composed of three fundamental building blocks: a sugar-phosphate "backbone" for each strand (the sugar is deoxyribose) bound together by reciprocal bases of adenine and thymine or guanine and cytosine. The combination of phosphate-sugar and a base is termed a nucleotide. The units of either A-T or G-C are called base pairs.

Genes are strung out long the length of each chromosome. Each gene is comprised of varying numbers of base pairs (see Table 1), but doing the math it's clear that there are many more base pairs than there are genes. In addition, the functional part of a gene, that is that part of a gene that is actually responsible for coding proteins represents under 10 % of the base pairs in the gene. The non-coding portion of DNA was referred to a "junk DNA", but recent studies have demonstrated that junk DNA plays a role in a variety of functions having clinical significance including disease and toxicity risk [3].

Fig. 3 Putting faces with names. Gregor Mendel, an Augustinian friar, is considered to be the father of modern genetics for his studies on inheritance at about the time of the American Civil War. Soon after (1869) Friedrich Miescher isolated nucleic acid from the nuclei of leukocytes. It wasn't until 1944 that Oswald Avery noted that genes were composed on DNA. In a little over 9 years, Watson and Crick, aided by information from the chemist and crystallographer, Rosalind Franklin, described the structure of DNA

Mutations and Variations in Genes

From a clinical standpoint, mutations and variations in genes play a big role in determining patients' risk of disease, how they respond to treatment or whether they're at high risk for certain drug toxicities or reactions. The fact that genes can

Table 1 The numbers game [15]

Number of human chromosomes	22 pairs + 2 sex chromosomes
Number of human genes	About 25, 000
Number of base pairs in human cells	About 3 billion
Number of base pairs in a gene (average)	3000 (largest 2,400,000)
Number of single nucleotide polymorphisms (human)	About 10 million
Frequency of SNPs	Once in every 300 base pairs
Number of genes on each chromosome	Varies by chromosome

change over time (maybe an extended period or acutely) in response to the environment in its broadest sense ultimately can influence phenotype. Changes in genes are called mutations and, by definition, a mutation is an alteration of the nucleotide sequence of a gene. Importantly, not all mutations result in disease, risk of disease or bad outcomes. The clinical importance of mutations varies. While all mutations are the consequence of nucleotide changes in sequence of a gene, some are subtle so show up infrequently and others are dramatic. Mutations which impact protein production in some way probably have the most impact clinically.

Mutations can be classified in a number of ways. Some describe their impact, while others are more descriptive relative to DNA morphology. Not all mutations are the same and not all have the same consequences. One way of classifying mutations is based on their impact on protein function. In this scheme, there are four possibilities:

1. Among the most common mutations are those associated with loss of function. Patients who have genetically-controlled enzymatic failures are representative of this category. One example are patients who don't have the enzyme to metabolize cancer chemotherapy drugs like methotrexate or 5-fluorouracil [6, 5, 18]. These type of drugs are toxic in their own right. You can imagine what happens when they're administered and continue to build up and stick around because the patient can't eliminate them. The levels of toxicity that affect patients in this category, like oral mucositis is horrific.
   Another example of diseases associated with this type of mutation are the diseases associated with inborn errors in metabolism [4]. A classic example of such a condition is phenylketonuria (PKU). Kids with PKU lack the enzyme phenylalanine hydroxylase which is critical to breaking down phenylalanine, a key component of proteins. As a result, if a child eats foods containing protein, phenylalanine accumulates and results in a range of symptoms and problems.
2. The opposite type of mutation may also occur in which there is a gain in function. As noted earlier, patients with Down syndrome have 3, not 2, #21 chromosomes.
3. Novel property mutations are those in which a specific gene change results in a clinical condition. Sickle cell anemia presents a good example [7, 13]. Sickle cell anemia is the most common blood disease in the United States and it affects thousands of patients worldwide. Due to a mutation of a single nucleotide (see SNPs below) the production of normal hemoglobin production does not occur and patients with the condition produce hemoglobin S.
4. Inappropriate gene expression characterized many of the genes identified with malignancies.

Mutations are also classified by changes that occur in placement or sequences of DNA within a chromosome, both within genes and non-gene-bearing parts of a chromosome as well. These types of mutations or variations in sequences are very common. The most common variations that occur happen at the level of a single base pair and involve a simple switch say, for example, for A-T to C-G [20]. Such changes, when they occur in at least 1 % of the population, are termed *single nucleotide polymorphisms* or SNPs (when you speak about them they're referred to as "snips"). There are over 10 million SNPs in the human genome; they occur about once in every 300 base-pairs.

For the purposes of the subject of this book, SNPs become valuable when they are linked to disease or toxicity risk. As you'll see in subsequent chapters, technology has evolved that allows investigators to detect SNPs in patients and to try to associate them with phenotypes. Since SNPs are sourced from DNA, they are very stable and present in all cells. Consequently, from a practical standpoint it is easy to collect DNA—blood, scraping cells from the cheek, saliva are all good sources of DNA [19].

Another type of mutation or variation is a consequence of structural differences in the sequence of base pairs in the genome. Stretches of DNA larger than 1000 base pairs (1 kb) that are different than the number of copies found in the normal population are referred to as *copy number variants* or CNVs [9]. CNVs can occur when there are deletions or additions to sections of a chromosome. The clinical significance of CNV relative to disease risk and treatment response is the subject of active investigations, but it is not hard to imagine how modifying sections of a chromosome might have an impact. So far, certain CNVs have been linked to risk of certain cancers, infectious diseases [10].

Epigenetics

In our discussion of mutations, we've learned that changes in DNA structure provide mechanisms that potentially impact the risk and course of a disease or disease response to treatment. There is an alternative way that gene expression can be modified that is totally independent of alterations in DNA structure—*epigenetics*. In particular, epigenetic pathways offer a conduit for the interaction and effect of environmental factors on host response and disease susceptibility.

Like mutations, epigenetic changes can affect gene expression (most often silencing gene expression) and protein transcription. But unlike heritable mutations, epigenetic modifications of gene expression provide a mechanism whereby environmental factors can influence phenotype. Epigenetic modifications result from four primary mechanisms which impact remodeling of chromatin:

- DNA methylation results from the addition of a methyl group in a specific site on DNA
- Histone modifications occur to the core structure of DNA (see structure of DNA above)

- Nucleosome positioning changes
- Non-coding RNA

As described above, DNA is wrapped around a histone core. Epigenetically driven acetylation or methylation modifies cellular responses by altering the production of transcription factors and subsequent gene expression [12]. The clinical importance of epigenetic changes is potentially significant [2] as abnormal epigenetic patterns can be identified in a range of diseases from asthma and chronic obstructive pulmonary disease to cancers. It seems likely that epigenetics may have an important in a number of oral diseases including periodontal disease and oral cancer.

Personalized Medicine

The concept of personalized or individualized medicine is not new [8, 11]. Clinicians (and patients) have long recognized that a 'one shoe fits all' approach to treatment is unrealistic. And yet clinical trial outcomes and many practice guidelines are designed around data based on analysis of a mean response to treatment in which a population of patients, defined by study inclusion criteria, is expected to react to treatment in the same way. As accumulating data confirms, such an approach is biologically naïve. With the coalescing advances in science and technology, we are fast approaching a point at which identification of risk and best treatment at the patient level is both clinically and economically feasible.

For all stakeholders—patients, clinicians, and payers—the advantages to individualizing treatment are significant and include determination of disease risk, assessment of a patient's likelihood to respond (or not respond) to a specific medication [1], and an appraisal of the probability of toxicities associated with treatment choices. An accurate probabilistic determination of these elements would enable a hierarchical guide to intervention and would be especially valuable when best therapy was based on preventive strategies.

Let's take a look at a non-oral very *hypothetical* example. Statins are used very broadly to reduce the risk of hyperlipidemia and consequent cardiac disease. And yet we know that not all patients are at equivalent risk of atherosclerotic heart disease and not all patients respond in the same way to equivalent doses of statins. Furthermore, it is also clear that statin use is not risk free. Occasionally some patients develop rhabdomyolysis, others liver damage. And, in 2010, $ 19 billion was spent on statins in the U.S. and not all statins are the same cost.

With an ideal individualized approach, patients would be screened for risk of atherosclerotic heart disease, response to each of the statin options, and for risk of statin-related toxicity. If a patient was found to be at risk for AHD, a list of statins would be produced with the top option being the agent with most likelihood of efficacy, lowest risk of toxicity and best price.

How would this approach apply to oral disease? This book is full of examples, but let's focus on one that is very real and very current. Palifermin is a growth factor that is approved for the prevention of oral mucositis (severe mouth sores) induced by chemotherapy administered prior to a hematopoietic stem cell transplant. In order for palifermin to be effective, it has to be given intravenously long before there are any visible changes in a patient's mouth or before the patient is symptomatic [14]. Of patients receiving this type of chemotherapy, about 40 % will develop mucositis, but 60 % escape this terrible complication. The 6 dose course of palifermin costs about $ 10,000. In the absence of a way to predict which patients will develop mucositis, the oncologist has two choices: treat all patients knowing that 6 out of 10 will receive the growth factor unnecessarily or don't treat any. But if there was a way to predict mucositis risk, the clinician could selectively treat only those patients who were likely to benefit. Aside from the clinical benefit of preventing mucositis in appropriate patients, targeting treatment would also save $ 60.000 spent on palifermin for those patients who didn't need it.

The actual translation of theoretical genomics to the clinic is now a reality. There are tests to assess the risk of diseases which are being actively marketed, including one for periodontal disease, and others to predict response to treatment and guide care. The field is expanding at a rapid rate; its impact will be felt throughout medicine. Oral diseases are among the most common maladies affecting humans. As described in the following chapters, personalized medicine and genomics will play a significant role in the future diagnosis and management of oral disease.

References

1. Alterovitz G, Tuthill C, Rios I, et al. Personalized medicine for mucositis: Bayesian networks identify unique gene clusters which predict the response to gamma-D-glutamyl-L-tryptophan (SCV-07) for the attenuation of chemoradiation-induced oral mucositis. Oral Oncol. 2011;47:951–5.
2. Arimondo PB, et al. Epigenetics. Biochim. 2012;94:2191–2.
3. Ayarpadikannan S, Kim HS. The impact of transposable elements in genome evolution and genetic instability and their implications in various diseases. Genomics Inform. 2014; 12:98–104.
4. Blau N, Hennermann JB, Langenbeck U, et al. Diagnosis, classification, and genetics of phenylketonuria and tetrahydrobiopterin (BH4) deficiencies. Mol Genet Metab. 2011;104(Suppl):S2–9.
5. Borras E, Dotor E, Arcusa A, et al. High-resolution melting analysis of the common c.1905+1G>A mutation causing dihydropyrimidine dehydrogenase deficiency and lethal 5-fluorouracil toxicity. Front Genet. 2013;17:312.
6. Castaldo P, Magi S, Nasti AA, et al. Clinical pharmacogenetics of methotrexate. Curr Drug Metab. 2011;12:278–86.
7. Fixler J, Styles L. Sickle cell disease. Pediatr Clin North Am. 2002;49:1193–210.
8. Ginsburg GS, Willard HF, editors. Essentials of genomic and personalized medicine 2010. San Diego: Academic; 2010.
9. Henrichsen CN, Chaignat E, Reymond A. Copy number variants, diseases and gene expression. Hum Mol Genet. 2009;18:R1–8.
10. Hollox EJ, Hoh BP. Human copy number variation and infectious disease. Hum Genet. 2014;133:1217–33.
11. Lesko LJ. Personalized medicine: elusive dream or imminent reality? Clin Pharmacol Ther. 2007;81:807–16.

12. Lod S, et al. The influence of epigenetics in relation to oral health. Int J Dent Hyg. 2013;12:48–54.
13. McCavit TL. Sickle cell disease. Pediatr Rev. 2012;33:195–204.
14. Nooka AK, Johnson HR, Kaufman JL, et al. Pharmacoeconomic analysis of palifermin to prevent mucositis among patients undergoing autologous hematopoietic stem cell transplantation. Biol Blood Marrow Transpl. 2014;20:852–7.
15. Palladino MA. Understanding the human genome project. 2nd ed. The benjamin cummings special topics in biology series. San Francisco: Pearson Publishers; 2006.
16. Portin P. The birth and development of the DNA theory of inheritance: sixty years since the discovery of the structure of DNA. J Genet. 2014;93:293–302.
17. Quackenbush J. The human genome. Watertown: Charlesbridge Publishers; 2011.
18. Rosmarin D, Palles C, Church D, et al. Genetic markers of toxicity for capecitabine and other fluorouracil-based regimens: investigation in the QUASAR2 study, systematic review, and meta-analysis. J Clin Oncol. 2014;32:1031–9.
19. Sonis S, Antin J, Tetaldi M, et al. SNP-based Bayesian networks can predict oral mucositis risk in autologous stem cell transplant recipients. Oral Dis. 2013;19:721–7.
20. Syvanen AC. Accessing genetic variation: genotyping single nucleotide polymorphisms. Nat Rev Genet. 2001;2:930–42.
21. Watson JD, Circk FH. Molecular structure of nucleic acids; a structure for deoxyribose nucleic acid. Nature. 1953;171:737–8.

Current and Evolving Technologies

J. Robert Chang, Enkhtsetseg Purev and Winston Patrick Kuo

Background on Genomic-Based Technologies

Genomic-based results can provide a powerful snapshot of an individual's state of health and disease and lead the way for diagnostic solutions in the field of personalized oral medicine. Identifying specific nucleic acids in either biological fluids or tissue samples from patients with a particular disease state can reflect both acute and chronic changes in diseased cells and tissue throughout the body and potentially inform downstream treatment decisions. The recent refinement of existing and current advancement of genomics-based technologies have allowed one to get a snapshot of a person's disease risks and status as revealed through DNA sequencing, DNA structural and gene expression analyses.

Genomic efforts over the past decade have identified an increasingly complex list of potential genomic-based biomarkers using large-scale approaches as illustrated in Table 1. The need for novel genomics-based diagnostic methods that are capable of measuring multiple analytes simultaneously are becoming critically important. Unlike in the past, these tests can be performed quickly at a relatively low cost, providing robust means of complementing current clinical best practices

---

W. P. Kuo (✉)
IES Diagnostics, Cambridge, MA 02139, USA
e-mail: wkuo@iesdiagnostics.com; wkuo@genetics.med.harvard.edu

Harvard Catalyst, Laboratory for Innovative Translational Technologies, Harvard Medical School, Boston, MA 02115, USA

Department of Developmental Biology, Harvard School of Dental Medicine, Boston, MA 02115, USA

J. R. Chang
OsteoInvent Inc., West Roxbury, MA 02132, USA

E. Purev
National Heart, Lung, and Blood Institute, National Institutes of Health, Bethesda, MD 20814, USA

S. T. Sonis (ed.), *Genomics, Personalized Medicine and Oral Disease*,
DOI 10.1007/978-3-319-17942-1_2

Table 1 Genomic-based biomarker types and corresponding methodology for biomarker discovery and validation. Genomic-based biomarker types include: (1) chromosomal translocations that are chromosomal rearrangements, involving the transfer of chromosomal segment to another chromosome or different region of the same chromosome; (2) DNA amplification, is the increased copy number of a chromosomal segment; (3) mutation that denotes a single nucleotide (point mutation) substitution, deletion and/or insertion; (4) epigenetics, which involves the control of genes without affecting their sequences (for example, DNA methylation leads to gene inactivation or gene silencing); and (5) gene expression denotes the steady state number (qualitative or quantitative) of genes in its transcript form at the time of the assay

Types of genomic-based biomarkers	Methodological approach	Application (biomarker discovery or validation)	Clinical application
Chromosomal translocations	Karyotyping	V	Yes
	FISH	V	Yes
	QRT-PCR	V	Yes
	CGH	BD	In development
	SOMA	BD	In development
DNA amplification	FISH	V	Yes
	aCGH	BD	In development
	CNV-Seq	BD	In development
	Digital karyotyping	BD	In development
	(BAC)-end Seq	BD	In development
Mutation	QRT-PCR	V	In development
	NGS	BD	In development
	High resolution melt	BD	In development
	Transcriptome-Seq	BD	In development
Epigenetics	Bisulfite sequencing	BD	In development
	Bisulfite pyrosequencing	BD	In development
	ChIP-Seq	BD	In development
	ChIP-chip	BD	In development
	Differential methylation hybridization	BD	In development
	SMRT	BD	In development
Gene Expression	Microarray	BD	Yes
	QRT-PCR	V	Yes
	RNA-seq	BD	In development
	Digital PCR	V	Yes
	NGS	BD/V	In development

Validation *V*, Biomarker discovery *BD*

in patient care and promises to play an important role in furthering personalized/precision medicine. To highlight the clinical utility of genomic-based studies, we will discuss the evolution of a large-scale breast cancer study that demonstrates the process from an exploratory phase to eventually, a commercially available diagnostic assay/test. Over the past decade, breast cancer research microarray studies have simultaneously assayed the expression profile of thousands of genes that have led to the classification and risk stratification based on molecular subtypes that include luminal A, luminal B, normal-like, HER2 and basal subtypes [1–3].

Table 2 List of a few FDA cleared genomic-based assays/tests utilizing different genomic-based technology types. To date, there are over 200 human and microbial genomic-based assay/tests available. (For complete list, see http://www.fda.gov/medicaldevices/productsandmedicalprocedures/invitrodiagnostics/ucm330711.htm)

Genomic-based technology type	Disease	Assay/test name	Company	References
Cytogenetics (Array-based)	Chromosomal abnormality	Affymetrix cytoScan® Dx assay	Affymetrix inc. (Santa Clara, CA)	[78–80]
Hybridization	Breast cancer	Prosigna™ breast cancer prognostic gene signature array	NanoString technologies (Seattle, WA)	[81]
Microarray	Occult primary cancer	Tissue of origin	Response genetics (Los Angeles, CA)	[35, 37, 82]
	Breast cancer	MammaPrint®	Agendia Inc. (Irvine, CA)	[83–85]
	Heart transplant monitoring	Allomap® molecular expression test	CareDx (Brisbane, CA)	[86–88]
QRT-PCR	Prostate cancer	Progensa™ PCA3 assay	GenProbe Inc. (San Diego, CA)	[89]
	Coagulation factors	Factor V leiden kit	Roche diagnostics Corp. (Nutley, NJ)	[90, 91]
	Drug metabolizing enzyme—warfarin	eSensor® warfarin sensitivity saliva test	GenMark diagnostics (Carlsbad, CA)	[92]
Sequencing	Cystic fibrosis	MiSeqDx cystic fibrosis clinical sequencing assay	Illumina, inc. (San Diego, CA)	[93]

These large-scale studies have provided pivotal information in which the final list of informative candidate biomarkers were narrowed down to a panel of a dozen to several dozen biomarker genes. The next phase is the development and commercialization of molecular assay/tests that contain these gene content that improve breast cancer risk stratification and support optimized treatment selection. As a result of these studies, through vigorous periods of biological validation, molecular diagnostic tests are now commercially available that include the OncotypeDx® assay (Genomic Health, Inc., Redwood City, CA, USA) and FDA-approved MammaPrint® (Agendia, The Netherlands) (Table 2), amongst others. The OncotypeDx® assay for example, uses quantitative reverse transcription polymerase chain reaction (QRT-PCR) to measure 16 cancer-related and 5 normalization gene transcripts from formalin-fixed paraffin-embedded (FFPE) tissue samples. This assay quantifies the likelihood of breast cancer recurrence in women with newly diagnosed early stage breast cancer [4, 5], and it can also be used to identify estrogen receptor (ER)-positive patients whose prognosis with hormonal therapy is favorable enough to waive adjuvant chemotherapy. The MammaPrint® on the other hand, measures a 70-gene panel from fresh frozen tissue samples to calculate a risk score for developing metastasis [6–8]. Other assays are available for use with fresh frozen or FFPE samples such as the Rotterdam Signature (Veridex Corp., Warren, NJ) [9], the

Mammostrat® test (Applied Genomics, Huntsville, AL) [10], and the Breast Cancer Recurrence assay (AviaraDX, San Diego, CA) [11].

Another important area in which genomic-based technologies have played a role is drug development. The identification of specific genetic alterations that can be targeted by specific drugs has become increasingly important in the current environment of personalized medicine in reference to therapies and biomarker-driven patient stratification. For example, this has led to regulatory approval for several successful anti-tumor drugs for specific mutation classes, where the U.S. Food and Drug Administration (FDA) approved "companion diagnostics" that use QRT-PCR-based assays to detect specific BRAF V600 mutations for which the drugs would be effective for patients with advanced stages of melanoma. These companion diagnostic assays proved to be a key factor for the approval and reimbursement policy of the following drugs; Vemurafenib, Dabrafenib, and Trametinib. Mutated BRAF has shown to be responsible for constitutive activation of the MAPK pathway and has been found in approximately 50% of melanoma cases. The drugs have significant anti-tumor activity but should be avoided in non-mutant patients due to the lack of efficacy. Unfortunately, the response is invariably followed by development of rapid tumor resistance, due to evolution or selection of mutations that either create alternative survival pathways or reactivate MAPK signaling [12]. It is clear cancer treatment, which harbors multiple or sequential mutations resulting in simultaneous activation of several pathways, will benefit from advanced molecular diagnostics and longitudinal therapeutic monitoring. This demand exists for patient stratification and selection in clinical trials and also for the downstream implementation of appropriate reimbursement rules for drugs and coupled tests in health care systems.

The above examples in the diagnostics and drug development has been made possible due to the technological advances in accuracy, sensitivity, specificity, rapid turnaround time, and streamlined workflow of analytic techniques such as microarrays, QRT-PCR, digital PCR (dPCR) and next generation sequencing (NGS) methods that have taken center stage as they transition into the clinical space, as indicated by growing number of FDA-approved in vitro diagnostics (IVD) based on genetic technologies. Examples of these tests are shown in Table 2. This is also attributable to the tremendous progress that has been achieved over the last 20 years in upstream sample preparation methods using nucleic acids (DNA, RNA, miRNA) for downstream genomics-based technologies.

For example, DNA sequencing has evolved from manual sequencing (based on the Sanger sequencing method) to NGS that provides detailed information related to the genome, transcriptome and epigenome. The enhanced NGS method allows sequencing of few hundred base pairs/day to where an individual's entire genome can be sequenced within a week for a cost of about $ 1000. This increased scalability allows evaluation of a patient's DNA not only for tumor-related mutations but would simultaneously assess the patient's drug response, susceptibility to side effects arising from therapy, and so forth. Furthermore, features have been incorporated into DNA sequencing to assess DNA structure (i.e. DNA methylation analysis) that would provide, not only the patient's allelic factors, but DNA structural features that would affect the disease and treatment outcome.

Although NGS field is evolving, other complementary genomic-based technologies have also seen significant progress that may alter future clinical applications. Oral and oropharyngeal cancer diagnostics, for example, can be tested for the presence of HPV as a prognostic tool for managing the disease. The virus is currently detected by either immunohistochemistry for HPV p16 or by *in situ* hybridization for HPV DNA. Implementing genetic assays using QRT-PCR, one can obtain a more robust (increased sensitivity and timeliness) results for managing such cancers. Similarly, comparative genomic hybridization (CGH) may be used to detect DNA translocations, gene duplications and/or insertions and deletions that may be associated with different diseases. In fact, one can envision monitoring systemic diseases by way of salivary diagnostics, which can be implemented using any of the commonly accepted and evolving genomic-based technologies. The adaptation of novel genomic-based technologies for clinical use can be seen by the growing number of genomic-based tests cleared by the FDA for commercialization (http://www.fda.gov/MedicalDevices/ProductsandMedicalProcedures/InVitroDiagnostics/ucm330711.htm). Recently there have been great advances utilizing genomic material in prenatal medicine exemplified by companies, such as, Sequenom (San Diego, CA) and oncology studies by companies such as Sysmex Inostics (Baltimore, MD) and TrovaGene (San Diego, CA) [13, 14]. Together, these genomic-based technologies have the potential to discover and utilize current and new biomarkers for clinical use (Table 1). This chapter will focus on an update of current and emerging downstream analytical genomic-based technologies as related to molecular diagnostics that are currently available, how such technologies are being utilized and what emerging technologies are awaiting entry into this robust and exciting field.

Genomic Isolation Methods

Many reliable nucleic acid extraction methods are streamlined, reliable and reproducible so that all downstream studies utilizing genomic-based technologies are more standardized and efficient. We will keep this section brief as is not the scope of this chapter; however the way in which nucleic acids are collected for evaluation is a critical step that should not be overlooked. As with all assays, the quality of the sample being analyzed plays a vital role in the quality of the downstream data. This is of particular importance when processing clinical samples: for example, RNA degradation due to its inherent structural weakness as well as the abundance of RNases that are found in biological systems. Sample preparation is a process that can be simplified by parsing into several steps: sample procurement, sample preservation and handling, nucleic acid isolation, library preparation for sequencing and sample processing. Certain aspects of the process can also be automated using robotics.

When collecting oral samples such as buccal cells, saliva or a tissue biopsy, the oral micro-environment is rich in enzymes that are capable of digesting nucleic acids, so implementing good laboratory practices (standardized operating protocols

Table 3 Commercially available nucleic isolation methods for saliva samples. Isolation of nucleic acids from saliva can be very challenging, there are a commercially available kits for research use only and available for diagnostics (FDA cleared) purposes. Methods for (a) DNA isolation and (b) RNA isolation from saliva are listed

Salivary collection method (DNA)	Trade name	Company	Saliva volume	Yield	Ref.
(a) Salivary oral DNA sample preparation methods					
Cotton Swab (manual)	BuccalAmp™	Epicentre® (an Illumina company) (madison, WI)	n/a	1–7 µg	[94, 95]
Cotton swab (chewable)	Salivette® with or without citric acid (fda approved for cortisol testing)	Sarstedt Inc. (nümbrecht Germany)	0.1–2.0 ml	6 µg	[77, 96, 97]
Saliva (with preservation reagent)	OrageneDx® (FDA cleared)	DNA Genotek Inc. (kanata, ontario, canada)	2.0 ml	~50 µg	[98–100]
Buccal swab (magnet beads)	NucleoMag® 96 Trace	Macherey-nagel inc. (düren, Germany)	n/a	0.4 µg/µl	[101]
(b) Salivary oral RNA sample preparation methods					
Saliva (with RNA preservation reagent)	OrageneRNA®	DNA Genotek Inc. (Kanata, Ontario, Canada)	2.0 ml	10–50 µg	[102]
	RNAProtect®	Qiagen GmbH (Hilden, Germany)	Scalable		[103–105]
Saliva (Phenol-Chloroform)	QIAzol®	Qiagen GmbH (Hilden, Germany)	200 µl	0.89–7.1 µg	[106]
	Trizol®	Life technologies (carlsbad, CA)	Scalable		[107]

for sample preparation) are needed when processing these sample types. Numerous methods and kits are available for nucleic acid isolation that include organic extraction (phenol:chloroform), spin columns, and magnetic beads (Tables 3a, b). Magnetic beads are especially useful in processing multiple samples due to the ease with which the beads can be incorporated into an automated system. As such, automated workstations using magnetic beads (Table 4) are able to minimize human error, improve consistency, and reliability. As technologies have improved from isolation of nucleic acids and downstream analytics, saliva has been targeted for diagnostics [15], prognostics [16] as well as biomarker discovery [15, 17, 18] due to the noninvasive and unpretentious nature of its collection. The ability to use saliva for diagnostic purposes has been greatly enhanced by various reagents/buffers that permits

Table 4 Nucleic acid sample preparation automation that complements downstream genomic-based technologies. Automation helps streamline sample processing in a high-throughput manner, thus allowing one analyze many samples in a short period of time. Quality of results are dependent on following standardized protocols

Automated work stations	Company	Samples/ Run(maximum)	Time/Run(minutes)	Ref.
MagNA pure compact	Roche diagnostics (Nutley, NJ)	96	20–45	[108, 109]
JANUS®	Perkin elmer (Waltham, MA)	96	120	[110]
SPRI-TE nucleic acid extractor	Beckman coulter (Indianapolis, IN)	10	N/A	[111]
NucliSENS® easyMAG®	bioMerieux Inc. (Durham, NC)	24	40–60	[112, 113]
EZ1 Advanced XL	Qiagen (Hilden, Germany)	14	20	[114]
QuickGene 810	Autogen Inc. (Holliston, MA)	8	10	[115]

for the preservation of nucleic acid found in the saliva such as OrageneDx®, which recently received FDA clearance for IVD use in the United States.

As for downstream NGS applications, library preparation is required from the DNA or the RNA (isolated from the samples) in order to provide sufficient amount of template for sequencing reactions. DNA fragmentation is often accomplished mechanically, for example hydrodynamic shearing and sonication or enzymatically by endonuclease digestion [19] while RNA fragmentation, in addition to mechanical and enzymatic approaches, can also be achieved chemically [19, 20]. With NGS sequencing, generating relatively uniform-sized fragments with numerous overlapping DNA reads assures the highest quality of data [21]. However, much sequencing bias has been observed with NGS [22–25], due, in some part to the fragmentation method used [26–28]. For instance, though mechanical fragmentation is often used to generate DNA library for NGS, similar DNA fragmentation bias was observed between sonication, nebulization and ultrasound methods due to the physiochemical nature of the DNA structure [28]. Upon fragmentation, DNA fragments are repaired to generate blunt ends, phosphorylated and ligated to platform-specific sequencing adaptor (usually barcoded). If small amount of nucleic acid was used to generate the library, optional amplification can be achieved using PCR [29], though low starting material may prevent detection of low copy transcripts [30, 31]. Considering the time and effort required to perform and analyze NGS data, assessing the quality and quantity of the library is highly desirable, which can be accomplished using traditional nucleic acid methods such as spectrophotometry, agarose and poly-acrylamide gel electrophoresis. Absolute quantification can also be performed using digital PCR (dPCR).

The FDA has been active in developing standards for isolation of genetic material as seen in their RNA sample preparation SOP (http://www.fda.gov/downloads/ScienceResearch/BioinformaticsTools/MicroarrayQualityControlProject/UCM126818.pdf) and assessing sample processing overview SOP (http://www.fda.gov/downloads/ScienceResearch/BioinformaticsTools/MicroarrayQualityControlProject/

UCM126825.pdf). Incorporating established standardized methods to ensure high quality would be advisable for obtaining consistent, clinically relevant data.

Genomics-Based Analytical Methods

Microarrays

Microarrays provide an unprecedented opportunity for comprehensive concurrent analysis of thousands of genes, DNA molecules or nucleic acids. Various methods such as optical, electrical, nanowire-based, magnetic, piezoelectric and mechanical transducers have been developed. The term "microarray" refers to the orderly arrangement, "array" of the probes of interest in a grid format used at the "micro" scale. The genomics context for the term "microarray" often refers to a device/platform where single-stranded DNA oligonucleotides (short sequences of nucleotides) or "oligos" are affixed to a solid surface. As many have experienced or read, a microarray is a platform for analyzing multiple genes simultaneously. As with DNA sequencing, microarray comes in many "shapes and sizes". Generally, nucleic acids are captured onto a platform (whether glass surface, beads, micro- and nanowells) and its presence is detected by several means, including hybridization to a set of characterized oligonucleotides, PCR amplification or even by semi-conductor approach. Furthermore, high-throughput microarrays have been incorporated into areas as gene expression profiling to slicing/fusion analyses, tiling/full genome coverage, DNA/RNA-protein interactions and comparative genomic hybridization (CGH). A more comprehensive overview of microarrays can be found in a recent publication by Trachtenberg et al. [32].

Microarrays have come a long way in reference to being a platform that can be utilized in the clinical setting. Kuo et al. was the first group to report a large-scale comprehensive cross-platform comparison of DNA microarrays [33]. Their results demonstrated that greater inter-platform consistency was observed in highly expressing genes than in low expressing genes [33]. When the same microarray experiments were performed in different laboratories, there was greater inter-laboratory variability than intra-laboratory variability, demonstrating users also play a role in generating different gene expression measurements [33]. The results suggested that there are many platforms available that provide good quality data, especially on highly expressed genes, and that, among these platforms, there is generally good agreement. These results were confirmed by another large-scale initiative called the MicroArray Quality Control (MAQC) project [34], spearheaded by the FDA.

After vigorous community criticism and evaluation, the FDA has cleared commercialization of some microarray platforms for clinical use. For example, Monzon et al. developed an Affymetrix GeneChip® test (Santa Clara, CA) for identifying the source of occult primary tumors [35] that can identify metastatic tumors found away from its tissue origin, thereby enabling diagnostics and targeted therapeutics for metastatic tumors (of previously unknown origin). In a multicenter validation study, the assay demonstrated high sensitivity (88 %) and specificity (99 %) in determining the source of the occult primary tumor [36]. The FDA approved the test, marketed under the trademark Response Dx: Tissue of Origin™ (Response Genetics, Los Angeles, CA) is cleared for both fresh frozen and paraffin-embedded tissue samples. Since commercialization, the test has proven to be effective for detecting metastatic cancers of the head and neck squamous carcinoma [37] and gynecological cancers [38], among others. Similar tests are demonstrating very encouraging results in providing better treatment and outcome for patients [39].

QRT-PCR

Traditional methods for nucleic acid quantification like Southern and Northern blots or traditional PCR are generally not accurate and have low sensitivity. In contrast, QRT-PCR platforms provide a robust means of quickly quantifying genes of interest and requires very small amounts of starting material. QRT-PCR is a widely adapted method for rapid quantification of known genomic content, mRNAs and miRNAs, in contrast to standard PCR, where the amplified product from designed primers specific to the gene of miRNA of interest is quantitated during the PCR reactions by a dye. In essence, this is achieved by fluorophores that fluoresces either by binding DNA (SYBR Green) or are released by Taq DNA polymerase exonuclease activity (probe-based including TaqMan® (Life Technologies, Carlsbad, CA) and locked nucleic acid (LNA™) (Exiqon, Vedbaek Denmark). The reactions are conducted using any of the numerous commercially available QRT-PCR platforms such as those from ViiA™ 7 system (Life Technologies, Carlsbad, CA), 480 LightCycler® (Roche Applied Sciences, Indianapolis, IN) to CFX96 Touch™ (Bio-Rad Laboratories, Hercules, CA) platform that are versatile such that you can mix and match reagents and primers from a variety of vendors to conduct both traditional SYBR green and probe-based assays.

Furthermore, with the flexibility of QRT-PCR, gene-specific DNA mutational analysis can be conducted. For example, Factor V Leiden Kit (Roche Diagnostics Corporation, Indianapolis, IN) is a FDA-cleared QRT-PCR assay on the LightCycler 1.2 (Roche Applied Sciences, Indianapolis, IN) platform that tests for the common point mutant variant of Factor V (http://www.accessdata.fda.gov/scripts/cdrh/cfdocs/cfPMN/pmn.cfm?ID=K033607) (Table 2).

Digital PCR

Digital PCR (dPCR) is a novel method for precise quantification of nucleic acids that uses similar assay reagents as those used in QRT-PCR measurements, but counts the total number of individual target molecules in a digital format [40, 41]. It is gaining acceptance in the field due to its superior sensitivity and precision providing absolute quantification of the low abundance RNA biomarkers. Recently a number of manufacturers have commercialized dPCR platforms using various approaches, all with the goal of providing improvements to legacy QRT-PCR methods [42–45].

Currently, there are three dPCR approaches: fluidic circuits, BEAMing and droplets. The three platforms differ in how the transcripts are partitioned. The integrated fluidic circuit platform for dPCR is commercialized by Fluidigm's BioMark™ HD (Fluidigm, South San Francisco, CA) system that utilizes a fluidic circuit model [46], where the sample is partitioned into hundreds of reaction panels (or circuits) consisting of nano-liter reaction volumes. Fluidigm's Digital Arrays are able to simultaneous process 48 and 96 samples simultaneously. The technology is flexible enough to accommodate, in addition to gene expression analysis [47, 48], SNP genotyping [49] and copy number variation analysis [50–52]. BEAMing digital PCR was developed in the laboratory of Dr. Vogelstein and is commercialized by Sysmex Inostics. BEAMing (beads, emulsion, amplification and magnetics) consists of beads coated with forward primers for the transcript of interest. Emulsion PCR is performed such that single transcript initiates PCR on a single bead. As the PCR cycles, newly formed DNA, in turn, are captured by another primer on the same bead, whereby localizing PCR amplification of a single template to a single bead. The PCR products are labeled and the beads are analyzed using flow cytometry [53, 54]. This technique has also been used to detect and quantify mutant DNAs in circulating tumor cells [55, 56] in addition to tumors [57, 58]. The third approach is droplet digital PCR (ddPCR) that involves partitioning the PCR reaction into tiny droplets where PCR occurs. Current systems are able to partition the PCR mastermix into 20,000 × 1.0 nl reactions by QX200™ Droplet Digital™ PCR system (Bio-Rad Laboratories, Hercules, CA) or millions of 1.0 pl reactions by the RainDrop® digital PCR system (RainDance Technologies, Billerica, MA) [59]. Using ddPCR, researchers are able to detect mutations at levels as low as 1:100,000 (mutant to wild type ratio) [60].

Next Generation Sequencing

NGS has already had a revolutionary impact on all fields of biology, and gives scientists the ability to economically and rapidly determine exact sequences of DNA and RNA molecules at the scale of the organism (such as an individual's entire genomic sequence), a single cell from an organism (such as a normal or cancerous liver cell), or even single DNA or RNA molecules (such as circulating DNA and RNA fragments in blood plasma). The NGS term is interchangeable with "high-throughput sequencing", "massively parallel sequencing", and "deep sequencing" that describe a rapid, inexpensive way to generate mega- and/or giga-base sequence reads with each run. Some of these methods are further evolution of the traditional Sanger sequencing methods, whereas others bring novel approaches to determining nucleic acid sequences. In contrast to other genomic-based platforms, NGS technologies provide an unprecedented opportunity to sequence thousands of genes concurrently to identify clinically relevant miRNAs [61] and genetic alterations including mutations [62]. This capability is being used to not only identify pathological mutations in the genes of individuals with inherited disorders (in which the DNA does not change much over the lifetime of the individual) but is also being used to rapidly and urgently identify important mutations in primary and metastatic tumors [63, 64].

In the 10 years since the first human genome was sequenced, there has been a million-fold drop in the cost of genome sequencing and 1000-fold increase in speed (10 years for the first genome to less than a week today). Different approaches have emerged to take advantage of this revolutionary technology platform. For instance, in a gene discovery mode, where it is unclear which genetic alterations are responsible for a given disease, it may be beneficial to sequence the entire genome of an individual or sets of individuals with the same disease. Alternatively, at the clinical end of the spectrum where it is clear that mutations in certain genes (such as cancer oncogenes) will cause disease, NGS technology may be used at lower cost and/or time to sequence specific limited sets of known genes to look for mutations in only these disease-specific subsets of genes in the human genome and may represent a more practical approach in a clinical setting. These are known as "Targeted NGS panels". In the following section, we will describe briefly the different NGS platforms that are currently commercially available.

NGS Platforms

Prior to the development of NGS, DNA sequencing was performed, predominantly, using the Sanger sequencing method, otherwise known as the "chain-termination method [65, 66]. DNA sequencing by the chain-termination method is achieved by performing four separate reactions in parallel with each reaction consisting of a nucleotide analog that inhibits DNA polymerase. The DNA fragments generated in each reaction is resolved adjacent to each other on a slab gel. As the fragments in each reaction are terminated due to an incorporated nucleotide analog, the identity of the last nucleotide in each fragment is known. By resolving them adjacent to each other, the positions of the nucleotides are identified with respect to each other. Initially, *de novo* synthesized DNA containing the inhibitory analog (at the 3' ends) was resolved in parallel using slab gel acrylamide electrophoresis. In the hands of trained personnel, the method is robust, providing accurate DNA sequences. However, typical sequencing (from sample prep to sequence deduction) required several days (including 12–16 h for resolving the DNA strands by electrophoresis) and the sequences generated are relatively short (< 1000 bases). Manual DNA sequencing gave way to automated sequencing, commercialized by Applied Biosystems, now part of Thermo Fisher Scientific. Coupling the nucleotide analogs to various fluorescent dyes allowed for a single reaction that was then resolved by capillary electrophoresis with each dye (detected using a laser sensor) representing the terminal

Table 5 Summary and technical descriptive details of commonly used sequencing technologies. The sequencing technologies are grouped into traditional, next generation sequencing (NGS) and third generation sequencing categories. The Sanger (traditional) method has transitioned from visual detection and manual sequencing to optical detection coupled with capillary electrophoresis whereas NGS platforms vary widely, that include Roche's 454, Illumina's HiSeq and Life Technologies' SOLiD® platforms. Third generation sequencing, single molecule approach, include Helicos (now bankrupt), Pacific BioSciences SMRT® system and Oxford Nanopore Technologies MinION™ and GridION™ platforms. There are technical differences between the different platforms. Third generation platforms have several technical advances over NGS in that (1) no template amplification is needed, (2) utilizes single molecule detection and (3) the technology allowed for direct sequencing of transcripts without converting to cDNA. The advantage of direct transcript sequencing is especially relevant in light of several studies showing the cDNA conversion can be inconsistent [116–120]. In addition, these platforms provide greater read lengths and shorter read time

	Platforms	Sequencing method	Detection technology	Read length limitation?	Template amplification required?	All solid-state materials?	Enzyme dependent?	Single molecule?	Tunable to arbitrary accuracy?	Whole genome sequencing	Transcriptome sequencing	Applicable for biomarker study?	Ref.
Traditional	Sanger	Dye terminator	Optical	Yes	Yes	No	No	No	No	No	No	No	[65, 67]
Next generation sequencing	Roche 454	Pyrosequencing	Optical	Yes	Yes	No	Yes	No	No	Yes	Indirect	Yes	[69, 121–124]
	HiSeq™	Sequencing by synthesis	Optical	Yes	Yes	No	Yes	No	No	Yes	Indirect	Yes	[71, 125–128]
	SOLiD®	Sequencing by Ligation	Optical	Yes	Yes	No	Yes	No	No	Yes	Indirect	Yes	[72, 129–132]
	Complete	Sequencing by ligation	Optical	Yes	Yes	No	Yes	No	No	Yes	Indirect	Yes	[133, 134]
	Ion TorrentTM	Ion semiconductor	Electronic	Yes	Yes	Yes	Yes	No	No	Yes	Indirect	Yes	[73, 135–137]
	Helicos	Virtual terminator	Optical	Yes	No	No	Yes	Yes	No	Yes	Direct	Yes	[138–143]

Table 5 (continued)

	Platforms	Sequencing method	Detection technology	Read length limitation?	Template amplification required?	All solid-state materials?	Enzyme dependent?	Single molecule?	Tunable to arbitrary accuracy?	Whole genome sequencing	Transcriptome sequencing	Applicable for biomarker study?	Ref.
Third generation sequencing	Pac Bio RS	Single molecule sequencing by synthesis	Optical	Yes	No	No	Yes	Yes	No	Yes	Indirect	Yes	[74, 75, 144–148]
	MinION™ /Grid-IONTM	Nanopore biosensor	Electronic	Yes	No	No	Yes	Yes	No	Yes	Indirect	Yes	[76, 149–151]

ends of *de novo* synthesized DNA strands [67, 68]. Though capillary electrophoresis allowed for a faster, user-friendly method for sequencing DNA, it took numerous collaborators over 13 years and billions of dollars to sequence 3.3 billion bases of the human genome (http://www.genome.gov/10001772#al-2). Table 5 summarizes an evolution of different sequencing technologies and their technical differences and similarities.

Pyrosequencing

Pyrosequencing is based on a method developed by Pal Nyren's group [69]. The method relies on pyrophosphates (PPi) that are released when nucleotides are incorporated during DNA synthesis [70]. Released PPi are detected indirectly as sulfurylase combines PPi with adenosine 5'-phosphosulfate to produce ATP. ATP is then used by luciferase to emit light that is captured by luminometer or CCD camera. In pyrosequencing, the sequence identity is captured by simply passing specific nucleotides through the reaction. Light is generated with each nucleotide that passes through the reaction and is a positive identifier of the sequence. The intensity of the light generated is directly proportional to the amount of PPi released. Therefore, two identical nucleotides adjacent to each other would emit twice as much light as a single nucleotide. In practice, as seen implemented by the 454 sequencing platform (Roche Applied Sciences, Indianapolis, IN), each DNA clone is captured on a single bead. Each bead is captured in a pico-titer plate where each well only accommodates a single bead. The plate is harnessed in a microfluidic chip to allow for uniform chemistry and individual nucleotides are passed through the chip with successive washes. Using this approach, a single run can generate approximately 1,000,000 reads in 23 h with average read length of ~700 bps. The 454 platform has been discontinued as of late 2013 and will stop their technical support as of mid-2016.

Sequencing by Synthesis

Sequencing by Synthesis (SBS) is a method that combines the DNA cluster technology developed Pascal Mayer and Laurent Farinelli together with a reversible dye-terminator chemistry originally conceptualized by Shankar Balasubramanian and David Klenerman [71]. The chemistry takes advantage of reversible DNA polymerase inhibitor nucleotides developed by the scientists at Solexa, now Illumina (Illumina, San Diego, CA). The DNA polymerase stalls after incorporating a single nucleotide that is coupled to a nucleotide-specific dye that acts to inhibit DNA polymerase. The DNA polymerase resumes the next nucleotide incorporation upon reversing the inhibition through cleavage of stereo-inhibitor, in this case, a dye. The dye is detected, thereby identifying the incorporated nucleotide. The chemistry was made "high-throughput" by combining with the DNA cluster technology. The DNA fragments are immobilized to a primer attached to a solid surface and clonally amplified, thereby creating a “DNA cluster” or a spot on a glass surface. The amplified DNA then serves as template for the step-wise DNA polymerase activity.

Sequencing by Ligation

The SOLiD system from Applied Biosystems (Life Technologies, Carlsbad, CA), now part of Thermo Fisher Scientific’s NGS portfolio, uses the ligation method for sequencing. The system was first developed in George Church’s laboratory [72]. The key to sequencing by ligation involves the generation of DNA library using Mmel restriction digest to generate paired “genomic” tags of 17–18 bps each. The unique tags are flanked by three adaptors that are used as a template for the anchor primers, required during sequencing. The DNA library amplification, similar to pyrosequencing, is performed on beads in an emulsion PCR reaction, with each bead representing a single DNA clone. The capture beads are used to isolate the clonal beads. The resulting beads with clonally amplified DNA fragments (tags) are spread on a solid surface where the sequencing is performed. Unlike other sequencing strategies, sequencing by ligation does not require DNA polymerase during sequencing, instead, a ligase that ligates the incoming nonamer to the anchor primer. The sequencing cycle then proceeds as follows: (1) hybridization of anchor primer; (2) ligation of degenerate nonamers (4 unique nonamers with fluorescent dye representing each); (3) four color imaging for sequence identification; and (4) stripping (denaturing) anchor primer:nonamer complex from the DNA template. The specificity and identity of the DNA sequence is determined as the nonamers are identical except for the single base (at the same parallel position). By cycling with nonamers with unique identifiers at various positions, the sequence of the template is identified. The resulting sequence has 99.99% accuracy and able to generate gigabytes of sequence in a single run.

Ion Semiconductor

The two by products of nucleotide incorporation during DNA synthesis are PPi and hydrogen ions. As discussed above, pyrosequencing uses PPi detection for sequencing DNA. Ion Torrent (Life Technologies, Carlsbad, CA), also part of Thermo Fisher Scientific’s NGS portfolio, on the other hand, developed a highly sensitive method of measuring H+ ions generated during DNA synthesis. As with pyrosequencing, single clonal beads are captured in each micro-volume wells and cycles of single nucleotides are passed through the device. Unlike pyrosequencing, a semiconductor layer under each well measures and converts the pH change to voltage when H+ ions are released [73].

"Third-Generation" Sequencing

Though NGS technologies have been in the market for a relatively short time, "third generation" sequencing technology has already been introduced to the public. Below we describe two of the most commonly used platforms.

Single-Molecule Real-Time

Single-Molecule Real-Time (SMRT) DNA sequencing technology (Pacific Biosciences, Menlo Park, CA) determines and captures the sequencing activity of the DNA polymerase to provide the sequence identity. In short, DNA polymerase:template complex is captured in a nano-photonic visualization chamber, coined zero-mode waveguide chamber. As each nucleotide is added to the new strand, a fluorophore that is coupled to each nucleotide is released, allowing photo-detection of the incorporated nucleotide in real time. The technology builds on the zero-mode waveguides first developed by Levene et al. [74]. They demonstrated that for single molecule visualization of an enzyme like DNA polymerase, the detection volume must be considerably small. In part, this allows for the micro-molar ligand concentration, as require for physiological enzymatic activity due to ligand diffusion to the enzyme. Additionally, the reduced volume is necessary so the released fluorophore may diffuse rapidly away from the enzyme. By capturing the DNA polymerase:DNA template complex to the bottom of a transparent chamber and allowing light to penetrate through the bottom 20–30 nm of the chamber, the resulting setup provides real time capture of enzymatic activity with very little noise [75]. By tagging the nucleotides with fluorophores that are cleaved when the nucleotide is incorporated, the sequence identity is revealed. The SMRT® technology to date has the longest read length (thousands of bps) compared to any current NGS technology.

Nanopore DNA Sequencing

Nanopore DNA sequencing is based on a nanopore biosensor technology, as devised by Kasianowicz et al. for nucleic acid sequencing [76]. In their work, Kasianowicz demonstrated how single stranded nucleic acids were able to traverse through ion channels embedded in a lipid bilayer. As a result of their work, Oxford Nanopore Technologies (Oxford, UK) was able to harness the concept and develop a new DNA sequencing platforms (GridIONTM and MinIONTM). The technology, as hypothesized by Kasianowicz et al., allows single stranded nucleic acid to traverse through an ion channel under a current. An unoccupied ion channel allows the current to flow through and a sensor coupled to the channel measures the current as nucleic acids traverse through (and inhibit the current in) the nanopore, with each nucleic acid blockage of the pore providing a unique "ion current signature," thereby providing real-time sequence reads that is limited only by the structure and length of the DNA.

Sequencing NGS Methods in Development

The future of sequencing looks even brighter. Current and new technologies such as tunnel-current based single molecule DNA sequencing, advanced hybridization technologies, integrating mass spectrometry and incorporating microfluidics to Sanger sequencing methods are all examples of what to expect in the future of DNA sequencing. Of course, as we see above, combining different strategies will expand the use of these advanced NGS technologies to beyond the scope of simple DNA sequencing.

NGS in the Clinic

In the new era of personalized/precision medicine, as discussed above, targeted NGS can detect heterogeneous gene mutations and interrogate relevant gene content with greater sensitivity to detect rare mutations in cell populations compared to whole genome sequencing and has proven to be a very useful tool in cancer diagnostics and drug development. To date, several NGS methods have been utilized for oncology, including RNA-Seq, whole genome sequencing, whole exome sequencing and targeted resequencing. Commercial targeted NGS cancer panels from Illumina, RainDance Technologies and Thermo Fisher Scientific require genomic DNA (gDNA) input of 10–250 ng and a practical turnaround time ranging from days to weeks. Of note, the United Kingdom has recently launched the AmpliSeqTM (Life Technologies, Carlsbad, CA) 46-gene sequencing tissue-based diagnostic test in their National Health Services [77] and several ongoing clinical trials are employing the AmpliSeqTM cancer panel for patient stratification in the move towards more personalized cancer care. Also, recently the FDA has cleared the use of MiSeqDx Cystic Fibrosis Clinical Sequencing Assay (Illumina, San Diego, CA) utilizing Illumina's HiSeq NGS platform as an IVD. It is the first NGS platform to receive FDA clearance and is used to sequence the cystic fibrosis conductance transmembrane regulator (CFTR) -coding and -noncoding regions of the chromosome. With growing number of studies showing the benefit of using mutational analysis as a prognostic test, it is likely that we will see more NGS assays becoming clinically available.

Conclusion

Human disease is poised to enter a new era of personalized health care, where diagnostic biomarkers act as a central hub in disease prevention, detection, and monitoring of therapeutic response. Nucleic acids can provide biomarkers for a variety of diseases in a non-invasive manner. Nucleic acid-based analytical methods have been rapidly evolving as demonstrated by the recent emergence of dPCR and NGS technologies. However, a critical problem has been poor reproducibility and the lack of standard isolation methods, standard biomarkers, or standard assay methods. Such improvements and engagement of the biotechnology and diagnostics industries will significantly reduce costs while increasing consistency and availability. Thus, we believe the evolution of these genomic-based technologies hold great promise as a new approach to biomarker discovery and precision medical diagnostics. Technology advancement is key to its clinical acceptance, and will rely upon emerging nucleic acid detection technologies (dPCR, NGS or both) as they co-evolve in the clinical space. Successful clinical application of genomic-based diagnostic assays will also require a close collaboration between industry, academia, regulatory agencies and access to patient samples.

References

1. Perou CM, Sorlie T, Eisen MB, et al. Molecular portraits of human breast tumours. Nature.2000;406(6797):747–52
2. Sorlie T, Perou CM, Tibshirani R, et al. Gene expression patterns of breast carcinomas distinguish tumor subclasses with clinical implications. Proc Natl Acad Sci U S A 2001;98(19):10869–74.
3. Sorlie T, Tibshirani R, Parker J, et al. Repeated observation of breast tumor subtypes in independent gene expression data sets. Proc Natl Acad Sci U S A. 2003;100(14):8418–23.
4. Habel LA, Shak S, Jacobs MK, et al. A population-based study of tumor gene expression and risk of breast cancer death among lymph node-negative patients. Breast Cancer Res. 2006;8(3):R25.
5. Paik S, Shak S, Tang G, et al. A multigene assay to predict recurrence of tamoxifen-treated, node-negative breast cancer. N Engl J Med. 2004;351(27):2817–26.
6. Buyse M, Loi S, van't Veer L, et al. Validation and clinical utility of a 70-gene prognostic signature for women with node-negative breast cancer. J Natl Cancer Inst. 2006;98(17):1183–92.
7. van de Vijver MJ, He YD, van't Veer LJ, et al. A gene-expression signature as a predictor of survival in breast cancer. N Engl J Med. 2002;347(25):1999–2009.
8. van 't VLJ, Dai H, van de Vijver MJ, et al. Gene expression profiling predicts clinical outcome of breast cancer. Nature. 2002;415(6871):530–6.
9. Foekens JA, Atkins D, Zhang Y, et al. Multicenter validation of a gene expression-based prognostic signature in lymph node-negative primary breast cancer. J Clin Oncol. 2006;24(11):1665–71.
10. Ring BZ, Seitz RS, Beck R, et al. Novel prognostic immunohistochemical biomarker panel for estrogen receptor-positive breast cancer. J Clin Oncol. 2006;24(19):3039–47
11. Ma XJ, Hilsenbeck SG, Wang W, et al. The HOXB13:IL17BR expression index is a prognostic factor in early-stage breast cancer. J Clin Oncol. 2006;24(28):4611–19.
12. Ascierto PA, Kirkwood JM, Grob JJ, et al. The role of BRAF V600 mutation in melanoma. J Transl Med. 2012;10:85.
13. Dennis Lo YM, Chiu RW. Prenatal diagnosis: progress through plasma nucleic acids. Nat Rev Genet. 2007;8(1):71–7.
14. Diaz LA Jr. The current clinical value of genomic instability. Semin Cancer Biol. 2005; 15(1):67–71.
15. Yakob M, Fuentes L, Wang MB, Abemayor E, Wong DT. Salivary biomarkers for detection of oral squamous cell carcinoma—current state and recent advances. Curr Or Health Rep. 2014;1(2):133–41.
16. Allegra E, Trapasso S, La Boria A, et al. Prognostic role of salivary CD44sol levels in the follow-up of laryngeal carcinomas. J Pathol Med. 2014;43(4):276–81.

17. Cuevas-Cordoba B, Santiago-Garcia J. Saliva: a fluid of study for OMICS. Omics. 2014;18(2):87–97.
18. Miller CS, Foley JD, 3rd, Floriano PN, et al. Utility of salivary biomarkers for demonstrating acute myocardial infarction. J Dent Res. 2014;93:72–9.
19. Spielmann N, Ilsley D, Gu J, et al. The human salivary RNA transcriptome revealed by massively parallel sequencing. Clin Chem. 2012;58(9):1314–21.
20. Lister R, O'Malley RC, Tonti-Filippini J, et al. Highly integrated single-base resolution maps of the epigenome in Arabidopsis. Cell. 2008;133(3):523–36.
21. Shendure J, Ji H. Next-generation DNA sequencing. Nat Biotechnol. 2008;26(10):1135–45.
22. Benjamini Y, Speed TP. Summarizing and correcting the GC content bias in high-throughput sequencing. Nucleic Acids Res. 2012;40(10):e72.
23. Cheung MS, Down TA, Latorre I, Ahringer J. Systematic bias in high-throughput sequencing data and its correction by BEADS. Nucleic Acids Res. 2011;39(15):e103.
24. Guo Y, Li J, Li CI, Long J, Samuels DC, Shyr Y. The effect of strand bias in Illumina short-read sequencing data. BMC Gen. 2012;13:666.
25. Taub MA, Corrada Bravo H, Irizarry RA. Overcoming bias and systematic errors in next generation sequencing data. Gen Med. 2010;2(12):87.
26. Grokhovsky SL, Il'icheva IA, Nechipurenko DY, et al. Sequence-specific ultrasonic cleavage of DNA. Biophys J. 2011;100(1):117–25.
27. Knierim E, Lucke B, Schwarz JM, Schuelke M, Seelow D. Systematic comparison of three methods for fragmentation of long-range PCR products for next generation sequencing. PloS one. 2011;6(11):e28240.
28. Poptsova MS, Il'icheva IA, Nechipurenko DY, et al. Non-random DNA fragmentation in next-generation sequencing. Sci Rep. 2014;4:4532.
29. Head SR, Komori HK, LaMere SA, et al. Library construction for next-generation sequencing: overviews and challenges. Biotechniques. 2014;56(2):61–4.(66 68 passim)
30. Bhargava V, Head SR, Ordoukhanian P, Mercola M, Subramaniam S. Technical variations in low-input RNA-seq methodologies. Sci Rep. 2014;4:3678.
31. Fu GK, Xu W, Wilhelmy J, et al. Molecular indexing enables quantitative targeted RNA sequencing and reveals poor efficiencies in standard library preparations. Proc Natl Acad Sci U S A 2014;111(5):1891–96.
32. Trachtenberg AJ, Robert JH, Abdalla AE, et al. A primer on the current state of microarray technologies. Methods Mol Biol. 2012;802:3–17.
33. Kuo WP, Liu F, Trimarchi J, et al. A sequence-oriented comparison of gene expression measurements across different hybridization-based technologies. Nat Biotechnol. 2006;24(7):832–40.
34. Shi L, Reid LH, Jones WD, et al. The MicroArray Quality Control (MAQC) project shows inter- and intraplatform reproducibility of gene expression measurements. Nat Biotechnol. 2006;24(9):1151–61.
35. Dumur CI, Lyons-Weiler M, Sciulli C, et al. Interlaboratory performance of a microarray-based gene expression test to determine tissue of origin in poorly differentiated and undifferentiated cancers. J Mol Diagn. 2008;10(1):67–77.
36. Monzon FA, Lyons-Weiler M, Buturovic LJ, et al. Multicenter validation of a 1550-gene expression profile for identification of tumor tissue of origin. J Clin Oncol. 2009;27(15):2503–8.
37. Lal A, Panos R, Marjanovic M, et al. A gene expression profile test to resolve head & neck squamous versus lung squamous cancers. Diagnc Pathol. 2013;8:44.
38. Lal A, Panos R, Marjanovic M, et al. A gene expression profile test for the differential diagnosis of ovarian versus endometrial cancers. Oncotarget. 2012;3(2):212–23.
39. Hainsworth JD, Rubin MS, Spigel DR, et al. Molecular gene expression profiling to predict the tissue of origin and direct site-specific therapy in patients with carcinoma of unknown primary site: a prospective trial of the Sarah Cannon research institute. J Clin Oncol. 2013;31(2):217–23.
40. Kalinina O, Lebedeva I, Brown J, Silver J. Nanoliter scale PCR with TaqMan detection. Nucleic Acids Res. 1997;25(10):1999–2004.

41. Vogelstein B, Kinzler KW. Digital PCR. Proc Natl Acad Sci U S A. 1999;96(16):9236–41.
42. Day E, Dear PH, McCaughan F. Digital PCR strategies in the development and analysis of molecular biomarkers for personalized medicine. Methods. 2013;59(1):101–7.
43. Hindson BJ, Ness KD, Masquelier DA, et al. High-throughput droplet digital PCR system for absolute quantitation of DNA copy number. Anal Chem. 2011;83(22):8604–10.
44. Warren L, Bryder D, Weissman IL, Quake SR. Transcription factor profiling in individual hematopoietic progenitors by digital RT-PCR. Proc Natl Acad Sci U S A. 2006;103(47):17807–12.
45. Zhong Q, Bhattacharya S, Kotsopoulos S, et al. Multiplex digital PCR: breaking the one target per color barrier of quantitative PCR. Lab Chip. 2011;11(13):2167–74.
46. Ramakrishnan R, Qin J, Jones RC, Weaver LS. Integrated Fluidic Circuits (IFCs) for digital PCR. Methods Mol Biol. 2013;949:423–31.
47. Valleron W, Laprevotte E, Gautier EF, et al. Specific small nucleolar RNA expression profiles in acute leukemia. Leukemia. 2012;26(9):2052–60.
48. Tehranchi R, Woll PS, Anderson K, et al. Persistent malignant stem cells in del(5q) myelodysplasia in remission. N Engl J Med. 2010;363(11):1025–37.
49. Lu X, Wang L, Chen S, et al. Genome-wide association study in Han Chinese identifies four new susceptibility loci for coronary artery disease. Nat Genet. 2012;44(8):890–4.
50. Sanders SJ, Ercan-Sencicek AG, Hus V, et al. Multiple recurrent de novo CNVs, including duplications of the 7q11.23 Williams syndrome region, are strongly associated with autism. Neuron. 2011;70(5):863–85.
51. Weinstein JA, Jiang N, White RA, 3rd, Fisher DS, Quake SR. High-throughput sequencing of the zebrafish antibody repertoire. Science. 2009;324(5928):807–10.
52. Whale AS, Cowen S, Foy CA, Huggett JF. Methods for applying accurate digital PCR analysis on low copy DNA samples. PloS one. 2013;8(3):e58177.
53. Dressman D, Yan H, Traverso G, Kinzler KW, Vogelstein B. Transforming single DNA molecules into fluorescent magnetic particles for detection and enumeration of genetic variations. Proc Natl Acad Sci U S A. 2003;100(15):8817–22.
54. Li M, Diehl F, Dressman D, Vogelstein B, Kinzler KW. BEAMing up for detection and quantification of rare sequence variants. Nat Methods. 2006;3(2):95–7.
55. Higgins MJ, Jelovac D, Barnathan E, et al. Detection of tumor PIK3CA status in metastatic breast cancer using peripheral blood. Clin Cancer Res. 2012;18(12):3462–9.
56. Taniguchi K, Uchida J, Nishino K, et al. Quantitative detection of EGFR mutations in circulating tumor DNA derived from lung adenocarcinomas. Clin Cancer Res. 2011;17(24):7808–15.
57. Diehl F, Schmidt K, Durkee KH, et al. Analysis of mutations in DNA isolated from plasma and stool of colorectal cancer patients. Gastroenterology. 2008;135(2):489–98.
58. Misale S, Yaeger R, Hobor S, et al. Emergence of KRAS mutations and acquired resistance to anti-EGFR therapy in colorectal cancer. Nature. 2012;486(7404):532–6.
59. Kiss MM, Ortoleva-Donnelly L, Beer NR, et al. High-throughput quantitative polymerase chain reaction in picoliter droplets. Anal Chem. 2008;80(23):8975–81.
60. Horoszewicz JS, Murphy GP. Prospective new developments in laboratory research and clinical trials in prostatic cancer. Cancer. 1990;66(5 Suppl):1083–5.
61. Burgos KL, Javaherian A, Bomprezzi R, et al. Identification of extracellular miRNA in human cerebrospinal fluid by next-generation sequencing. RNA. 2013;19(5):712–22.
62. Kinde I, Wu J, Papadopoulos N, Kinzler KW, Vogelstein B. Detection and quantification of rare mutations with massively parallel sequencing. Proc Natl Acad Sci U S A. 2011;108(23):9530–5.
63. Cronin M, Ross JS. Comprehensive next-generation cancer genome sequencing in the era of targeted therapy and personalized oncology. Biomark Med. 2011;5(3):293–305.
64. Lipson D, Capelletti M, Yelensky R, et al. Identification of new ALK and RET gene fusions from colorectal and lung cancer biopsies. Nat Med. 2012;18(3):382–4.
65. Sanger F, Nicklen S, Coulson AR. DNA sequencing with chain-terminating inhibitors. Proc Natl Acad Sci U S A. 1977;74(12):5463–7.
66. Sanger F, Nicklen S, Coulson AR. DNA sequencing with chain-terminating inhibitors. 1977. Biotechnology. 1992;24:104–8.

67. Smith LM. High-speed DNA sequencing by capillary gel electrophoresis. Nature. 1991; 349(6312):812–3.
68. Swerdlow H, Gesteland R. Capillary gel electrophoresis for rapid, high resolution DNA sequencing. Nucleic Acids Res. 1990;18(6):1415–9.
69. Ronaghi M, Karamohamed S, Pettersson B, Uhlen M, Nyren P. Real-time DNA sequencing using detection of pyrophosphate release. Anal Biochem. 1996;242(1):84–9.
70. Nyren P. Enzymatic method for continuous monitoring of DNA polymerase activity. Anal Biochem. 1987;167(2):235–8.
71. Bentley DR, Balasubramanian S, Swerdlow HP, et al. Accurate whole human genome sequencing using reversible terminator chemistry. Nature. 2008;456(7218):53–9.
72. Shendure J, Porreca GJ, Reppas NB, et al. Accurate multiplex polony sequencing of an evolved bacterial genome. Science. 2005;309(5741):1728–32.
73. Rothberg JM, Hinz W, Rearick TM, et al. An integrated semiconductor device enabling non-optical genome sequencing. Nature. 2011;475(7356):348–52.
74. Levene MJ, Korlach J, Turner SW, Foquet M, Craighead HG, Webb WW. Zero-mode waveguides for single-molecule analysis at high concentrations. Science. 2003;299(5607):682–6.
75. Korlach J, Bjornson KP, Chaudhuri BP, et al. Real-time DNA sequencing from single polymerase molecules. Methods Enzymol. 2010;472:431–55.
76. Kasianowicz JJ, Brandin E, Branton D, Deamer DW. Characterization of individual polynucleotide molecules using a membrane channel. Proc Natl Acad Sci U S A. 1996;93(24):13770–3.
77. Phalane KG, Kriel M, Loxton AG, et al. Differential expression of host biomarkers in saliva and serum samples from individuals with suspected pulmonary tuberculosis. Mediat Inflamm. 2013;2013:981984.
78. Burnside RD, Spudich L, Rush B, Kubendran S, Schaefer GB. Secondary complex chromosome rearrangement identified by chromosome analysis and FISH subsequent to detection of an unbalanced derivative chromosome 12 by SNP array analysis. Cytogenet Genome Res. 2014;142(2):129–33.
79. Puiggros A, Puigdecanet E, Salido M, et al. Genomic arrays in chronic lymphocytic leukemia routine clinical practice: are we ready to substitute conventional cytogenetics and fluorescence in situ hybridization techniques? Leuk Lymphoma. 2013;54(5):986–95.
80. Yu YP, Michalopoulos A, Ding Y, Tseng G, Luo JH. High fidelity copy number analysis of formalin-fixed and paraffin-embedded tissues using Affymetrix Cytoscan HD chip. PloS One. 2014;9(4):e92820.
81. Nielsen T, Wallden B, Schaper C, et al. Analytical validation of the PAM50-based prosigna breast cancer prognostic gene signature assay and nCounter Analysis system using formalin-fixed paraffin-embedded breast tumor specimens. BMC Cancer. 2014;14:177.
82. Handorf CR, Kulkarni A, Grenert JP, et al. A multicenter study directly comparing the diagnostic accuracy of gene expression profiling and immunohistochemistry for primary site identification in metastatic tumors. Am J Surg Pathol. 2013;37(7):1067–75.
83. Mook S, Schmidt MK, Weigelt B, et al. The 70-gene prognosis signature predicts early metastasis in breast cancer patients between 55 and 70 years of age. Ann Oncol. 2010;21(4):717–22.
84. Bedard PL, Mook S, Piccart-Gebhart MJ, Rutgers ET, Van't Veer LJ, Cardoso F. MammaPrint 70-gene profile quantifies the likelihood of recurrence for early breast cancer. Expert Opin Med Diagn. 2009;3(2):193–205.
85. Glas AM, Floore A, Delahaye LJ, et al. Converting a breast cancer microarray signature into a high-throughput diagnostic test. BMC Gen. 2006;7:278.
86. Deng MC, Eisen HJ. Mehra MR, et al. Noninvasive discrimination of rejection in cardiac allograft recipients using gene expression profiling. Am J Transplant. 2006;6(1):150–60.
87. Pham MX, Teuteberg JJ, Kfoury AG, et al. Gene-expression profiling for rejection surveillance after cardiac transplantation. N Engl J Med. 2010;362(20):1890–900.
88. Starling RC, Pham M. Valantine H, et al. Molecular testing in the management of cardiac transplant recipients: initial clinical experience. J Heart Lung Transplant. 2006;25(12):1389–95.

89. Gittelman MC, Hertzman B. Bailen J, et al. PCA3 molecular urine test as a predictor of repeat prostate biopsy outcome in men with previous negative biopsies: a prospective multicenter clinical study. J Urol. 2013;190(1):64–9.
90. Nauck M, Marz W, Wieland H. Evaluation of the roche diagnostics LightCycler-Factor V leiden mutation detection kit and the LightCycler-Prothrombin mutation detection kit. Clin Biochem. 2000;33(3):213–6.
91. Svensson AM, Chou LS, Meadows C, et al. Implementation of a cost-effective unlabeled probe high-resolution melt assay for genotyping of Factor V Leiden. Genetic Test Mol Biomark. 2011;15(4):207–13.
92. Maurice CB, Barua PK, Simses D, Smith P, Howe JG, Stack G. Comparison of assay systems for warfarin-related CYP2C9 and VKORC1 genotyping. Clin Chim Acta. 2010;411(13–14):947–54.
93. Grosu DS, Hague L. Chelliserry M, et al. Clinical investigational studies for validation of a next-generation sequencing in vitro diagnostic device for cystic fibrosis testing. Expert Rev Mol Diagn. 2014;14(5):605–22.
94. Keller MA, Martinez J. Baradet TC, et al. Fibrinogen Philadelphia, a hypodysfibrinogenemia characterized by abnormal polymerization and fibrinogen hypercatabolism due to gamma S378P mutation. Blood. 2005;105(8):3162–8.
95. Raliou M, Wiencis A. Pillias AM, et al. Nonsynonymous single nucleotide polymorphisms in human tas1r1, tas1r3, and mGluR1 and individual taste sensitivity to glutamate. Am J Clin Nutr. 2009;90(3):789S–99S.
96. Durdiakova J, Kamodyova N, Ostatnikova D, Vlkova B, Celec P. Comparison of different collection procedures and two methods for DNA isolation from saliva. Clin Chem Lab Med. 2012;50(4):643–7.
97. Keijzer H, Endenburg SC, Smits MG, Koopmann M. Automated genomic DNA extraction from saliva using the QIAxtractor. Clin Chem Lab Med. 2010;48(5):641–3.
98. Matthews AM, Kaur H, Dodd M, et al. Saliva collection methods for DNA biomarker analysis in oral cancer patients. Br J Oral Maxillofac Surg. 2013;51(5):394–8.
99. Nishita DM, Jack LM. McElroy M, et al. Clinical trial participant characteristics and saliva and DNA metrics. BMC Med Res Methodol. 2009;9:71.
100. Rylander-Rudqvist T, Hakansson N, Tybring G, Wolk A. Quality and quantity of saliva DNA obtained from the self-administrated oragene method–a pilot study on the cohort of Swedish men. Cancer Epidemiol, Biomark Prev. 2006;15(9):1742–45.
101. Witt S, Neumann J, Zierdt H, Gebel G, Roscheisen C. Establishing a novel automated magnetic bead-based method for the extraction of DNA from a variety of forensic samples. Forens Sci Int Genet. 2012;6(5):539–47.
102. Patel RS, Jakymiw A. Yao B, et al. High resolution of microRNA signatures in human whole saliva. Arch Oral Biol. 2011;56(12):1506–13.
103. Jones TH, Muehlhauser V. Effect of handling and storage conditions and stabilizing agent on the recovery of viral RNA from oral fluid of pigs. J Virol Methods. 2014;198:26–31.
104. Park NJ, Li Y, Yu T, Brinkman BM, Wong DT. Characterization of RNA in saliva. Clin Chem. 2006;52(6):988–94.
105. Vermeersch P, Marien G, Bossuyt X. A case of pseudoparaproteinemia on capillary zone electrophoresis caused by geloplasma. Clin Chem. 2006;52(12):2309–11.
106. Pandit P, Cooper-White J, Punyadeera C. High-yield RNA-extraction method for saliva. Clin Chem. 2013;59(7):1118–22.
107. Nwokeji ED, Rascati KL, Brown CM, Eisenberg A. Influences of attitudes on family physicians' willingness to prescribe long-acting opioid analgesics for patients with chronic nonmalignant pain. Clin Ther. 2007;29:2589–602.
108. Barkai G, Ari-Even Roth D, Barzilai A, et al. Universal neonatal cytomegalovirus screening using saliva—Report of clinical experience. J Clin Virol. 2014;60(4):361–6.
109. Esona MD, McDonald S, Kamili S, Kerin T, Gautam R, Bowen MD. Comparative evaluation of commercially available manual and automated nucleic acid extraction methods for rotavirus RNA detection in stools. J Virol Methods. 2013;194(1–2):242–9.

110. Li S, Liu H. Jia Y, et al. An automatic high-throughput single nucleotide polymorphism genotyping approach based on universal tagged arrays and magnetic nanoparticles. J Biomed Nanotechnol. 2013;9(4):689–98.
111. DeAngelis MM, Wang DG, Hawkins TL. Solid-phase reversible immobilization for the isolation of PCR products. Nucleic Acids Res. 1995;23(22):4742–3.
112. Griesemer SB, Holmberg R. Cooney CG, et al. Automated, simple, and efficient influenza RNA extraction from clinical respiratory swabs using TruTip and epMotion. J Clin Virol. 2013;58(1):138–43.
113. Kalina WV, Douglas CE, Coyne SR, Minogue TD. Comparative assessment of automated nucleic acid sample extraction equipment for biothreat agents. J Clin Microbiol. 2014;52(4):1232–4.
114. Fujii K, Inokuchi S, Kitayama T, Nakahara H, Mizuno N, Sekiguchi K. A comparison of DNA extraction using AutoMate Express and EZ1 advanced XL from liquid blood, bloodstains, and semen stains. J Forensic Sci. 2013;58(4):981–8.
115. Pereira JC, Chaves R, Bastos E, Leitao A, Guedes-Pinto H. An efficient method for genomic DNA extraction from different molluscs species. Int J Mol Sci. 2011;12(11):8086–95.
116. Roy SW, Irimia M. When good transcripts go bad: artifactual RT-PCR 'splicing' and genome analysis. Bioessays. 2008;30(6):601–5.
117. Bustin SA, Nolan T. Pitfalls of quantitative real-time reverse-transcription polymerase chain reaction. J Biomol Tech. 2004;15(3):155–66.
118. Roberts JD, Preston BD, Johnston LA, Soni A, Loeb LA, Kunkel TA. Fidelity of two retroviral reverse transcriptases during DNA-dependent DNA synthesis in vitro. Mol Cell Biol. 1989;9(2):469–76.
119. Gubler U. Second-strand cDNA synthesis: mRNA fragments as primers. Methods Enzymol. 1987;152:330–5.
120. Spiegelman S, Burny A, Das MR et al. DNA-directed DNA polymerase activity in oncogenic RNA viruses. Nature. 1970;227(5262):1029–31.
121. Cancer Genome Atlas Network. Comprehensive molecular portraits of human breast tumours. Nature. 2012;490(7418):61–70.
122. Lopez-Rios F, Angulo B, Gomez B, et al. Comparison of testing methods for the detection of BRAF V600E mutations in malignant melanoma: pre-approval validation study of the companion diagnostic test for vemurafenib. PloS one. 2013;8(1):e53733.
123. Shukla R, Upton KR, Munoz-Lopez M, et al. Endogenous retrotransposition activates oncogenic pathways in hepatocellular carcinoma. Cell. 2013;153(1):101–11.
124. Tenedini E, Bernardis I, Artusi V, et al. Targeted cancer exome sequencing reveals recurrent mutations in myeloproliferative neoplasms. Leukemia. 2014;28(5):1052–9.
125. Arinami T, Ohtsuki T, Ishiguro H, et al. Genomewide high-density SNP linkage analysis of 236 Japanese families supports the existence of schizophrenia susceptibility loci on chromosomes 1p, 14q, and 20p. Am J Hum Genet. 2005;77(6):937–44.
126. Newman AM, Bratman SV, To J, et al. An ultrasensitive method for quantitating circulating tumor DNA with broad patient coverage. Nat Med. 2014;20(5):548–54.
127. Tragante V, Barnes MR, Ganesh SK, et al. Gene-centric meta-analysis in 87,736 individuals of European ancestry identifies multiple blood-pressure-related loci. Am J Hum Genet. 2014;94(3):349–60.
128. Treutlein B, Brownfield DG, Wu AR et al. Reconstructing lineage hierarchies of the distal lung epithelium using single-cell RNA-seq. Nature. 2014;509(7500):371–5.
129. Stark MS, Woods SL, Gartside MG, et al. Frequent somatic mutations in MAP3K5 and MAP3K9 in metastatic melanoma identified by exome sequencing. Nat Genet. 2012;44(2):165–9.
130. Ju YS, Kim JI, Kim S, et al. Extensive genomic and transcriptional diversity identified through massively parallel DNA and RNA sequencing of eighteen Korean individuals. Nat Genet. 2011;43(8):745–52.
131. Kegel A, Betts-Lindroos H, Kanno T, et al. Chromosome length influences replication-induced topological stress. Nature. 2011;471(7338):392–6.

132. Wilhelm BT, Briau M, Austin P, et al. RNA-seq analysis of 2 closely related leukemia clones that differ in their self-renewal capacity. Blood. 2011;117(2):e27–e38.
133. Peters BA, Kermani BG, Sparks AB, et al. Accurate whole-genome sequencing and haplotyping from 10 to 20 human cells. Nature. 2012;487(7406):190–5.
134. Drmanac R, Sparks AB, Callow MJ, et al. Human genome sequencing using unchained base reads on self-assembling DNA nanoarrays. Science. 2010;327(5961):78–81.
135. Woyach JA, Furman RR, Liu TM, et al. Resistance mechanisms for the Bruton's tyrosine kinase inhibitor ibrutinib. N Engl J Med. 2014;370(24):2286–94.
136. Palles C, Cazier JB, Howarth KM, et al. Germline mutations affecting the proofreading domains of POLE and POLD1 predispose to colorectal adenomas and carcinomas. Nat Genet. 2013;45(2):136–44.
137. Biankin AV, Waddell N, Kassahn KS, et al. Pancreatic cancer genomes reveal aberrations in axon guidance pathway genes. Nature. 2012;491(7424):399–405.
138. van den Oever JM, Balkassmi S, Johansson LF, et al. Successful noninvasive trisomy 18 detection using single molecule sequencing. Clin Chem. 2013;59(4):705–9.
139. Ozsolak F, Milos PM. Single-molecule direct RNA sequencing without cDNA synthesis. Wiley Interdiscip Rev RNA. 2011;2(4):565–70.
140. Bowers J, Mitchell J, Beer E, et al. Virtual terminator nucleotides for next-generation DNA sequencing. Nat Methods. 2009;6(8):593–5.
141. Gupta PK. Single-molecule DNA sequencing technologies for future genomics research. Trends Biotechnol. 2008;26(11):602–11.
142. Harris TD, Buzby PR, Babcock H, et al. Single-molecule DNA sequencing of a viral genome. Science. 2008;320(5872):106–9.
143. Braslavsky I, Hebert B, Kartalov E, Quake SR. Sequence information can be obtained from single DNA molecules. Proc Natl Acad Sci U S A. 2003;100(7):3960–4.
144. Roberts RJ, Carneiro MO, Schatz MC. The advantages of SMRT sequencing. Genome Biol. 2013;14(6):405.
145. Loomis EW, Eid JS, Peluso P, et al. Sequencing the unsequenceable: expanded CGG-repeat alleles of the fragile X gene. Genome Res. 2013;23(1):121–8.
146. Pugh TJ, Weeraratne SD, Archer TC, et al. Medulloblastoma exome sequencing uncovers subtype-specific somatic mutations. Nature. 2012;488(7409):106–10.
147. Flusberg BA, Webster DR, Lee JH, et al. Direct detection of DNA methylation during single-molecule, real-time sequencing. Nat Methods. 2010;7(6):461–5.
148. Eid J, Fehr A, Gray J, et al. Real-time DNA sequencing from single polymerase molecules. Science. 2009;323(5910):133–8.
149. Wallace EV, Stoddart D, Heron AJ, et al. Identification of epigenetic DNA modifications with a protein nanopore. Chem Commun (Camb). 2010;46(43):8195–7.
150. Clarke J, Wu HC, Jayasinghe L, Patel A, Reid S, Bayley H. Continuous base identification for single-molecule nanopore DNA sequencing. Nat Nanotechnol. 2009;4(4):265–70.
151. Howorka S, Cheley S, Bayley H. Sequence-specific detection of individual DNA strands using engineered nanopores. Nat Biotechnol. 2001;19(7):636–9.

The Oral Microbiome and Its Relationship to Genomics and Oral Disease

Ingrid Glurich, Sanjay Shukla, Amit Acharya, Geoffrey S. Ginsburg and Murray H. Brilliant

What Metagenomics Has Revealed about the Oral Microbiome

This chapter will provide an overview of advances in the understanding of the oral microbiome and its contributions to health and disease. The chapter has been organized to first highlight emerging insights emanating from the study of the oral microbiota as constituents of the human mega-microbiome. Next, a brief overview of 'omics' technologies most commonly used to explore microbiota and host interaction is provided. Progress in defining a 'core' microbiome associated with oral health is then explored, followed by characterization of microbiota in selected oral and systemic disease states. Finally, implications for diagnostics, therapeutics and personalized care associated with this rapidly expanding field are briefly examined. A compilation of pivotal resources advancing understanding of the oral microbiome is presented at the end of the chapter.

M. H. Brilliant (✉) · S. Shukla
Center for Human Genetics, Marshfield Clinic Research Foundation, Marshfield, WI 54449, USA
e-mail: brilliant.murray@mcrf.mfldclin.edu

I. Glurich
Institute for Oral and Systemic Health, Marshfield Clinic Research Foundation, Marshfield, WI 54449, USA

A. Acharya
Institute for Oral and Systemic Health, Marshfield Clinic Research Foundation, Marshfield, WI 54449, USA

G. S. Ginsburg
Center for Genomic Medicine, Duke Translational Medicine Institute, Durham, NC 27710, USA

S. T. Sonis (ed.), *Genomics, Personalized Medicine and Oral Disease*,
DOI 10.1007/978-3-319-17942-1_3

Concepts and Advances in Microbiome and Metagenomics

The collective microbial load of a human being is estimated to quantitatively exceed the total volume of human cells in that individual by an order of magnitude [95]. Collectively microbial genomic material (or metagenome) is estimated to include >19,000 microbial phylotypes [48]. The number of species in the oral cavity varies by site, with over 16,000 species identified in subgingival plaque alone [40]. This bacterial load expands the genomic content of any given human by approximately 100 fold and adds systemic attributes that were not genetically endowed by the human genome [102]. By discerning and adapting relational balances through immune mechanisms with resident microbial communities in the context of shifting environmental dynamics, the host establishes an equilibrium which promotes health.

The study of human microbiome over the last few years has opened up new horizons to query, understand and explain the complex interactions between human host and the microbial life it harbors, and how these interactions may keep one relatively healthy or contribute to diseases. Newer 'omic' technologies such as metagenomics, (meta)/transcriptomics, proteomics, and metabolomics are providing insights into the highly interdependent metabolic networks that exist between humans and their microbial constituents. In the past decade, application of 'omic' technologies has facilitated investigation and characterization of microbial capacity to adapt to focal environments that exist throughout the body. With the advent of deep sequencing technology, an important thrust in metagenomic research has been to explore whether a 'core' microbiome associated with health can be defined broadly for human oral microbiome. Challenges associated with addressing definition of a 'core' are addressed in this chapter in Sect. 1.3.

Emerging insights into microbiome structure and function facilitated by the availability of multifaceted metagenomic technologies are affecting paradigm shifts and new insight into the role of the oral cavity in overall health, including which organisms are implicated in health and disease. Importantly, these culture-independent technologies have revealed the presence of a subpopulation of microorganisms, constituting ~50% of the oral taxa associated with periodontal disease and endodontic periodontitis, which are uncultivatable and which were first identified via metagenomic study. (reviewed by [69, 89] Certain members of uncultivatable bacterial phyla are now being associated with periodontal disease and cariogenesis, but virtually nothing is known about these organisms or their relative contribution to oral or systemic disease, or how to create *in vitro* environmental constructs to support their growth and characterization. Based on presentations at the 2014 American Association for Dental Research (AADR) meeting in Charlotte NC, some laboratories are reporting success in defining and systematically creating supportive environments to culture representative organisms from phyla that are associated with disease based on metagenomic data. The extent to which currently uncultivatable organisms contribute to oral health and disease (or interact with traditional periodontal or cariogenic pathogens or oncogenesis) is presently unknown, but may become more apparent as microbiota associated with health and various disease states are further delineated using 'omic' approaches. Thus, these organisms represent virtually uncharted territory based on current understanding of oral and systemic health and disease, but have broadened the landscape of organisms with pathogenic potential while introducing new challenges with respect to clinical management of their presence in the context of oral disease.

Recent evidence further suggests that there is a network between the oral microbiomes and other distant microbiomes within the host. This networking may impact constituent population profiles within these disparate microbiomes. In the context of some disease states, immune deficiencies or dysfunction of the host may further amplify dysbiosis (microbial imbalance) and establish disease chronicity or advancement. Moreover, emerging evidence suggests that the physiological environment created in the context of some diseases and its impact on immune status of the affected individual may strongly modulate the composition of the oral microbiome. Thus, the oral microbiome in the presence of other underlying pathological states (e.g. Crohn's disease [23] and Type 2 diabetes mellitus [13, 83, 105] may be distinct from the microbiome discerned in disease-free subjects. The scenarios in these two disease states are discussed in detail later in this chapter in Sect. 2.0.

Recent reports have recognized a potential role for host-microbial interactions in the oral cavity in the evolution of drug resistance mechanisms. Approximately 3 % of genes identified in microbiomes derived from dental plaque are associated with antibiotic resistance [81, 98] A recent study of dental plaque derived from skeletons excavated from a medieval monastic site dating back to ca. 950–1200CE, demonstrated genetically-encoded, broad-spectrum antibiotic resistance predating the era of therapeutic antibiotics. Detected resistance genes included broad spectrum efflux pumps, genes encoding resistance to aminoglycosides, bacitracin, bacteriocins and macrolides, plasmid-encoded transposons for efflux pump genes, including one with high homology to CTn5 of *Clostridium difficile* [93]. Employing shotgun and targeted DNA sequencing, the investigators successfully characterized the microbiome of calculus from these ancient individuals. This proteomic study revealed evidence of: (1) bacterial carriage of virulence factors including those facilitating hemagglutination, adhesion and tissue invasion and transmission elements including plasmids, transposons and phages which support mechanisms for horizontal gene transfer; (2) enrichment for human proteins associated with innate immune responses, inflammation and mechanisms of host defense, secreted proteins of neutrophilic origin deposited at the junctions of the epithelial pocket to kill bacteria harboring in the plaque otherwise protected from phagocytosis, salivary proteins and immunoglobulin heavy and light chains; (3) morphological evidence of bone and attachment loss providing evidence of periodontal disease. Interestingly, ancient and modern root morphologies were structurally distinct, both functionally and in composition, with the former exhibiting high collagen content. Nine bacterial phyla dominated the medieval dental calculus, and reflect phyla represented in studies performed within the past decade. At the species level, organisms associated with periodontal disease (PD), cariogenesis, upper respiratory infection and endocarditis were also prominent in the ancient calculus [93].

Another important emerging paradigm shift recognizes that the 'oral microbiome' actually represents distinct microbiomes defined by the surface with which they are associated. Studies sampling microbiomes from different surfaces in the oral cavity report different constituent communities and variable quantitative representation across the distinct microbiomes. Thus, microbiome characterization from saliva, tongue cultures, dental plaque, endodontic samples collected from root canals, tonsils or cheek swabs show variability in their microbial composition, contributing considerable challenge to establishing a definition of a healthy 'core' microbiome in association with the oral cavity. These differences are presumably due to differences in availability of oxygen and nutrients, and protective properties of saliva, among others, in different parts of oral environment. Complicating this even more, approaches to sampling have also yielded different constituency profiles depending on the depth of penetration of the periodontal pocket during sample collection [28]. Progress in definition of the healthy oral 'core' microbiome is discussed further in Sect. 1.3.

Metagenomic studies exploring dynamic aspects of the microbiome associated with aging have recently been undertaken, largely in the context of cariogenic susceptibility with advancing age, and are reviewed in Sect. 1.4.2. These studies have shown that shifts in microbiota occur across the human lifespan.

Quantitative Approaches to Microbiome Analysis

Overview of Technologies Applied to Metagenomic Microbiome Research

Traditional technologies used to characterize microbial pathogens associated with PD have included culture and microscopy techniques, enzyme or immunoassays, PCR and DNA-DNA hybridization. Whereas culture and microscopy techniques provided insights into microbiota present in the oral cavity under conditions of health and disease and supported accurate identification of cultivatable organisms, including evidence of antimicrobial resistance, these techniques focused only on detection and quantitation of a core subset of microbes traditionally associated with oral diseases that can be cultivated. Application of PCR technology and checkerboard DNA-DNA hybridization technology has facilitated detection and quantitation to organisms undetected by other technologies.

Analysis of the full spectrum of constituent microorganisms comprising microbiomes has only become possible with the advent of high throughput metagenomic approaches and these have radically redefined the role of microorganisms in oral and systemic disease pathogenesis. Indeed, advances in sequencing technologies to study metagenome, bioinformatics tools to analyze microbiome, delineation of host genetic variation and its impact on microbial pathogenesis has generated a new study domain coined 'Infectogenomics' by Kellam and Weiss [49] that facilitates exploration of the host-pathogen interaction at a metagenomic level. Such studies are shifting paradigms defining dynamic relationships between the host, and the microbial inhabitants that establish and occupy environmental niches in the host in the context of constantly shifting environmental dynamics. Further, the emerging deep sequencing technologies have expanded 'omic' study to encompass meta-genomic, -transcriptomic, -proteomic, and -metabolomic investigation of microbial communities. 'Omic' studies are providing new insight into the dynamics of microbial interaction with the host and environment, how these interactions may contribute to shifts in the community composition, disease emergence and progression. These technologies have contributed to new global perspectives on the functional importance of microbiomes in maintenance of health and potential for contribution to specific disease processes. Their advantages over traditional approaches include: (1) rapid or high throughput capacity for many technologies with availability of annotated resources to support interpretation, (2) capacity to study of organisms that cannot be cultivated, (3) capacity for deep sequencing, analysis of metabolic dynamics and functional characteristics, (4) support for definition of potential pathogenic mechanisms, and (5) broadened perspective into impact of global community dynamics as opposed to characterization of only those few constituent members traditionally linked with oral and/or systemic disease. 'Omic' technologies currently employed to study microbiomes are discussed briefly below.

16S rRNA gene based sequencing technology supports screening and classification of microbiota through detection of the unique genetic configurations of the resident microorganisms in the hypervariable regions of prokaryotic16S ribosomal RNA (RNA). This approach focuses on aligning DNA primers with highly conserved ribosomal RNA sequences common to bacterial species, PCR amplification and sequencing to support examination of known variable regions that allow discrimination of the microorganism and evolutionary changes that may have occurred over time. Thus, 16S rRNA sequencing is a cost effective approach for detailed characterization of microbial diversity associated with microbiomes. Some limitations associated with this approach include PCR bias and sensitivity to contamination.

Another sequencing approach is pyrosequencing, which involves synthetic sequencing and relies on detection of pyrophosphate signal released prior to incorporation of the subsequent nucleotide onto a single stranded template. Determination of which of four possible nucleotide is incorporated is dictated by the relative intensity of signal released as the preceding nucleotide in the sequence is degraded. Limitations associated with pyrosequencing are that this method does not produce full length 16S sequences but instead produces relatively short sequence stretches ranging between 300 to 500 nucleotides which may make alignment for genome assembly challenging. Advantages of the technique are that it is less prone to bias because no cloning is involved and further, it is more nimble at highlighting biodiversity.

Custom array-based approaches which selectively target the most prevalent species in the oral cavity based on Sanger sequence data are a more targeted alternative to genomic analysis. An example includes the Human Oral Microbe Identification Microarray (HOMIM) hybridization assay, which surveys 300 of the most prevalent oral species and has been extensively validated. The technology supporting the array is the 16S rRNA approach. One limitation of this approach is that the assay is constrained to detecting only constituents that correspond to the preselected primers included in the array. On comparison with 16SrRNA pyrosequencing with HOMIM high concordance was demonstrated. Pyrosequencing was superior in detecting less common genera whose contribution was deemed minor. HOMIM was reported to be an acceptable array approach for developing broad-scope microbiome profiling and may be a cost effective alternative screening tool for personalized medicine approaches [1]. A more comprehensive review and comparative analysis of these technologies is presented by Ahn et al. [2].

Transcriptomic Approaches

Metatranscriptomic approaches could be used to determine the metabolic environment associated with oral microbiota in the context of health and disease. For example, a recent study by Jorth et al. [43] demonstrated that despite variability in composition of microbiota across the spectrum of periodontal health and disease, comparison of gene expression of 16,000 genes in periodontal microbial communities consistently correlated with differences in metabolic patterns that associated with, and were shared by, patients based on absence or presence of periodontal disease.

Similarly, one can study the transcriptome of oral tissues in response to changing microbiota from diseased and healthy states. These comparisons will help to identify the role of regulatory molecules, especially those contributing to inflammatory processes. A few studies have focused on transcriptomic analysis of whole genome expression data in the context of oral disease to assist with accurate phenotypic classification of disease subtypes. For example, a study by Kebschull et al. [47], used the whole genome expression data from gingival tissues to help with differentiation of patients into chronic and aggressive PD phenotypes. These investigators were able to distinguish several distinct gene expression signatures that differentiated between phenotypes. The authors posited that transcriptomics may represent a new approach to aid in disease classification, thus reducing 'noise' associated with misclassification errors in designing case control studies. Further, Mans et al. [60] have shown that transcriptomic study of epithelial cells subjected to single and complex microbial exposures provided a better understanding of how bacterial-bacterial interactions and bacterial-host interactions modulate the overall host response.

Transcriptomic analysis could be potentially used in the evaluation of gene expression patterns in saliva for creation of diagnostic tools. An example of this application includes use of mRNA or miRNA expression arrays to define signatures associated with oral cancer detection followed by quantitative polymerase chain reaction to validate genes whose signals denoted a differential expression pattern. In a study of oral squamous cell carcinoma, Li et al. [57] successfully defined transcriptomic signatures of *IL8, IL1B, DUSP1, HA3, OAZ1, S200P* and *SAT* as potentially diagnostic RNA biomarkers.

Proteomics

Proteomic analysis involves protein isolation, digestion and separation utilizing protein identification technology (such as MudPIT , which involves using micro elution of proteins separated on cationic exchange columns) followed by mass spectrometric analysis of separated proteins and quantitation using quantitative PCR approaches [51]. Proteomic approaches are particularly useful to the analysis of proteins produced by the host or microbes in the context of host-microbial interaction and creation of environments conducive to microbial establishment and survival. For example, high throughput quantitative proteomic analysis can help to define proteins associated with bacterial adaptation and survival and virulence factor production during shifts in environmental conditions that support pathogenesis.

The utility of this approach is illustrated in a study by Klein et al. [51] who studied the protein profiles of *S. mutans* during attachment and establishment of pro-cariogenic biofilms in a mixed-species *in-vitro* model simulating colonization patterns and an environmental milieu associated with cariogenesis *in vivo.* Based on protein expression patterns, the investigators discerned a role for *S. mutans* genes critical to establishment of a pro-cariogenic biofilm including: *gtfB, gtfC dexA, ftf, gbpB, manL, glgP, atpD, fabM, groES,* and *nox.* These genes play central roles in adaptation of *S. mutans* to an increasingly stressful and acidic environment established by other co-habitating microbes, increased metabolic capacity for glycogen storage polymers and lipoteichic acid, and capacity for glucan synthesis, remodeling and binding. Other biofilm model systems are defined in more detail in Sect. 1.2.5.

Metabolomics

Metabolomic analysis involves metabolite identification by application of ultra-high performance liquid chromatographic separation of metabolites with a basic pH followed by tandem mass spectrometry (UHPLC/MS/MS) and UHPLC/MS/MS adapted for metabolites with an acidic pH followed by gas chromatography/mass spectrometry [7]. For example, a study which applied metabolomic analysis to saliva collected from subjects with PD revealed high levels of macromolecular degradation commensurate with bacterial metabolic activity, compared to saliva of healthy subjects. The investigators noted that increased enzymatic breakdown of lipids, proteins and polysaccharides by bacteria in subjects with PD generated a favorable environmental energy balance in which oral pathogens could thrive and exacerbate pathogenic processes [7].

In vitro Models Supporting 'Omic' Analysis

In vitro biofilm model systems mimic early oral biofilm development *in vitro.* An artificial environment is created that reproduces *in vivo* environmental conditions to enrich for microorganisms which then establish the biofilm on scaffold. Several approaches have been employed (reviewed by Edlund et al. [24]), including:

1. chemostats which recreate the environmental conditions thereby permitting observation of dynamics of microbial community formation and responsive shifts associated with environmental perturbations, such as pH shifts, changes in relative O_2 content, iron or nutrient availability, and others ([54] review);

2. constant depth film fermenters, which reproduce the nutrient milieu that attracts specific subsets of microorganism with capacity to thrive under the given conditions and exist symbiotically with other microorganisms within the simulated environment; [50]
3. saliva conditioned flow cells, which expose microbial communities to host factors found in saliva which bathes microbiomes in the oral cavity and artificial mouths which simulate conditions in the oral cavity to the greatest extent possible; [27]
4. dental plaque microcosm model: an 'artificial mouth' plaque culture system in which bacteria are cultured in the presence of 'plaque enriched' saliva collected from saliva of donors who have abstained from any oral hygiene for 24 h [96].

In vitro simulation has exhibited capacity to establish stable oral biofilms which accurately reflect taxonomic carriage and proportions detected *in vivo*. Biofilms incorporating one hundred operational taxonomic units (OTUs), reflecting 60–80 % of the OTU contained in the original inoculum, have been achieved. Notably, uncultivated human oral taxa are constituents of these in vitro biofilms, accounting for approximately 33 % of the 16S rRNA genetic diversity detected in these simulated biofilms [24]. This technology holds great promise from a personalized medicine perspective because it facilitates creation of a biofilm from pooled saliva to which biofilms from individuals can be compared to establish the degree of individual variance from the composite biofilm [24]. Further, environmental analysis can support proteomic and metabolomic characterization. Environmental simulation may also hold promise for characterizing microbial response on exposure to antimicrobial compounds in the context of defining evolving mechanisms of drug resistance.

An example of application of this technology can be found in a recent study by Langfeldt [56] who collected samples from membrane filters supported by splints placed adjacent to teeth and sutured to the human gingiva for 14 days. Microbial content on the filter was examined by metagenomic analysis longitudinally over 14 days. The authors reported highly variable colonization profiles over time representing up to 8 phyla. The authors identified three distinctive patterns of microbial clustering based on relative abundance of dominant species including a *Prevotella* cluster, a *Proteobacteria* cluster and a *Streptococcus* cluster. Multivariate analysis demonstrated a strong correlation for mutual exclusion between the *Prevotella* and *Streptococcus* cluster. The investigators postulated that disease susceptibility risk was likely associated with both disease cluster prevalence and host inflammatory status [56].

Progress in Defining a 'Healthy' or 'Core' Oral Microbiome

In response to the NIH Core database initiative, definition of a 'core microbiome' was attempted for 18 microbiomes in over 240 individuals by analyzing 16S rRNA sequences amplified for V1-3 and V3-5 from among 12 hypervariable regions which support organism classification, followed by pyrosequencing approaches which yield functional insight. Reports by the Human Microbiome Project Consortium and several of its members concluded that definition of such a core was confounded by high rates of inter individual variability in abundance and diversity of microbial constituents across the microbiomes [32, 39, 40]. The oral microbiome, reconstructed based on sample collection from oral sites/samples including: gums, cheek, tongue, throat and saliva, exhibited the greatest number of 'core' operational taxonomic units (OTUs), (defined as OTUs having shared representation across 95% of individuals tested) [40]. These investigators further reported that although broad representation was noted across oral sites at the genus level, selective site-specificity prevailed at the sub-genus levels. Finally, quantitative differences by orders of magnitude were prevalent across individuals for OTUs commonly represented across all individuals. By contrast, a study by Zaura et al. [103] that undertook metagenomic microbiome characterization of three unrelated individuals after sampling various oral niches, reported that despite diversity across individuals, a discernable pattern existed that suggested that the majority of the OTUs represented a shared 'core' of organisms common to all individuals tested. Further, principal component analysis allowed differentiation of the oral niche from which OTUs originated. Predominant taxa were members of Firmicutes, Proteobacteria, Actinobacteria and Bacteroidetes, with 75% of OTUs and 65% of unique sequences common to 2 to 3 of the oral sub-microbiomes. By applying taxonomic 'binning' and cluster analysis, Alcaraz et al. [4] similarly analyzed bacterial diversity in the context of a 'healthy core phenotype' following metagenomic analysis of plaque from patients with and without dental caries. These investigators delineated a pattern which discerned a 'core' of genera that distinguished healthy individuals from those with caries. Notably, a shift in diversity in association with disease phenotypes has been a recurrent observation across metagenomic studies.

Evidence suggests that microbes grow in an organized manner in micro-environmental strata that are defined by physiological conditions that dictate which organisms can survive within a given strata based on intrinsic properties. These properties include tolerance for acidic environments, tolerance to varying amount of O_2, and relative abundance of nutrients required by the individual organism, capacity for symbiotic interaction or host evasion, among other factors. Within a site-specific microbiome, core organisms have the capacity to create environments that either selectively promote, or exclude, survival of specific types of organisms within a given strata. For example, constituency of the subgingival pocket microbiome may be dynamic and is defined by different organisms within each strata as one proceeds from the juncture of the periodontal pocket at the gum line and proceeds downward towards the root and the apical surface [19]. External factors posited to contribute to high inter-individual microbiome diversity include diet, exposures in early childhood, host genetics, comorbid presence, and the host's immune and systemic inflammatory response status.

Whereas constituent carriage of microorganisms varied across individuals, metagenomic carriage of metabolic pathways based on metabolomic investigations appear to remain stable across all microbiomes despite the variability in microbiome community constituency [39]. 'Core' pathways identified consistently across the microbiomes included ribosomal and translational capacity, nucleotide charging, ATP synthesis and glycolysis.

Progress in Microbiome Characterization in Disease States

Periodontal Disease and Endodontic Periodontal Disease

An ever-increasing number of studies published over the past several years have presented outcomes of metagenomic analyses of the oral microbiome in context of periodontal disease. Representative studies supporting key observations are presented here. Predominant phyla reported with good consistency across various metagenomic approaches to define oral microbiota in gingival plaque include the following phyla: Firmicutes, Spirochaetes, Bacteroidetes, Proteobacteria, Actinobacteria, and Fusobacteria [1]. Recent studies have focused on defining 'PD signatures' through analysis of differences in microbial representation and relative abundance when compared to healthy controls. Notably, discernment of PD signatures has produced variable results [33, 59]. For example, the aforementioned study by Liu et al. [59] did not report differences in phyla representation in the PD and healthy individuals, but did detect differences at the genus level as follows: *Prevotella* and TM7 represented the most predominant genera present in PD samples with low representation of *Fusobacterium* and *Porphyromonas*, whereas *Streptococcus* and *Actinomyces* were dominant in healthy individuals. By contrast, Griffen et al. [33] also defined distinctive profiles in health and PD but their study differed with respect to predominant phyla reported (which included Spirochaetes, Synergistetes and Bacteroidetes in PD and Proteobacteria in healthy controls). At a genus level, they defined prevalent PD-associated profiles to include: *Fusobacterium, Treponema, Prevotella, Leptotrichea, Porphyromonas, Filofactor* and *Synergistes* while predominant profiles in health included *Streptococcus, Acinetobacter, Moraxcella, Haemophlus, Granicatella*, and *Actinomyces*. Diaz [19] posited that differences in reported representation of microbiota in health and disease are potentially attributable to discrepant PCR amplification using variable primer sets. Additionally, as reported by Griffen et al. [33], variable results were demonstrated based on (1) selection of hypervariable region target for 16s rRNA amplification, and (2) pocket depth at which samples were collected.

Six studies to date have applied metagenomic analysis to endodontic infection generally in the context of primary infection of root canals. Prominent phyla detected included Bacteriodetes, Firmicutes, Porphyromonas and Selemonas and Proteobacteria [37, 38, 58, 66, 70, 79]. Hong et al. [37] applied metagenomic analysis to characterize microbiota associated with primary versus persistent endodontic infection. No differences were noted either in bacterial diversity or relative abundance of microbiota, with Bacteriodetes representing the predominant phyla detected in their study.

A recent study by Carneiro et al. [12], employed application of stable complementary isotope-labeling to achieve quantitation of gingival crevicular fluid (GCF) proteins following their separation by SDS PAGE gels. Protein content was then analyzed by mass spectrometry in order to achieve quantitative definition of the (GCF) proteome in PD compared to that of subjects with no PD. Employing this proteomic approach, the investigators detected significantly elevated levels of 50 host proteins and 16 microbial proteins in CGF from subjects with PD not previously reported. Among these, were proteins of high relevance to PD including host proteins associated with inflammation, innate and adaptive immune response, defensins, cytokine regulatory proteins, matrix components, and proteinases, among others. Bacterial proteins from the microbiota included virulence factor-associated proteins such as ATP- dependent DNA helicase 2 subunit, cadhedrin 6, chromodomain-helicase DNA binding protein 7, Complement C2, C region of the Ig alpha 2 chain, latent transforming growth factor beta binding protein 3, mucin 19, membrane-associated phosphotydinositol transfer protein 1, multidrug resistant protein 3 and *P. gingivalis* virulence factor OMP85. In addition, 13 proteins found only in GCF obtained from healthy periodontal sites, were identified. Such approaches hold promise for identification of distinctive biomarkers that might have clinical application in assessment of PD status since they closely reflect underlying pathological processes paralleling disease activity and components reflecting host-pathogen interactions.

Cariogenesis

Metagenomic study attempting delineation of signature microbiomes in association with cariogenic phenotypes has been relatively successful. However, several studies have reported variable constituency of the 'caries-associated' microbiome depending on severity of caries status and which clinical samples were analyzed [31, 42, 100]. Notably, microbiome constituency remains dynamic across the human lifespan, responding to external stimuli such as diet, host genetics and immune status. For example, Cephas et al. [15] undertook a comparative metagenomic study of salivary samples collected from edentulous infants and their mothers or primary care givers. The number of OTUs in the adult saliva approached nearly double the number OTUs present in infant saliva and represented a total of 397 genera and 1033 species across all individuals. The adult saliva exhibited high diversity and differed from infant saliva by 28 genera with 27/28 exhibiting higher prevalence. *Streptococcus* was the predominant genera in infants (62% in infants *vs.* 20% in adults). Predominant genera common to both infants and adults included *Veillonella, Neisseria, Haemophilus, Rothia,* and *Fusobacterium*. Genera predominant only in infants included *Gemella, Granulicatella* and *Leptotrichia,* whereas *Treponema, Oribacterium* and *Actinomyces* were predominant only in adults. Notably, predominant genera in adults not found in infants have been associated with PD.

Carcinogenesis

Based on epidemiological studies which have reported high prevalence of PD in subjects in the context of various cancers, contribution of the chronic infectious and consequent hyperinflammatory processes have been posited to contribute to creation of a physiological state that increases risk for carcinogenesis [2]. Evidence is found in a study by Michaud et al. [62] that modeled cancer risk in subjects with and without PD. Following adjustment for other risk factors including smoking and diet, patients with a history of PD were projected to be at increased risk for cancer compared to those with no PD. Analysis of strength of the evidence to date undertaken by Fitzpatrick and Katz [26] concluded that a preponderance of studies supported higher risk for oral, esophageal, gastric and pancreatic cancer in individuals impacted by PD. In contrast, studies defining links between PD and lung, prostate and hematological cancers association to date report equivocal results or remain understudied.

Whereas pathophysiological factors linking PD and cancer may vary with cancers, some common themes are noteworthy. A review by Pendyala et al. [67] indicted multifaceted inflammatory processes as strong contributors to pathological processes that impact both PD and cancer, identifying specific mediators common to both PD and cancer. Mediators of inflammatory processes include cytokines, chemokines, acute phase proteins, innate immune signaling molecules and reactive oxygen and nitrogen species with potential to damage cellular DNA. Notably, pathogens associated with carcinogenesis may thrive in subgingival pockets in the microbiome environment favoring PD. For example, presence of *Helicobacter pylori,* implicated in gastric and pancreatic cancers, was identified as a constituent of the subgingival biofilm microbiota co-aggregated with *Fusobacterium* species [67].

Oral Cancer

Symptomology associated with oral cancers such as gingival squamous cell carcinoma closely parallels key features of severe PD. Fitzpatrick and Katz (2010) [26] reviewed studies examining oral criteria measured in the oral cavity associated with development of various cancers. Tooth loss and/or PD were recurrent risk factors for carcinogenesis.

A study by Bebek et al. [8] demonstrated hypermethylation of four genes associated with head and neck squamous cell carcinoma in tumors. Notably, the aberrant methylation of *MDR1* was associated with an altered microbiome signature, specifically when compared to normal mucosa, in which prominence of two species, (*Enterobacteriaceae* and *Tenericutes),* significantly correlated with focal nodal metastases. Similarly, a study by Pushalkar et al [68] which conducted clonal analysis in tumor and normal tissue demonstrated distinctive oral microbiota in each tissue type, lending further support to the premise that changes in the host at the level of the mucosa were associated with shifts in microbial community. High levels of copy number variation (CNV) and changes in differential gene expression patterns in gingival buccal cancers have been described by Ambatipudi et al. [5]. Further studies are warranted to define whether epigenetic changes causing shifts in host genetic expression are associated with development of signature patterns of microbiota in the context of oral cancers.

A preliminary study by Xu et al. [99] examined the oral microbiome in the context of cancer treatment in a small number of subjects. Chemoradiation treatment in patients with nasopharyngeal cancer of the oral-esophygeal epithelial lining is associated with significant side effects including oral mucositis, gingivitis, oral candidiasis, cellulitis and 'radiation caries'. To better understand how microbiota are impacted by chemoradiation in the context of this cancer, Xu et al. [99] conducted metagenomic analyses of samples collected at baseline (pre-treatment), and at two time points post treatment (month 7 and 12) and compared microbiome profiles to those of healthy controls. Although the diversity indices did not differ across samples, relative abundance of microbiota in patients receiving chemotherapy at baseline showed statistically significant differences compared to controls and post treatment time points. Notably, changes among treated patients were highly heterogeneous, underlining the need for a personalized approach in treatment planning. Whereas all 13 phyla most frequently encountered in the oral cavity were detected including Firmicutes, Bacteroidetes, Actinobacteria, Proteobacteria, Fusobacteria, Spirochaetes and from the TM7 phyla of uncultivated organisms, levels of Firmicutes were depleted in cancer patients, whereas Proteobacteria were enriched. Whereas infectious side effects did not occur in their study subjects, further investigation is warranted to determine whether specific discernable shifts in relative representation and abundance of microbiota are associated with establishment of post treatment oral infections.

'Infectogenomics': Relationship between Host Genetic Factors and Oral Microbiota

Shifts in microbiome community structure that are conducive to pathological development may be driven by some combination of environmental stimuli, microbial constitution and host immune dysfunction. In examining evidence supporting a role for periodontal infectogenomics, Nibali et al. [64] have provided examples of genetic factors that impacted on the relative capacity for pathogenic invasion, proliferation and clearance. Immune dysfunction may have genetic underpinnings at the level of the host or may be induced through stimuli such as microbial encounters. Potential manifestations of immunologically-mediated pathology stimulated by microbial processes or inappropriate host response may include induction of chronic inflammatory states driven by the chemokine network with local or systemic impact, induction of autoimmunity, potentially driven by collateral damage to host tissue or aberrant immunological tolerance, or breakdowns in innate immune signaling at the level of the oral cavity resulting in inappropriate or ineffective stimulation of immunological response pathways of the adaptive immune system. Igari et al. [41] posited that inflammatory mediator production stimulated by PD enter the blood stream and induce a systemic inflammatory state that may exacerbate other comorbid conditions sensitive to inflammatory stimuli and deregulate other biological processes (e.g., induction of preterm labor). The magnitude of this inflammatory response is a function of both host factors and microbial stimulus. Further, it is well documented that oral microbes do not remain sequestered in the mouth but gain systemic access by a variety of mechanisms [35, 41]. Further, Han and Wang [35] noted that systemic extravasation is largely limited to subspecies of bacteria supported by evolutionary acquisition of virulence factors which support survival in the hostile extra- oral environment facilitated to some extent, by host factors. An example of this is the capacity of some pathogens to be transported and released intact from phagocytic cells.

Several recent studies have examined microbiomes in the context of host diseases and report altered microbiome structures and dysbiosis. Whether the dysbiosis is a function of immunological dysfunction or arises in response to local changes in the microenvironment due to the underlying disease state remains to be elucidated. Two examples follow:

Inflammatory Bowel Disease (IBD) Recent studies have examined the genetic basis of immune dysfunction present in the context of inflammatory bowel disease. In depth genetic analysis by Jostins et al. [44] defined 163 susceptibility loci in association with (IBD), most common to both Crohn's disease and ulcerative colitis that contribute to disease emergence. The potential causal genes all impact on the host capacity to respond to infectious processes associated with IBD, in particular, at the level of mucosal immune responses to microbes localized at the level of the epithelial cell surfaces of luminal surfaces along the GI tract. Among genes exhibiting high Bonferroni-significant selection were genes impacting on cytokine production including IL17 that plays a central role both in defense against infectious processes as well as autoimmunity. The most significant signal was noted for *NOD2* that occurs in a cluster of candidate genes associated with capacity of the host to respond to mycobacterial infection. IBD is in some cases, associated with oral manifestations. An interesting study by Docktor et al. [23] explored compositional diversity of the oral microbiome in the context of inflammatory bowel diseases including Crohn's disease (CD) and ulcerative colitis (UC). Applying the 16S rRNA Human Oral Microbe Identification Microarray (HOMIM) technology to tongue and buccal samples, these investigators demonstrated a loss in microbial diversity with statistically significant decrease in level of signal associated with two phyla, *Fusobacteria* and *Firmicutes,* in subjects with CD compared to healthy controls and patients with UC. Notably, loss of *Fusobacteria* and *Firmicutes* has also been reported at the level of the intestinal microbiome [90]. By contrast, an increased quantitative signal was associated with Spirochaete, Synergistetes and Bacterodetes phyla in oral microbiome samples collected from the CD patients. Docktor et al postulated that dysbiosis, prominently associated with a loss in bacterial diversity, correlated with loss of important commensal organisms that promote a healthy local environment and permitted the establishment of pathogenic organisms in a less functional microbiome. The authors posited that the aberrant immune response in cases of CD that are associated with pathological manifestations in the oral mucosa in some patients, may impact on microbiome profiles found in the intestinal tract of these patients. In the context of CD, aberrant immunological response has been implicated and manifests through stimulation of enhanced cytokine production at the level of the oral epithelial lining. Detection of high titers of antibodies specific to *Saccharomyces cerevisiae* has been reported (Docktor et al. [23]).

Type 2 diabetes mellitus (T2DM) The bidirectional exacerbative impact of periodontal disease and poor glycemic control in the context of T2DM has been recognized [61, 85]. Further, non-surgical periodontal treatment contributes to improved glycemic control in patients with T2DM([53]; systematic review/meta analysis by Corbella et al. [16]), allegedly by attenuation of host hyper-inflammatory responses in the context of PD [55].

Three independent studies published in 2013 utilized different approaches to explore the subgingival plaque in the context of T2DM. Castrillon et al.[14] conducted a PCR-based examination of relative prevalence of traditional periodontal pathogens *A. actinomycetemcomitans* (green complex) and the three 'red complex' periodontal pathogens, *P. gingivalis, T. forsythia* and *T. denticola,* in subgingival plaque of patients with T1 or T2DM and non-diabetic subjects with and without periodontitis. The authors postulated that subgingival microbiota differed with diabetic status. All four pathogens were found at higher frequency in patients with periodontitis without T2DM compared to non-diabetic periodontally healthy subjects. High prevalence of *A. actinomycetemcomitans* was noted among patients with T2DM compared to non- diabetic patients with periodontal disease and low prevalence in the absence of periodontal disease. Patients with diabetes also exhibited higher levels of attachment loss. Conversely, non-diabetic patients with periodontitis displayed higher frequency of *P. gingivalis*.

An independent report by Casarin et al. [13] examined subgingival microbiome diversity among uncontrolled diabetic subjects and non-diabetic subjects, where both groups presented with severe, generalized chronic periodontitis and no diabetes. These investigators reported significant differences in distribution of constituent members within their microbiomes. Whereas non-diabetic subjects exhibited higher prevalence of *Porphyromonas, Filfactor, Eubacterium, Syergistetes, Tannerella* and *Treponema* genera, diabetic patients exhibited increased levels of the following genera: *TM7, Aggregatibacter, Neisserian, Gemella, Eikenella, Selenomonas, Actinomyces, Capnacytophaga, Fusobacterium, Veillonella* and *Streptococcus*. At a species level, *F. nucleotum, V. parvula, V. dispar* and *E. corrodens*, were encountered at significantly higher frequency in diabetic patients compared to non-diabetic subjects. Notably, *TM7* genus clones which represent a class of non-cultivable bacteria, were detected exclusively in diabetic subjects in this study. These authors similarly reported higher prevalence of *P. gingivalis* and *T. forrsythia* in non-diabetic patients, as reported by Castrillon et al. [14].

The third study by Zhou et al. [105] compared subgingival microbiomes of periodontally healthy subjects with, and without, DM and subjects with T2DM with, and without, periodontal disease. These investigators posited that T2DM contributes to an altered subgingival plaque composition both in the absence or presence of periodontal disease. The investigators reported that sequences represented organisms from 16 phyla, 27 classes, 48 orders, and 85 families, 126 genera, and 1141 species. Further, these investigators defined 3 prominent genera associated with health: *Prevotella, Pseudomonas* and *Tannerella.* Nine species level operational taxonomic units (OTUs) distinguished diabetic from non-diabetic subjects in the absence of periodontal disease while 6 OTUs differed significantly between diabetics and non-diabetics with T2DM. The results of this study differed from the other two studies in that these investigators reported *P. gingivalis*, *Treponemia medium*, *Tanneralla forsythia, Synergistacease, Porpyromonas endodontailis* and *Filifactor alocis* in the context of patients with diabetes and periodontitis, while A. *actinomycetemcomitans* was not reported as a significant organism in any of the groups. *Selenomonas* was associated with periodontitis on diabetes negative background in contrast to the report by Castrillon et al. [14]. PD- associated OTUs found in both diabetes positive and negative subjects included *T. denticola*, and *P. intermedia.* Variability of these outcomes could stem from a variety of factors. The individuals in the latter study differed by race from the those in the former study and differences could reflect inter-individual variability, small numbers of subjects included in study groups, sampling technique, analysis platform, glycemic control among diabetic subjects, location of the teeth chosen for sample collection, among others. Data from these three studies further reinforces that there are formidable challenges to defining core microbiome associated with health including a high degree of inter-individual variation associated with the composition of flora defining the microbiome constituents. Importantly however, all of these studies underline dynamic shifts in microbiome patterns across the T2DM disease trajectory.

T2DM increases in prevalence with advancing age. This phenomenon has been attributed to distinctive epigenetic changes (changes in methylation patterns) to a risk gene. In the context of T2DM, two independent studies reported definition of hypomethylation of the *FTO* gene in subjects who developed T2DM over the course of their lifetime compared to non-diabetic population controls, and further it was demonstrated that epigenetic changes at this locus were causal. [10, 85]. Thus, host genetics simultaneously contribute to establishment of T2DM and mediate shifts in microbiota constituency towards organisms with capacity to thrive deep in periodontal pockets in a hyperglycemic environment.

Characterization of Genetics Supporting Disease Emergence

Genetics and Periodontal Disease (PD)

The induction of PD has been viewed from both the perspective of a largely pathogen-driven infectious process and from the vantage of heightened host susceptibility due to compromised immune and inflammatory responses. Traditional approaches have focused largely on genetic polymorphisms with potential functional relevance such as single nucleotide polymorphisms (SNPs) occurring in genes associated with immune response or chemokines regulating inflammatory processes, which may affect host capacity to effectively interact with microbial presence. Meta analyses of conventional candidate-driven approaches have reported strongest support for candidate single nucleotide polymorphisms (SNPs) summarized in Table 1.

The availability of metagenomic technology such as genome wide association study (GWAS) of PD in the context of the oral microbiome has permitted a shift to studies supporting hypothesis generation which may be missed by the former approach [88]. To date, 4 GWAS studies have permitted validity testing of previously proposed candidate SNPs while further exploring other potential genetic underpinnings of PD which may have remained unexplored to date because functional relevance was unknown or unrecognized [22, 73, 78, 84]. Interestingly, while strong putative association has been associated with several SNPs, none have achieved the genome wide significance threshhold of $p < 5 \times 10^{-8}$.

Table 1 Validated and putative candidate genes associated with PD or PD and coronary heart disease(CHD)

Candidate gene/ region	Methodological approach	Relevant function	Impact	Reference
IL1A	Candidate approach	Proinflammatory cytokine	Chronic PD	[46, 65, 73]
IL1B	Candidate approach	Proinflammatory cytokine	Chronic PD	[46, 65, 73]
Fc-gamma-RIIIB NA1/NA2	Candidate approach	Fc gamma receptor	Aggressive & Chronic PD	[20, 82]
IL10[a]	Candidate approach/GWAS/ candidate (meta analysis) GWAS	Cytokine: immune response regulation	Aggressive & Chronic PD Chronic PD Aggressive PD	[3, 73, 104]
ANRIL[a]	GWAS GWAS Meta analysis	Anti sense RNA regulator;VAMP3 regulator; cardiolipin inducer	Aggressive PD Aggressive PD Aggressive PD CHD	[71, 25, 73, 76]
CAMPTA1[a]	DNA-DNA checkerboard titration Genomic expression Arrays	Calmodulin-binding transcription activator 1: increased PD pathogen colonization	Non-PD CHD	[11, 21]
COX2	GWAS Candidate approach Candidate approach	Cyclooxegenase 2 Inflammatory mediator	Aggressive PD Chronic PD Protective	[36, 71, 97]
ERC2 & gene region	GWAS GWAS		Chronic PD Chronic PD	[22, 84]
9p21.3[a]	GWAS Meta analysis GWAS		Aggressive PD CHD Aggressive PD and CHD	[25, 71, 75]
Additional putative candidates not validated (p value $<5 \times 10^6$)				
Divaris et al. [21, 22]	*KVNK1, PKN2, FXOB38, UHRF2, TBC1D1, CLIC5, CSMD3, FOS, ODZ2, GRID1, KIAA1715 IL33, RUNX2, TRPS1, NIN, NPY, NCR2,CELF2*, region between *WNT5A* and *ERC2*, region between *EMR1* and *VAV1*			
Teumer et al. [84]	*EPHA3, RAB6C, C9orf150, IQSEC1, ERC2, CAMK4, MFSD1, LBP, ETS2*, and *FAM180A*			
Schaffer et al. [78]	*LAMA2, HAS2, CDH2, ESR1, OSBPL10*, HSP90AB2P and GVINP1 pseudogene regions, region near SEL1 and FHOD3, region between SOS2 and *NIN*			

Bolded genes/regions show have been validated in another study
[a] Have been validated in both heart disease and PD.

GWAS studies have also been conducted in larger populations to identify potential genome wide association.

Table 1 shows genes which exhibit strong p values in these studies although genome-wide significance was not achieved. Notably putative candidate gene status was noted in more than one study (either by GWAS analysis or candidate SNP approach) for genes listed. Further, many mutations occurred within regulatory/intronic regions (Vaithilingam et al. [88]).

An interesting study by Ernst el al. [25] validated association of 4 SNPs occurring within *9p21.3* in subjects with generalized aggressive PD. This finding was particularly significant since this region has demonstrated strong putative association with coronary heart disease in previous GWAS studies [25, 71]. Further putative associations reported between chronic PD and coronary artery disease (CAD) include *ANRIL* [25, 71, 73], IL10 [3, 73, 104] and *CAMPTA1* [21].

In a follow-up study, Divaris et al. [22] reported on a meta-analysis of data derived from GWAS performed in two separate cohorts examining genetic association in the context of chronic periodontitis. Validated and additional putative candidate genes/regions are reported in Table 1. An increase in estimated heritable variance associated with severe chronic PD (present in 17 % of their study population) from 18 to 52 % was reported with smoking as an interactive variable.

The GWAS study by Teumer et al. [84] similarly focused on chronic periodontitis in two separate populations. Putative SNPs with the strongest association reported in their study are summarized in Table 1. Following data modeling (which included imputation of HapMap, autosomal and X-chromosomal genotypes, and indels), these investigators reported that cumulative interactive effects of all common SNPs on mean attachment loss contributed 23 % of the estimated variance, and could explain 34 %, if subjects >60 years of age were excluded.

The latest GWAS study by Shaffer et al.[78] reported that their study validated other candidate loci for PD previously associated in candidate approaches including: *LAMA2*, *HAS2*, *CDH2, ESR1*, and a genomic region on chromosome 14q21-22 between *SOS2* and *NIN*. The study further nominated new candidates including *OSBPL10*, a lipid receptor that has shown association with hyperlipidemia, two loci near pseudogenes *HSP 90AB2P* and *GVINP1* and two additional loci near *SEL1L* and *FHOD3* [78].

Interestingly, among the four GWAS studies to date, only few common risk alleles have been observed across more than one GWAS study or replicate risk alleles reported in previous studies (e.g. *IL10*). Inability to observe genome-wide significance may be associated with factors such as heterogeneity among populations, sample size, and variability in inter-individual variability in microbiome constituency across the spectrum of PD.

Is there Genetic Predisposition Supporting Cariogenesis?

Applying a heritability analysis approach, 740 multigenerational families were genotyped for 72 ancestry informative SNPs [91]. Family members were classified into the following phenotypes following oral examination: (1) no decay, (2) 'white spots' (pre decay state), (3) fillings, (4) missing due to cavitation, (5) hypoplasticity, (6) localization of the decay and (7) missing due to non caries-related variables. Strong heritability patterns that attributed between 54–70 and 35–60 % of variability to a genetic component for primary and permanent dentition, respectively, were observed.

Studies defining caries-associated candidate genes have identified some genetic predisposition due to SNPs in genes relating to tooth structure, the innate immune response and taste receptor genes which may predispose to diets that promote cariogenesis. Putative genetic associations to cariogenesis are summarized in Table 2.

Two GWAS studies exploring genetic underpinnings in cariology have been reported. Schaffer et al. (2011) [77] explored heritability factors contributing to cariogenesis risk in primary teeth in children. Although putative SNPs did not achieve GWAS significance, strong association was nonetheless observed for loci at or near genes with plausible functional roles that could contribute to caries risk (see Table 2). Other SNPs showed variable association depending on adequacy of fluoride treatment (see Table 2). A second GWAS study carried out [92] in 5 distinct adult cohorts, advanced several putative genes as candidates contributing to caries risk including genomic loci in the vicinity of genes with plausible functional roles in caries development (shown in Table 2). In 2011, Shaffer et al. [77] defined a secondary caries phenotype characterized by caries specifically affecting maxillary incisors. Notably, in a murine model, *ISL1* was specifically associated with incisor development. The relationship between this mutation and incisor caries remains to be explored. Whereas these mutations showed strong associations (p value $< 10^{-7}$), none achieved genome-wide significance. Future meta-analysis and testing for genetic and genetic/environmental factor interactions will further increase understanding of the genetic mechanisms interacting with the oral microbiomes.

Employing the HuMiChip 1.0, a geochip designed to measure metabolism of microbiota, Yang et al. [101], demonstrated a functional gene structure that differentiated individuals with caries from individuals without caries. These investigators observed a higher level of conservation of non-core genes in healthy individuals compared to caries-active individuals who exhibited a loss of genetic diversity in three distinct metabolic pathways thus creating a 'cariogenic signature' detectable in saliva. They proposed that, when applied as a screening tool, this functional microarray exhibited greater sensitivity to detect caries-active individuals than any other biomarker currently available. Biomarkers associated with the caries-active phenotype included altered levels of diaminopimelate epimerase, (functions in the amino acid synthesis pathway and is critical to bacterial cell wall biosynthesis), prephenate dehydrogenase (oxidative decarboxylation and is critical to tyrosine synthesis, N-acetlymuramoyl-L-alanine amidase (glycan synthesis/metabolism critical to cell wall autolysis). Study outcomes await replication in larger populations.

Advances in Proteomics in Assessing the Microbiome in Health and Disease States

Determination of disease presence in the oral cavity is largely dependent on clinical examination in the absence of other available diagnostic tools. Among 'omic' studies, genomics has been predominantly applied to microbiome analysis with the

Table 2 Putative candidate genes associated with cariogenesis

Candidate gene/region	Approach	Relevant function	Impact	Reference
AMELX	Candidate approach	Encodes amelogenin	p values for TT genotype subject (fluoride vs non fluoride) for Amelix SNPs: rs17878486: 0.003 rs5933871: 0.001 rs5934997: 0.000	[45]*
TUFT1	Candidate approach	Encodes tuftelin	significant interaction between S. mutans and tuftelin. (r square = 0.268)	[80]*
CD14	Candidate approach	Microbial pattern recognition	CD14-260 TT genotype in children with caries protective against abscess/fistula p = 0.005	[17]*
TAS2R38	Candidate approach	Taste receptor	CA and CAT haplotype for TAS238 is associated with caries risk in primary dentition (p = 0.03 and 0.02, respectively)	[94]*
RPS6KA2	GWAS	Kinase; inflammatory mediator gene regulation	Permanent dentition caries	[92]
PTK2B	GWAS	Kinase; inflammatory mediator gene regulation	Permanent dentition caries	[92]
RHOU	GWAS	In *WNT* signaling cascade: (tooth development)	Permanent dentition caries	[92]
FZD1	GWAS	In *WNT* signaling cascade: (tooth development)	Permanent dentition caries	[92]
TLR2	GWAS	Immune response to oral pathogens	Permanent dentition caries	[92]
ADMTS3	GWAS	Tooth development	Permanent dentition caries	[92]
ISL1	GWAS	Incisor development	Permanent dentition caries	[92]
ACTN2	GWAS	Regulation of tooth enamel formation	Primary dentition caries	[77]
MTR	GWAS	Methionine and homocysteine production	Primary dentition caries	[77]
EDARADD	GWAS	Tooth development	Primary dentition caries	[77]
MPPED2	GWAS	Expressed in epithelial cells during microbial challenge	Primary dentition caries	[77]
LPO	GWAS	Encodes salivary enzymes	Primary dentition caries	[77]
TF1P11	GWAS	Enamel synthesis	Caries, low fluoride	[77]
EPHA7	GWAS	Tooth development	Caries, adequate fluoride	[77]
ZMPSTE24	GWAS	Mandibular development	Caries, adequate fluoride	[77]

expectation that these approaches would define clear definitions of community profiles associated with health and disease states, with mixed success. Instead, what has emerged in many studies are similar profiles exhibiting subtle shifts in relative representation, generally with modest changes in diversity or relative abundance. Further, the choice of sample analyzed has presented researchers with varied profiles in the same disease state. To complicate matters further, these studies have revealed that approximately 50% of the microbiome constituents have never been cultured and their contribution to oral health and disease remains unexplored and begs the question of whether major pathogens responsible for PD remain to be defined.

As illustrated by the study of Yang et al. [101] discussed in the preceding section, multidimensional approaches to include proteomic and/or metabolomics investigation are just beginning to provide increased granularity and perspective into the complex nuances of host-microbe and inter-microbe interaction and the consequences of these interactions systemically and locally within the microbiome. However, such complementary multi-omic approaches to microbiome analyses are just beginning to gain traction and a recent meta-analysis identified only a dozen studies that had applied proteomic approaches to characterizing periodontitis and only one metabolomic study at the time of publication (Trinidade F et al. [86]). Advances in proteomic technology are likely to advance microbiome analysis from this perspective. Examples include: (1) Protein topography and Migration Analysis Platform (PROTOMAP0 which merges application of shotgun proteomics to proteins separated by SDS-PAGE and (2) Matrix assisted laser desorption/ionization (MALDI) which can be applied to rapid protein detection in mixtures [34]. These authors review a list of some periodontal proteins characterized by application of structural proteomic study, interactive proteomic study or functional proteomic study in the context of PD or clinical treatment of PD. Potentially some of these may represent candidates for incorporation into panels for use as clinical screening tools designed to have high sensitivity and specificity for evaluating PD disease status.

Three recent proteomic studies of PD that illustrate potential of proteomic approach include a study by Tsuchida et al. [87], who described a tandem mass tags approach to accomplish quantitative proteomic analysis to discover potential proteomic biomarkers associated with PD in gingival crevicular fluid. Of 619 proteins, the investigators reported that metalloproteinase 9 (MMP9) and neutrophil gelatinase-associated lipocalin (LCN2) levels were higher in patients with PD compared to healthy subjects. Both of these proteins have previously been implicated in progression of PD. A study by Ngo et al. [63] subjected GCF to MALDI-TOF mass spectrometry analysis and demonstrated that GCF mass spectra data could be modeled to predict attachment loss at a site with 97% specificity. Further, a study by Carneiro et al. [12] applied stable isotope-labeled ICAT and mTRAQ in mass spectroscopy studies to examine the GCF proteome in subjects with and without PD and the results were validated using ELISA. The investigators found 180 proteins common to both healthy subjects and those with PD, and 26 and 32 proteins present only in healthy subjects or those with PD, respectively. Other proteins quantified for the first time in GCF associated with PD included host and bacterial proteins and virulence factor OMP85. These authors reported largely novel host ($n=50$)- or microbial pathogen- associated ($n=16$) proteins present in significantly elevated levels over those measured in healthy subjects and predict that some of these have applicability to the clinical setting.

Implications for Diagnostic and Therapeutic Approaches

Perturbations in the microbiome or changes in the host interaction with the microbial load may shift the balances within microbiomes and the global ecosystem from one supporting health to one supporting disease. These shifts underscore the necessity for a personalized approach to diagnosis and clinical management due to dynamic relationships between the host and its microbiome which has shown substantial inter-individual variability both in the state of homeostasis or dysbiosis as determined by host factors. However, 'omic' studies are just beginning to provide insights into patterns associated with health and disease. Definitions of heritability will help identify genes vested in health maintenance or emergence of disease. Functional analysis of these genes will provide clues to translational products and metabolic pathways they impact. Once patterns associated with health and disease are defined, the potential to harness this knowledge to screen individuals for profiles associated with healthy states or detect subtle perturbations that may portend risk for disease emergence, offer the potential for personalizing approaches for maintaining oral health, evaluating disease risk, or detecting disease emergence [52]. This approach may inform development of interventions tailored specifically to individual treatment to restore a healthy equilibrium. For example, research is pointing to a profile of core microorganisms associated with oral health that deter establishment of cariogenic bacterial strains. Core commensal strains may prove useful as 'probiotics' to maintain a healthy balance in the oral cavity, thus discouraging establishment of microbiota with cariogenic potential [9].

Further, studies suggest that presence and number of risk factors for periodontitis and tooth loss are directly correlated with number of annual visits to the dentist for prophylactic treatment. For example, the Michigan Personalized Prevention Study by Giannobile et al. [29] demonstrated significantly higher rates of tooth loss in patients with one or more of the following risk factors: smoking, diabetes, and presence of a specified pattern of IL-1 SNPs as determined by genetic testing, if they visited the dentist for prophylaxis once yearly compared to twice yearly compared to low risk patients whose rate was the same whether they were seen once or twice annually. In a follow-up editorial, Giannobile et al. [30] challenge the paradigm of twice annual visits suggesting that this should be personalized to patients depending on risk, suggesting that one visit annually may be sufficient for low risk patients whereas high risk patients with multiple risk factors may require more than 2 prophylactic visits annually to maintain periodontal health. Based on increasing information with respect to maintenance of health and risk emerging from microbiomic study, these authors advocate application of the new knowledge to risk stratification and an approach to patient care with embraces 'the four P's: prediction, prevention, personalization and participatory health care on the part of the patient.

Whereas there is much heterogeneity in the interaction between the host and associated microbiomes across individuals, and the enormous complexity in the interactions between the host, microbiota, and environment, it becomes highly critical to carefully standardize and define approaches that analyze these complex relationships in order to create capacity to clearly discern any global patterns through meta-analysis of metadata generated by these approaches without factoring in additional confounding contributed by experimental artifact. Historically, lack of consensus on standardized definitions of periodontal disease or definition of chronic periodontitis have contributed significant heterogeneity to study data creating serious challenges in comparing and testing validity of data over time. 'Omic' approaches show promise for more accurate classification of disease phenotypes [47]. With metagenomic approaches to oral microbiome analysis still burgeoning, critical analysis of currently available data and how these could inform standardization of metagenomic and future 'omic' approaches while reducing technology-associated confounding, could immensely accelerate progress in this emerging discipline in the future and potentially reduce the number of conflicting reports. A critical editorial by Schaefer et al. [74] highlights important considerations that provide a good starting point for planning and advancing future initiatives.

Metagenomic and other 'Omic' Resources

Finally, a summary of important resources that have been developed over the past decade to facilitate 'omic' study of the host-microbiota interaction are important to cite. Quantum leaps in advancing understanding of species diversity and composition within disparate microbiome environments in the human body have been facilitated by the National Institutes of Health's (NIH) sponsorship of three important initiatives and have generated several invaluable resources which have rapidly advanced this field of study. These include the:

Human Microbiome Project (HMP)

Launched in 2008, HMP has enabled characterization of various microbiomes throughout the body at the sequence level with the objective of defining and compiling reference genomes of species represented within these local microbial communities. http://commonfund.nih.gov/hmp/index.

Human Oral Microbiome Database (HOMD: http://www.homd.org/)

Curated by Dewhirst et al. [18], this singularly important resource represents a compilation of prokaryotic species associated with the oral cavity with capacity to link sequence data to phenotypic, phylogenetic, clinical and other available data characterized for each organism. This resource contains over 600 validated taxa of which only ~25 % are named, 8 % have been cultivated but remain unnamed and 68 % represent uncultivated phylotypes about which little is known.

Core Human Oral Microbiome Database (CORE: http://microbiome.osu.edu) and other 16s rRNA gene reference resources

CORE is a database which stores phylogenetically-curated 16S rDNA sequences associated with the oral microbiome. Comprehensive representation of the oral microbiome was achieved by alignment of 668 full length 16S sequences [32]. The Core database represents the most accurate and validated curation to date and consists of a compilation of 1043 sequences representing 152 genus level and 636 species level OTUs detected in the oral cavity. Average genus and species-level divergence reported within OTUs is 7.3 % (SD 5.5 %) and 1.3 % (SD 0.8 %), respectively [32]. Because of the extremely diverse number of taxa found in the oral cavity, this resource was designed as an aid in achievement of taxonomic assignment and as a framework to study community divergence utilizing the highly detailed phylogenetic tree. An example of an oral genus-level phylogenetic tree configured based on CORE database data, shown in Fig. 1, (from [32]). Performance of the Oral CORE database for identification of clinical sequences exceeded that of GenBank, 16s rRNA gene reference resource, HOMD and RDP (http://rdp.cme.msu.edu/). (Griffen et al. [32] [39, 40]). An additional available 16s rRNA gene reference resource is Silva (http://www.arb-silva.de) [2].

OralCard

OralCard represents comprehensively curated published data on the oral proteome inclusive of both human and microbial data. This resource is accessible at http://bioinformatics.ua.pt/oralcard [6].

Key Terminology

Cariogenesis: the process leading to cavitation of tooth enamel (i.e. formation of dental caries)

Core Microbiome: composition of the microbiome associated with the state of health

Genomics: the study of genes and their function

Gingivitis: a mild self-limited infection of the gums of lower severity than periodontal disease

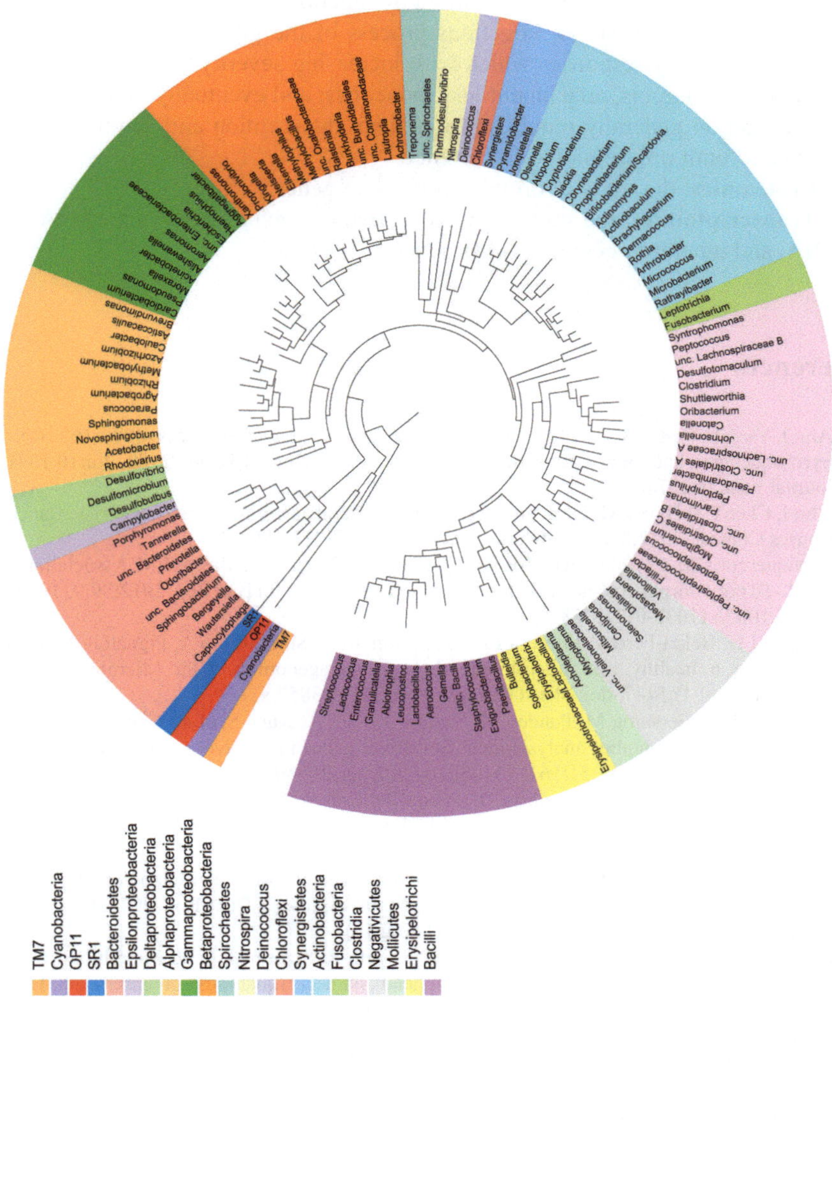

Fig. 1 Circular maximum likelihood phylogenetic tree at level of genus. ([32]. CC BY 2.5; with permission, under the Creative Commons Attribution License)

Metabolomics: the study of products arising from metabolic processes

Metamicrobiome-the collective communities of microorganisms organized into environmental niches throughout the body. The term microbiome is often used interchangeably with **microbiota**

Omics-technology applied to the study of the various molecules of human or microbial origin in the body classified by function

Periodontal disease: an infectious process of the gums exacerbated by host inflammatory responses which with increasing severity leads to bleeding and painful gums, attachment loss, bone loss, and eventually, tooth loss. This condition generally requires professional intervention and careful oral hygiene to limit progression or recurrence

Proteomics: the study of protein products of genes

Transcriptomics: the study of RNA molecules, including mRNA, rRNA, tRNA, and other non-coding RNA produced

References

1. Ahn J, Yang L, Paster BJ, Ganly I, Morris L, Pei Z, et al. Oral microbiome profiles: 16S rRNA pyrosequencing and microarray assay comparison. PLoS One. 2011;6:e22788. doi:10.1371/journal.pone.0022788.
2. Ahn J, Chen CY, Hayes RB. Oral microbiome and oral and gastrointestinal cancer risk. Cancer Causes Control. 2012;23:399–404. doi:10.1007/s10552-011-9892-7.
3. Albuquerque CM, Cortinhas AJ, Morinha FJ, Leitão JC, Viegas CA, Bastos EM. Association of the IL-10 polymorphisms and periodontitis: a meta-analysis. Mol Biol Rep. 2012;39:9319–29. doi:10.1007/s11033-012-1738-1.
4. Alcaraz LD, Belda-Ferre P, Cabrera-Rubio R, Romero H, Simón-Soro A, Pignatelli M, et al. Identifying a healthy oral microbiome through metagenomics. Clin Microbiol Infect. 2012;18(Suppl 4):54–7. doi:10.1111/j.1469-0691.2012.03857.x.
5. Ambatipudi S, Gerstung M, Pandey M, Samant T, Patil A, Kane S, et al. Genome-wide expression and copy number analysis identifies driver genes in gingivobuccal cancers. Genes Chromosom Cancer. 2012;51:161–73. doi:10.1002/gcc.20940..
6. Arrais JP, Rosa N, Melo J, Coelho ED, Amaral D, Correia MJ, et al. OralCard: a bioinformatic tool for the study of oral proteome. Arch Oral Biol. 2013;58:762–72. doi:10.1016/j.archoralbio.2012.12.012. (Epub 2013 Feb 8).
7. Barnes VM, Ciancio SG, Shibly O, Xu T, Devizio W, Trivedi HM, et al. Metabolomics reveals elevated macromolecular degradation in periodontal disease. J Dent Res. 2011;90:1293–7. doi:10.1177/0022034511416240. (Epub 2011 Aug 19).
8. Bebek G, Bennett KL, Funchain P, Campbell R, Seth R, Scharpf J, et al. Microbiomic subprofiles and MDR1 promoter methylation in head and neck squamous cell carcinoma. Hum Mol Genet. 2012;21:1557–65. doi:10.1093/hmg/ddr593.
9. Belda-Ferre P, Alcaraz LD, Cabrera-Rubio R, Romero H, Simón-Soro A, Pignatelli M, et al. The oral metagenome in health and disease. ISME J. 2012;6:46–56. doi:10.1038/ismej.2011.85.
10. Bell CG, Finer S, Lindgren CM, Wilson GA, Rakyan VK, Teschendorff AE, et al. Integrated genetic and epigenetic analysis identifies haplotype-specific methylation in the *FTO* type 2 diabetes and obesity susceptibility locus. PLoS One. 2010;5:e14040. doi:10.1371/journal.pone.0014040.

11. Bochenek G, Häsler R, El Mokhtari NE, König IR, Loos BG, Jepsen S, Rosenstiel P, Schreiber S, Schaefer AS. The large non-coding RNA ANRIL, which is associated with ath-ero*scl*erosis, periodontitis and several forms of cancer, regulates ADIPOR, VAMP3 and C11ORF10. Hum Mol Genet. 2013;22(22):4516–27. doi:10.1093/hmg/ddt299.
12. Carneiro LG, Nouh H, Salih E Quantitative gingival crevicular fluid proteome in health and periodontal disease using stable-isotope chemistries and mass spectrometry. J Clin Periodontol. 2014. doi:10.1111/jcpe.12262. (Epub ahead of print)
13. Casarin RC, Barbagallo A, Meulman T, Santos VR, Sallum EA, Nociti FH, et al. Subgingival biodiversity in subjects with uncontrolled type-2 diabetes and chronic periodontitis. J Periodontal Res. 2013;48:30–6. doi:10.1111/j.1600-0765.2012.01498.x.
14. Castrillon CA, Hincapie JP, Yepes FL, Roldan N, Moreno SM, Contreras A, et al. Occurrence of red complex microorganisms and Aggregatibacter actinomycetemcomitans in patients with diabetes. J Investig Clin Dent. 2013. doi:10.1111/jicd.12051. (Epub ahead of print).
15. Cephas KD, Kim J, Mathai RA, Barry KA, Dowd SE, Meline BS, et al. Comparative analysis of salivary bacterial microbiome diversity in edentulous infants and their mothers or primary care givers using pyrosequencing. PLoS One. 2011;6:e23503. doi:10.1371/journal.pone.0023503.
16. Corbella S, Francetti L, Taschieri S, De Siena F, Fabbro MD. Effects of periodintal treatment on glycemic control of patients with diabetes: a systematic review and meta analysis. J Diabetes Investig. 2013;4:502–9. doi:10.1111/jdi.12088. (Epub 2013 Apr 18)
17. De Soet JJ, van Gemert-Schriks MCM, Laine ML, van Amerongen WE, Morre SA, Winkehoff AJ. Host and microbiological factors related to dental caries development. Caries Res. 2008;42:340–7. doi:10.1159/000151329.
18. Dewhirst FE, Chen T, Izard J, Paster BJ, Tanner AC, Yu WH, et al. The human oral microbiome. J Bacteriol. 2010;192:5002–17. doi:10.1128/JB.00542-10.
19. Diaz PI. Microbial diversity and interactions in subgingival biofilm communities. Front Oral Biol. 2012;15:17–40. doi:10.1159/000329669.
20. Dimou NL, Nikolopoulos GK, Hamodrakas SJ, Bagos PG. Fcgamma receptor polymorphisms and their association with periodontal disease: a meta-analysis. J Clin Periodontol. 2010;37:255–65. doi:10.1111/j.1600-051X.2009.01530.x.
21. Divaris K, Monda KL, North KE, Olshan AF, Lange EM, Moss K, et al. Genome-wide association study of periodontal pathogen colonization. J Dent Res. 2012;91(Suppl 7):21–8.
22. Divaris K, Monda KL, North KE, Olshan AF, Reynolds LM, Hsueh WC, et al. Exploring the genetic basis of chronic periodontitis: a genome-wide association study. Hum Mol Genet. 2013;22:2312–24. doi:10.1093/hmg/ddt065.
23. Docktor MJ, Paster BJ, Abramowicz S, Ingram J, Wang YE, Correll M, et al. Alterations in diversity of the oral microbiome in pediatric inflammatory bowel disease. Inflamm Bowel Dis. 2012;18:935–42. doi:10.1002/ibd.21874.
24. Edlund A, Yang Y, Hall AP, Guo L, Lux R, He X, et al. An in vitro biofilm model system maintaining a highly reproducible species and metabolic diversity approaching that of the human oral microbiome. Microbiome. 2013;1:25. doi:10.1186/2049-2618-1-25.
25. Ernst FD, Uhr K, Teumer A, Fanghänel J, Schulz S, Noack B, et al. Replication of the association of chromosomal region 9p21.3 with generalized aggressive periodontitis (gAgP) using an independent case-control cohort. BMC Med Genet. 2010;11:119. doi:10.1186/1471-2350-11-119.
26. Fitzpatrick SG, Katz J. The association between periodontal disease and cancer: a review of the literature. J Dent. 2010;38:83–95. doi:10.1016/j.jdent.2009.10.007.
27. Foster JS, Kolenbrander PE. Development of a multispecies oral bacterial community in a saliva-conditioned flow cell. Appl Environ Microbiol. 2004;70:4340–8.
28. Ge X, Rodriguez R, Trinh M, Gunsolley J, Xu P. Oral microbiome of deep and shallow dental pockets in chronic periodontitis. PLoS One. 2013;8:e65520. doi:10.1371/journal.pone.0065520.
29. Giannobile WV, Braun TM, Caplis AK, Doucette-Stamm L, Duff GW, Kornman KS. Patient stratification for preventive care in dentistry. J Dent Res. 2013;92:694–701.

30. Giannobile WV, Kronman KS, Williams RC. Personalized medicine enters dentistry. JADA. 2013;144:874–6.
31. Gomar-Vercher S, Cabrera-Rubio R, Mira A, Montiel-Company JM, Almerich-Silla JM. Relationship of children's salivary microbiota with their caries status: a pyrosequencing study. Clin Oral Investig. 2014. doi:10.1007/s00784-014-1200-y.
32. Griffen AL, Beall CJ, Firestone ND, Gross EL, Difranco JM, Hardman JH, et al. CORE: a phylogenetically-curated 16S rDNA database of the core oral microbiome. PLoS One. 2011;6:e19051. doi:10.1371/journal.pone.0019051.
33. Griffen AL, Beall CJ, Campbell JH, Firestone ND, Kumar PS, Yang ZK, et al. Distinct and complex bacterial profiles in human periodontitis and health revealed by 16S pyrosequencing. ISME J. 2012;6:1176–85. doi:10.1038/ismej.2011.191.
34. Grover HS, Kapoor S, Saksena N. Periodontal Proteomics: Wonders Never Cease! Int J Proteomics. 2013. doi:10.1155/2013/850235.
35. Han YW, Wang X. Mobile microbiome: oral bacteria in extra-oral infections and inflammation. J Dent Res. 2013;92:485–91. doi:10.1177/0022034513487559. (Epub 2013 Apr 26).
36. Ho YP, Lin YC, Yang YH, Ho KY, Wu YM, Tsai CC. Cycooxygenase-2 Gene 765 single nucleotide polymorphism as a protective factor against periodontitis in Taiwanese. J Clin Periodontol. 2008;35:1–8. doi:10.1111/j.1600-051X.2007.01167.x.
37. Hong BY, Lee TK, Lim SM, Chang SW, Park J, Han SH, et al. Microbial analysis in primary and persistent endodontic infections by using pyrosequencing. J Endod. 2013;39:1136–40. doi:10.1016/j.joen.2013.05.001.
38. Hsiao WW, Li KL, Liu Z, Jones C, Fraser-Ligett CM, Fouad AF. Microbial transformation form normal oral microbiota to acute endodontic infections. BMC Genomics. 2012;13:345. doi:10.1186/1471-2164-13-345.
39. Human Microbiome Project Consortium. Structure, function and diversity of the healthy human microbiome. Nature. 2012;486:207–14. doi:10.1038/nature11234.
40. Huse SM, Ye Y, Zhou Y, Fodor AA. A core human microbiome as viewed through 16S rRNA sequence clusters. PLoS One. 2012;7:e34242. doi:10.1371/journal.pone.0034242.
41. Igari K, Kudo T, Toyofuku T, Inoue Y, IwaiT. Association between periodontitis and the development of systemic diseases. Oral Biol Dent. 2014;2:4. doi:10.7243/2053-5775-2-4.
42. Jiang W, Ling Z, Lin X, Chen Y, Zhang J, Yu J, et al. Pyrosequencing analysis of oral microbiota shifting in various caries states in childhood. Microb Ecol. 2014;67:962–9. doi:10.1007/s00248-014-0372-y.
43. Jorth P, Turner KH, Gumus P, Nizam N, Buduneli N, Whitely M. Metatranscriptomics of the human oral microbiome during health and disease. MBio. 2014;5:e01012–14. doi:10.1128/mBio.01012-14.
44. Jostins L, Ripke S, Weersma RK, Duerr RH, McGovern DP, Hui KY, et al. Host-microbe interactions have shaped the genetic architecture of inflammatory bowel disease. Nature. 2012;491:119–24. doi:10.1038/nature11582.
45. Kang SW, Yoon I, Lee HW, Cho J. Association between AMELIX polymorphisms and dental caries in Koreans. Oral Dis. 2011;17:399–406. doi:10.1111/j.1601-0825.2010.01766.x. (Epub 2010 Nov 29).
46. Karimbux NY, Saraiya VM, Elangovan S, Allareddy V, Kinnunen T, Kornman KS, et al. Interleukin-1 gene polymorphisms and chronic periodontitis in adult whites: a systematic review and meta-analysis. J Periodontol. 2012;83:1407–19. doi:10.1902/jop.2012.110655.
47. Kebschull M, Demmer RT, Grün B, Guarnieri P, Pavlidis P, Papanou PN. Ginigival tissue transcriptomes identify distinct periodontitis phenotypes. J Dent Res. 2014;93:459–68. doi:10.1177/0022034514527288. (Epub 2014 Mar 19).
48. Keijser BJF, Zaura E, Huse SM, van der Vossen JMBM, Schuren F;HJ, Montijn RC, et al. Pyrosequencing analysis of the oral microflora of healthy adults. J Det Res. 2008;87:1016–20.
49. Kellam P, Weiss RA. Infectogenomics: insights from the host genome into infectious diseases. Cell. 2006;124:695–7.
50. Kinniment SL, Wimpenny JW, Adams D, Marsh PD. Development of a steady-state oral microbiofilm community using the constant depth film fermenter. Microbiology. 1996;142(pt 3):631–8..

51. Klein MI, Xiao J, Lu B, Delahunty CM, Yates JR, III, Koo H. *Streptococcus mutans* protein synthesis during mixed species biofilm development by high throughput quantitative proteomics. PLoS One. 2012;7(9):e45795. doi:10.1371/journal.pone.0045795. (Epub 2012 Sep 25).
52. Kornman KS, Polverini PJ. Clinical application of genetics to guide prevention and treatment of oral diseases. Clin Genet. 2014;86:44–9. doi:10.1111/cge.12396.
53. Koromantzos PA, Makrilakis K, Dereka X, Katsulambros N, Vrotsos IA, Madianos PN. A randomized, controlled trial on the effect of non-surgical periodontal therapy in patients with type 2 diabetes. Part I: effect on periodontal status and glycaemic control. J Clin Periodontol. 2011;38:142–7. doi:10.1111/j.1600-051X.2010.01652.x. (Epub 2010 Nov 29).
54. Kuboniwa M, TRibble GD, Hendrickson EL, Amano A, Lamont RJ, Hacket M. Insights into the virulence of oral biofilms: discoveries from proteomics. Expert Rev Proteomics. 2012;9:311–23. doi:10.1586/epr.12.16.
55. Lalla E, Papapanou PN. Diabetes mellitus and periodontitis: a tale of two common interrelated diseases. Nat Rev Endocrinol. 2011;7:738–48. doi:10.1038/nrendo.2011.106.
56. Langfeldt D, Neulinger SC, Heuer W, Staufenbiel I, Künzel S, Baines JF, et al. Composition of microbial oral biofilms during maturation in young healthy adults. PLoS One. 2014;9:e87449. doi:10.1371/journal.pone.0087449.
57. Li Y, John MA St, Zhou X, Kim Y, Sinha U, Jordan RC, et al. Salivary transcriptome diagnostics for oral cancer detection. Clin Cancer Res. 2004;15:8442–50.
58. Li L, Hsaio WW, Nandakumar R, Barbuto SM, Mongodin EF, Paster BJ, Fraser-Liggett CM, et al. Analyzing endodontic infections by deep coverage pyrosequencing. J Dent Res. 2010;89:980–4. doi:10.1177/0022034510370026. (Epub 2010 Jun 2).
59. Liu B, Faller LL, Klitgord N, Mazumdar V, Ghodsi M, Sommer DD, et al. Deep sequencing of the oral microbiome reveals signatures of periodontal disease. PLoS One. 2012;7:e37919. doi:10.1371/journal.pone.0037919.
60. Mans JJ, von Lackum K, Dorsey C, Willis S, Wallet SM, Baker HV, Lamont RJ, Handfield M. The degree of microbiome complexity influences the epithelial response to infection. BMC Genomics. 2009;10:380. doi:10.1186/1471-2164-10-380.
61. Mealey BL, Oates TW. American academy of periodontoloty. Diabetes mellitus and periodontal diseases. J Periodontol. 2006;77:1289–303.
62. Michaud DS, Liu Y, Meyer M, Giovannucci E, Joshipura K. Periodontal disease, tooth loss and cancer risk in male health professionals: a prospective cohort study. Lancet Oncol. 2008;9:550–8. doi:10.1016/S1470-2045(08)70106-2.
63. Ngo LH, Darby IB, Veith PD, Locke AG, Reynolds EC. Mass spectrometric analysis of gingival crevicular fluid biomarkers can predict periodontal disease progression. J Periodontal Res. 2013;48:331–41. doi:10.1111/jre.12012.
64. Nibali L, Donas N, Henderson B. Periodontal Infectogenomics. J Med Microbiol. 2009; 58:1269–74. doi:10.1099/jmm.0.012021-0.
65. Nikolopoulos GK, Dimou NL, Hamodrakas SJ, Bagos PG. Cytokine gene polymorphisms in periodontal disease: a meta-analysis of 53 studies including 4178 cases and 4590 controls. J Clin Periodontol. 2008;35:754–67. doi:10.1111/j.1600-051X.2008.01298.x.
66. Ózok AR, Persoon IF, Huse SM, Keijser BJ, Wesselink PR, Crielaard W, et al. Ecology of the microbiome of the infected root canal system: a comparison between apical and coronal root segments. Int Endod J. 2012;45(6):530–41. doi:10.1111/j.1365-2591.2011.02006.x. (Epub 2012 Jan 17).
67. Pendyala G, Joshi S, Chaudhari S, Gandhage D. Links demystified: periodontitis and cancer. Dent Res J (Isfahan). 2013;10:704–12.
68. Pushalkar S, Ji X, Li Y, Estilo C, Yegnanarayana R, Singh B, et al. Comparison of oral microbiota in tumor and non-tumor tissues of patients with oral squamous cell carcinoma. BMC Microbiol. 2012;12:144. doi:10.1186/1471-2180-12-144.
69. Rôças IN, Neves MA, Provenzano JC, Siqueira JF, Jr. Susceptibility of as-yet-uncultivated and difficult-to-culture bacteria to chemomechanical procedures. J Endod. 2014;40:33–7. doi:10.1016/j.joen.2013.07.022.

70. Santos AL, Siqueira JF, Jr, Rôças IN, Jesus EC, Rosado AS, Tiedje JM. Comparing the bacterial diversity of acute and chronic dental root canal infections. PLoS One. 2011;6(11):e28088. doi:10.1371/journal.pone.0028088. (Epub 2011 Nov 21)
71. Schaefer AS, Richter GM, Groessner-Schreiber B, Noack B, Nothnagel M. El Mokhtari NE et al. Identification of a shared genetic susceptibility locus for coronary heart disease and periodontitis. PLoS Genet. 2009;5(2):e1000378. doi:10.1371/journal.pgen.1000378. (Epub 2009 Feb 13).
72. Schaefer AS, Richter GM, Nothnagel M, Laine ML, Noack B, Glas J, et al. COX-2 is associated with periodontitis in Europeans. J Dent Res. 2010;89:384–8. doi:10.1177/0022034509359575. (Epub 2010 Feb 22).
73. Schaefer AS, Bochenek G, Manke T, Nothnagel M, Graetz C, et al. Validation of reported genetic risk factors for periodontitis in a large replication study. J Clin Periodontol. 2013;40:563–72. doi:10.1111/jcpe.12092. (Epub 2013 Apr 16.)
74. Schäfer AS, Jepsen S, Loos BG. Periodontal genetics: a decade of genetic association studies mandates better study designs. J Clin Periodontol. 2011;38:103–7. doi:10.1111/j.1600-51X.2010.01653x.
75. Schunkert H, Gotz A, Braund P, McGinnis R, Tregouet DA, Mangino M, et al. Repeated replication and a prospective meta analysis of the association between chromosome 9p21.3 and coronary artery disease. Circulation. 2008;117:1675–84. doi:10.1161/CIRCULATIONAHA.107.730614. (Epub 2008 Mar 24.)
76. Schunkert H, Konig IR, Kathiresan S, Reilly MP, Assimes TL, Holm H, et al. Large scale association analysis identifies 13 new susceptibility loci for coronary artery disease. Nat Genet. 2011;43:333–8. doi:10.1038/ng.784.
77. Shaffer JR, Wang X, Feingold E, Lee M, Begum F, Weeks DE, et al. Genome-wide association scan for childhood caries implicates novel genes. J Dent Res. 2011;90:1457–62. doi:10.1177/0022034511422910.
78. Shaffer JR, Polk DE, Wang X, Feingold E, Weeks DE, Lee MK, et al. Genome-wide association study of periodontal health measured by probing depth in adults ages 18–49 years. G3 (Bethesda). 2014;4:307–14. doi:10.1534/g3.113.008755.
79. Siqueira JF, Jr, Alves FR, Rocas IN. Pyrosequencing analysis of the apical root canal microbiota. J Endod. 2011;37:1499–503. doi:10.1016/j.joen.2011. 08.012. (Epub 2011 Sep 16).
80. Slayton RL, Cooper ME, Marazita ML. Tuftelin, mutans streptococci and dental caries susceptibility. J Dent Res. 2005;84:711–4.
81. Sommer MO, Dantes G, Church GM. Functional characterization of the antibiotic resistance reservoir in the human microflora. Science. 2009;325:1128–31. doi:10.1126/science.1176950.
82. Song GG, Lee YH. Associations between FCGR2A rs1801274, FCGR3A rs396991, FCGR3B NA1/NA2 polymorphisms and periodontitis: a meta-analysis. Mol Biol Rep. 2013;40:4985–93. doi:10.1007/s11033-013-2599-y.
83. Taylor GW, Burt BA, Becker MP, Genco RJ, Shlossman M, Knowler WC, et al. Severe periodontitis and risk for poor glycemic control in patients with non insulin dependent diabetes mellitus. J Periodontol. 1996;67(Suppl. 10):1085–93.
84. Teumer A, Holtfreter B, Völker U, Petersmann A, Nauck M, Biffar R, et al. Genome-wide association study of chronic periodontitis in a general German population. J Clin Periodontol. 2013;40:977–85. doi:10.1111/jcpe.12154.
85. Toperoff G, Aran D, Kark JD, Rosenberg M, Dubnikov T, et al. Genome-wide survey reveals predisposing diabetes thype 2-related DNA methylation variations in human peripheral blood. Hu Molec Genet. 2012;21:371–83. doi:10.1093/hmg/ddr472. (Epub 2011 Oct 12.)
86. Trinidade F, Oppenheim FG, Helmerhorst EJ, Amado F, Gomes PS, Vitorino R. Uncovering the molecular networks in periodontitis. Proteomics Clin Appl. 2014 8(9-10):748–61. doi:10.1002/prca.201400028.
87. Tsuchida S, Satoh M, Kawashima Y, Sogawa K, Kado S, Sawai S, et al. Application of quantitative proteomic analysis using tandem mass tags for discovery and identification of novel biomarkers in periodontal disease. Proteomics. 2013;13:2339–50. doi:10.1002/pmic.201200510.

88. Vaithilingam RD, Safii SH, Baharuddin NA, Ng CC, Cheong SC, Bartold PM, et al. Moving into a new era of periodontal genetic studies: relevance of large case-control samples using severe phenotypes for genome-wide association studies. J Periodontal Res. 2014. doi:10.1111/jre.12167. [Epub ahead of print]
89. Wade WG. Has the use of molecular methods for the characterization of the human oral microbiome changed our understanding of the role of bacteria in the pathogenesis of periodontal disease? J Clin Periodontol. 2011;38(Suppl. 11):7–16. doi:10.1111/j.1600-051X.2010.01679.x.
90. Walker AW, Sanderson JD, Churcher C, Parkes GC, Hudspith BN, Rayment N, et al. High-throughput clone library analysis of the mucosa-associated microbiota reveals dysbiosis and differences between inflamed and non-inflamed regions of the intestine in inflammatory bowel disease. BMC Microbil. 2011;11:7. doi:10.1186/1471-2180-11-7.
91. Wang X, Shaffer JR, Weyant RJ, Cuenco KT, DeSensi RS, Crout R, et al. Genes and their effects on dental caries may differ between primary and permanent dentitions. Caries Res. 2010;44:277–84. doi:10.1159/000314676.
92. Wang X, Shaffer JR, Zeng Z, Begum F, Vieira AR, Noel J, et al. Genome-wide association scan of dental caries in the permanent dentition. BMC Oral Health. 2012;12:57. doi:10.1186/1472-6831-12-57.
93. Warinner C, Rodrigues JF, Vyas R, Trachsel C, Shved N, Grossmann J, et al. Pathogens and host immunity in the ancient human oral cavity. Nat Genet. 2014;46:336–44. doi:10.1038/ng.2906.
94. Wendell S, Wang X, Brown M, Cooper ME, DeSensi RS, Weyant RJ, et al. Taste genes associated with dental caries. J Dent Res. 2010;89:1198–202. doi:10.1177/0022034510381502. (Epub 2010 Sep 21).
95. Whitman WB, Coleman DC, Wiebe WJ. Prokaryotes: the unseen majority. Proc Natl Acad Sci USA. 1998;95:6578–83.
96. Wong L, Sissons C. A comparison of human dental plaque microcosm biofilms grown in an undefined medium and a chemically defined artificial saliva. Arch Orla Biol. 2001;46:477–86.
97. Xie CJ, Xiao LM, Fan WH, Xuan DY, Zhang JC. Common single nucleotide polymorphisms in cyclooxygenase-2 and risk for severe chronic periodontitis in a Chinese population. J Clin Periodontol. 2009;36:198–203. doi:10.1111/j.1600-051X.2008.01366.x.
98. Xie G, Chain PS, Lo CC, Liu KL, Gans J, Merritt J, Qi F. Community and gene composition of a human dental plaque microbiota obtained by metagenomic sequencing. Mol Oral Microbiol. 2010;25:391–405. doi:10.1111/j.2041-1014.2010.00587.x.
99. Xu Y, Teng F, Huang S, Lin Z, Yuan X, Zeng X, et al. Changes of saliva microbiota in nasopharyngeal carcinoma patients under chemoradiation therapy. Arch Oral Biol. 2014;59:176–86. doi:10.1016/j.archoralbio.2013.10.011.
100. Yang F, Zeng X, Ning K, Liu KL, Lo CC, Wang W, et al. Saliva microbiomes distinguish caries-active from healthy human populations. ISME J. 2012;6:1–10. doi:10.1038/ismej.2011.71.
101. Yang F, Ning K, Chang X, Yuan X, Tu Q, Yuan T, et al. Saliva microbiota carry caries-specific functional gene signatures. PLoS One. 2014;9:e76458. doi:10.1371/journal.pone.0076458.
102. Zarco MF, Vess TJ, Ginsburg GS. The oral microbiome in health and disease and the potential impact on personalized dental medicine. Oral Dis. 2012;18:109–20. doi:10.1111/j.1601-0825.2011.01851.x.
103. Zaura E, Keijser BJ, Huse SM, Crielaard W. Defining the healthy "core microbiome" of oral microbial communities. BMC Microbiol. 2009;9:259. doi:10.1186/1471-2180-9-259.
104. Zhong Q, Ding C, Wang M, Sun Y, Xu Y. Interleukin-10 gene polymorphisms and chronic/aggressive periodontitis susceptibility: a meta-analysis based on 14 case-control studies. Cytokine. 2012;60:47–54. doi:10.1016/j.cyto.2012.05.014.
105. Zhou M, Rong R, Munro D, Zhu C, Gao X, Zhang Q, et al. Investigation of the effect of type 2 diabetes mellitus on subgingival plaque microbiota by high-throughput 16S rDNA pyrosequencing. PLoS One. 2013;8:e61516. doi:10.1371/journal.pone.0061516.

Host Genomics and Response to Infectious Agents

Manuela Moraru and Carlos Vilches

Herpes simplex virus type 1 (HSV-1) is a human pathogen that causes one of the most wide-spread infections of orofacial skin and mucous membranes. It infects most individuals in early life, the primary infection often lacking clinical manifestations. However, after viral replication at the entry site, HSV-1 is transported by sensory neuron fibers to cell bodies in central ganglia, where the virus establishes life-long latency. Reactivation of quiescent virus results in recurrent disease, most often at the site of primary infection.

In the United States, the overall age-adjusted HSV-1 seroprevalence was estimated at 60%, [1, 2] but only about half of individuals experience clinically relevant herpetic infection [3, 4]. Furthermore, the recurrence frequency varies largely among the symptomatic individuals, ranging from few episodes in a lifetime to more than one monthly flare/outbreak. Susceptibility to clinically manifest reactivation of dormant HSV-1 has been shown to depend on the virus itself, environmental factors and host genetics.

The Virus

HSV-1 (or Human herpesvirus 1; family: *Herpesviridae*; subfamily: *Alphaherpesvirinae*; genus: *Simplexvirus*) is a large (150–200 nm), spherical, enveloped virus, comprising four major structures: the core containing viral DNA, an icosahedral capsid, the tegument and the envelope. The large HSV-1 genome (~150,000 base pairs of double stranded DNA) includes more than eighty genes organized into a

C. Vilches (✉) · M. Moraru
Immunogenetics & Histocompatibility, Instituto de Investigación Sanitaria Puerta de Hierro, Manuel de Falla1, 28222 Majadahonda, Spain
e-mail: carlos.vilches@yahoo.com

M. Moraru
e-mail: manuelamoraru@yahoo.com

S. T. Sonis (ed.), *Genomics, Personalized Medicine and Oral Disease*,
DOI 10.1007/978-3-319-17942-1_4

long (L) and a short (S) region, each containing a unique region (U_L and U_S, respectively) flanked by inverted repeat sequences. Compared to RNA viruses, the HSV-1 genome is fairly stable; however, differences in virulence and pathogenesis have been observed between different strains [5].

HSV-1 productive infection follows a stepwise sequence, starting with virus entry into the host cell, followed by viral gene expression and DNA replication, and ending with virion assembly and exit. Interactions of HSV-1 envelope glycoproteins B (gB), gC, gD and the gH/gL complex with host cell surface molecules allow virus attachment, its fusion with the plasmatic membrane, and, eventually, nucleocapsid and tegument proteins passage to cytoplasm. Virus entry stage ends with nucleocapsid delivery to the nucleus, where gene transcription begins. This latter process also follows a chronological sequence where three major gene groups are distinguished: immediate early (IE), early and late genes. IE genes are transcribed without prior HSV-1 protein synthesis because their promoters recruit the host cell transcriptional machinery. IE-gene encoded proteins promote the transcription of early genes and a subset of late genes, which guide synthesis of virus DNA and structural proteins in the productive stage of the infection. HSV-1 DNA synthesis generates numerous progeny genomes in each infected cell; when viral nucleocapsids are assembled, they are loaded with viral DNA and exit host nucleus to cytoplasm. Finally, virions egress the infected cell, either after its lysis or taking advantage of cellular secretory mechanisms [6].

Pathogenesis and Clinical Course

The mainstay of all members of *Herperviridae* family is their capacity to establish latent, lifetime infection. Additionally, Alphaherpesviruses are able to invade and replicate in the Central Nervous System (CNS), the site of HSV latency. HSV-1 transmission from one host to another requires direct contact of damaged skin or mucosae of an individual with HSV-1 virions shed by another one. When the virus surpasses anatomical barriers of a susceptible subject, it replicates at the site of inoculation. It follows the virus (or capsid fragments) uptake into the sensory nerve fibers in the epithelium and transport to the neuronal body, where, after further viral replication, latency establishes. This initial stage of infection is usually asymptomatic and self-limiting, although in exceptional cases it can evolve in life-threatening herpetic encephalitis, sepsis-like syndrome, eczema herpeticum, a serious complication of atopic dermatitis, or Herpes simplex keratitis, a major cause of blindness in developed countries [7–12].

Unlike other chronic infections, the virus does not replicate during latency, when only very limited viral gene transcription is detected. Certain signals (e.g., UV-light exposure, fever, stress), through as yet incompletely understood mechanisms, can cause viral reactivation, followed by transport from neuronal bodies to periphery. Reactivation is characterized by either asymptomatic virion excretion or clinically relevant manifestations lasting for about 1 week, with keratinocyte damage presenting as typical vesicular lesions and subsequent ulcers, which can undergo secondary bacterial infection. Reactivations occur at variable intervals, and most often affect the perioral area. Less frequently found than herpes labialis is the occurrence of gingivostomatitis (difficult to distinguish clinically from aphtas of other aetiologies), which usually affects the tongue, gingival, buccal or palate mucosae [13]. Outside the oral area, potential sites of cutaneous recurrent herpetic lesions include, with variable (albeit lower) frequency, face, ears, neck, trunk or limbs (including the rare herpetic paronychia) [13]. Genital and ocular herpetic lesions are more common, but still far less frequent than herpes labialis [1, 12]. Erythema multiforme, which may affect the oral cavity, is also a rare complication of herpetic infections [14].

Host Immune Response to HSV-1 and Virus Evasion from Host Immunosurveillance

Innate Cytokine Response to Virus Molecular Patterns

Virus survival inside host cells implies the development of a large array of strategies to overcome the host blocking its replication. To this end, HSV-1 selectively degrades or inhibits the proper function of host proteins and limits their synthesis (reviewed in [15]). In primary infection, when HSV-1 replicates prior to maturation of host adaptive immunity, evasion from innate immunity should be crucial for its survival. Later on, additional strategies to overcome both innate and acquired host immune responses facilitate virus latency and reactivation.

As outlined previously, the first step in HSV-1 life cycle is entry into host cells, taking advantage of intrinsic cellular mechanisms. Conversely, the trigger for an immune response to any pathogen is its recognition by host immune cells. To this end, host pattern recognition receptors (PRRs) recognize pathogen constitutive and conserved molecular motifs termed Pathogen-associated molecular patterns (PAMPs). HSV-1 glycoproteins, DNA and RNA are all PAMPs that can be sensed by host PRRs, mainly Toll-like receptors (TLRs) and cytosolic DNA and RNA sensors. These PRRs are expressed in epithelial and fibroblast cells at the site of primary infection and also in different innate immune cell types.

PRR-dependent pathways (and also PRR-independent ones, see following section) converge to trigger proinflammatory cytokine responses, particularly interferon (IFN) production. Of the three types of IFN, types I and III (IFN-α/β and IFN-λ, respectively) are produced within the first hours after infection by a large array of cellular types, whereas type II (i.e. IFN-γ) is secreted in a subsequent step of the immune response, mainly upon NK- and T-lymphocyte sensing of infected cells [16]. IFNs, largely considered a first line defence against virus and, probably, the most critical for innate response, inhibit viral replication and control the transcription of several host genes, leading to global repression of protein synthesis in infected cells. Furthermore, they constitute potent stress signals that enhance antigen presentation, shaping acquired immune response to HSV-1 infection [17].

Type I IFN induction requires the activation of several transcription factors, including interferon regulatory factor (IRF)-3 and nuclear factor-κB (NF-κB). An additional autocrine feedback enhancing mechanism is initiated in epithelial and fibroblast cells through IFN receptors, triggering a phosphorylation cascade of Janus kinases (Jak) and signal transducer of activated transcription (Stat) molecules, which eventually leads to the transcription of IFN-stimulated genes and subsequent anti-viral responses (reviewed in [18]).

Considering the importance of these early host defence mechanisms, it is not surprising that several HSV-1 encoded proteins alter TLR-dependent and -independent IFN-mediated antiviral pathways at different levels. For instance, ICP34.5 and US11-encoded proteins interfere with host cell response to double stranded (ds) RNA (a marker of virus infection), consequently preventing both infected cell death and NF-κB activation [19–23]; UL41 RNase and IE protein ICP27 reduce Jak1, Stat1 and Stat2 supplies, hence limiting the signalling through type I IFN receptors [24–26]; and ICP0 blocks both TLR-triggered activation of NF-κB and IRF signalling, thus limiting the ensuing production of type I IFN and its downstream outcomes [21, 27–30].

Cellular Immunity

Cytotoxic cells, including CD8 +T lymphocytes (CTL) and natural killer (NK) cells, are the main effectors of cellular immunity to viral infection. NK cells can recognize and kill infected cells without prior sensitization, and they also secrete soluble factors, notably IFN-γ and TNF-α, potent anti-viral effectors and immune cell activators. NK-cell activation depends on the balance of signals elicited by activating and inhibitory receptors [31–33]. NK cells, besides recognizing PAMPs, sense pathogen- or stress-induced molecules of infected cells through specific activating receptors (e.g. Natural Cytotoxicity Receptors and NKG2D, which recognises some polymorphic ligands) [34–36]. Furthermore, they are strongly stimulated upon recognition of immune complexes and antibody-coated cells through receptors for the Fc of IgG (FcγR), like CD16 [37]. Also essential for NK cell regulation and surveillance of infection is, as explained in more detail in Diversity of NK-cell receptors, recognition through inhibitory receptors of self Major Histocompatibility Complex class I molecules (MHC, called HLA in humans). Supporting a crucial role for NK cells in control of HSV-1 infection are several primary immunodeficiencies affecting either the number or the function of this lymphocyte subset, whereby the clinical picture is dominated by recurrent herpetic infections (reviewed in [38, 39]).

In addition to the evolutionarily ancient pathogen recognition by germ-line encoded PRRs and other stress signal receptors on innate cells, adaptive immune lymphocytes use clonotypic receptors to recognize and activate in response to either complete viral antigens (B cells) or (T-cells) pathogen-derived peptides on the surface of antigen presenting cells (APC). Although acquired immunity requires longer periods of time to mature, it confers highly-specific, long-lasting protection against pathogens. HSV-1 activates APCs (essentially dendritic cells, DCs) at the infection site, either directly or through type I IFN release by infected cells. DCs then carry viral antigens to draining lymph nodes, where acquired immune responses are primed [17, 40].

In lymph nodes, antigen-specific CD8+ T lymphocytes recognize viral peptides crosspresented by DCs on HLA class I molecules. After clonal expansion and maturation, these T lymphocytes specifically recognize and kill infected cells, and secrete IFNγ, thus controlling both the virus clearance at the infection sites and reactivation from latency [41–43]. Differentiation of antigen-specific CTLs and generation of memory CD8+ T lymphocytes require functional CD4+ T helper cells, which, in turn, depend on antigen presentation and co-stimulation by DCs [43–45]. CD4+ helper and regulatory T lymphocytes are also critical to promote and control, through cell-cell interaction and cytokine secretion, efficacious humoral and inflammatory responses to infection. Effective antigen-specific humoral and cellular antiviral responses thus relay on viral peptide presentation to T-lymphocytes of the CD8 and CD4 subsets by HLA molecules, which belong to two classes.

HLA class I molecules, expressed on the surface of nearly all nucleated human cells, specialize in presentation to $CD8^+$ T-lymphocytes of peptides derived from proteins synthesized by the cell expressing the HLA molecule (i.e. this reflects the cell health state). In contrast, HLA class II molecules, whose expression is restricted to, roughly, professional APCs (e.g. Langerhans and other DCs) and B-lymphocytes, can additionally present peptides derived from proteins taken from the extracellular medium (e.g. endocytosed virions and dead cells); i.e., through HLA class II molecules, APCs show CD4+ T-lymphocytes the antigens they have encountered in their environment.

Classical HLA class I proteins (HLA-A, -B and -C) are membrane-bound heterodimers composed of a highly polymorphic α-chain, non-covalently associated with the smaller β_2-microglobulin, which form a binding grove that harbours peptides of 8–11 amino acids in length for T-cell receptor recognition. Roughly similar structures are seen in HLA-class II molecules (DR, DQ and DP), albeit each of these is composed of alpha and beta-chains of comparable size, and they present peptides of greater and more variable length. Complex intracellular structures generate and transport peptides of appropriate length to HLA class I and class II molecules [46].

Consistently with the key role of antigen presentation in the response to HSV-1, the virus has evolved several mechanisms to circumvent T cell recognition of infected cells. For instance, viral protein ICP47, encoded by the immediate-early gene US12, efficiently binds the human transporter for antigenic peptides (TAP), hence interfering with delivery of virus peptides to HLA class I molecules [47]. Other levels of antigen presentation targeted by HSV-1 include the blockage of DC maturation [48] and ICP0-mediated down-modulation of CD83, an adhesion molecule involved in T-cell/DC interaction [49, 50]. Furthermore, viral protein ICP22 inhibits CD4+ T lymphocyte activation by B-cells [51]. Additionally, by a yet unknown mechanism, HSV-1 down-regulates CD1 proteins (a family of HLA-like lipid-antigen-presenting molecules), hence subverting CD1-restricted T cells response to infection [52, 53]. Though CD1 proteins have not yet formally been involved in HSV-1-antigen presentation, murine CD1-restricted T cells (also called NKT cells) are somehow involved in the control of HSV-1 infection [54].

Humoral Immunity

Antibody response is a hallmark of adaptive immunity. Following a primary IgM response, CD4+ T cells promote further maturation of the humoral response, and the generation of HSV-1 antigen-specific plasma cells and B memory cells. This results in a secondary response of circulating specific antibodies of higher affinity, primarily of the IgG isotype, which act as effectors of the acquired immune response during HSV-1 reactivations. In addition to direct virus neutralization, antibodies are also required for classical pathway of complement-mediated cell lysis and for antibody-dependent cellular cytotoxicity (ADCC) through FcγR, as detailed in Diversity of cellular receptors for immunoglobulin G. It is therefore not unexpected that HSV-1 glycoproteins gE and gI have evolved to form a decoy FcγR domain expressed on both the viral envelope and the infected cell membrane [55]. The decoy FcγR may compete with functional FcγR of effector immune cells, and partially protect the infected cells from ADCC; in addition, it generates an antibody bipolar bridging effect (the antibody uselessly binding to the infected cell through both the antigen-binding site and the Fc). Bipolar bridging, besides hindering ADCC, blocks C1q binding, thus inhibiting immunocomplex recognition and complement classical pathway activation [56, 57]. Furthermore, viral glycoprotein C targets C3b, blocking both C3 and C5 convertases assembly and anaphylatoxin release, hence inhibiting all pathways of complement activation [58–60]. Of note, perhaps due in part to subversion mechanisms, symptomatic reactivations of HSV-1 (and other herpesviruses) take place almost invariably in presence of specific antibodies, showing ineffectiveness of humoral response alone to eradicate the virus or completely block its replication.

Genetics, Genomics and Susceptibility to HSV-1 Infection

Lessons from Mouse Experimental HSV-1 Infection

Differences between individuals in susceptibility to particular infections, and in their response to these, have been observed since centuries ago. It followed the observation that these inter-individual variations in susceptibility to disease depend on both the infectious agent and the host. The study of infectious disease under complex genetic control in humans is complicated by a variety of aspects including population heterogeneity, environmental factors, low or incomplete penetrance, and phenotypic differences among infected individuals. Furthermore, genetic factors contributing to pathogen susceptibility most often act at different stages of the infection process. Mouse models using both reverse- and forward-genetics approaches have been of great help for geneticists searching for disease susceptibility loci. In the case of HSV-1 infection, mice are not natural hosts, and reactivation from latency is less frequent in them than in humans, thus limiting the utility of murine models for studying this phenomenon. However, progress has been made in studying mouse susceptibility to HSV-1 primary infection, the establishment of latency and pathogenesis. Alike other infectious diseases, mouse models confirmed that virus-specific factors including virus strain, the site of inoculation and the amount of viral inoculum influence the infection outcome. Furthermore, initial forward genetics studies found that natural resistance to HSV-1 primary infection is also mouse strain-dependent [61]; differences manifested as reduced pathogenicity and increased survival, determined by a delay in HSV-1 spreading to CNS and latency establishment, processes dependent on type I IFN production and NK cell activation [62–66]. Several susceptibility loci on mouse chromosome 6 were subsequently found responsible for these differences [63, 64, 67]. Additionally, a very recent study has extended knowledge on genetic factors influencing disease severity by identifying a locus on mouse chromosome 16 which modifies both susceptibility to die of herpetic encephalitis and the severity of ocular disease [68]. Indeed, predisposition to develop severe keratitis is another example where mouse models can be of great help. Early studies identified a region around immunoglobulin heavy-chain locus (IGH) on mouse chromosome 12 involved in keratitis development [69], which has been confirmed in two more recent studies [68, 70]. Additional susceptibility loci for herpetic keratitis were also identified on chromosomes 4, 5, 13, and 14 [70]. However, while these approaches are promising, they have so far identified no susceptibility gene.

Reverse genetics allows direct studying of specific genes involved in resistance to infection through manipulation of mouse genome. Mouse knock-outs for several genes implicated in innate and adaptive immunity elucidated most of the known host defence mechanisms against HSV-1 infection, including the crucial role of interferons in the control of primary HSV-1 infection [71–73], opening the way towards defining the first genetic cause of human susceptibility to devastating, primary HSV-1 infection: Herpes Simplex Encephalitis (HSE).

TLR-IFN Pathways and Herpes Simplex Encephalitis

As outlined above, IFNs are key players in limiting HSV-1 infection in both mice and humans, and PRR recognition of pathogen structures is an important trigger of IFN production. Since PRR discovery in the early nineties, an increasing number of HSV-1-derived structures have been found to activate these receptors. Thereby, cell surface-expressed TLR2 recognizes both HSV-1 gB and the gH/gL complex [74, 75]. Furthermore, endosomal TLR3 and TLR9 trigger cellular activation in response to HSV-1 dsRNA intermediates and unmethylated DNA containing CpG motifs, respectively [76–78]. TLR7 and TLR8 sense virus-derived single-stranded RNA, but they have not yet been implicated in HSV-1 recognition. Additionally, RIG-1 (retinoic acid-inducible gene 1) and MDA-5 (melanoma differentiation factor 5) sense cytosolic HSV-1 RNA [79, 80]. Finally, a growing number of new cytosolic DNA sensors are being reported to recognize intracellular HSV-1 structures (reviewed in [81, 82]). However, more research is warranted to provide an integrating model linking all these findings.

Parallel to advance in PRR characterization, great progress has been made in identifying single-gene inborn defects causing, or predisposing to, herpetic encephalitis. Investigations with both murine models and human patients have contributed to defining genetic defects related to type I IFN as primary immunodeficiencies capable of causing HSE. The first report on genetic mutations causing isolated, childhood HSE described two unrelated patients with impaired type I and III IFN production and different autosomal recessive mutations in the gene coding for the endoplasmic reticulum protein UNC93B1, implicated in the TLR3, TLR7 and TLR9 traffic [83–85]. This was soon followed by descriptions of another three unrelated patients with autosomal dominant [86] and recessive mutations in the *TLR3* gene [77], and pediatric patients with history of HSE and mutations affecting several proteins downstream the TLR3-IFN pathway (namely TRAF3, TRIF/TICAM1 and TBK1) [87–89]. While these studies provide strong evidence that human TLR3 is involved in host defence against HSV-1 primary infection in the CNS, the immunological alterations responsible for many other patients experiencing HSE remain to be established.

Besides immune alterations predisposing exclusively to HSE, a minority of patients with other primary immunodeficiencies affecting the IFN activation pathway also develop HSE, among other recurrent infections caused by different pathogens. Such is the case with *NEMO* and *STAT1* mutations, which affect NF-κB activation, and response to type I IFNs, respectively. Also affecting the latter pathway, the only *TYK2* mutation described to date did not associate with HSE, but with HSV-1 cutaneous infection (reviewed in [90, 91]).

Interestingly, only a minority of pediatric patients and relatives with an identified genetic defect causing HSE suffered episodes of herpes labialis [86]. Furthermore, very few experienced HSE recurrences [7, 92], arguing against a crucial role of the TLR3 pathway in immunity to HSV-1 outside CNS and in latter stages of infection (i.e. latency control and reactivations). A single study reported that a leucine-412 to phenylalanine TLR3 polymorphism associates with a reduced NK-cell responsiveness through TLR3 activation, and with susceptibility to frequent recurrences of herpes labialis [93]. However, only thirty-seven individuals were studied and the statistical analysis was not standard, since chromosomes, instead of individuals, were apparently counted, therefore these results need confirmation in larger series. Likewise, no consensus about the involvement of TRL3 polymorphism in HSV-2 infection has so far been reached [94, 95].

HLA Polymorphism

As detailed in Cellular immunity, host ability to mount adequate cellular and humoral adaptive responses to viral infection depends completely on the capacity of CD8 and CD4 T-lymphocytes to detect virus peptides presented by HLA class I and class II molecules, respectively. Although HLA molecules can present a wide array of peptides with different sequences, each individual HLA molecule has a defined preference for peptides with certain sequence motifs, imposed by the biochemical features of the peptide-binding groove. Furthermore, HLA molecules are the most polymorphic in the human proteome, and their polymorphic residues concentrate around the peptide-binding grove, this way providing a variety of alleles capable of binding very diverse sets of microbial peptides. In human communities, balancing selection has favoured heterozygosis through persistence of multiple alleles at each HLA locus, thus providing nearly infinite opportunities for antigen recognition. In turn, the genomes of many pathogens evolve under the pressure of the immune response, to avoid encoding the sequence motifs better presented by the more prevalent HLA alleles.

HLA molecules, because of their extreme polymorphism and their central role in the control of the adaptive and the NK-cell-mediated immune responses, are good candidates to influence the course of HSV-1 infection. The HLA locus on chromosome 6 is in fact more frequently associated with protection or susceptibility to disease and variation in the immune response than any other region in the human genome. During the last 40 years, several studies evaluated contribution of HLA allotypes to the clinical course of HSV-1 infection. Though results were hardly reproducible in different populations, these studies deserve attention ([96] and references 16–27 therein). Low reproducibility of results obtained by each study could reflect varying selection pressures exerted by the pathogen strains, and their balance with the effects of other diseases affecting different populations, besides the diverse HLA allelic and haplotypic backgrounds of these. In addition, or alternatively, study design could have biased the results of at least some of them; possible confounding factors include the number and selection of the studied individuals (e.g. use of asymptomatically HSV-1 infected negative controls); the HLA typing methods (reliable genotyping methods were implemented only in the last two decades); and, ultimately, the information available on HLA diversity at the moment of the results divulgation. Our group addressed recently those issues in a new study of HLA polymorphism, using DNA typing methods for all assessed loci (including HLA-C and DRB1, whose serological typing was very imprecise), and well characterized negative controls. This study identified some new potential associations and its results were partially consistent with some of the previously proposed ones [96].

The general picture that emerges from studies on association of HLA with HSV-1 infection is that class II polymorphism appears not to influence substantially the course of the infection; whilst different HLA-A, -B and -C allele groups may associate with risk (e.g. A19-associated alleles, and the C*15-B*51 haplotype) or protection (e.g., B*18 and B*35) from symptomatic infection, even though none stands out clearly as a strong predictor of its clinical course. The most obvious reason for an HLA allele to be protective would be its capacity to bind immunodominant virus peptides with high affinity. However, polymorphism also determines additional functional variability that influences the capacity of an HLA molecule to present efficaciously virus antigens, such as its expression levels [97]; its speed of assembly (from protein synthesis to migration to the surface of a fully folded, peptide-loaded molecule) [98]; or the degree of dependency of its expression on a fully functional peptide-loading complex (targeted by HSV-1 and other herpesviruses). Another manner in which HLA diversity could influence immunity to pathogens is, as discussed in the following section, through the role of HLA class I molecules, as cell health markers, in the regulation of NK cells.

Diversity of NK-cell Receptors for HLA Class I

Human NK cells use several families of surface receptors to survey abnormal expression of HLA class I molecules. Two such families recognize conserved or little polymorphic HLA-sequence motifs: heterodimers of CD94 and either NKG2A (inhibitory) or NKG2C (activating) recognize the scarcely polymorphic HLA-E molecule; and the B1 member of the Leukocyte Immunoglobulin-Like Receptor family (i.e. LILRB1), which is expressed by NK and T cell subsets among other leukocyte lineages, interacts with sequences conserved in HLA-A, -B, -C, and -G). A third family, Killer-cell Ig-like Receptors (KIR), has expanded and diversified during human evolution to encode a repertoire of approximately fifteen inhibitory and activating KIR that recognize specifically subsets of HLA molecules [99]. This enables NK-cell precursors to differentiate into clones expressing diverse KIR, which monitor separately the expression of the different HLA class I molecules. Such NK-cell clone repertoires are believed to have evolved for counteracting pathogens that tamper selectively with HLA expression—i.e., which subvert expression of the HLA molecules that present efficaciously their antigens, whilst respecting others. In support of this view is the fact that convergent evolution has generated in rodents and other mammals an analog repertoire made up of Ly49 receptors, which pertain to a protein superfamily totally unrelated with human KIR [100]. Of note, one such receptor, mouse Ly49H, turned out to correspond to the Cmv1 locus of resistance to murine cytomegalovirus and recognize a decoy, MHC-like, viral molecule [101, 102].

The members of the KIR family diverge in several structural and functional aspects, including expression frequency and levels, and capacity to bind polymorphic distinct sets of HLA ligands with variable avidity, and to transmit inhibitory signals with variable strength. Furthermore, KIRs are enormously diverse, owing to conspicuous copy-number variation, allelic polymorphism, existence of several chimeric recombinant genes with mixed features, and highly variable haplotypic combinations of *KIR* genes and alleles [99, 103, 104]. The combination of functional diversity and genetic polymorphism makes humans to possess repertoires of NK cells differently calibrated to sense pathologic variations in HLA expression. This has prompted much interest on KIR as potential susceptibility/resistance genes in infection and other human health conditions. Well studied is the influence of KIR and HLA diversities on HIV infection, in which delayed or accelerated progress of the disease associates with combined presence in the genome of genes encoding certain polymorphic KIR of inhibitory or activating function, and HLA-B ligand molecules bearing specific sequence motifs recognised by those KIR [105].

In HSV-1 infection, we reported that presence in the genome of the KIR2DL2 gene, which encodes an inhibitory receptor for multiple HLA-C molecules, together with asparagine in residue 80 of the HLA-C α-chain (present in most KIR2DL2 ligands) associated with symptomatic HSV-1 infection and with recurrent reactivations [96]. Because KIR2DL2 recognizes HLA-C ligands with higher affinity than its KIR2DL3 allotype, we speculate that its association with a poorer course of the infection could be due with this molecule being a poorer sensor (precisely for its higher affinity) of viral subversion of HLA-C expression. Or that, in the absence of HLA-C downregulation, its inhibitory KIR2DL2 receptor sets too high a threshold for NK activation in response to other stimuli triggered by HSV-1-infected cells. Nevertheless, we can not formally exclude that association of KIR2DL2 with herpetic disease could be related to its nearly complete linkage disequilibrium with the neighbor KIR2DS2 gene, which encodes an activating homolog receptor; controversial is, however, whether this homologue also recognizes HLA-C alleles overrepresented among clinically infected individuals [106, 107].

In contrast with KIR receptors, the CD94/NKG2 family of C-type lectin-like receptors encoded in the NK complex (NKC) on chromosome 12 is largely conserved [108]. However, CNV of the gene coding for the activating receptor NKG2C affects about one third of the population, and ~4 % of individuals lack the gene altogether [96, 109, 110]. NKG2C is expressed at high levels on subsets of cytotoxic cells (NK cells and CTL). Such subsets are found in only a fraction of individuals infected by cytomegalovirus (human herpesvirus 5), and they can help control the infection. Zygosity for *NKG2C* modulates the expression and function of the receptor [111], but we did not find any relation between *NKG2C* deletion and the clinical course of the herpetic infection [96].

LILRB1, a receptor for multiple HLA class I molecules, also recognizes with even higher affinity an HLA-decoy molecule of human cytomegalovirus [112, 113], and it is encoded by a polymorphic gene [114]. Other members of the LILR family (encoded close to *KIR* genes on chromosome 19) are expressed mainly in myelomonocytic cells and display allelic polymorphism and, in some cases, CNV [115]. Their possible implication in HSV-1 infection is unexplored.

Diversity of Cellular Receptors for Immunoglobulin G

A family of receptors for the Fc fraction of IgG (FcγR) links humoral and cellular immune responses, and confers high specificity to innate and acquired cellular responses against invading pathogens. Because virtually every lineage of effector leukocytes expresses one or more FcγR, the effects of the interactions of these with antigen-antibody immune complexes are vast and pleiotropic, including ADCC, phagocytosis, cytokine production, B cell homeostasis, immunocomplex clearance and antigen presentation [116].

FcγRs belong to the immunoglobulin superfamily and in humans they are subdivided in three types that diverge in structure, affinity for IgG and expression pattern—high-affinity FcγRI (or CD64); low-affinity FcγRII (or CD32, including three structurally related receptors: CD32A, CD32B and CD32C); and intermediate-affinity FcγRIII (or CD16, comprising CD16A and CD16B). CD32B is the single inhibitory receptor of the family. The complexity of the FcγR family is paralleled by IgG Fc diversity, and the different IgG subclasses (and allotypes) bind with variable avidity to each FcγR [117].

Genetic polymorphism in the *FCGR* locus on the long arm of chromosome 1 further increases the intricate tuning of immune responses mediated by these receptors. Relevant for HSV-1 infection is the functional dimorphism in *FCGR3A* (coding for activating CD16A), because it is the most widely FcγR expressed on NK cells, triggering potent activating signals upon antibody-coated target-cell recognition. A valine for phenylalanine change at position 158 of CD16A increases the receptor affinity for IgG1 and IgG3, thereby potentially modulating the intensity of NK-cell mediated ADCC against HSV-1 infected cells [118–120].

We have recently reported that the valine-158 allotype of CD16A (of higher affinity for IgG) is underrepresented among individuals susceptible to develop clinically relevant HSV-1 reactivations, the dose of allele CD16A-158V correlating significantly with the probability of symptom-free infection [96]. Apparent protection from recurrences could be directly related to enhanced ADCC against HSV-1 infected cells in NK cells expressing CD16A-158V. Increased CD16A-IgG avidity might also surpass the viral FcR decoy subversion mechanism.

Likewise, genetic polymorphisms determine expression of activating CD32C (a non-functioning gene in most humans) or inhibitory CD32B on NK cells (normally restricted to B lymphocytes and myeloid cells) in a minority of individuals. Those polymorphisms could potentially impact on cellular response against IgG-coated infected cells [121–123]. Additional CD32B functional polymorphisms have been described and they might also influence NK-cell mediated ADCC [124, 125]. Further studies should elucidate the complex interplay between viral and host FcγR, and IgG polymorphism, and its contribution to the immune response against HSV-1 infection.

Genome-Wide Association Studies, Cold Sore Susceptibility Gene 1 and Apolipoprotein E—A Connection with Alzheimer Disease?

Exploring human susceptibility to HSV-1 was initially restricted to candidate-gene studies (where the genes-to-analyze are selected based on their believed mechanistic relevance to the disease pathogenesis) and, more recently, was approached using whole-genome scans.

A single, family-based, genome-wide association study (GWAS) has so far tried to identify candidate genes responsible for the differences in the frequency of herpes labialis (4). This study identified a susceptibility region for HSV-1 recurrence on chromosome 21, where six different genes mapped. Subsequent studies found open reading frame 91 of chromosome 21 the locus of interest, and proposed for it the name 'cold sore susceptibility gene 1' (*CSSG-1*) [126]. Five *CSSG-1* genotypes were defined, and they associated with mean cold sore annual frequencies ranging 1.16–2.03. *CSSG-1* codes for an intracytoplasmatic protein of yet unknown function, thus more research is needed before this finding can be understood and translated to clinical practice. Furthermore, considering the sample size of this first GWAS (431 individuals from 39 families) and that it failed to detect any of the candidate genes previously identified in other studies, its results warrant confirmation. Interestingly, mouse chromosome 16 has recently been shown to bear a susceptibility locus for HSE and herpetic keratitis severity, which maps close to (but is different from) a murine homologue of the human *CSSG-1* gene [68]. The meaning of this co-incidence is unknown.

Also of possible interest is the fact that *CSSG-1* is a neighbour to several genes involved in susceptibility to Alzheimer disease (AD), including that encoding amyloid-β precursor protein (APP), and that this is but one of several connections between AD and HSV-1 infection [127, 128]. The most widely accepted risk factor for non-familial Alzheimer disease, allele ε4 of apolipoprotein E (ApoE), has also been related to increased HSV-1 neuroinvasiveness, frequently recurrent herpes labialis [127, 129], and other viral infection outcomes, including hepatitis C and human immunodeficiency viruses [130–132]. ApoE, critical for triglyceride-rich lipoprotein catabolism and a major component of very-low density lipoproteins [133], facilitates lipid-antigen presentation by CD1 molecules [134], and its variations could skew CD1 repertoire. Given that several RNA viruses depend on ApoE for cell-to-cell passage [135–138], and that gB of HSV-1 has also been shown to bind ApoE [139], it is also tempting to speculate that a similar mechanism might contribute to HSV-1 infectivity. As HSV-1 has been directly implicated in the pathogenesis of AD, assessment of epistatic interactions between *APOE* and *CSSG1* may shed light on the genic risk of both AD and HSV-1 infection, and on the relationships between the two diseases.

Perspectives

There is ample evidence that genomic diversity, besides environmental factors, plays a crucial role in the final outcome of many virus-host interactions, which manifests as resistance or susceptibility to disease; and that much of that diversity pertains to immune response genes. The extremely variable control of HSV-1 by different humans has intrigued physicians and scientists since decades ago, but it remains only partially explained. Advances in genomics have greatly modified our understanding of host genetics contribution to the course of many infectious diseases. Genomic studies on HSV-1 infection are lagged in this regard, and further research is needed to confirm the only candidate gene identified in the single GWAS performed so far, and to understand the mechanism by which it modifies susceptibility to the disease.

Although GWAS represent a valuable tool for identifying disease-associated candidate genes, most available methods only spot single nucleotide polymorphisms of clear Mendelian inheritance; in other words, they are not well suited for studying more complex forms of variation, including deletions, insertions, frameshifts, copy-number variation (CNV), and rare allelic variants. Furthermore, some very polymorphic families of highly homologous immune-response genes (e.g. the genes encoding HLA, immunoglobulins, FcγR and KIR), are poorly or not at all represented among the genetic tags of many GWAS platforms, and demand other methodological approaches. Also of note, protein expression levels do not necessarily mirror genetic variation, genomic complexity extending beyond nucleotide sequence variation; examples of these are regulation of gene expression through epigenetic mechanisms (inheritable covalent modifications of the chromatin that regulate tightly its function) and non-coding-RNAs, issues equally not covered by GWAS. Nevertheless the rapid expansion, development and refinement of genomic tools (e.g., the human genome and HapMap projects, massive sequencing techniques) and genome-wide genotyping platforms available at affordable prices, provide valuable opportunities for disease-association studies that might soon shed new light on susceptibility to HSV-1.

Hypothesis-driven research on HSV-1 infection has identified critical elements that help the immune system fight the virus (Figs. 1, 2). Experimental mouse infection and mutations that lead to human herpetic encephalitis illustrate the crucial role of interferons and sensors of innate immunity that trigger their production, therefore genetic variations affecting these pathways might in the future be found relevant also for common forms of HSV-1 infection. Consistent and plausible are as well data supporting the essential contribution of cytotoxic lymphocytes of the NK and T-CD8 subsets to immunity against HSV-1. Polymorphic molecules that regulate these cells (e.g. NK-cell receptors, FcγR and HLA molecules) are therefore relevant candidates to become HSV-1 disease markers. Further research should clarify the precise role of KIR in this context, and the contribution of their genetic diversity to the risk of clinical HSV-1 reactivations. New approaches should also confirm the tantalizing evidence of HLA class I polymorphism as a determinant of HSV-1 infection outcome, such as typing of HLA alleles at high resolution and evaluation of patients complete HLA genotypes (instead of frequencies of individual HLA alleles in a population), along with in-silico tools [140] that enable prediction of T-cell epitopes presented by different HLA alleles from a given pathogen antigen.

Immunoglobulin genes, one example of polymorphic genes relevant for infectious diseases for which there are no available tags included in GWAS platforms, have also been hypothesised to contribute to variability in the response to HSV-1 [141]. Human immunoglobulin allotypes, determined by polymorphic residues on the heavy or light chains, may influence FcγR-binding affinity. Moreover, the HSV-1 FcγR decoy discriminates between two major IgG1 allotypes [142]. Further supporting the interest of investigations of these genes in human HSV-1 infection are the susceptibility marker identified close to mouse immunoglobulin heavy-chain genes locus, and our own finding of a FcγR dimorphism being associated with lack of symptomatic infection [69, 96]. Single-gene candidate approaches should clarify the possible involvement of immunoglobulin-gene diversity (either by itself or through epistatic interactions with other immunogenetic polymorphisms) in the cellular response against HSV-1 infection and its clinical course.

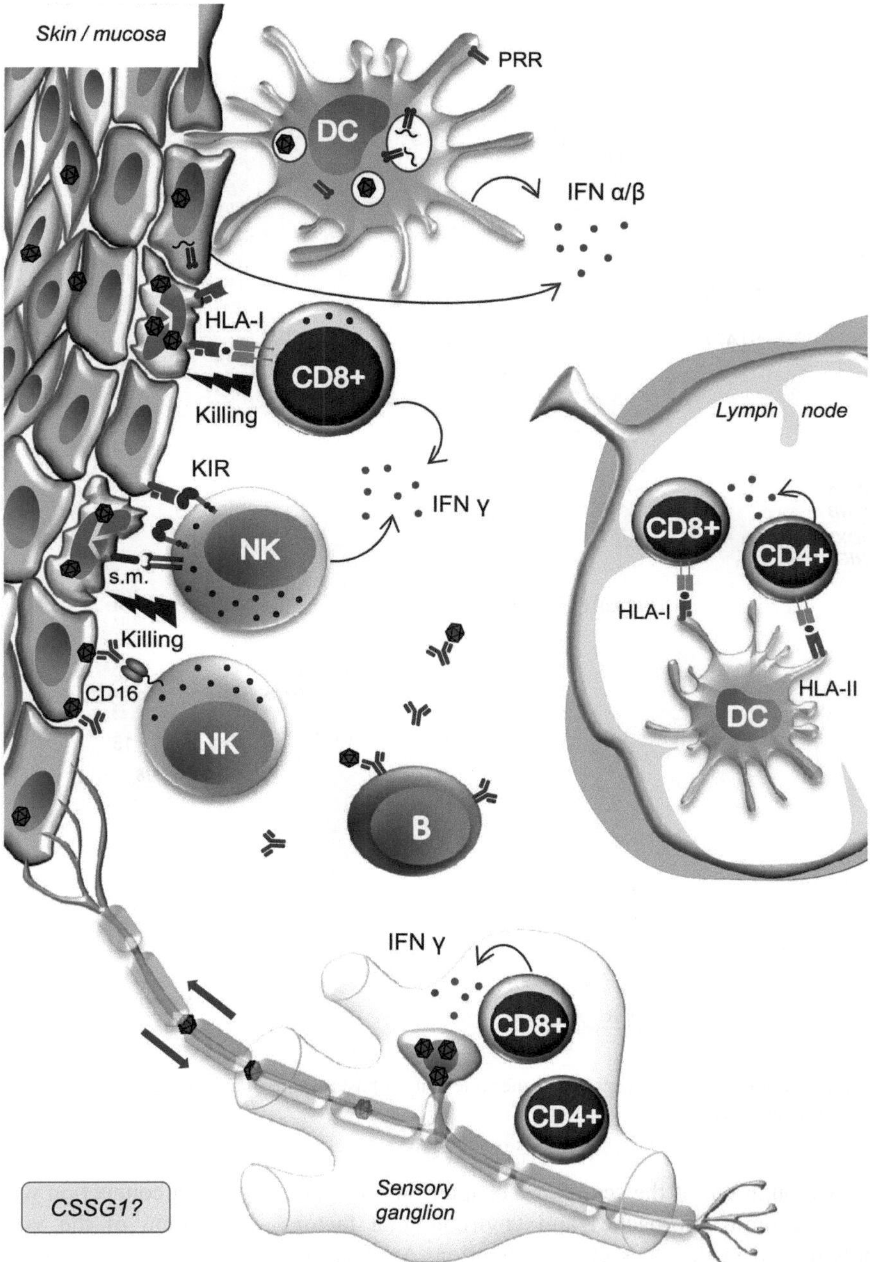

Fig. 1 Scheme of the immune response to HSV-1 infection. Only some of the relevant involved molecules are represented. Elements are not shown to scale. S.m.: stress- or virus-induced self molecule; see text for other abbreviations

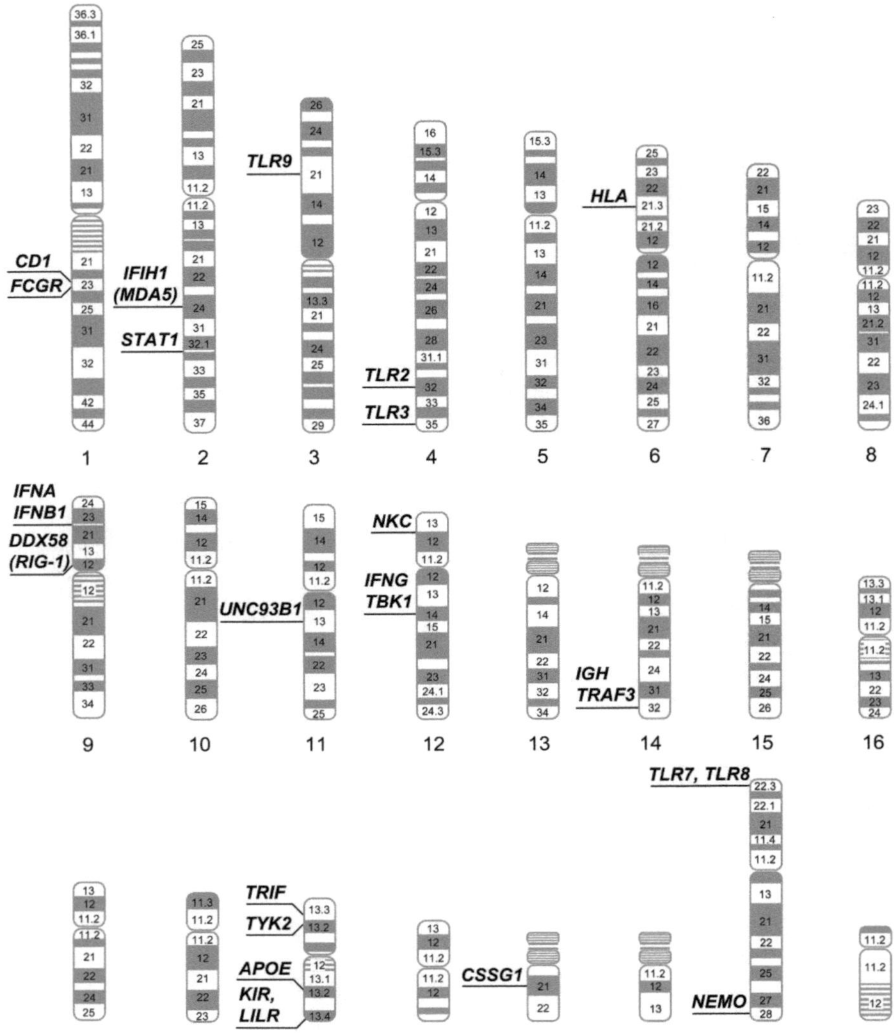

Fig. 2 Chromosomal localization of genes relevant for host response against HSV-1 infection. See text for abbreviations

Although out of the scope of a chapter on host genetics, but in connection with both genomic approaches and environmental factors, exploration of the virus polymorphism deserves greater attention. Application of novel genomic methods is warranted to shed light on the hypothesis of whether, and to which extent, genetic diversity of different HSV-1 strains is responsible for clinical variability of the herpetic infection; in other words, which specific HSV-1 genes or polymorphic variants of them increase the virus pathogenicity.

A substantial part of the resources of a human immune system is devoted to control the latent and very prevalent infections by different herpesviruses, which can shape profoundly the receptor repertoire of different lymphocyte lineages. A largely unexplored field is how the imprint exerted by one virus conditions the response to a different one, and to which extent this is influenced by the host genomics. As hinted in Diversity of NK-cell receptors, human cytomegalovirus shapes the NK-cell repertoire by inducing, in some individuals, a variable expansion of cells expressing a defined profile of activating and inhibitory receptors [143]. The extent of this expansion is possibly modulated by unknown genetic determinants, and it is likely that the expanded population might modify qualitative or quantitatively the NK-cell response to other viruses, like HSV-1.

Genetic susceptibility to infection has classically assumed to fall into one of two categories. On one hand, primary immunodeficiencies in which a known or putative mutation of a single gene with Mendelian inheritance causes frequent or severe infections by different pathogens, many of them exotic or opportunistic, but also ones prevalent in healthy subjects. On the other hand, susceptibility to common pathogens in otherwise healthy immunocompetent subjects is generally assumed to be governed by common polymorphisms of multiple genes, whose interactions would calibrate the degree of resistance to the infection. However, a third paradigm has been recognised in recent years, as increasing numbers of newly characterized non-conventional primary immunodeficiencies, caused by a monogenic mutation, confer susceptibility to severe infection by a specific, sometimes common, pathogen, such as pneumococcus, mycobacteria, Epstein-Barr virus or, notably, HSV-1. In fact, this novel model of immunodeficiency has been built on disclosure of several inborn defects predisposing to herpetic encephalitis, caused by isolated mutations in the TLR/IFN pathways, as described above [144]. Therefore, although it cannot be formally excluded that variations of a predominant gene might be responsible for common forms of herpetic infection, the different forms of predisposition to HSV-1 infection appear to fit into all three models of genetic susceptibility.

Likewise, different forms of susceptibility to Epstein-Barr virus (human herpesvirus 4, which latently infects most individuals but triggers infectious mononucleosis in a minority of them) can also fall into these three models of predisposition [145–147], as might be the case with other chronic infections affecting the perioral area, for which strategies of research on HSV-1 infection could be a model. Understanding the molecular basis for differences in predisposition to infectious diseases should provide useful insight into their pathogenesis and eventually improve their control.

Acknowledgements We thank Dr. Miguel López-Botet (Universitat Pompeu Fabra, Barcelona, Spain) for comments on the manuscript and continued scientific advice. Manuela Moraru is currently supported by a grant (121532) from Fundació La Marató TV3, Barcelona. Our studies on genetics of the immune response to HSV-1 were successively supported by grants from *Instituto de Salud Carlos III* and *Plan Nacional de I + D*, Spain (SAF2010-22153-C03-03, BFU2005-04622, and FIS01/0381).

References

1. Xu F, Sternberg MR, Kottiri BJ, et al. Trends in herpes simplex virus type 1 and type 2 seroprevalence in the United States. Jama. 2006;296:964–73.
2. Schillinger JA, Xu F, Sternberg MR, et al. National seroprevalence and trends in herpes simplex virus type 1 in the United States, 1976–1994. Sex Transm Dis. 2004;31:753–60.
3. Whitley RJ, Kimberlin DW, Roizman B. Herpes simplex viruses. Clin Infect Dis. 1998;26:541–53; quiz 54–5.
4. Hobbs MR, Jones BB, Otterud BE, Leppert M, Kriesel JD. Identification of a herpes simplex labialis susceptibility region on human chromosome 21. J Infect Dis. 2008;197:340–6. doi:10.1086/525540.
5. Watson G, Xu W, Reed A, et al. Sequence and comparative analysis of the genome of HSV-1 strain McKrae. Virology 2012;433:528–37.
6. Taylor TJ, Brockman MA, McNamee EE, Knipe DM. Herpes simplex virus. Front Biosci. 2002;7:d752–64.
7. Abel L, Plancoulaine S, Jouanguy E, et al. Age-dependent Mendelian predisposition to herpes simplex virus type 1 encephalitis in childhood. J Pediatr. 2010;157:623–9.
8. Chase RA, Pottage JC Jr., Haber MH, Kistler G, Jensen D, Levin S. Herpes simplex viral hepatitis in adults: two case reports and review of the literature. Rev Infect Dis. 1987;9:329–33.
9. Frederick DM, Bland D, Gollin Y. Fatal disseminated herpes simplex virus infection in a previously healthy pregnant woman. A case report. J Reprod Med. 2002;47:591–6.
10. Leung DY. Why is eczema herpeticum unexpectedly rare? Antiviral Res. 2013;98:153–7.
11. Whitley R, Arvin A, Prober C, et al. Predictors of morbidity and mortality in neonates with herpes simplex virus infections. The national institute of allergy and infectious diseases collaborative antiviral study group. N Engl J Med 1991;324:450–4.
12. Liesegang TJ. Herpes simplex virus epidemiology and ocular importance. Cornea 2001;20: 1–13.
13. Arduino PG, Porter SR. Herpes simplex virus type 1 infection: overview on relevant clinico-pathological features. J Oral Pathol Med 2008;37:107–21. doi:10.1111/j.600–0714.2007.00586.x.
14. Sokumbi O, Wetter DA. Clinical features, diagnosis, and treatment of erythema multiforme: a review for the practicing dermatologist. Int J Dermatol 2012;51:889–902.
15. Roizman B, Taddeo B. The strategy of herpes simplex virus replication and takeover of the host cell. In: Arvin Aea, Editor. Human herpesviruses. Cambridge: Cambridge University Press; 2007. pp. 163–73.
16. Durbin RK, Kotenko SV, Durbin JE. Interferon induction and function at the mucosal surface. Immunol Rev. 2013;255:25–39.
17. Pollara G, Jones M, Handley ME, et al. Herpes simplex virus type-1-induced activation of myeloid dendritic cells: he roles of virus cell interaction and paracrine type I IFN secretion. J Immunol. 2004;173:4108–19.
18. Mossman KL, Ashkar AA. Herpesviruses and the innate immune response. Viral Immunol. 2005;18:267–81.
19. Leib DA, Machalek MA, Williams BR, Silverman RH, Virgin HW. Specific phenotypic restoration of an attenuated virus by knockout of a host resistance gene. Proc Natl Acad Sci U. S. A. 2000;97:6097–101.
20. Ishioka K, Ikuta K, Sato Y, et al. Herpes simplex virus type 1 virion-derived US11 inhibits type 1 interferon-induced protein kinase R phosphorylation. Microbiol Immunol. 2013;57:426–36.
21. Mossman KL, Smiley JR. Herpes simplex virus ICP0 and ICP34.5 counteract distinct interferon-induced barriers to virus replication. J Virol. 2002;76:1995–8.
22. Taddeo B, Luo TR, Zhang W, Roizman B. Activation of NF-kappaB in cells productively infected with HSV-1 depends on activated protein kinase R and plays no apparent role in blocking apoptosis. Proc Natl Acad Sci U. S. A. 2003;100:12408–13. (Epub 2003 Oct 6).

23. Xing J, Wang S, Lin R, Mossman KL, Zheng C. Herpes simplex virus 1 tegument protein US11 downmodulates the RLR signaling pathway via direct interaction with RIG-I and MDA-5. J Virol. 2012;86:3528–40.
24. Suzutani T, Nagamine M, Shibaki T, et al. The role of the UL41 gene of herpes simplex virus type 1 in evasion of non-specific host defence mechanisms during primary infection. J Gen Virol. 2000;81:1763–71.
25. Chee AV, Roizman B. Herpes simplex virus 1 gene products occlude the interferon signaling pathway at multiple sites. J Virol. 2004;78:4185–96.
26. Johnson KE, Knipe DM. Herpes simplex virus-1 infection causes the secretion of a type I interferon-antagonizing protein and inhibits signaling at or before Jak-1 activation. Virology. 2010;396:21–9.
27. Melroe GT, DeLuca NA, Knipe DM. Herpes simplex virus 1 has multiple mechanisms for blocking virus-induced interferon production. J Virol. 2004;78:8411–20.
28. van Lint AL, Murawski MR, Goodbody RE, et al. Herpes simplex virus immediate-early ICP0 protein inhibits Toll-like receptor 2-dependent inflammatory responses and NF-kappaB signaling. J Virol. 2010;84:10802–11.
29. Lin R, Noyce RS, Collins SE, Everett RD, Mossman KL. The herpes simplex virus ICP0 RING finger domain inhibits IRF3- and IRF7-mediated activation of interferon-stimulated genes. J Virol 2004;78:1675–84.
30. Daubeuf S, Singh D, Tan Y, et al. HSV ICP0 recruits USP7 to modulate TLR-mediated innate response. Blood. 2009;113:3264–75. doi:10.1182/blood-2008-07-168203. (Epub 2008 Oct 24.)
31. Moretta L, Pietra G, Montaldo E, et al. Human NK cells: from surface receptors to the therapy of leukemias and solid tumors. Front Immunol. 2014;5:87.
32. Trinchieri G. Biology of natural killer cells. Adv Immunol. 1989;47:187–376.
33. Lanier LL. NK cell recognition. Annu Rev Immunol. 2005;23:225–74.
34. Fernandez-Messina L, Reyburn HT, Vales-Gomez M. Human NKG2D-ligands: cell biology strategies to ensure immune recognition. Front Immunol. 2012;3:299.
35. Bahram S, Inoko H, Shiina T, Radosavljevic M. MIC and other NKG2D ligands: from none to too many. Curr Opin Immunol. 2005;17:505–9.
36. Sivori S, Carlomagno S, Pesce S, Moretta A, Vitale M, Marcenaro E. TLR/NCR/KIR: Which one to use and when? Front Immunol. 2014;5:105. doi:10.3389/fimmu.2014.00105. eCollection 2014.
37. Bryceson YT, March ME, Ljunggren HG, Long EO. Activation, coactivation, and costimulation of resting human natural killer cells. Immunol Rev. 2006;214:73–91.
38. Jouanguy E, Gineau L, Cottineau J, Beziat V, Vivier E, Casanova JL. Inborn errors of the development of human natural killer cells. Curr Opin Allergy Clin Immunol. 2013;13:589–95.
39. Orange JS. Natural killer cell deficiency. J Allergy Clin Immunol. 2013;132:515–25.
40. Allan RS, Waithman J, Bedoui S, et al. Migratory dendritic cells transfer antigen to a lymph node-resident dendritic cell population for efficient CTL priming. Immunity. 2006;25:153–62.
41. Cose SC, Jones CM, Wallace ME, Heath WR, Carbone FR. Antigen-specific CD8 + T cell subset distribution in lymph nodes draining the site of herpes simplex virus infection. Eur J Immunol. 1997;27:2310–6.
42. Khanna KM, Bonneau RH, Kinchington PR, Hendricks RL. Herpes simplex virus-specific memory CD8 + T cells are selectively activated and retained in latently infected sensory ganglia. Immunity. 2003;18:593–603.
43. Frank GM, Lepisto AJ, Freeman ML, Sheridan BS, Cherpes TL, Hendricks RL. Early CD4(+) T cell help prevents partial CD8(+) T cell exhaustion and promotes maintenance of Herpes Simplex Virus 1 latency. J Immunol. 2010;184:277–86.
44. Shedlock DJ, Shen H. Requirement for CD4 T cell help in generating functional CD8 T cell memory. Science. 2003;300:337–9.
45. Sun JC, Bevan MJ. Defective CD8 T cell memory following acute infection without CD4 T cell help. Science. 2003;300:339–42.

46. Germain RN. Antigen processing and presentation. In: WEP, Editor. Fundamental immunology. Phladelphia: Lippincott-Raven; 1999. pp. 287–340.
47. Hill A, Jugovic P, York I, et al. Herpes simplex virus turns off the TAP to evade host immunity. Nature. 1995;375:411–5.
48. Kobelt D, Lechmann M, Steinkasserer A. The interaction between dendritic cells and herpes simplex virus-1. Curr Top Microbiol Immunol. 2003;276:145–61.
49. Kruse M, Rosorius O, Kratzer F, et al. Mature dendritic cells infected with herpes simplex virus type 1 exhibit inhibited T-cell stimulatory capacity. J Virol. 2000;74:7127–36.
50. Kummer M, Turza NM, Muhl-Zurbes P, et al. Herpes simplex virus type 1 induces CD83 degradation in mature dendritic cells with immediate-early kinetics via the cellular proteasome. J Virol. 2007;81:6326–38. (Epub 2007 Apr 11.)
51. Barcy S, Corey L. Herpes simplex inhibits the capacity of lymphoblastoid B cell lines to stimulate CD4 + T cells. J Immunol. 2001;166:6242–9.
52. Raftery MJ, Winau F, Kaufmann SH, Schaible UE, Schonrich G. CD1 antigen presentation by human dendritic cells as a target for herpes simplex virus immune evasion. J Immunol. 2006;177:6207–14.
53. Yuan W, Dasgupta A, Cresswell P. Herpes simplex virus evades natural killer T cell recognition by suppressing CD1d recycling. Nat Immunol. 2006;7:835–42. (Epub 2006 Jul 16.)
54. Grubor-Bauk B, Arthur JL, Mayrhofer G. Importance of NKT cells in resistance to herpes simplex virus, fate of virus-infected neurons, and level of latency in mice. J Virol. 2008;82:11073–83. doi:10.1128/JVI.00205–08. (Epub 2008 Jul 9.)
55. Johnson DC, Frame MC, Ligas MW, Cross AM, Stow ND. Herpes simplex virus immunoglobulin G Fc receptor activity depends on a complex of two viral glycoproteins, gE and gI. J Virol. 1988;62:1347–54.
56. Dubin G, Socolof E, Frank I, Friedman HM. Herpes simplex virus type 1 Fc receptor protects infected cells from antibody-dependent cellular cytotoxicity. J Virol. 1991;65:7046–50.
57. Frank I, Friedman HM. A novel function of the herpes simplex virus type 1 Fc receptor: participation in bipolar bridging of antiviral immunoglobulin G. J Virol. 1989;63:4479–88.
58. Friedman HM, Cohen GH, Eisenberg RJ, Seidel CA, Cines DB. Glycoprotein C of herpes simplex virus 1 acts as a receptor for the C3b complement component on infected cells. Nature. 1984;309:633–5.
59. Fries LF, Friedman HM, Cohen GH, Eisenberg RJ, Hammer CH, Frank MM. Glycoprotein C of herpes simplex virus 1 is an inhibitor of the complement cascade. J Immunol. 1986;137:1636–41.
60. Kostavasili I, Sahu A, Friedman HM, Eisenberg RJ, Cohen GH, Lambris JD. Mechanism of complement inactivation by glycoprotein C of herpes simplex virus. J Immunol. 1997;158:1763–71.
61. Lopez C. Genetics of natural resistance to herpesvirus infections in mice. Nature. 1975;258:152–3.
62. Halford WP, Balliet JW, Gebhardt BM. Re-evaluating natural resistance to herpes simplex virus type 1. J Virol 2004;78:10086–95.
63. Lundberg P, Welander P, Openshaw H, et al. A locus on mouse chromosome 6 that determines resistance to herpes simplex virus also influences reactivation, while an unlinked locus augments resistance of female mice. J Virol. 2003;77:11661–73.
64. Pereira RA, Scalzo A, Simmons A. Cutting edge: a NK complex-linked locus governs acute versus latent herpes simplex virus infection of neurons. J Immunol. 2001;166:5869–73.
65. Zawatzky R, Hilfenhaus J, Marcucci F, Kirchner H. Experimental infection of inbred mice with herpes simplex virus type 1. I. Investigation of humoral and cellular immunity and of interferon induction. J Gen Virol. 1981;53:31–8.
66. Zawatzky R, Gresser I, DeMaeyer E, Kirchner H. The role of interferon in the resistance of C57BL/6 mice to various doses of herpes simplex virus type 1. J Infect Dis. 1982;146:405–10.
67. Lopez C. Resistance to HSV-1 in the mouse is governed by two major, independently segregating, non-H-2 loci. Immunogenetics. 1980;11:87–92.

68. Thompson RL, Williams RW, Kotb M, Sawtell NM. A forward phenotypically driven unbiased genetic analysis of host genes that moderate herpes simplex virus virulence and stromal keratitis in mice. PLoS One. 2014;9:e92342.
69. Foster CS, Tsai Y, Monroe JG, et al. Genetic studies on murine susceptibility to herpes simplex keratitis. Clin Immunol Immunopathol. 1986;40:313–25.
70. Norose K, Yano A, Zhang XM, Blankenhorn E, Heber-Katz E. Mapping of genes involved in murine herpes simplex virus keratitis: identification of genes and their modifiers. J Virol. 2002;76:3502–10.
71. Vollstedt S, Arnold S, Schwerdel C, et al. Interplay between alpha/beta and gamma interferons with B, T, and natural killer cells in the defense against herpes simplex virus type 1. J Virol. 2004;78:3846–50.
72. Smith PM, Wolcott RM, Chervenak R, Jennings SR. Control of acute cutaneous herpes simplex virus infection: T cell-mediated viral clearance is dependent upon interferon-gamma (IFN-gamma). Virology. 1994;202:76–88.
73. Cantin E, Tanamachi B, Openshaw H, Mann J, Clarke K. Gamma interferon (IFN-gamma) receptor null-mutant mice are more susceptible to herpes simplex virus type 1 infection than IFN-gamma ligand null-mutant mice. J Virol. 1999;73:5196–200.
74. Cai MS, Li ML, Zheng CF. Herpesviral infection and Toll-like receptor 2. Protein Cell. 2012;3:590–601.
75. Leoni V, Gianni T, Salvioli S, Campadelli-Fiume G. Herpes simplex virus glycoproteins gH/gL and gB bind Toll-like receptor 2, and soluble gH/gL is sufficient to activate NF-kappaB. J Virol. 2012;86:6555–62.
76. Rasmussen SB, Sorensen LN, Malmgaard L, et al. Type I interferon production during herpes simplex virus infection is controlled by cell-type-specific viral recognition through Toll-like receptor 9, the mitochondrial antiviral signaling protein pathway, and novel recognition systems. J Virol. 2007;81:13315–24. (Epub 2007 Oct 3.)
77. Zhang SY, Jouanguy E, Ugolini S, et al. TLR3 deficiency in patients with herpes simplex encephalitis. Science. 2007;317:1522–7.
78. Weber F, Wagner V, Rasmussen SB, Hartmann R, Paludan SR. Double-stranded RNA is produced by positive-strand RNA viruses and DNA viruses but not in detectable amounts by negative-strand RNA viruses. J Virol 2006;80:5059–64.
79. Melchjorsen J, Rintahaka J, Soby S, et al. Early innate recognition of herpes simplex virus in human primary macrophages is mediated via the MDA5/MAVS-dependent and MDA5/MAVS/RNA polymerase III-independent pathways. J Virol. 2010;84:11350–8.
80. Rasmussen SB, Jensen SB, Nielsen C, et al. Herpes simplex virus infection is sensed by both Toll-like receptors and retinoic acid-inducible gene- like receptors, which synergize to induce type I interferon production. J Gen Virol. 2009;90:74–8. doi:10.1099/vir.0.005389–0.
81. Unterholzner L. The interferon response to intracellular DNA: why so many receptors? Immunobiology. 2013;218:1312–21.
82. Paludan SR, Bowie AG. Immune sensing of DNA. Immunity. 2013;38:870–80.
83. Casrouge A, Zhang SY, Eidenschenk C, et al. Herpes simplex virus encephalitis in human UNC-93B deficiency. Science. 2006;314:308–12. (Epub 2006 Sep 14.)
84. Brinkmann MM, Spooner E, Hoebe K, Beutler B, Ploegh HL, Kim YM. The interaction between the ER membrane protein UNC93B and TLR3, 7, and 9 is crucial for TLR signaling. J Cell Biol. 2007;177:265–75.
85. Kim YM, Brinkmann MM, Paquet ME, Ploegh HL. UNC93B1 delivers nucleotide-sensing toll-like receptors to endolysosomes. Nature. 2008;452:234–8. doi:10.1038/nature06726. (Epub 2008 Feb 27).
86. Guo Y, Audry M, Ciancanelli M, et al. Herpes simplex virus encephalitis in a patient with complete TLR3 deficiency: TLR3 is otherwise redundant in protective immunity. J Exp Med. 2011;208:2083–98.
87. Sancho-Shimizu V, Perez de Diego R, Lorenzo L, et al. Herpes simplex encephalitis in children with autosomal recessive and dominant TRIF deficiency. J Clin Invest. 2011;121:4889–902.

88. Perez de Diego R, Sancho-Shimizu V, Lorenzo L, et al. Human TRAF3 adaptor molecule deficiency leads to impaired Toll-like receptor 3 response and susceptibility to herpes simplex encephalitis. Immunity. 2010;33:400–11.
89. Herman M, Ciancanelli M, Ou YH, et al. Heterozygous TBK1 mutations impair TLR3 immunity and underlie herpes simplex encephalitis of childhood. J Exp Med. 2012;209:1567–82.
90. Sancho-Shimizu V, Perez de Diego R, Jouanguy E, Zhang SY, Casanova JL. Inborn errors of anti-viral interferon immunity in humans. Curr Opin Virol. 2011;1:487–96.
91. Bustamante J, Boisson-Dupuis S, Jouanguy E, et al. Novel primary immunodeficiencies revealed by the investigation of paediatric infectious diseases. Curr Opin Immunol. 2008;20:39–48.
92. Spiegel R, Miron D, Yodko H, Lumelsky D, Habib A, Horovitz Y. Late relapse of herpes simplex virus encephalitis in a child due to reactivation of latent virus: clinicopathological report and review. J Child Neurol. 2008;23:344–8. doi:10.1177/0883073807309243. (Epub 2008 Jan 29.)
93. Yang CA, Raftery MJ, Hamann L, et al. Association of TLR3-hyporesponsiveness and functional TLR3 L412F polymorphism with recurrent herpes labialis. Hum Immunol. 2012;73:844–51.
94. Svensson A, Tunback P, Nordstrom I, Padyukov L, Eriksson K. Polymorphisms in Toll-like receptor 3 confer natural resistance to human herpes simplex virus type 2 infection. J Gen Virol. 2012;93:1717–24.
95. Bochud PY, Magaret AS, Koelle DM, Aderem A, Wald A. Polymorphisms in TLR2 are associated with increased viral shedding and lesional rate in patients with genital herpes simplex virus Type 2 infection. J Infect Dis. 2007;196:505–9. (Epub 2007 Jun 29.)
96. Moraru M, Cisneros E, Gomez-Lozano N, et al. Host genetic factors in susceptibility to herpes simplex type 1 virus infection: contribution of polymorphic genes at the interface of innate and adaptive immunity. J Immunol. 2012;188:4412–20.
97. Apps R, Qi Y, Carlson JM, et al. Influence of HLA-C expression level on HIV control. Science. 2013;340:87–91.
98. Rizvi SM, Salam N, Geng J, et al. Distinct assembly profiles of HLA-B molecules. J Immunol. 2014;192:4967–76.
99. Vilches C, Parham P. KIR: diverse, rapidly evolving receptors of innate and adaptive immunity. Annu Rev Immunol. 2002;20:217–51.
100. Raulet DH, Held W, Correa I, Dorfman JR, Wu MF, Corral L. Specificity, tolerance and developmental regulation of natural killer cells defined by expression of class I-specific Ly49 receptors. Immunol Rev. 1997;155:41–52.
101. Arase H, Mocarski ES, Campbell AE, Hill AB, Lanier LL. Direct recognition of cytomegalovirus by activating and inhibitory NK cell receptors. Science. 2002;296:1323–6.
102. Smith HR, Heusel JW, Mehta IK, et al. Recognition of a virus-encoded ligand by a natural killer cell activation receptor. Proc Natl Acad Sci U. S. A. 2002;99:8826–31.
103. Pyo CW, Guethlein LA, Vu Q, et al. Different patterns of evolution in the centromeric and telomeric regions of group A and B haplotypes of the human killer cell Ig-like receptor locus. PLoS One. 2010;5:e15115.
104. Jiang W, Johnson C, Jayaraman J, et al. Copy number variation leads to considerable diversity for B but not A haplotypes of the human KIR genes encoding NK cell receptors. Genome Res. 2012;22:1845–54.
105. Martin MP, Carrington M. Immunogenetics of HIV disease. Immunol Rev. 2013;254:245–64.
106. Moesta AK, Graef T, Abi-Rached L, Older Aguilar AM, Guethlein LA, Parham P. Humans differ from other hominids in lacking an activating NK cell receptor that recognizes the C1 epitope of MHC class I. J Immunol. 2010;185:4233–7.
107. David G, Djaoud Z, Willem C, et al. Large spectrum of HLA-C recognition by killer Ig-like receptor (KIR)2DL2 and KIR2DL3 and restricted C1 SPECIFICITY of KIR2DS2: dominant impact of KIR2DL2/KIR2DS2 on KIR2D NK cell repertoire formation. J Immunol. 2013;191:4778–88.

108. Shum BP, Flodin LR, Muir DG, et al. Conservation and variation in human and common chimpanzee CD94 and NKG2 genes. J Immunol. 2002;168:240–52.
109. Miyashita R, Tsuchiya N, Hikami K, et al. Molecular genetic analyses of human NKG2C (KLRC2) gene deletion. Int Immunol. 2004;16:163–8.
110. Hikami K, Tsuchiya N, Yabe T, Tokunaga K. Variations of human killer cell lectin-like receptors: common occurrence of NKG2-C deletion in the general population. Genes Immun. 2003;4:160–7.
111. Muntasell A, Lopez-Montanes M, Vera A, et al. NKG2C zygosity influences CD94/NKG2C receptor function and the NK-cell compartment redistribution in response to human cytomegalovirus. Eur J Immunol. 2013;43:3268–78.
112. Colonna M, Navarro F, Bellon T, et al. A common inhibitory receptor for major histocompatibility complex class I molecules on human lymphoid and myelomonocytic cells. J Exp Med. 1997;186:1809–18.
113. Yang Z, Bjorkman PJ. Structure of UL18, a peptide-binding viral MHC mimic, bound to a host inhibitory receptor. Proc Natl Acad Sci U. S. A. 2008;105:10095–100.
114. Davidson CL, Li NL, Burshtyn DN. LILRB1 polymorphism and surface phenotypes of natural killer cells. Hum Immunol. 2010;71:942–9.
115. Lopez-Alvarez MR, Jones DC, Jiang W, Traherne JA, Trowsdale J. Copy number and nucleotide variation of the LILR family of myelomonocytic cell activating and inhibitory receptors. Immunogenetics. 2014;66:73–83.
116. Nimmerjahn F, Ravetch JV. Fcgamma receptors as regulators of immune responses. Nat Rev Immunol. 2008;8:34–47.
117. van Sorge NM, van der Pol WL, van de Winkel JG. FcgammaR polymorphisms: Implications for function, disease susceptibility and immunotherapy. Tissue Antigens. 2003;61:189–202.
118. Wu J, Edberg JC, Redecha PB, et al. A novel polymorphism of FcgammaRIIIa (CD16) alters receptor function and predisposes to autoimmune disease. J Clin Invest. 1997;100:1059–70.
119. Sondermann P, Huber R, Oosthuizen V, Jacob U. The 3.2-A crystal structure of the human IgG1 Fc fragment-Fc gammaRIII complex. Nature. 2000;406:267–73.
120. Radaev S, Motyka S, Fridman WH, Sautes-Fridman C, Sun PD. The structure of a human type III Fcgamma receptor in complex with Fc. J Biol Chem. 2001;276:16469–77. (Epub 2001 Jan 31.)
121. Metes D, Ernst LK, Chambers WH, Sulica A, Herberman RB, Morel PA. Expression of functional CD32 molecules on human NK cells is determined by an allelic polymorphism of the FcgammaRIIC gene. Blood. 1998;91:2369–80.
122. van der Heijden J, Breunis WB, Geissler J, de Boer M, van den Berg TK, Kuijpers TW. Phenotypic variation in IgG receptors by nonclassical FCGR2C alleles. J Immunol. 2012;188:1318–24.
123. Mueller M, Barros P, Witherden AS, et al. Genomic pathology of SLE-associated copy-number variation at the FCGR2C/FCGR3B/FCGR2B locus. Am J Hum Genet. 2013;92:28–40.
124. Floto RA, Clatworthy MR, Heilbronn KR, et al. Loss of function of a lupus-associated FcgammaRIIb polymorphism through exclusion from lipid rafts. Nat Med. 2005;11:1056–8. (Epub 2005 Sep 18.)
125. Su K, Li X, Edberg JC, Wu J, Ferguson P, Kimberly RP. A promoter haplotype of the immunoreceptor tyrosine-based inhibitory motif-bearing FcgammaRIIb alters receptor expression and associates with autoimmunity. II. Differential binding of GATA4 and Yin-Yang1 transcription factors and correlated receptor expression and function. J Immunol. 2004;172:7192–9.
126. Kriesel JD, Jones BB, Matsunami N, et al. C21orf91 genotypes correlate with herpes simplex labialis (cold sore) frequency: description of a cold sore susceptibility gene. J Infect Dis. 2011;204:1654–62.
127. Itzhaki R, Wozniak M. Susceptibility to herpes simplex labialis conferred by the gene encoding apolipoprotein E. J Infect Dis. 2008;198:624–5. (author reply 5–6. doi:10.1086/590213.)

128. Burgos JS, Ramirez C, Sastre I, Valdivieso F. Effect of apolipoprotein E on the cerebral load of latent herpes simplex virus type 1 DNA. J Virol. 2006;80:5383–7.
129. Burgos JS, Ramirez C, Sastre I, Bullido MJ, Valdivieso F. ApoE4 is more efficient than E3 in brain access by herpes simplex virus type 1. Neuroreport. 2003;14:1825–7.
130. Toniutto P, Fabris C, Fumo E, et al. Carriage of the apolipoprotein E-epsilon4 allele and histologic outcome of recurrent hepatitis C after antiviral treatment. Am J Clin Pathol. 2004;122:428–33.
131. Wozniak MA, Itzhaki RF, Faragher EB, James MW, Ryder SD, Irving WL. Apolipoprotein E-epsilon 4 protects against severe liver disease caused by hepatitis C virus. Hepatology. 2002;36:456–63.
132. Corder EH, Robertson K, Lannfelt L, et al. HIV-infected subjects with the E4 allele for APOE have excess dementia and peripheral neuropathy. Nat Med. 1998;4:1182–4.
133. Willnow TE. The low-density lipoprotein receptor gene family: multiple roles in lipid metabolism. J Mol Med (Berl). 1999;77:306–15.
134. van den Elzen P, Garg S, Leon L, et al. Apolipoprotein-mediated pathways of lipid antigen presentation. Nature. 2005;437:906–10.
135. Agnello V, Abel G, Elfahal M, Knight GB, Zhang QX. Hepatitis C virus and other flaviviridae viruses enter cells via low density lipoprotein receptor. Proc Natl Acad Sci U. S. A. 1999;96:12766–71.
136. Finkelshtein D, Werman A, Novick D, Barak S, Rubinstein M. LDL receptor and its family members serve as the cellular receptors for vesicular stomatitis virus. Proc Natl Acad Sci U. S. A. 2013;110:7306–11.
137. Chang KS, Jiang J, Cai Z, Luo G. Human apolipoprotein e is required for infectivity and production of hepatitis C virus in cell culture. J Virol. 2007;81:13783–93. (Epub 2007 Oct 3.)
138. Faustino AF, Carvalho FA, Martins IC, et al. Dengue virus capsid protein interacts specifically with very low-density lipoproteins. Nanomedicine. 2014;10:247–55.
139. Huemer HP, Menzel HJ, Potratz D, et al. Herpes simplex virus binds to human serum lipoprotein. Intervirology. 1988;29:68–76.
140. Stranzl T, Larsen MV, Lundegaard C, Nielsen M. NetCTLpan: pan-specific MHC class I pathway epitope predictions. Immunogenetics. 2010;62:357–68.
141. Pandey JP. Immunoglobulin genes and immunity to herpes simplex virus type 1. J Infect Dis. 2012;206:143–4.
142. Atherton A, Armour KL, Bell S, Minson AC, Clark MR. The herpes simplex virus type 1 Fc receptor discriminates between IgG1 allotypes. Eur J Immunol. 2000;30:2540–7.
143. Muntasell A, Vilches C, Angulo A, Lopez-Botet M. Adaptive reconfiguration of the human NK-cell compartment in response to cytomegalovirus: a different perspective of the host-pathogen interaction. Eur J Immunol. 2013;43:1133–41.
144. Casanova JL, Abel L. Primary immunodeficiencies: a field in its infancy. Science. 2007; 317:617–9.
145. Veillette A, Perez-Quintero LA, Latour S. X-linked lymphoproliferative syndromes and related autosomal recessive disorders. Curr Opin Allergy Clin Immunol. 2013;13:614–22.
146. Li FY, Chaigne-Delalande B, Su H, Uzel G, Matthews H, Lenardo MJ. XMEN disease: a new primary immunodeficiency affecting Mg2 + regulation of immunity against Epstein-Barr virus. Blood. 2014;123:2148–52.
147. Fernandez-Flores A. Epstein-Barr virus in cutaneous pathology. Am J Dermatopathol. 2013;35:763–86.

Developmental Anomalies – Clefts

Elizabeth J. Leslie and Mary L. Marazita

List of Abbreviations

CL =	Cleft lip
CLP =	Cleft lip plus cleft palate
CL/P =	Cleft lip with or without cleft palate
CP =	Cleft palate
GWAS =	Genome-wide association study
OFC =	Orofacial cleft

Introduction

Congenital developmental anomalies present both opportunities and challenges for personalized medicine. Since such anomalies develop before birth, prevention (one of the major goals of personalized medicine) is less pertinent. Instead, other opportunities are likely to be more attainable, for example, the use of personalized medicine approaches to improve treatment, prognosis, long-term outcomes, and prevention of associated health complications. There are a large number of congenital anomalies that involve the oral, facial, and craniofacial complex. Here we will

M. L. Marazita (✉)
Center for Craniofacial and Dental Genetics, Department of Oral Biology, School of Dental Medicine, Department of Human Genetics, Graduate School of Public Health, Clinical and Translational Science Institute, and Department of Psychiatry, School of Medicine, University of Pittsburgh, Suite 500 Bridgeside Point, 100 Technology Dr., Pittsburgh, PA 15219, USA
e-mail: marazita@pitt.edu

E. J. Leslie
Center for Craniofacial and Dental Genetics, Department of Oral Biology, School of Dental Medicine, University of Pittsburgh, Pittsburgh, PA 15219, USA
e-mail: ejl40@pitt.edu

S. T. Sonis (ed.), *Genomics, Personalized Medicine and Oral Disease*,
DOI 10.1007/978-3-319-17942-1_5

focus on orofacial clefts (OFCs), the most common of those anomalies, as a model for other congenital anomalies with respect to personalized medicine.

OFCs comprise any cleft, i.e. break or gap, in orofacial structures including the lips, palate, eyes, ears, nose, cheeks and forehead with about 15 different types observed and annotated [142]) . Aside from cleft lip (CL) and cleft palate (CP) most OFCs are extremely rare. Cleft lip (CL) and cleft palate (CP) are among the most common birth defects in all populations worldwide [106, 107] , with population and ethnic differences in birth prevalence. Interest in the etiology of these birth defects goes back centuries as does formal scientific interest [17, 35, 91, 145]. There is now general consensus that CL, CP and CL with or without CP (CL/P) represent complex human traits with both environmental and genetic components contributing to susceptibility. OFC represents a major success story in the genetics of complex human traits in that about 18 genes have been discovered for CL/P at genome-wide significance levels, that can account for about 55 % of the heritability of the trait, and many of which now have substantial replication and even functional verification in some cases.

In this chapter, we will focus on CL/P, the most common OFC and thus a priority for personalized medicine. We will review development of the orofacial complex, CL and CP epidemiology and co-morbidities, specific clinical and sub-clinical phenotypes, and genetic studies. All of these areas are necessary components of the foundational knowledge required to realize the promise of personalized medicine.

Orofacial Development

Development of the lip and palate requires growth and fusion of multiple embryonic structures, and coordination of a complex series of events including cell growth, migration, differentiation, and apoptosis. Disruptions of development at any stage of the process can result in OFCs at birth, i.e., there can be defects in the many steps requiring fusion, or there can be disruption of timing and/or positioning of the processes and/or palatal shelves (e.g. in Robin sequence in which micrognathia prevents the tongue from dropping). There is excellent video of facial development available at http://www.youtube.com/watch?v=wFY_KPFS3LA.

Normal development of the lip and palate occurs very early in embryogenesis. The lip forms first and is complete by week 6, followed by the palate which is complete around week 12 [62, 143]. By the fourth week of embryonic development, a variety of tissues are in place: the frontonasal prominence, paired maxillary processes, and paired mandibular processes surround the oral cavity. By the fifth week, the nasal pits have fused to form the paired medial and lateral nasal processes. By the end of the sixth week of normal development, the lip is formed, i.e. the medial nasal processes have merged with the maxillary processes to form the upper lip and primary palate. During the sixth-seventh weeks, there are bilateral outgrowths from the maxillary processes which grow down on either side of the tongue to become the palatal shelves. The palatal shelves initially grow vertically along the sides of the developing tongue but later elevate into a horizontal position as the tongue flattens (reviewed by [50]). Continued growth leads to the palatal shelves meeting at the midline followed by fusion along the medial edge epithelia. Successful fusion of the secondary palate results in complete separation of the nasal and oral cavities.

Orofacial Cleft Epidemiology and Co-Morbidities

Epidemiology

The epidemiology of OFC has a number of striking features. There are both ethnic and gender differences in the birth prevalence of CL/P. Native Americans and Asians have the highest rates (close to 2/1000 live births), Caucasians intermediate (about 1/1000) and African-derived populations the lowest rates (about 1/2500) [26, 30, 106, 107]. There is a 2:1 male:female ratio for CL/P, but an approximate 1:2 male:female ratio for CP. Furthermore, OFC can be unilateral or bilateral; interestingly the majority (about 2/3) of unilateral clefts are on the left side.

About 70–80 % of CL/P and 50 % of CP [66]) are considered "nonsyndromic", i.e. isolated anomalies with no other apparent cognitive or structural abnormalities. The "nonsyndromic" designation is therefore arbitrary and to some extent reflects our current lack of certainty about OFC etiologies [67] . Many of the genetic variants or mutations causing syndromic forms of OFC have been identified (see Online Inheritance in Man database at www.ncbi.nlm.nih.gov/omim, and Box 1 in Dixon et al. [38]).

Epidemiological data support a role for environmental risk factors in the development of orofacial clefts. Maternal smoking has been consistently associated with an increased risk of clefting, with a population-attributable risk estimated as high as 20 % and an odds-ratio of 1.3 for CLP [137]. While alcohol is an established teratogen [167], evidence supporting a role for maternal alcohol use increasing OFC risk has been inconsistent [109]. However, some support for maternal alcohol consumption comes from an association between clefting and genetic variants in the alcohol dehydrogenase gene *ADH1C* [70]. Moreover, a recent study demonstrated that the combination of *ADH1C* variants with reduced enzymatic activity and heavy maternal alcohol use increased the risk for orofacial clefts [18]. However, a role for alcohol may be confounded by other risk factors such as nutrition, smoking, or stress that can be associated with alcohol consumption in some contexts.

Nutrition during pregnancy has been suggested as another contributing factor based on observational and interventional studies using folate supplements as a preventive measure [159]. The beneficial effect of folate use, however, remains controversial and has not been consistently replicated [159, 168] . Other nutrients, including cholesterol [124], zinc [108], and general multivitamin use [63] have also been studied, but need to be expanded to larger populations. Finally, other exposures to teratogens and environmental toxins have also been associated with increased risk of clefting [1] such as retinoic acid, valproic acid, and phenytoin. A more comprehensive review of environmental risk factors for orofacial clefts is provided by [128] and [150].

Genetic Epidemiology

Interest in the causes of OFC goes back thousands of years and continues to the present. Historically there have been a number of folklore explanations of how OFC arises, and how to prevent occurrence during pregnancy; some of which survive to the present day [33]. Although such folkloric explanations have since been discounted, it is notable that many cultures also felt that OFC is familial, e.g. "in the blood" [25, 33], which we still agree with today. The first published observational study of OFC inheritance was in 1757 [145] of a family with several affected members. 100 years later Darwin [35] cited a paper by Sproule [140], mentioning "the transmission during a century of hare-lip with a cleft-palate" in his discussion of variation in plants and animals. Additional pre-1900 publications and OFC pedigrees were summarized by Rischbieth [99, 131].

Today we consider OFC to be heterogeneous, with single genes of major effect potentially modified by polygenic background and environmental/behavioral factors. Statistical segregation analyses of OFCs in a number of ethnicities bear out this consensus for US and European Caucasians (e.g. [59, 92]), Asian and mixed Asian (e.g. [29, 93]), and others (see review in [90]) Also consistent with these results were analyses based on evaluations of recurrence risk in OFC families that in the aggregate estimated that from two to fifteen genes of major effect are likely to be involved in OFC [27, 136]

Co-Morbidities

Affected individuals initially face difficulties feeding and also experience speech, hearing, and dental problems. Although clefts can be surgically repaired, patients often undergo multiple craniofacial and dental surgeries, as well as speech and hearing therapy. Surgical repair of CL is done around 2–3 months of age, with CP closure from 6–12 months (see www.acpa-cpf.org/team_care/ for more details) The complications of OFCs in early life are particularly devastating in developing countries where access to medical care may be limited [160]. In developed countries, routine surgical treatment, with ongoing orthodontia, speech and other therapies, is very successful in ameliorating OFC anomalies but there is still a significant financial burden for individuals with OFC, their families, and society [10, 158].

Despite medical interventions where available, individuals affected with an OFC can experience lifelong psychosocial effects from the malformation. In fact, individuals born with a cleft have increased incidence of mental health problems and higher mortality rates at all stages of life [28, 158]. OFC is also associated with a higher risk of various cancer types, including breast, brain, and colon cancers, in the individual with a cleft as well as their family members [12, 37, 100, 170].

Orofacial Cleft Phenotypes

Accurate phenotyping is essential for successful studies of etiology, and is particularly important for human genetic studies and for applications of etiologic research to personalized medicine. Along with the explosion of research tools developed during the Human Genome Project [76] there has also been an increase in bioninformatic tools and resources related to phenotyping. Examples include the PhenomicDB [54, 55], the GEN2PHEN project [157], the PhenX toolkit [58], and FaceBase [60] which is specific to OFC and other craniofacial anomalies.

OFCs are a heterogeneous group of disorders with a wide range of expression and severity, affecting the structure of the face and oral cavity. There are three general categories of phenotypes that are felt to represent the range of expression in individuals with OFC and their families: the clinically relevant birth defects (CL, CP, see Fig. 1), microforms (see Fig. 2) and subclinical phenotypic features (see Fig. 3). The clinically relevant phenotypes include those that affect the lip only (CL, Fig. 1a, b), those affecting the palate alone (CP, Fig. 1c), and those affecting the lip plus palate (CLP, Fig. 1d). Further, CL and CLP share a defect of the primary palate, motivating the inclusion of CL and CLP into a common group—cleft lip with or without cleft palate (CL/P) [44, 46]. However, note that epidemiological [51] and

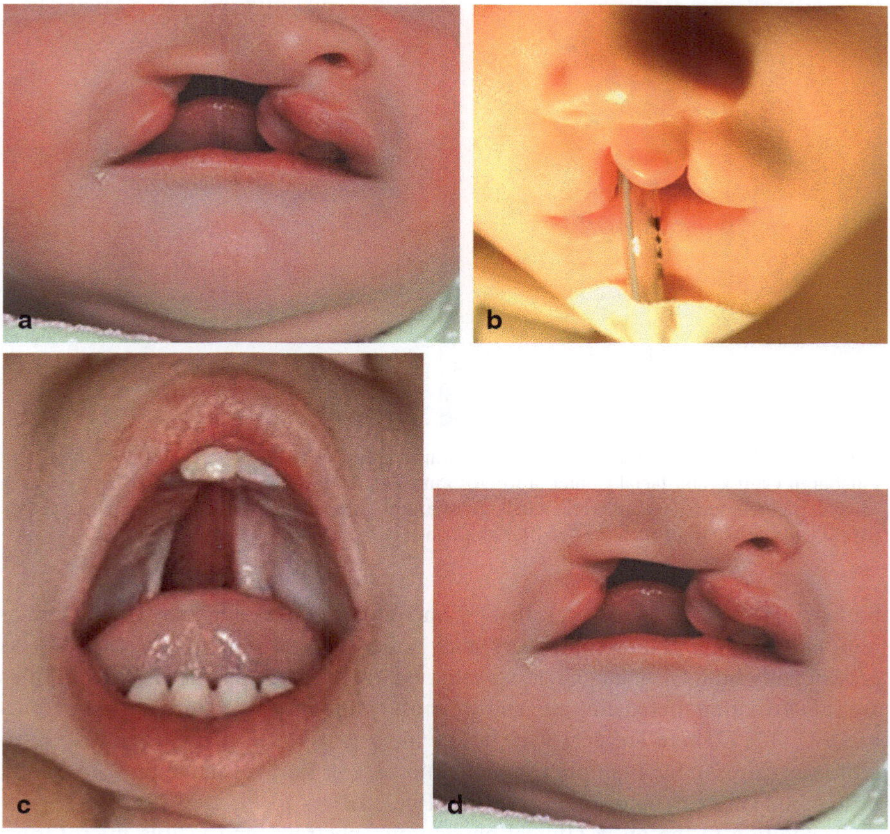

Fig. 1 Examples of overt types of OFC. Photographs courtesy of M. Ford, Children's Hospital of Pittsburgh Cleft Craniofacial Center. **a** Unilateral incomplete cleft lip. **b** Bilateral incomplete cleft lip. **c** *Right* unilateral complete cleft lip plus cleft palate. **d** Cleft of the hard and soft palates

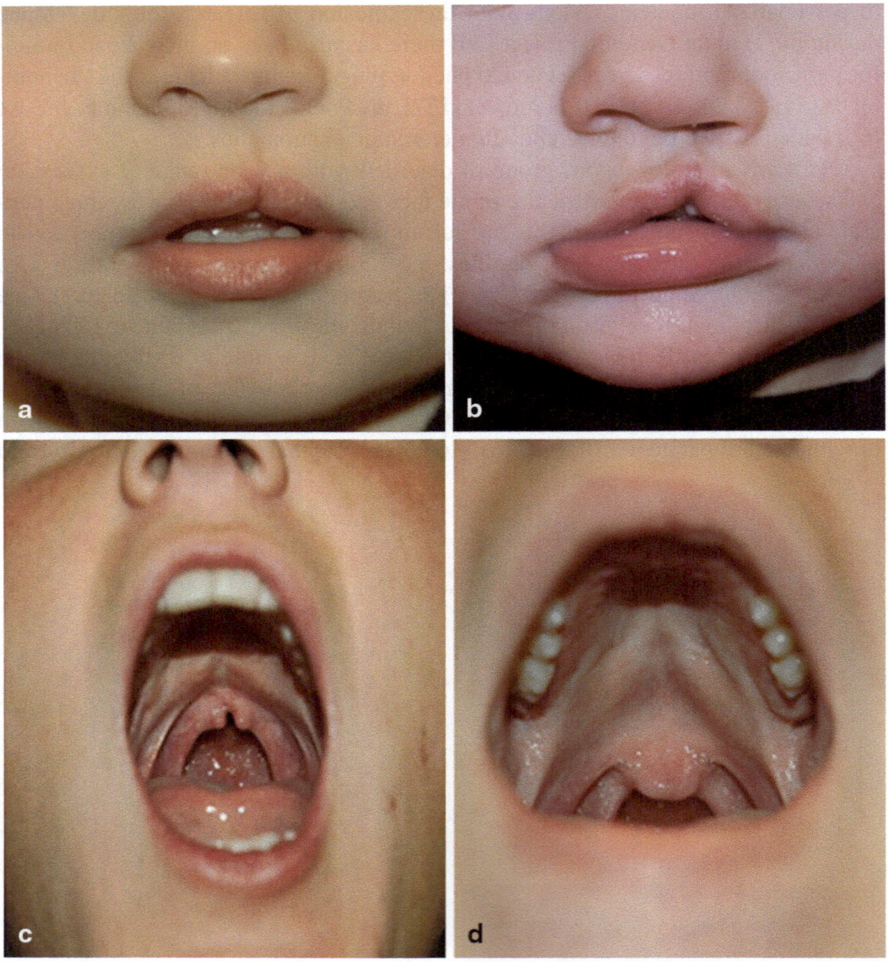

Fig. 2 Microform lip and palate defects Photographs courtesy of M. Ford, Children's Hospital of Pittsburgh Cleft Craniofacial Center. **a** Microform lip defect: *Left* unilateral notch and groove. **b** Microform lip defect: *Left* unilateral notch, groove, and slumped nares. **c** Microform palate defect: Bifid Uvula. **d** Microform palate defedt: Submucous cleft palate

biological [85, 127] data suggest that CL and CLP may have separate genetic etiologies. Nonetheless, common pathways may underlie the etiologies of each group, as occasionally both CL/P and CP are present within the same pedigree. Such families are said to exhibit mixed clefting, and are most commonly noted in syndromic forms of OFC [128]. There are also very mild expressions of CL and CP known as microforms (Fig. 2), for example congenital notches or grooves in the lip (Fig. 2a, b), bifid uvula (Fig. 2c), and submucous CP (Fig. 2d).

In addition to the overt clinical phenotypic spectrum, and visible microforms, there is increasing research on sub-clinical phenotypic features, i.e. features within the range of normal variability that are seen at increased frequency in individuals

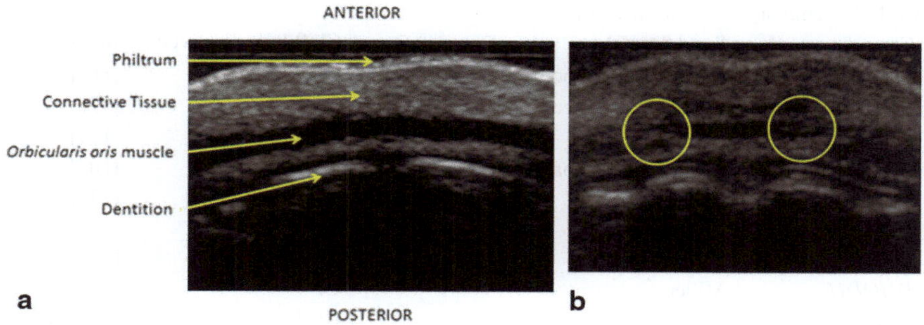

Fig. 3 Examples of sub-clinical phenotypes. **a** Normal Orbicularis oris muscle visualized via high-resolution ultrasound in a cross-section through the *upper* lip. Note the wide, uniform appearance of the muscle and contrast to the breaks seen in Fig. 3b. **b** Oribicularis oris muscle with subclinical (i.e. not externally visible) defects that appear as breaks in the muscle. This image shows bilateral breaks (*circled*) which notably are located where overt clefts of the lip would be

with OFC or their relatives, versus controls with no family history of OFC[161]. The earliest studies were on features related to laterality, such as handedness (Rintala, 1985#84; [163], while more recent studies have implicated a range of sub-clinical phenotypes such as *orbicularis oris* muscle (OO) defects (see Fig. 3a, b; [34, 97, 111, 135]), dental anomalies [4, 151], lip dermatoglyphics [112], facial measurements [101, 161, 162, 165, 166], brain variants on MRI [118–121].

Such sub-clinical features could represent the mildest physical expression of OFC risk genes (e.g. OO defects, Fig. 3), and/or pleiotropic effects of the risk genes (e.g. lip dermatoglyphics). Further, these features may clarify the lack of typical Mendelian patterns seen in OFC families, "missing" heritability from GWAS studies [89], and OFC discordance in MZ twins. Interestingly, a recent study of Danish twins found essentially identical recurrence risks for offspring of either the affected or unaffected twin in discordant MZ pairs [51, 53]. Furthermore, examination of such phenotypes are beginning to blur the historical distinction between CL/P and CP in some cases, e.g. in a small study there was a significant proportion of CP cases with OO defects [164].

Orofacial Cleft Genetics

Lessons Learned from Genetic Studies of Syndromic Orofacial Clefts

Orofacial clefts are designated as syndromic based on the presence of additional physical or cognitive abnormalities. There are at least 275 described orofacial clefting syndromes (http://www.ncbi.nlm.nih.gov/OMIM) that are caused by mutation of a single genetic locus, chromosomal abnormalities, or teratogens. Most clefting syndromes are rare, affecting only one in several hundred thousand live births. Approximately 75 % of the described syndromes have a known genetic cause (summarized in Table 1). With advances in genomic technologies, identification of genes

Table 1 Summary of syndromic forms of OFC

Gene	Syndrome(s)	Cleft type observed	Reference
BCOR	Oculofaciocardiodental	CP	[113]
CDH1	Familial gastric cancer and CLP	CL/P	[47]
CHD7	CHARGE	CP	[152]
CHRNG	Lethal and Escobar multiple pterygium	CP	[105]
COL11A1, COL11A2	Stickler type 2	CP	[139]
COL2A1	Stickler type 1	CP	[139]
DHODH	Miller	CP	[1]
FGFR1	Hartsfield, Kallmann	CL/P	[39, 138]
FGFR2	Apert, Crouzon	CP	[129, 169]
FLNA	Otopalatodigital types 1 and 2	CP	[132]
FOXE1	Bamforth-Lazarus	CP	[5]
GLI2	Holoprosencephaly	CL/P	[134]
GLI3	Oro-facial-digital	CL/P	[65]
GRHL3	Van der Woude	CL/P	[123]
IRF6	Popliteal pterygium, Van der Woude	CL/P	[73]
KDM6A	Kabuki	CL/P	[78]
KMT2D (MLL2)	Kabuki	CL/P	[114]
MID1	Opitz G/BBB	CL/P	[126]
MSX1	Tooth agenesis with or without cleft	CL/P	[146]
NIPBL	Cornelia de Lange	CP	[74, 144]
OFD1	Oro-facial-digital type 1	CL/P	[42]
PHF8	Siderius X-linked mental retardation	CL/P	[77]
PTCH1	Gorlin	CL/P	[57, 64]
PVRL1	CLP ectodermal dysplasia	CL/P	[141]
RIPK4	Bartsocas-Papas	CL/P	[71, 103]
SATB2	Isolated cleft palate	CP	[43]
SF3B4	Nager	CP	[11, 32]
SHH	Holoprosencephaly	CL/P	[133]
SIX3	Holoprosencephaly	CL/P	[154]
SOX9	Campomelic dysplasia, Pierre Robin sequence	CP	[9, 45, 153]
TBX1	DiGeorge	CP	[122]
TBX22	X-linked cleft palate and ankyloglossia	CP	[19]
TCOF1	Treacher Collins	CP	[56]
TFAP2A	Branchio-oculo-facial	CL/P	[102]
TGFBR1, TGFBR2	Loeys-Dietz	CP	[84]
TGIF	Holoprosencephaly	CL/P	[49]
TP63	Ankyloblepharon-ectodermal dysplasia-clefting, Ectrodactyly-ectodermal dysplasia-clefting	CL/P	[24, 98]
TWIST1	Saethre-Chotzen	CP	[41, 61]
WNT3	Tetra-amelia with CLP	CL/P	[116]

CL cleft lip
CP cleft palate
CL/P cleft lip with or without cleft palate

causing these syndromes has been very successful [38] and has been further facilitated by advances in sequencing technology. For example, whole exome sequencing has recently identified the genes causing Kabuki syndrome [114], Miller syndrome [115], Bartsocas-Papas syndrome [71, 103], and Van der Woude syndrome [123] .

Van der Woude syndrome (VWS; MIM #119300) is the most common orofacial clefting syndrome. With a prevalence of 1/34,000 live births [20], it accounts for approximately 2 % of all CL/P cases. Individuals with VWS have at least one of the following three anomalies: congenital, typically bilateral, paramedian lower-lip pits or mounds with a sinus tract leading from a mucous gland of the lip; cleft lip (CL); or cleft palate (CP) [147]. VWS and its allelic disorder, popliteal pterygium syndrome (PPS; MIM #119500), are caused by mutations is the transcription factor, *Interferon Regulatory Factor 6* (*IRF6*) [73]. To date, several hundred mutations in IRF6 have been reported to cause these disorders; mutations are found throughout the gene but are enriched in the DNA-binding domain [36, 82].

Mutations in *IRF6* are only present in 65–75 % of all cases of VWS. Prior to the identification of *IRF6* as the gene for VWS, [72] described a Finnish family in which ten family members were affected with CP and one had CLP. Of these eleven individuals, one had lip pits and two others had a "wave-like" lower lip. The phenotypes in this family are reminiscent of VWS, but linkage to *IRF6 at* 1q32-q41 was excluded (multipoint LOD scores < -13.0 for markers across this region). The locus was subsequently mapped to a 30-cM region on 1p32–p36 [72] . In 2014, Peyrard-Janvid and colleagues [123] performed exome sequencing on this family and identified a heterozygous pathogenic variant in the gene *Grainy-head like three* (*GRHL3*). A cohort of 44 additional individuals with clinical features of VWS who lacked a pathogenic variant in *IRF6* was sequenced, which identified pathogenic variants in seven of these families. The authors concluded that pathogenic variants in *GRHL3* appear to account for approximately 5 % of all cases of VWS. In these eight VWS families with *GRHL3* mutations, the full range of VWS-associated orofacial clefts and lip pits was observed. However, the affected individuals were significantly more likely to have cleft palate, less likely to have cleft lip, and less likely to have lip pits.

VWS and PPS are especially interesting as there are few single gene disorders or syndromes in which family members have CL/P or CP. This type of "mixed" clefting can also occur with mutation of *MSX1* [146], *TP63*, and *FGFR1*, and can be seen in individuals with 22q11.2 deletion syndrome, fetal alcohol syndrome, Kabuki syndrome and CHARGE syndrome.

Lessons Learned from Genetic Studies of Nonsyndromic Orofacial Clefts

Orofacial Cleft Candidate Gene Studies Early molecular genetics advances and the development of improved genetic markers (e.g. restriction fragment length polymorphisms and microsatellites) were important for genetic studies of OFCs, allowing for family studies of candidate genes and for genome scans. Prior studies of candidate genes for OFC and other traits were done with non-DNA based genetic markers, such as HLA or the ABO blood group, and employed statistical methods such as association analysis in case/control series or linkage analysis in multiplex families or affected relative pairs.

The first published OFC association studies examined HLA alleles [16, 155] based on the observed susceptibility to cortisone-induced CP in some mouse strains that was associated primarily with genotypes at the H2 locus [15]. Although several studies were performed in multiple populations, no overall positive associations were found between *HLA* and CL/P or CP. The first publication reporting a positive association with OFC was a population-based study examining CL/P and a TaqI restriction site polymorphism in the *transforming growth factor alpha* locus (*TGFA*; [3]). Linkage approaches were also applied in tests for candidate genes, first to HLA (again with no positive results for either CP or CL/P [148]). The first positive linkage finding with OFC was with *F13A* on chromosome 6p [40], which utilized a subset of the multiplex Danish families first documented by Fogh-Andersen in 1942 [44].

Throughout the 1970's, 1980's, and 1990's, there were many other candidate gene studies (see reviews of those studies [70, 109]). Many genes or regions had positive results in one or more of the early studies, however few have been consistently positive across all studies, primarily due to lack of adequate sample size. In addition to *TGFA* and *F13A*, at least one positive linkage or association with OFC was reported during this time period for the following genes/regions: Interferon regulatory factor 6 (*IRF6*, chromosome 1q32.3–q41), homeobox 7 (*MSX1*; chromosome 4p16), anonymous markers on 4q31, transforming growth factor beta-3 (*TGFB3*; 14q24), retinoic acid receptor alpha (*RARA*; 17q21), proto-oncogene *BCL3* plus nearby anonymous markers (19q13).

In addition to these candidate gene linkage and association studies, the candidate gene investigations that drove many early studies searching for the genetic etiology of OFCs often included resequencing to identify coding mutations causing OFCs. Genes sequenced in this approach included *BMP4*, *FGF10*, *FGF8*, *FGFR1*, *FGFR2*, *FGFR3*, *FOXE1*, *GLI2*, *JAG2*, *LHX8*, *MSX1*, *MSX2*, *NUDT6*, *PAX9*, *PTCH1*, *PVR*, *PVRL1*, *PVRL2*, *RYK*, *SATB2*, *SKI*, *SPRY2*, *TBX10*, *TGFB3*, and *TP63* [38, 68]. Although many rare, coding variants have been reported, the majority of these variants are likely to be rare polymorphisms [80]. Candidate gene resequencing efforts suggest that rare variants in *MSX1* and members of the FGF signaling pathway, as well as de novo variants in *FGF8* and *TP63*, contribute risk to OFCs [79, 130]. Recently, there has been a resurgence of candidate gene resequencing studies following the genome-wide studies described below. In these studies, one or more genes in the regions of association are sequenced to provide some evidence that those genes are causal for OFCs. Briefly, these studies have provided evidence for *GREM1* [2], *MAFB* [6], and *ARHGAP29* [81], and have examined *PAX7* [21], *VAX1* [110], *ABCA4* [6] and NOG [2].

Orofacial Cleft Genome-Wide Linkage Studies The first genome-wide linkage scans for CL/P were in English affected affected sib pairs [125] and then in multiplex Chinese families [94]. Multiple other genome-wide linkage scans were performed, each noting a number of positive signals; however, due to limited sample size few individual study results reached the standard levels of genome-wide significance (LOD scores ≥3.2 [75] for the typical 400 microsatellite-marker panel). It was not until a consortium of research groups pooled their studies that the first genome-wide significant results for CL/P were obtained for regions on 1q32, 2p, 3q27–28, 9q21, 14q21–24 and 16q24 [95, 96] Follow-up fine mapping of these regions showed significant results for SNPs in *IRF6,* previously associated in candidate gene studies [127] and later also identified by genome-wide association studies [6, 14, 88], and in *FOXE1*, which was later confirmed and strengthened with the results in other populations [83, 86, 104, 117, 149].

Orofacial Cleft Genome-Wide Association Studies (GWAS) GWAS are commonly used for studying common, complex traits and disorders. Most of the studied disorders have adult onset, making OFC one of the few congenital/developmental anomalies to have been studied using the GWAS method. To date there have been four independent GWAS for CL/P [6, 14, 48, 88], one for CP [7] , one in consanguineous CL/P families [23], and a meta-analysis of the two largest CL/P studies [85]. These GWAS have been extremely successful in that they identified multiple genome-wide significant associations with CL/P, and for CP identified potential gene-environment interactions (Table 2).

The first two successful OFC GWAS [14, 48] used Caucasian CL/P cases and controls and identified a novel region on chromosome 8q24 with extremely strong evidence of association. In addition, the Birnbaum study confirmed an association with *IRF6* (1q32.3–q41; which had prior positive candidate gene [13, 69, 127] and linkage analysis [96] results). A third study, with an increased sample size and replication populations [88] confirmed 8q24 and *IRF6*, and identified two additional loci: 17q22 near *NOG* and 10q25.3 near *VAX1.*

8q24 and *IRF6* were also confirmed by a nuclear-trio based GWAS of Caucasians and Asians [6] that was part of the trans-NIH GENEVA study [31]. In the latter study, there were marked differences in the strength of association seen in the Caucasian and Asian trios, apparently due to lower minor allele frequencies in Asians. Notably the statistical evidence for *IRF6* variants was strongest in the Asians, while 8q24 was strongest in the Caucasians. The GENEVA study also identified at least two novel loci (on 1p22 in the *ABCA4* gene and 20q12 near *MAFB*) reaching genome-wide significance, with stronger signals in the Asians compared to the Caucasians.

A meta-analysis, combining the GENEVA study and the Mangold *et al.* studies identified several additional risk loci [85]—three loci in Caucasians (2p21, 13q31, and 15q22) and three in Asians (1p36, 3p11, and 8q21). Based on previous evidence that CL and CLP could have distinct genetic etiologies [52, 127], Mangold *et al.* also performed separate analyses for CL and CLP. This analysis provided an important advance in our understanding of CL vs. CLP as it demonstrated that the 13q31 locus was exclusively associated with NSCLP. It has been estimated that the CL/P GWAS loci contribute to about 55% of the overall population attributable risk for CL/P. Unlike many other common complex human traits studied by GWAS [89], the results from CL/P GWAS are potentially capable of explaining a substantial portion of the variation in CL/P.

Table 2 Summary of genomic regions with strong statistical support from genomic-wide linkage and association studies

Associated locus	Candidate gene in region	Lowest reported p-value (association)	Associated phenotype	References for genome-wide significance
1p36.13	*PAX7*	7.0×10^{-10}(meta)	CL/P	[85]
1p22	*ARHGAP29*	3.1×10^{-12}(meta)	CL/P	[6, 87]
1q32.2	*IRF6*	9.1×10^{-15}	CL/P	[6, 14, 88, 96]
2p13	*TGFA*	3.25 (HLOD)	CL/P	[96]
2p21	*THADA*	1.1×10^{-8}(meta)	CL/P	[85]
3p11	*EPHA3*	3.9×10^{-8}(meta)	CL/P	[85]
3q12	*COL8A1/FILIPIL*	4.49×10^{-5}	CL/P	[8]
3q27–28	*TP63*	4.13 (HLOD)	CL/P	[96]
8q21.3	*DCAF4L2*	1.9×10^{-8}(meta)	CL/P	[8]
8q22.3	*BAALC*	2.0×10^{-7}	CP with multivitamins	[7]
8q24	Gene Desert	5.1×10^{-35}(meta)	CL/P	[6, 13, 48, 88]
9q22.2	*GADD45G*	3.0×10^{-5}	CL/P	[8]
9q22.33	*FOXE1*	2.0×10^{-9}	CL/P and CP	[86, 96]
9q31.1	*SMC2*	1.53×10^{-8}	CP with maternal alcohol	[7]
10q25.3	*VAX1*	4.0×10^{-11}	CL/P	[6, 88, 22]
12q14	*TBK1*	7.86×10^{-8}	CP with maternal smoking	[7]
13q31.2	*SPRY2*	9×10^{-11}(meta)	CLP	[85]
14q21–24	*PAX9, TGFB3, BMP4*	4.18 (HLOD)	CL/P	[96]
15q13.3	*GREM1*	1×10^{-6}	CL/P	[88]
15q22	*TPM1*	8×10^{-7}	CL/P	[85]
16q24	*CRIPSLD2*	3.56 (HLOD)	CL/P	[87]
17p13.1	*NTN1*	6.0×10^{-9}	CL/P	[85]
17q22	*NOG*	1.1×10^{-8}(meta)	CL/P	[88]
18q22	*ZNF236*	6.75×10^{-8}	CP with maternal smoking	[7]
20q12	*MAFB*	1.6×10^{-11}	CL/P	[6]

As with other etiological investigations of OFC, there has been a dearth of GWAS of CP. In the European CL/P GWAS [14, 88], the replication panel of SNPs for four most significant loci was also tested in CP trios and showed no statistically significant results, implying little or no overlap in the findings for CL/P versus CP. The first GWAS of CP was performed in 2011 [7] and found no genome-wide significant signals. Using gene by environment models incorporating three common exposures during pregnancy: maternal smoking, alcohol consumption, and multivitamin supplementation, several significant results were obtained including *MLLT3* and *SMC2* (chromosome 9) with alcohol consumption, *TBK1* (chromosome 12) and *ZNF236* (chromosome 18) with maternal smoking, and *BAALC* (chromosome 8) with multivitamin supplementation [7].

Orofacial Clefts and Other Developmental Anomalies: Future Directions and Challenges for Personalized Medicine

As we have reviewed, there has been substantial progress in understanding the etiology of OFCs, and therefore the field is poised to consider opportunities for personalized medicine applications.

OFC has been a major success in applying genome-wide approaches to a common, complex disorder, given that there are multiple genome-wide significant regions identified (Table 2). Of the significant regions, four (*IRF6* on 1q32–41, 8q24, 17q22, and 10q25.3) have been estimated to account for almost 55 % of the variation in CL/P, representing one of the highest proportions achieved for any common, complex disorder. In addition to progress in understanding the genetics of OFCs, there has also been substantial progress is understanding the range of phenotypic expression that can now be correlated with specific genetic signals. Given the wide range of overt and sub-clinical phenotypes that are now known to aggregate in OFC families, phenotyping is predicted to be key in future studies to understand the expression of OFC risk genes, and to move to personalized applications.

The challenge is to continue to move from the essentially population-based statistical genetic and phenotypic results to truly understanding the etiology of OFC in specific individuals and their families. The positive genome-wide results need further study in order to identify the functional variants (both common and rare) in OFC risk genes and/or their regulatory regions. Sequencing projects are now underway to begin to identify additional variants in the genome-wide significant regions. With the wealth of data that the sequencing and genotyping projects will bring, it will be imperative to maximize the analysis of the data by pathway analyses and other methods to detect interactions (GxG and GxE) and more complex interplay between etiologic variants, the environment, and phenotypes. Enhanced inclusion of phenotyping will be important in understanding the function of risk variants and pathways that are identified.

Successful completion of variant identification is likely to lead to new ways to designate phenotype patterns in terms of the responsible risk gene. Using *IRF6* as an example, mutations in the gene cause VWS, common polymorphic variants are associated with nonsyndromic CL/P (particularly CL alone), and rare variants are currently under scrutiny. Perhaps lip pits and sub-clinical lip pits (i.e. lip dermatoglyphic patterns) will become the defining feature of the presence of *IRF6* risk variants, and greatly enhance our ability to estimate specific recurrence risks. Further, understanding variants can also lead to treatment and personalized medicine implications.

Other opportunities include extending OFC studies to additional clinical phenotypes and ethnicities, to bring personalized medicine approaches to as many individuals and families touched by OFC as possible. There has been substantial progress in identifying risk genes for CL/P; similar approaches should be extended to isoloated CP as well, requiring a concerted effort to identify substantial numbers of nonsyndromic CP individuals and families. Progress to date has been concentrated in studies of Caucasian and Asian OFC families; it will be important to broaden studies to other ethnicities, especially to African-derived populations in order to better understand the notable ethnic differences in birth prevalence. There are two studies that have tested the loci significant in Caucasians and Asians, with no significant findings to date [21, 156].

As additional progress is made, OFC categorizations should be re-visited to develop a less arbitrary classification than non-syndromic versus syndromic, in favor of a gene-and/or phenotype- centric system. Animal models will continue to be key in expression and functional studies following statistically successful human studies; genomics resources such as FaceBase [60] hope to provide improved integration of animal model and human genetic results for OFC.

In conclusion, OFCs and other congenital developmental anomalies present multiple challenges and opportunities for application of a personalized medicine approach. Recent genetic and phenotypic studies provide necessary preliminary knowledge, but require substantial investment in additional research to move to such a paradigm.

Reference

1. Abbott BD. The etiology of cleft palate: a 50-year search for mechanistic and molecular understanding. Birth defects Res B Dev Reprod Toxicol. 2010;89(4):266–74. doi:10.1002/bdrb.20252.
2. Al Chawa T, Ludwig KU, Fier H, Potzsch B, Reich RH, Schmidt G, Braumann B, Daratsianos N, Bohmer AC, Schuencke H, Alblas M, Fricker N, Hoffmann P, Knapp M, Lange C, Nothen MM, Mangold E. Nonsyndromic cleft lip with or without cleft palate: increased burden of rare variants within Gremlin-1, a component of the bone morphogenetic protein 4 pathway. Birth Defects Res A Clin Mol Teratol. 2014. doi:10.1002/bdra.23244
3. Ardinger HH, Buetow KH, Bell GI, Bardach J, VanDemark DR, Murray JC. Association of genetic variation of the transforming growth factor- alpha gene with cleft lip and palate. Am J Hum Genet. 1989;45(3):348–53.
4. Aspinall A, Raj S, Jugessur A, Marazita M, Savarirayan R, Kilpatrick N. Expanding the cleft phenotype: the dental characteristics of unaffected parents of Australian children with non-syndromic cleft lip and palate. Int J Paedtr Dent. Br Paedodontic Soc. Int Assoc Dent Child. 2013. doi:10.1111/ipd.12072.
5. Bamforth JS, Hughes IA, Lazarus JH, Weaver CM, Harper PS. Congenital hypothyroidism, spiky hair, and cleft palate. J Med. Genet. 1989;26(1):49–51.
6. Beaty TH, Murray JC, Marazita ML, Munger RG, Ruczinski I, Hetmanski JB, Liang KY, Wu T, Murray T, Fallin MD, Redett RA, Raymond G, Schwender H, Jin SC, Cooper ME, Dunnwald M, Mansilla MA, Leslie E, Bullard S, Lidral AC, Moreno LM, Menezes R, Vieira AR, Petrin A, Wilcox AJ, Lie RT, Jabs EW, Wu-Chou YH, Chen PK, Wang H, Ye X, Huang S, Yeow V, Chong SS, Jee SH, Shi B, Christensen K, Melbye M, Doheny KF, Pugh EW, Ling H, Castilla EE, Czeizel AE, Ma L, Field LL, Brody L, Pangilinan F, Mills JL, Molloy AM, Kirke PN, Scott JM, Arcos-Burgos M, Scott AF. A genome-wide association study of cleft lip with and without cleft palate identifies risk variants near MAFB and ABCA4. Nat Genet. 2010;42(6):525–9. doi:10.1038/ng.580.
7. Beaty TH, Ruczinski I, Murray JC, Marazita ML, Munger RG, Hetmanski JB, Murray T, Redett RJ, Fallin MD, Liang KY, Wu T, Patel PJ, Jin SC, Zhang TX, Schwender H, Wu-Chou YH, Chen PK, Chong SS, Cheah F, Yeow V, Ye X, Wang H, Huang S, Jabs EW, Shi B, Wilcox AJ, Lie RT, Jee SH, Christensen K, Doheny KF, Pugh EW, Ling H, Scott AF. Evidence for gene-environment interaction in a genome wide study of nonsyndromic cleft palate. Genet Epidemiol. 2011;35(6):469–78. doi:10.1002/gepi.20595.

8. Beaty TH, Taub MA, Scott AF, Murray JC, Marazita ML, Schwender H, Parker MM, Hetmanski JB, Balakrishnan P, Mansilla MA, Mangold E, Ludwig KU, Noethen MM, Rubini M, Elcioglu N, Ruczinski I. Confirming genes influencing risk to cleft lip with/without cleft palate in a case-parent trio study. Hum Genet. 2013;132(7):771–81. doi:10.1007/s00439-013-1283-6.
9. Benko S, Fantes JA, Amiel J, Kleinjan DJ, Thomas S, Ramsay J, Jamshidi N, Essafi A, Heaney S, Gordon CT, McBride D, Golzio C, Fisher M, Perry P, Abadie V, Ayuso C, Holder-Espinasse M, Kilpatrick N, Lees MM, Picard A, Temple IK, Thomas P, Vazquez MP, Vekemans M, Roest Crollius H, Hastie ND, Munnich A, Etchevers HC, Pelet A, Farlie PG, Fitzpatrick DR, Lyonnet S. Highly conserved non-coding elements on either side of SOX9 associated with Pierre Robin sequence. Nat Genet. 2009;41(3):359–64. doi:10.1038/ng.329.
10. Berk NW, Marazita ML. The costs of cleft lip and palate: personal and societal implications. In: Wyszynski DF, editor. Cleft lip and palate: from origin to treatment. Oxford: Oxford University Press; 2002.
11. Bernier FP, Caluseriu O, Ng S, Schwartzentruber J, Buckingham KJ, Innes AM, Jabs EW, Innis JW, Schuette JL, Gorski JL, Byers PH, Andelfinger G, Siu V, Lauzon J, Fernandez BA, McMillin M, Scott RH, Racher H, Consortium FC, Majewski J, Nickerson DA, Shendure J, Bamshad MJ, Parboosingh JS. Haploinsufficiency of SF3B4, a component of the pre-mRNA spliceosomal complex, causes Nager syndrome. Am J Hum Genet. 2012;90(5):925–33. doi:10.1016/j.ajhg.2012.04.004.
12. Bille C, Knudsen LB, Christensen K. Changing lifestyles and oral clefts occurrence in Denmark. Cleft Palate Craniofac J Off Publ Am Cleft Palate Craniofac Assoc. 2005;42(3):255–9. doi:03-139 [pii] 10.1597/03-139.1.
13. Birnbaum S, Ludwig KU, Reutter H, Herms S, de Assis NA, Diaz-Lacava A, Barth S, Lauster C, Schmidt G, Scheer M, Saffar M, Martini M, Reich RH, Schiefke F, Hemprich A, Potzsch S, Potzsch B, Wienker TF, Hoffmann P, Knapp M, Kramer FJ, Nothen MM, Mangold E. IRF6 gene variants in Central European patients with non-syndromic cleft lip with or without cleft palate. Eur J Oral Sci. 2009a;117(6):766–9. doi:10.1111/j.1600-0722.2009.00680.x.
14. Birnbaum S, Ludwig KU, Reutter H, Herms S, Steffens M, Rubini M, Baluardo C, Ferrian M, Almeida de Assis N, Alblas MA, Barth S, Freudenberg J, Lauster C, Schmidt G, Scheer M, Braumann B, Berge SJ, Reich RH, Schiefke F, Hemprich A, Potzsch S, Steegers-Theunissen RP, Potzsch B, Moebus S, Horsthemke B, Kramer FJ, Wienker TF, Mossey PA, Propping P, Cichon S, Hoffmann P, Knapp M, Nothen MM, Mangold E. Key susceptibility locus for nonsyndromic cleft lip with or without cleft palate on chromosome 8q24. Nat Genet. 2009b;41(4):473–7. doi:10.1038/ng.333.
15. Bonner JJ, Tyan ML. Backcross test demonstrates the linkage of glucocorticoid-induced cleft palate susceptibility to H-2. Teratology. 1982;26(2):213–6. doi:10.1002/tera.1420260214.
16. Bonner JJ, Terasaki PI, Thompson P, Holve LM, Wilson L, Ebbin AJ, Slavkin HC. HLA phenotype frequencies in individuals with cleft lip and/or cleft palate. Tissue Antigens. 1978;12(3):228–32.
17. Boo-Chai K. An ancient Chinese text on a cleft lip. Plast Reconstr Surg. 1966;38(2):89–91.
18. Boyles AL, DeRoo LA, Lie RT, Taylor JA, Jugessur A, Murray JC, Wilcox AJ. Maternal alcohol consumption, alcohol metabolism genes, and the risk of oral clefts: a population-based case-control study in Norway, 1996–2001. Am J Epidemiol. 2010;172(8):924–31. doi:10.1093/aje/kwq226.
19. Braybrook C, Doudney K, Marcano AC, Arnason A, Bjornsson A, Patton MA, Goodfellow PJ, Moore GE, Stanier P. The T-box transcription factor gene TBX22 is mutated in X-linked cleft palate and ankyloglossia. Nat Genet. 2001;29(2):179–83. doi:10.1038/ng730.
20. Burdick AB. Genetic epidemiology and control of genetic expression in van der Woude syndrome. J Craniofac Genet Dev Biol Suppl. 1986;2:99–105.
21. Butali A, Mossey PA, Adeyemo WL, Jezewski PA, Onwuamah CK, Ogunlewe MO, Ugboko VI, Adejuyigbe O, Adigun AI, Abdur-Rahman LO, Onah, II, Audu RA, Idigbe EO, Mansilla MA, Dragan EA, Petrin AL, Bullard SA, Uduezue AO, Akpata O, Osaguona AO, Olasoji HO, Ligali TO, Kejeh BM, Iseh KR, Olaitan PB, Adebola AR, Efunkoya E, Adesina OA, Oluwatosin OM, Murray JC. Genetic studies in the nigerian population implicate an MSX1 mutation in complex oral facial clefting disorders. Cleft Palate Craniofac J. 2011;48(6):646–53. doi:10.1597/10-133.

22. Butali A, Suzuki S, Cooper ME, Mansilla MA, Cuenco K, Leslie EJ, Suzuki Y, Niimi T, Yamamoto M, Ayanga G, Erkhembaatar T, Furukawa H, Fujiwawa K, Imura H, Petrin AL, Natsume N, Beaty TH, Marazita ML, Murray JC. Replication of GWAS identified candidate genes in four independent populations confirm the role of common and rare variants in PAX7 and VAX1 in the etiology of non-syndromic CL(P). Am J Med Genet A. 2013;161A:965–72.
23. Camargo M, Rivera D, Moreno L, Lidral AC, Harper U, Jones M, Solomon BD, Roessler E, Velez JI, Martinez AF, Chandrasekharappa SC, Arcos-Burgos M. GWAS reveals new recessive loci associated with non-syndromic facial clefting. Eur J Med Genet. 2012;55(10):510–4. doi:10.1016/j.ejmg.2012.06.005.
24. Celli J, Duijf P, Hamel BC, Bamshad M, Kramer B, Smits AP, Newbury-Ecob R, Hennekam RC, Van Buggenhout G, van Haeringen A, Woods CG, van Essen AJ, de Waal R, Vriend G, Haber DA, Yang A, McKeon F, Brunner HG, van Bokhoven H. Heterozygous germline mutations in the p53 homolog p63 are the cause of EEC syndrome. Cell. 1999;99(2):143–53.
25. Cheng LR. Asian-American cultural perspectives on birth defectrs: focus on cleft palate. Cleft Palate Craniofac J. 1990;27:294–300.
26. Christensen K. The 20th century Danish facial cleft population—epidemiological and genetic-epidemiological studies. Cleft Palate Craniofac J. 1999;36(2):96–104.
27. Christensen K, Mitchell LE. Familial recurrence-pattern analysis of nonsyndromic isolated cleft palate—a Danish Registry study. Am J Hum Genet. 1996;58(1):182–90.
28. Christensen K, Juel K, Herskind AM, Murray JC. Long term follow up study of survival associated with cleft lip and palate at birth. BMJ. 2004;328(7453):1405. (Clinical research ed).
29. Chung CS, Ching GH, Morton NE. A genetic study of cleft lip and palate in Hawaii. II. Complex segregation analysis and genetic risks. Am J Hum Genet. 1974;26(2):177–88.
30. Cooper ME, Ratay JS, Marazita ML. Asian oral-facial cleft birth prevalence. Cleft Palate Craniofac J. 2006;43(5):580–9.
31. Cornelis MC, Agrawal A, Cole JW, Hansel NN, Barnes KC, Beaty TH, Bennett SN, Bierut LJ, Boerwinkle E, Doheny KF, Feenstra B, Feingold E, Fornage M, Haiman CA, Harris EL, Hayes MG, Heit JA, Hu FB, Kang JH, Laurie CC, Ling H, Manolio TA, Marazita ML, Mathias RA, Mirel DB, Paschall J, Pasquale LR, Pugh EW, Rice JP, Udren J, van Dam RM, Wang X, Wiggs JL, Williams K, Yu K, Consortium G. The Gene, Environment Association Studies consortium (GENEVA): maximizing the knowledge obtained from GWAS by collaboration across studies of multiple conditions. Genet Epidemiol. 2010;34(4):364–72. doi:10.1002/gepi.20492.
32. Czeschik JC, Voigt C, Alanay Y, Albrecht B, Avci S, Fitzpatrick D, Goudie DR, Hehr U, Hoogeboom AJ, Kayserili H, Simsek-Kiper PO, Klein-Hitpass L, Kuechler A, Lopez-Gonzalez V, Martin M, Rahmann S, Schweiger B, Splitt M, Wollnik B, Ludecke HJ, Zeschnigk M, Wieczorek D. Clinical and mutation data in 12 patients with the clinical diagnosis of Nager syndrome. Hum Genet. 2013;132(8):885–98. doi:10.1007/s00439-013-1295-2.
33. Daack-Hirsch S. Filipino explanatory models of cleft lip with or without cleft palate. Cleft Palate Craniofac J. 2010;47(2):122–133.
34. Dado DV, Kernahan DA. Anatomy of the orbicularis oris muscle in incomplete unilateral cleft lip based on histological examination. Ann Plast Surg. 1985;15(2):90–8.
35. Darwin C. The variation of animals and plants under domestication. vol. 1. 1875. (Second Edition, Revised edn. John Murray, Albermarle Street, London).
36. de Lima RL, Hoper SA, Ghassibe M, Cooper ME, Rorick NK, Kondo S, Katz L, Marazita ML, Compton J, Bale S, Hehr U, Dixon MJ, Daack-Hirsch S, Boute O, Bayet B, Revencu N, Verellen-Dumoulin C, Vikkula M, Richieri-Costa A, Moretti-Ferreira D, Murray JC, Schutte BC. Prevalence and nonrandom distribution of exonic mutations in interferon regulatory factor 6 in 307 families with Van der Woude syndrome and 37 families with popliteal pterygium syndrome. Genet Med Off J Am Coll Med Genet. 2009;11(4):241–7. doi:10.1097/GIM.0b013e318197a49a.
37. Dietz A, Pedersen DA, Jacobsen R, Wehby GL, Murray JC, Christensen K. Risk of breast cancer in families with cleft lip and palate. Ann Epidemiol. 2012; 22(1):37–42. doi:10.1016/j.annepidem.2011.09.003.

38. Dixon MJ, Marazita ML, Beaty TH, Murray JC. Cleft lip and palate: understanding genetic and environmental influences. Nat Rev Genet. 2011;12(3):167–78. doi:10.1038/nrg2933.
39. Dode C, Levilliers J, Dupont JM, De Paepe A, Le Du N, Soussi-Yanicostas N, Coimbra RS, Delmaghani S, Compain-Nouaille S, Baverel F, Pecheux C, Le Tessier D, Cruaud C, Delpech M, Speleman F, Vermeulen S, Amalfitano A, Bachelot Y, Bouchard P, Cabrol S, Carel JC, Delemarre-van de Waal H, Goulet-Salmon B, Kottler ML, Richard O, Sanchez-Franco F, Saura R, Young J, Petit C, Hardelin JP. Loss-of-function mutations in FGFR1 cause autosomal dominant Kallmann syndrome. Nat Genet. 2003;33(4):463–5. doi:10.1038/ng1122.
40. Eiberg H, Bixler D, Nielsen LS, Conneally PM, Mohr J. Suggestion of linkage of a major locus for nonsyndromic orofacial cleft with F13A and tentative assignment to chromosome 6. Clin Genet. 1987;32(2):129–32.
41. el Ghouzzi V, Le Merrer M, Perrin-Schmitt F, Lajeunie E, Benit P, Renier D, Bourgeois P, Bolcato-Bellemin AL, Munnich A, Bonaventure J. Mutations of the TWIST gene in the Saethre-Chotzen syndrome. Nat Genet. 1997;15(1):42–6. doi:10.1038/ng0197-42.
42. Ferrante MI, Giorgio G, Feather SA, Bulfone A, Wright V, Ghiani M, Selicorni A, Gammaro L, Scolari F, Woolf AS, Sylvie O, Bernard L, Malcolm S, Winter R, Ballabio A, Franco B. Identification of the gene for oral-facial-digital type I syndrome. Am J Hum Genet. 2001;68(3):569–76.
43. FitzPatrick DR, Carr IM, McLaren L, Leek JP, Wightman P, Williamson K, Gautier P, McGill N, Hayward C, Firth H, Markham AF, Fantes JA, Bonthron DT. Identification of SATB2 as the cleft palate gene on 2q32-q33. Hum Mol Genet. 2003;12(19):2491–501. doi:10.1093/hmg/ddg248ddg248 [pii].
44. Fogh-Andersen P. Inheritance of harelip and cleft palate. Thesis, Nyt nordisk forlag. Copenhagen: Arnold Busck; 1942.
45. Foster JW, Dominguez-Steglich MA, Guioli S, Kwok C, Weller PA, Stevanovic M, Weissenbach J, Mansour S, Young ID, Goodfellow PN, et al. Campomelic dysplasia and autosomal sex reversal caused by mutations in an SRY-related gene. Nature. 1994;372(6506):525–30. doi:10.1038/372525a0.
46. Fraser FC. Thoughts on the etiology of clefts of the palate and lip. Acta Genet Stat Med. 1955;5(4):358–69.
47. Frebourg T, Oliveira C, Hochain P, Karam R, Manouvrier S, Graziadio C, Vekemans M, Hartmann A, Baert-Desurmont S, Alexandre C, Lejeune Dumoulin S, Marroni C, Martin C, Castedo S, Lovett M, Winston J, Machado JC, Attie T, Jabs EW, Cai J, Pellerin P, Triboulet JP, Scotte M, Le Pessot F, Hedouin A, Carneiro F, Blayau M, Seruca R. Cleft lip/palate and CDH1/E-cadherin mutations in families with hereditary diffuse gastric cancer. J Med Genet. 2006;43(2):138–42. doi:10.1136/jmg.2005.031385.
48. Grant SF, Wang K, Zhang H, Glaberson W, Annaiah K, Kim CE, Bradfield JP, Glessner JT, Thomas KA, Garris M, Frackelton EC, Otieno FG, Chiavacci RM, Nah HD, Kirschner RE, Hakonarson H. A genome-wide association study identifies a locus for nonsyndromic cleft lip with or without cleft palate on 8q24. J Pediatr. 2009;155(6):909–13. doi:S0022-3476(09)00575-7 [pii] 10.1016/j.jpeds.2009.06.020.
49. Gripp KW, Wotton D, Edwards MC, Roessler E, Ades L, Meinecke P, Richieri-Costa A, Zackai EH, Massague J, Muenke M, Elledge SJ. Mutations in TGIF cause holoprosencephaly and link NODAL signalling to human neural axis determination. Nat Genet. 2000;25(2):205–8. doi:10.1038/76074.
50. Gritli-Linde A. Molecular control of secondary palate development. Dev Biol. 2007; 301(2):309–26.
51. Grosen D, Bille C, Pedersen JK, Skytthe A, Murray JC, Christensen K. Recurrence risk for offspring of twins discordant for oral cleft: a population-based cohort study of the Danish 1936–2004 cleft twin cohort. Am J Med Genet A. 2010a;152A(10):2468–74. doi:10.1002/ajmg.a.33608.
52. Grosen D, Chevrier C, Skytthe A, Bille C, Molsted K, Sivertsen A, Murray JC, Christensen K. A cohort study of recurrence patterns among more than 54,000 relatives of oral cleft cases in Denmark: support for the multifactorial threshold model of inheritance. J Med Genet. 2010b;47(3):162–8. doi:10.1136/jmg.2009.069385.

53. Grosen D, Bille C, Petersen I, Skytthe A, Hjelmborg J, Pedersen JK, Murray JC, Christensen K. Risk of oral clefts in twins. Epidemiology. 2011;22(3):313–9. doi:10.1097/EDE.0b013e3182125f9c.
54. Groth P, Pavlova N, Kalev I, Tonov S, Georgiev G, Pohlenz HD, Weiss B. PhenomicDB: a new cross-species genotype/phenotype resource. Nucleic Acids Res. 2007;35(Database issue):D696–9. doi:10.1093/nar/gkl662.
55. Groth P, Leser U, Weiss B. Phenotype mining for functional genomics and gene discovery. Methods Mol Biol. 2011;760:159–73. doi:10.1007/978-1-61779-176-5_10.
56. Group TCC. Positional cloning of a gene involved in the pathogenesis of Treacher Collins syndrome. The Treacher Collins Syndrome Collaborative Group. Nat Genet. 1996;12(2):130–6. doi:10.1038/ng0296-130.
57. Hahn H, Wicking C, Zaphiropoulous PG, Gailani MR, Shanley S, Chidambaram A, Vorechovsky I, Holmberg E, Unden AB, Gillies S, Negus K, Smyth I, Pressman C, Leffell DJ, Gerrard B, Goldstein AM, Dean M, Toftgard R, Chenevix-Trench G, Wainwright B, Bale AE. Mutations of the human homolog of Drosophila patched in the nevoid basal cell carcinoma syndrome. Cell. 1996;85(6):841–51.
58. Hamilton CM, Strader LC, Pratt JG, Maiese D, Hendershot T, Kwok RK, Hammond JA, Huggins W, Jackman D, Pan H, Nettles DS, Beaty TH, Farrer LA, Kraft P, Marazita ML, Ordovas JM, Pato CN, Spitz MR, Wagener D, Williams M, Junkins HA, Harlan WR, Ramos EM, Haines J. The PhenX Toolkit: get the most from your measures. Am J Epidemiol. 2011;174(3):253–60. doi:10.1093/aje/kwr193.
59. Hecht JT, Yang P, Michels VV, Buetow KH. Complex segregation analysis of nonsyndromic cleft lip and palate. Am J Hum Genet. 1991;49(3):674–81.
60. Hochheiser H, Aronow BJ, Artinger K, Beaty TH, Brinkley JF, Chai Y, Clouthier D, Cunningham ML, Dixon M, Donahue LR, Fraser SE, Hallgrimsson B, Iwata J, Klein O, Marazita ML, Murray JC, Murray S, de Villena FP, Postlethwait J, Potter S, Shapiro L, Spritz R, Visel A, Weinberg SM, Trainor PA. The FaceBase Consortium: a comprehensive program to facilitate craniofacial research. Dev Biol. 2011;355(2):175–82. doi:10.1016/j.ydbio.2011.02.033.
61. Howard TD, Paznekas WA, Green ED, Chiang LC, Ma N, Ortiz de Luna RI, Garcia Delgado C, Gonzalez-Ramos M, Kline AD, Jabs EW. Mutations in TWIST, a basic helix-loop-helix transcription factor, in Saethre-Chotzen syndrome. Nat Genet . 1997;5(1):36–41. doi:10.1038/ng0197-36.
62. Jiang R, Bush JO, Lidral AC. Development of the upper lip: morphogenetic and molecular mechanisms. Dev Dyn Off Publ Am Assoc Anat. 2006;235(5):1152–66.
63. Johnson CY, Little J. Folate intake, markers of folate status and oral clefts: is the evidence converging? Int J Epidemiol. 2008;37(5):1041–58. doi:10.1093/ije/dyn098.
64. Johnson RL, Rothman AL, Xie J, Goodrich LV, Bare JW, Bonifas JM, Quinn AG, Myers RM, Cox DR, Epstein EH, Jr., Scott MP. Human homolog of patched, a candidate gene for the basal cell nevus syndrome. Science. 1996;272(5268):1668–71. (New York, NY).
65. Johnston JJ, Sapp JC, Turner JT, Amor D, Aftimos S, Aleck KA, Bocian M, Bodurtha JN, Cox GF, Curry CJ, Day R, Donnai D, Field M, Fujiwara I, Gabbett M, Gal M, Graham JM, Hedera P, Hennekam RC, Hersh JH, Hopkin RJ, Kayserili H, Kidd AM, Kimonis V, Lin AE, Lynch SA, Maisenbacher M, Mansour S, McGaughran J, Mehta L, Murphy H, Raygada M, Robin NH, Rope AF, Rosenbaum KN, Schaefer GB, Shealy A, Smith W, Soller M, Sommer A, Stalker HJ, Steiner B, Stephan MJ, Tilstra D, Tomkins S, Trapane P, Tsai AC, Van Allen MI, Vasudevan PC, Zabel B, Zunich J, Black GC, Biesecker LG. Molecular analysis expands the spectrum of phenotypes associated with GLI3 mutations. Hum Mutat. 2010;31(10):1142–54. doi:10.1002/humu.21328.
66. Jones MC. Etiology of facial clefts: prospective evaluation of 428 patients. Cleft Palate J. 1988;25(1):16–20.
67. Jones MC. Facial clefting. Etiology and developmental pathogenesis. Clin Plast Surg. 1993; 20(4):599–606.
68. Jugessur A, Murray JC. Orofacial clefting: recent insights into a complex trait. Curr Opin Genet Dev. 2005;15(3):270–8. doi:S0959-437X(05)00051-1[pii]10.1016/j.gde.2005.03.003.

69. Jugessur A, Rahimov F, Lie RT, Wilcox AJ, Gjessing HK, Nilsen RM, Nguyen TT, Murray JC. Genetic variants in IRF6 and the risk of facial clefts: single-marker and haplotype-based analyses in a population-based case-control study of facial clefts in Norway. Genet Epidemiol. 2008;32(5):413–24. doi:10.1002/gepi.20314.
70. Jugessur A, Farlie PG, Kilpatrick N. The genetics of isolated orofacial clefts: from genotypes to subphenotypes. Oral Dis. 2009;15(7):437–53. doi:10.1111/j.1601-0825.2009.01577.x.
71. Kalay E, Sezgin O, Chellappa V, Mutlu M, Morsy H, Kayserili H, Kreiger E, Cansu A, Toraman B, Abdalla EM, Aslan Y, Pillai S, Akarsu NA. Mutations in RIPK4 cause the autosomal-recessive form of popliteal pterygium syndrome. Am J Hum Genet. 2012;90(1):76–85. doi:10.1016/j.ajhg.2011.11.014.
72. Koillinen H, Wong FK, Rautio J, Ollikainen V, Karsten A, Larson O, Teh BT, Huggare J, Lahermo P, Larsson C, Kere J. Mapping of the second locus for the Van der Woude syndrome to chromosome 1p34. Eur J Hum Genet. 2001;9(10):747–52. doi:10.1038/sj.ejhg.5200713.
73. Kondo S, Schutte BC, Richardson RJ, Bjork BC, Knight AS, Watanabe Y, Howard E, de Lima RL, Daack-Hirsch S, Sander A, McDonald-McGinn DM, Zackai EH, Lammer EJ, Aylsworth AS, Ardinger HH, Lidral AC, Pober BR, Moreno L, Arcos-Burgos M, Valencia C, Houdayer C, Bahuau M, Moretti-Ferreira D, Richieri-Costa A, Dixon MJ, Murray JC. Mutations in IRF6 cause Van der Woude and popliteal pterygium syndromes. Nat Genet. 2002;32(2):285–9.
74. Krantz ID, McCallum J, DeScipio C, Kaur M, Gillis LA, Yaeger D, Jukofsky L, Wasserman N, Bottani A, Morris CA, Nowaczyk MJ, Toriello H, Bamshad MJ, Carey JC, Rappaport E, Kawauchi S, Lander AD, Calof AL, Li HH, Devoto M, Jackson LG. Cornelia de Lange syndrome is caused by mutations in NIPBL, the human homolog of Drosophila melanogaster Nipped-B. Nat Genet. 2004;36(6):631–5. doi:10.1038/ng1364.
75. Lander E, Kruglyak L. Genetic dissection of complex traits: guidelines for interpreting and reporting linkage results. Nat Genet. 1995;11(3):241–7. doi:10.1038/ng1195-241.
76. Lander ES, Linton LM, Birren B, Nusbaum C, Zody MC, Baldwin J, Devon K, Dewar K, Doyle M, FitzHugh W, Funke R, Gage D, Harris K, Heaford A, Howland J, Kann L, Lehoczky J, LeVine R, McEwan P, McKernan K, Meldrim J, Mesirov JP, Miranda C, Morris W, Naylor J, Raymond C, Rosetti M, Santos R, Sheridan A, Sougnez C, Stange-Thomann N, Stojanovic N, Subramanian A, Wyman D, Rogers J, Sulston J, Ainscough R, Beck S, Bentley D, Burton J, Clee C, Carter N, Coulson A, Deadman R, Deloukas P, Dunham A, Dunham I, Durbin R, French L, Grafham D, Gregory S, Hubbard T, Humphray S, Hunt A, Jones M, Lloyd C, McMurray A, Matthews L, Mercer S, Milne S, Mullikin JC, Mungall A, Plumb R, Ross M, Shownkeen R, Sims S, Waterston RH, Wilson RK, Hillier LW, McPherson JD, Marra MA, Mardis ER, Fulton LA, Chinwalla AT, Pepin KH, Gish WR, Chissoe SL, Wendl MC, Delehaunty KD, Miner TL, Delehaunty A, Kramer JB, Cook LL, Fulton RS, Johnson DL, Minx PJ, Clifton SW, Hawkins T, Branscomb E, Predki P, Richardson P, Wenning S, Slezak T, Doggett N, Cheng JF, Olsen A, Lucas S, Elkin C, Uberbacher E, Frazier M, Gibbs RA, Muzny DM, Scherer SE, Bouck JB, Sodergren EJ, Worley KC, Rives CM, Gorrell JH, Metzker ML, Naylor SL, Kucherlapati RS, Nelson DL, Weinstock GM, Sakaki Y, Fujiyama A, Hattori M, Yada T, Toyoda A, Itoh T, Kawagoe C, Watanabe H, Totoki Y, Taylor T, Weissenbach J, Heilig R, Saurin W, Artiguenave F, Brottier P, Bruls T, Pelletier E, Robert C, Wincker P, Smith DR, Doucette-Stamm L, Rubenfield M, Weinstock K, Lee HM, Dubois J, Rosenthal A, Platzer M, Nyakatura G, Taudien S, Rump A, Yang H, Yu J, Wang J, Huang G, Gu J, Hood L, Rowen L, Madan A, Qin S, Davis RW, Federspiel NA, Abola AP, Proctor MJ, Myers RM, Schmutz J, Dickson M, Grimwood J, Cox DR, Olson MV, Kaul R, Raymond C, Shimizu N, Kawasaki K, Minoshima S, Evans GA, Athanasiou M, Schultz R, Roe BA, Chen F, Pan H, Ramser J, Lehrach H, Reinhardt R, McCombie WR, de la Bastide M, Dedhia N, Blocker H, Hornischer K, Nordsiek G, Agarwala R, Aravind L, Bailey JA, Bateman A, Batzoglou S, Birney E, Bork P, Brown DG, Burge CB, Cerutti L, Chen HC, Church D, Clamp M, Copley RR, Doerks T, Eddy SR, Eichler EE, Furey TS, Galagan J, Gilbert JG, Harmon C, Hayashizaki Y, Haussler D, Hermjakob H, Hokamp K, Jang W, Johnson LS, Jones TA, Kasif S, Kaspryzk A, Kennedy S, Kent WJ, Kitts P, Koonin EV, Korf I, Kulp D,

Lancet D, Lowe TM, McLysaght A, Mikkelsen T, Moran JV, Mulder N, Pollara VJ, Ponting CP, Schuler G, Schultz J, Slater G, Smit AF, Stupka E, Szustakowski J, Thierry-Mieg D, Thierry-Mieg J, Wagner L, Wallis J, Wheeler R, Williams A, Wolf YI, Wolfe KH, Yang SP, Yeh RF, Collins F, Guyer MS, Peterson J, Felsenfeld A, Wetterstrand KA, Patrinos A, Morgan MJ, de Jong P, Catanese JJ, Osoegawa K, Shizuya H, Choi S, Chen YJ, International Human Genome Sequencing C. Initial sequencing and analysis of the human genome. Nature. 2001;409(6822):860–921. doi:10.1038/35057062.

77. Laumonnier F, Holbert S, Ronce N, Faravelli F, Lenzner S, Schwartz CE, Lespinasse J, Van Esch H, Lacombe D, Goizet C, Phan-Dinh Tuy F, van Bokhoven H, Fryns JP, Chelly J, Ropers HH, Moraine C, Hamel BC, Briault S. Mutations in PHF8 are associated with X linked mental retardation and cleft lip/cleft palate. J Med Genet. 2005;42(10):780–6. doi:10.1136/jmg.2004.029439.
78. Lederer D, Grisart B, Digilio MC, Benoit V, Crespin M, Ghariani SC, Maystadt I, Dallapiccola B, Verellen-Dumoulin C. Deletion of KDM6A, a histone demethylase interacting with MLL2, in three patients with Kabuki syndrome. Am J Hum Genet. 2012;90(1):119–24. doi:10.1016/j.ajhg.2011.11.021.
79. Leoyklang P, Siriwan P, Shotelersuk V. A mutation of the p63 gene in non-syndromic cleft lip. J Med Genet. 2006;43(6):e28. doi:43/6/e28 [pii] 10.1136/jmg.2005.036442.
80. Leslie E, Murray J. Evaluating rare coding variants as contributing causes to non-syndromic cleft lip and palate. Clin Genet. 2013. doi:10.1111/cge.12018.
81. Leslie EJ, Mansilla MA, Biggs LC, Schuette K, Bullard S, Cooper M, Dunnwald M, Lidral AC, Marazita ML, Beaty TH, Murray JC. Expression and mutation analyses implicate ARHGAP29 as the etiologic gene for the cleft lip with or without cleft palate locus identified by genome-wide association on chromosome 1p22. Birth Defects Res A Clin Mol Teratol. 2012a. doi:10.1002/bdra.23076.
82. Leslie EJ, Standley J, Compton J, Bale S, Schutte BC, Murray JC. Comparative analysis of IRF6 variants in families with Van der Woude syndrome and popliteal pterygium syndrome using public whole-exome databases. Genet Med Off J Am Coll Med Genet. 2012b. doi:10.1038/gim.2012.141.
83. Letra A, Menezes R, Govil M, Fonseca RF, McHenry T, Granjeiro JM, Castilla EE, Orioli IM, Marazita ML, Vieira AR. Follow-up association studies of chromosome region 9q and nonsyndromic cleft lip/palate. Am J Med Genet A. 2010;152A(7):1701–10. doi:10.1002/ajmg.a.33482.
84. Loeys BL, Chen J, Neptune ER, Judge DP, Podowski M, Holm T, Meyers J, Leitch CC, Katsanis N, Sharifi N, Xu FL, Myers LA, Spevak PJ, Cameron DE, De Backer J, Hellemans J, Chen Y, Davis EC, Webb CL, Kress W, Coucke P, Rifkin DB, De Paepe AM, Dietz HC. A syndrome of altered cardiovascular, craniofacial, neurocognitive and skeletal development caused by mutations in TGFBR1 or TGFBR2. Nat Genet. 2005;37(3):275–81. doi:10.1038/ng1511.
85. Ludwig KU, Mangold E, Herms S, Nowak S, Reutter H, Paul A, Becker J, Herberz R, Alchawa T, Nasser E, Bohmer AC, Mattheisen M, Alblas MA, Barth S, Kluck N, Lauster C, Braumann B, Reich RH, Hemprich A, Potzsch S, Blaumeiser B, Daratsianos N, Kreusch T, Murray JC, Marazita ML, Ruczinski I, Scott AF, Beaty TH, Kramer FJ, Wienker TF, Steegers-Theunissen RP, Rubini M, Mossey PA, Hoffmann P, Lange C, Cichon S, Propping P, Knapp M, Nothen MM. Genome-wide meta-analyses of nonsyndromic cleft lip with or without cleft palate identify six new risk loci. Nat Genet. 2012. doi:10.1038/ng.2360.
86. Ludwig KU, Bohmer AC, Rubini M, Mossey PA, Herms S, Nowak S, Reutter H, Alblas MA, Lippke B, Barth S, Paredes-Zenteno M, Munoz-Jimenez SG, Ortiz-Lopez R, Kreusch T, Hemprich A, Martini M, Braumann B, Jager A, Potzsch B, Molloy A, Peterlin B, Hoffmann P, Nothen MM, Rojas-Martinez A, Knapp M, Steegers-Theunissen RP, Mangold E. Strong association of variants around FOXE1 and orofacial clefting. J Dent Res. 2014;93(4):376–81. doi:10.1177/0022034514523987.
87. Mangold E, Reutter H, Birnbaum S, Walier M, Mattheisen M, Henschke H, Lauster C, Schmidt G, Schiefke F, Reich RH, Scheer M, Hemprich A, Martini M, Braumann B, Krimmel

M, Opitz C, Lenz JH, Kramer FJ, Wienker TF, Nothen MM, Diaz Lacava A. Genome-wide linkage scan of nonsyndromic orofacial clefting in 91 families of central European origin. Am J Med Genet A. 2009;149A(12):2680–94. doi:10.1002/ajmg.a.33136.
88. Mangold E, Ludwig KU, Birnbaum S, Baluardo C, Ferrian M, Herms S, Reutter H, de Assis NA, Chawa TA, Mattheisen M, Steffens M, Barth S, Kluck N, Paul A, Becker J, Lauster C, Schmidt G, Braumann B, Scheer M, Reich RH, Hemprich A, Potzsch S, Blaumeiser B, Moebus S, Krawczak M, Schreiber S, Meitinger T, Wichmann HE, Steegers-Theunissen RP, Kramer FJ, Cichon S, Propping P, Wienker TF, Knapp M, Rubini M, Mossey PA, Hoffmann P, Nothen MM. Genome-wide association study identifies two susceptibility loci for nonsyndromic cleft lip with or without cleft palate. Nat Genet. 2010;42(1):24–6. doi:10.1038/ng.506.
89. Manolio TA, Collins FS, Cox NJ, Goldstein DB, Hindorff LA, Hunter DJ, McCarthy MI, Ramos EM, Cardon LR, Chakravarti A, Cho JH, Guttmacher AE, Kong A, Kruglyak L, Mardis E, Rotimi CN, Slatkin M, Valle D, Whittemore AS, Boehnke M, Clark AG, Eichler EE, Gibson G, Haines JL, Mackay TF, McCarroll SA, Visscher PM. Finding the missing heritability of complex diseases. Nature. 2009;461(7265):747–53. doi:nature08494 [pii] 10.1038/nature08494.
90. Marazita ML. Segregation analyses. In: Wyszynski DF, editor. Cleft lip and palate: from origin to treatment. New York: Oxford University Press; 2002. pp. 222–233.
91. Marazita ML. The evolution of human genetic studies of cleft lip and cleft palate. Annu Rev Genomics Hum Genet. 2012. doi:10.1146/annurev-genom-090711-163729.
92. Marazita ML, Spence MA, Melnick M. Genetic analysis of cleft lip with or without cleft palate in Danish kindreds. Am J Med Genet. 1984;19(1):9–18.
93. Marazita ML, Hu DN, Spence MA, Liu YE, Melnick M. Cleft lip with or without cleft palate in Shanghai, China: evidence for an autosomal major locus. Am J Hum Genet. 1992;51(3):648–53.
94. Marazita ML, Field LL, Cooper ME, Tobias R, Maher BS, Peanchitlertkajorn S, Liu YE. Genome scan for loci involved in cleft lip with or without cleft palate, in Chinese multiplex families. Am J Hum Genet. 2002;71(2):349–64.
95. Marazita ML, Murray JC, Lidral AC, Arcos-Burgos M, Cooper ME, Goldstein T, Maher BS, Daack-Hirsch S, Schultz R, Mansilla MA, Field LL, Liu YE, Prescott N, Malcolm S, Winter R, Ray A, Moreno L, Valencia C, Neiswanger K, Wyszynski DF, Bailey-Wilson JE, Albacha-Hejazi H, Beaty TH, McIntosh I, Hetmanski JB, Tuncbilek G, Edwards M, Harkin L, Scott R, Roddick LG. Meta-analysis of 13 genome scans reveals multiple cleft lip/palate genes with novel loci on 9q21 and 2q32–35. Am J Hum Genet. 2004;75(2):161–73.
96. Marazita ML, Lidral AC, Murray JC, Field LL, Maher BS, Goldstein McHenry T, Cooper ME, Govil M, Daack-Hirsch S, Riley B, Jugessur A, Felix T, Morene L, Mansilla MA, Vieira AR, Doheny K, Pugh E, Valencia-Ramirez C, Arcos-Burgos M. Genome scan, -mapping, and candidate gene analysis of non-syndromic cleft lip with or without cleft palate reveals phenotype-specific differences in linkage and association results. Hum Hered. 2009;68(3):151–70. doi:000224636 [pii] 10.1159/000224636.
97. Martin RA, Hunter V, Neufeld-Kaiser W, Flodman P, Spence MA, Furnas D, Martin KA. Ultrasonographic detection of orbicularis oris defects in first degree relatives of isolated cleft lip patients. Am J Med Genet. 2000;90(2):155–61.
98. McGrath JA, Duijf PH, Doetsch V, Irvine AD, de Waal R, Vanmolkot KR, Wessagowit V, Kelly A, Atherton DJ, Griffiths WA, Orlow SJ, van Haeringen A, Ausems MG, Yang A, McKeon F, Bamshad MA, Brunner HG, Hamel BC, van Bokhoven H. Hay-Wells syndrome is caused by heterozygous missense mutations in the SAM domain of p63. Hum Mol Genet. 2001;10(3):221–9.
99. Melnick M. Cleft lip and palate etiology and its meaning in early 20th century England: Galton/Pearson vs. Bateson; polygenically poor protoplasm vs. Mendelism. J Craniofac Genet Dev Biol. 1997;17(2):65–79.

100. Menezes R, Marazita ML, Goldstein McHenry T, Cooper ME, Bardi K, Brandon C, Letra A, Martin RA, Vieira AR. AXIS inhibition protein 2, orofacial clefts and a family history of cancer. J Am Dent Assoc. 2009;140(1):80–4. doi:140/1/80[pii].
101. Miller SF, Weinberg SM, Nidey NL, Defay DK, Marazita ML, Wehby GL, Moreno Uribe LM. Exploratory genotype-phenotype correlations of facial form and asymmetry in unaffected relatives of children with non-syndromic cleft lip and/or palate. J Anat. 2014;224(6):688–709. doi:10.1111/joa.12182.
102. Milunsky JM, Maher TA, Zhao G, Roberts AE, Stalker HJ, Zori RT, Burch MN, Clemens M, Mulliken JB, Smith R, Lin AE. TFAP2A mutations result in branchio-oculo-facial syndrome. Am J Hum Genet. 2008;82(5):1171–7. doi:10.1016/j.ajhg.2008.03.005.
103. Mitchell K, O'Sullivan J, Missero C, Blair E, Richardson R, Anderson B, Antonini D, Murray JC, Shanske AL, Schutte BC, Romano RA, Sinha S, Bhaskar SS, Black GC, Dixon J, Dixon MJ. Exome sequence identifies RIPK4 as the Bartsocas-Papas syndrome locus. Am J Hum Genet. 2012;90(1):69–75. doi:10.1016/j.ajhg.2011.11.013.
104. Moreno LM, Mansilla MA, Bullard SA, Cooper ME, Busch TD, Machida J, Johnson MK, Brauer D, Krahn K, Daack-Hirsch S, L'Heureux J, Valencia-Ramirez C, Rivera D, Lopez AM, Moreno MA, Hing A, Lammer EJ, Jones M, Christensen K, Lie RT, Jugessur A, Wilcox AJ, Chines P, Pugh E, Doheny K, Arcos-Burgos M, Marazita ML, Murray JC, Lidral AC. FOXE1 association with both isolated cleft lip with or without cleft palate, and isolated cleft palate. Hum Mol Genet. 2009;18(24):4879–96. doi:ddp444 [pii]10.1093/hmg/ddp444.
105. Morgan NV, Brueton LA, Cox P, Greally MT, Tolmie J, Pasha S, Aligianis IA, van Bokhoven H, Marton T, Al-Gazali L, Morton JE, Oley C, Johnson CA, Trembath RC, Brunner HG, Maher ER. Mutations in the embryonal subunit of the acetylcholine receptor (CHRNG) cause lethal and Escobar variants of multiple pterygium syndrome. Am J Hum Genet. 2006;79(2):390–5. doi:10.1086/506256.
106. Mossey P. Epidemiology underpinning research in the aetiology of orofacial clefts. Orthod Craniofac Res. 2007;10(3):114–20. doi:OCR398 [pii] 10.1111/j.1601-6343.2007.00398.x.
107. Mossey PA, Little J, Munger RG, Dixon MJ, Shaw WC. Cleft lip and palate. Lancet. 2009;374(9703):1773–85. doi:10.1016/S0140-6736(09)60695-4.
108. Munger RG, Tamura T, Johnston KE, Feldkamp ML, Pfister R, Carey JC. Plasma zinc concentrations of mothers and the risk of oral clefts in their children in Utah. Birth Defects Res A Clin Mol Teratol. 2009;85(2):151–5. doi:10.1002/bdra.20516.
109. Murray JC. Gene/environment causes of cleft lip and/or palate. Clin Genet. 2002;61(4): 248–56.
110. Nasser E, Mangold E, Tradowsky DC, Fier H, Becker J, Boehmer AC, Herberz R, Fricker N, Barth S, Wahle P, Nowak S, Reutter H, Reich RH, Lauster C, Braumann B, Kreusch T, Hemprich A, Potzsch B, Hoffmann P, Kramer FJ, Knapp M, Lange C, Nothen MM, Ludwig KU. Resequencing of VAX1 in patients with nonsyndromic cleft lip with or without cleft palate. Birth Defects Res A Clin Mol Teratol. 2012;94(11):925–33. doi:10.1002/bdra.23078.
111. Neiswanger K, Weinberg SM, Rogers CR, Brandon CA, Cooper ME, Bardi KM, Deleyiannis FW, Resick JM, Bowen A, Mooney MP, de Salamanca JE, Gonzalez B, Maher BS, Martin RA, Marazita ML. Orbicularis oris muscle defects as an expanded phenotypic feature in nonsyndromic cleft lip with or without cleft palate. Am J Med Genet A. 2007;143(11):1143–9.
112. Neiswanger K, Chirigos KW, Klotz CM, Cooper ME, Bardi KM, Brandon CA, Weinberg SM, Vieira AR, Martin RA, Czeizel AE, Castilla EE, Poletta FA, Marazita ML. Whorl patterns on the lower lip are associated with nonsyndromic cleft lip with or without cleft palate. Am J Med Genet A. 2009;149A(12):2673–9. doi:10.1002/ajmg.a.33089.
113. Ng D, Thakker N, Corcoran CM, Donnai D, Perveen R, Schneider A, Hadley DW, Tifft C, Zhang L, Wilkie AO, van der Smagt JJ, Gorlin RJ, Burgess SM, Bardwell VJ, Black GC, Biesecker LG. Oculofaciocardiodental and Lenz microphthalmia syndromes result from distinct classes of mutations in BCOR. Nat Genet. 2004;36(4):411–6. doi:10.1038/ng1321.

114. Ng SB, Bigham AW, Buckingham KJ, Hannibal MC, McMillin MJ, Gildersleeve HI, Beck AE, Tabor HK, Cooper GM, Mefford HC, Lee C, Turner EH, Smith JD, Rieder MJ, Yoshiura K, Matsumoto N, Ohta T, Niikawa N, Nickerson DA, Bamshad MJ, Shendure J. Exome sequencing identifies MLL2 mutations as a cause of Kabuki syndrome. Nat Genet. 2010a;42(9):790–3.
115. Ng SB, Buckingham KJ, Lee C, Bigham AW, Tabor HK, Dent KM, Huff CD, Shannon PT, Jabs EW, Nickerson DA, Shendure J, Bamshad MJ. Exome sequencing identifies the cause of a mendelian disorder. Nat Genet. 2010b;42(1):30–5. doi:ng.499 [pii] 10.1038/ng.499.
116. Niemann S, Zhao C, Pascu F, Stahl U, Aulepp U, Niswander L, Weber JL, Muller U. Homozygous WNT3 mutation causes tetra-amelia in a large consanguineous family. Am J Hum Genet. 2004;74(3):558–63. doi:10.1086/382196.
117. Nikopensius T, Kempa I, Ambrozaityte L, Jagomagi T, Saag M, Matuleviciene A, Utkus A, Krjutskov K, Tammekivi V, Piekuse L, Akota I, Barkane B, Krumina A, Klovins J, Lace B, Kucinskas V, Metspalu A. Variation in FGF1, FOXE1, and TIMP2 genes is associated with nonsyndromic cleft lip with or without cleft palate. Birth Defects Res A Clin Mol Teratol. 2011;91(4):218–25. doi:10.1002/bdra.20791.
118. Nopoulos P, Berg S, Canady J, Richman L, Van Demark D, Andreasen NC. Abnormal brain morphology in patients with isolated cleft lip, cleft palate, or both: a preliminary analysis. Cleft Palate Craniofac J. 2000;37(5):441–6.
119. Nopoulos P, Berg S, VanDemark D, Richman L, Canady J, Andreasen NC. Increased incidence of a midline brain anomaly in patients with nonsyndromic clefts of the lip and/or palate. J Neuroimaging. 2001;11(4):418–24.
120. Nopoulos P, Berg S, Canady J, Richman L, Van Demark D, Andreasen NC. Structural brain abnormalities in adult males with clefts of the lip and/or palate. Genet Med. 2002;4(1):1–9.
121. Nopoulos P, Langbehn DR, Canady J, Magnotta V, Richman L. Abnormal brain structure in children with isolated clefts of the lip or palate. Arch Pediatr Adolesc Med. 2007;161(8):753–8. doi:161/8/753 [pii] 10.1001/archpedi.161.8.753.
122. Packham EA, Brook JD. T-box genes in human disorders. Hum Mol Genet. 2003;12(Spec No. 1):R37–44.
123. Peyrard-Janvid M, Leslie EJ, Kousa YA, Smith TL, Dunnwald M, Magnusson M, Lentz BA, Unneberg P, Fransson I, Koillinen HK, Rautio J, Pegelow M, Karsten A, Basel-Vanagaite L, Gordon W, Andersen B, Svensson T, Murray JC, Cornell RA, Kere J, Schutte BC. Dominant mutations in GRHL3 cause Van der Woude Syndrome and disrupt oral periderm development. Am J Hum Genet. 2014;94(1):23–32. doi:10.1016/j.ajhg.2013.11.009.
124. Porter FD. Cholesterol precursors and facial clefting. J Clin Invest. 2006;116(9):2322–5. doi:10.1172/JCI29872.
125. Prescott NJ, Lees MM, Winter RM, Malcolm S. Identification of susceptibility loci for nonsyndromic cleft lip with or without cleft palate in a two stage genome scan of affected sib-pairs. Hum Genet. 2000;106(3):345–50.
126. Quaderi NA, Schweiger S, Gaudenz K, Franco B, Rugarli EI, Berger W, Feldman GJ, Volta M, Andolfi G, Gilgenkrantz S, Marion RW, Hennekam RC, Opitz JM, Muenke M, Ropers HH, Ballabio A. Opitz G/BBB syndrome, a defect of midline development, is due to mutations in a new RING finger gene on Xp22. Nat Genet. 1997;17(3):285–91. doi:10.1038/ng1197-285.
127. Rahimov F, Marazita ML, Visel A, Cooper ME, Hitchler MJ, Rubini M, Domann FE, Govil M, Christensen K, Bille C, Melbye M, Jugessur A, Lie RT, Wilcox AJ, Fitzpatrick DR, Green ED, Mossey PA, Little J, Steegers-Theunissen RP, Pennacchio LA, Schutte BC, Murray JC. Disruption of an AP-2alpha binding site in an IRF6 enhancer is associated with cleft lip. Nat Genet. 2008;40(11):1341–7. doi:ng.242 [pii] 10.1038/ng.242.
128. Rahimov F, Jugessur A, Murray JC. Genetics of nonsyndromic orofacial clefts. Cleft Palate Craniofac J Off Publ Am Cleft Palate Craniofac Assoc. 2012;49(1):73–91. doi:10.1597/10-178.
129. Reardon W, Winter RM, Rutland P, Pulleyn LJ, Jones BM, Malcolm S. Mutations in the fibroblast growth factor receptor 2 gene cause Crouzon syndrome. Nat Genet. 1994;8(1):98–103. doi:10.1038/ng0994-98.

130. Riley BM, Mansilla MA, Ma J, Daack-Hirsch S, Maher BS, Raffensperger LM, Russo ET, Vieira AR, Dode C, Mohammadi M, Marazita ML, Murray JC. Impaired FGF signaling contributes to cleft lip and palate. Proc Nat Acad Sci USA. 2007;104(11):4512–7. doi:10.1073/pnas.0607956104.
131. Rischbieth H. Hare-lip and cleft palate. In: Pearson K, editor. Treasury of human inheritance, part IV. London: Dulau; 1910. pp. 79–123.
132. Robertson SP, Twigg SR, Sutherland-Smith AJ, Biancalana V, Gorlin RJ, Horn D, Kenwrick SJ, Kim CA, Morava E, Newbury-Ecob R, Orstavik KH, Quarrell OW, Schwartz CE, Shears DJ, Suri M, Kendrick-Jones J, Wilkie AO, Group OP-sDCC. Localized mutations in the gene encoding the cytoskeletal protein filamin A cause diverse malformations in humans. Nat Genet. 2003;33(4):487–91. doi:10.1038/ng1119.
133. Roessler E, Belloni E, Gaudenz K, Jay P, Berta P, Scherer SW, Tsui LC, Muenke M. Mutations in the human Sonic Hedgehog gene cause holoprosencephaly. Nat Genet. 1996;14(3):357–60. doi:10.1038/ng1196-357.
134. Roessler E, Du YZ, Mullor JL, Casas E, Allen WP, Gillessen-Kaesbach G, Roeder ER, Ming JE, Ruiz i Altaba A, Muenke M. Loss-of-function mutations in the human GLI2 gene are associated with pituitary anomalies and holoprosencephaly-like features. Proc Natl Acad Sci USA. 2003;100(23):13424–9. doi:10.1073/pnas.2235734100.
135. Rogers CR, Weinberg SM, Smith TD, Deleyiannis FW, Mooney MP, Marazita ML. Anatomical basis for apparent subepithelial cleft lip: a histological and ultrasonographic survey of the orbicularis oris muscle. Cleft Palate Craniofac J. 2008;45(5):518–24. doi:07-115 [pii] 10.1597/07-115.1.
136. Schliekelman P, Slatkin M. Multiplex relative risk and estimation of the number of loci underlying an inherited disease. Am J Hum Genet. 2002;71(6):1369–85.
137. Shi M, Wehby GL, Murray JC. Review on genetic variants and maternal smoking in the etiology of oral clefts and other birth defects. Birth Defects Res C Embryo Today. 2008;84(1):16–29.
138. Simonis N, Migeotte I, Lambert N, Perazzolo C, de Silva DC, Dimitrov B, Heinrichs C, Janssens S, Kerr B, Mortier G, Van Vliet G, Lepage P, Casimir G, Abramowicz M, Smits G, Vilain C. FGFR1 mutations cause Hartsfield syndrome, the unique association of holoprosencephaly and ectrodactyly. J Med Genet. 2013;50(9):585–92. doi:10.1136/jmedgenet-2013-101603.
139. Snead MP, Yates JR. Clinical and Molecular genetics of Stickler syndrome. J Med Genet. 1999;36(5):353–9.
140. Sproule J. Hereditary nature of hare-lip. Br Med J. 1863;1:412.
141. Suzuki K, Hu D, Bustos T, Zlotogora J, Richieri-Costa A, Helms JA, Spritz RA. Mutations of PVRL1, encoding a cell-cell adhesion molecule/herpesvirus receptor, in cleft lip/palate-ectodermal dysplasia. Nat Genet. 2000;25(4):427–30. doi:10.1038/78119.
142. Tessier P. Anatomical classification of facial, cranio-facial and latero-facial clefts. J Maxillofac Surg. 1976;4(2):69–92. doi:http://dx.doi.org/10.1016%2FS0301-0503%2876%2980013-6.
143. Thomason H, Dixon M. Craniofacial defects and cleft lip and palate. Encycl Life Sci. 2009. doi:10.1002/9780470015902.a0020915.
144. Tonkin ET, Wang TJ, Lisgo S, Bamshad MJ, Strachan T. NIPBL, encoding a homolog of fungal Scc2-type sister chromatid cohesion proteins and fly Nipped-B, is mutated in Cornelia de Lange syndrome. Nat Genet. 2004;36(6):636–41. doi:10.1038/ng1363.
145. Trew CJ. Sistens plura exempla palati deficientis. Nova Acta Phys Med Acad Caesarae Leopold Carol. 1757;1:445–7.
146. van den Boogaard MJ, Dorland M, Beemer FA, van Amstel HK. MSX1 mutation is associated with orofacial clefting and tooth agenesis in humans. Nat Genet. 2000;24(4):342–3.
147. Van Der Woude A. Fistula labii inferioris congenita and its association with cleft lip and palate. Am J Hum Genet. 1954;6(2):244–56.
148. Van Dyke DC, Goldman AS, Spielman RS, Zmijewski CM. Segregation of HLA in families with oral clefts: evidence against linkage between isolated cleft palate and HLA. Am J Med Genet. 1983;15(1):85–8. doi:10.1002/ajmg.1320150111.

149. Venza M, Visalli M, Venza I, Torino C, Tripodo B, Melita R, Teti D. Altered binding of MYF-5 to FOXE1 promoter in non-syndromic and CHARGE-associated cleft palate. J Oral Pathol Med Off Publ Int Assoc Oral Pathol Am Acad Oral Pathol. 2009;38(1):18–23. doi:10.1111/j.1600-0714.2008.00726.x.
150. Vieira AR. Genetic and environmental factors in human cleft lip and palate. Front Oral Biol. 2012;16:19–31. doi:10.1159/000337521.
151. Vieira AR, McHenry TG, Daack-Hirsch S, Murray JC, Marazita ML. A genome wide linkage scan for cleft lip and palate and dental anomalies. Am J Med Genet A. 2008;146A(11):1406–13. doi:10.1002/ajmg.a.32295.
152. Vissers LE, van Ravenswaaij CM, Admiraal R, Hurst JA, de Vries BB, Janssen IM, van der Vliet WA, Huys EH, de Jong PJ, Hamel BC, Schoenmakers EF, Brunner HG, Veltman JA, van Kessel AG. Mutations in a new member of the chromodomain gene family cause CHARGE syndrome. Nat Genet. 2004;36(9):955–7. doi:10.1038/ng1407.
153. Wagner T, Wirth J, Meyer J, Zabel B, Held M, Zimmer J, Pasantes J, Bricarelli FD, Keutel J, Hustert E, Wolf U, Tommerup N, Schempp W, Scherer G. Autosomal sex reversal and campomelic dysplasia are caused by mutations in and around the SRY-related gene SOX9. Cell. 1994;79(6):1111–20.
154. Wallis DE, Roessler E, Hehr U, Nanni L, Wiltshire T, Richieri-Costa A, Gillessen-Kaesbach G, Zackai EH, Rommens J, Muenke M. Mutations in the homeodomain of the human SIX3 gene cause holoprosencephaly. Nat Genet. 1999;22(2):196–8. doi:10.1038/9718.
155. Watanabe T, Ohishi M, Tashiro H. Population and family studies of HLA in Japanese with cleft lip and cleft palate. Cleft Palate J. 1984;21(4):293–300.
156. Weatherley-White RC, Ben S, Jin Y, Riccardi S, Arnold TD, Spritz RA. Analysis of genomewide association signals for nonsyndromic cleft lip/palate in a Kenya African Cohort. Am J Med Genet A. 2011;155A(10):2422–5.
157. Webb AJ, Thorisson GA, Brookes AJ, Consortium GP. An informatics project and online "Knowledge Centre" supporting modern genotype-to-phenotype research. Hum Mutat. 2011;32(5):543–50. doi:10.1002/humu.21469.
158. Wehby GL, Cassell CH. The impact of orofacial clefts on quality of life and healthcare use and costs. Oral Dis. 2010;16(1):3–10. doi:10.1111/j.1601-0825.2009.01588.x.
159. Wehby GL, Murray JC. Folic acid and orofacial clefts: a review of the evidence. Oral Dis. 2010;16(1):11–19. doi:10.1111/j.1601-0825.2009.01587.x.
160. Wehby GL, Castilla EE, Goco N, Rittler M, Cosentino V, Javois L, McCarthy AM, Bobashev G, Litavecz S, Mariona A, Dutra G, Lopez-Camelo JS, Orioli IM, Murray JC. Description of the methodology used in an ongoing pediatric care interventional study of children born with cleft lip and palate in South America [NCT00097149]. BMC Pediatr. 2006;6:9. doi:1471-2431-6-9 [pii] 10.1186/1471-2431-6-9.
161. Weinberg SM, Maher BS, Marazita ML. Parental craniofacial morphology in cleft lip with or without cleft palate as determined by cephalometry: a meta-analysis. Orthod Craniofac Res. 2006a;9(1):18–30.
162. Weinberg SM, Naidoo S, Govier DP, Martin RA, Kane AA, Marazita ML. Anthropometric precision and accuracy of digital three-dimensional photogrammetry: comparing the Genex and 3dMD imaging systems with one another and with direct anthropometry. J Craniofac Surg. 2006b;17(3):477–83.
163. Weinberg SM, Neiswanger K, Martin RA, Mooney MP, Kane AA, Wenger SL, Losee J, Deleyiannis F, Ma L, De Salamanca JE, Czeizel AE, Marazita ML. The Pittsburgh Oral-Facial Cleft study: expanding the cleft phenotype. Background and justification. Cleft Palate Craniofac J. 2006c;43(1):7–20.
164. Weinberg SM, Brandon CA, McHenry TH, Neiswanger K, Deleyiannis FW, de Salamanca JE, Castilla EE, Czeizel AE, Vieira AR, Marazita ML. Rethinking isolated cleft palate: evidence of occult lip defects in a subset of cases. Am J Med Genet A. 2008a;146A(13):1670–5. doi:10.1002/ajmg.a.32291.
165. Weinberg SM, Neiswanger K, Richtsmeier JT, Maher BS, Mooney MP, Siegel MI, Marazita ML. Three-dimensional morphometric analysis of craniofacial shape in the unaffected relatives of individuals with nonsyndromic orofacial clefts: a possible marker for genetic susceptibility. Am J Med Genet A. 2008b;146A(4):409–20.

166. Weinberg S, Naidoo, SD, Bardi, KM, Brandon, CA, Neiswanger, K, Resick, JM, Martin, RA, and Marazita, ML. Face shape of unaffected parents with cleft affected offspring: combining three-dimensional surface imaging and geometric morphometrics. Orthod Craniofac Res. 2009;12(4):271–81.
167. West JR, Blake CA. Fetal alcohol syndrome: an assessment of the field. Exp Biol Med. 2005;230(6):354–6.
168. Wilcox AJ, Lie RT, Solvoll K, Taylor J, McConnaughey DR, Abyholm F, Vindenes H, Vollset SE, Drevon CA. Folic acid supplements and risk of facial clefts: national population based case-control study. BMJ. 2007;334(7591):464. doi:10.1136/bmj.39079.618287.0B. (Clinical Research Edition).
169. Wilkie AO, Slaney SF, Oldridge M, Poole MD, Ashworth GJ, Hockley AD, Hayward RD, David DJ, Pulleyn LJ, Rutland P, et al. Apert syndrome results from localized mutations of FGFR2 and is allelic with Crouzon syndrome. Nat Genet. 1995;9(2):165–72. doi:10.1038/ng0295-165.
170. Zhu JL, Basso O, Hasle H, Winther JF, Olsen JH, Olsen J. Do parents of children with congenital malformations have a higher cancer risk? A nationwide study in Denmark. Br J Cancer. 2002;87(5):524–8. doi:10.1038/sj.bjc.6600488.

Caries

John R. Shaffer and Mary L. Marazita

Dental Caries: Public Health and Pathogenesis

Dental caries (tooth decay) is among the most widespread of chronic diseases worldwide, affecting both children and adults of all ages, and in some cases leading to serious co-morbidities. Though improvements in disease prevention over the last 50 years had led to a decrease in caries prevalence in many populations, dental caries continues to be a major public health problem. Disease burden is extremely skewed between wealthy and poor societies. Moreover, even within comparatively affluent societies, oral health disparities persist, with the most severe cases of disease, barriers to accessing oral health care, and negative social consequences occurring in vulnerable subgroups such as racial minorities, residents of rural areas, children, and those living in poverty. For these reasons, dental caries is a focal issue for alleviating public health disparities.

Though used somewhat interchangeably, the term dental caries technically refers to the symptoms of cariogenesis, a disease process characterized by the dissolution of the tooth surface caused by metabolic processes of micro-organisms in the surface biofilm. Under the classic Keyes model [85], illustrated in Fig. 1, disease occurs at the intersection of three key elements: a susceptible tooth surface, the presence of cariogenic bacteria in the attached biofilm, and the availability of fermentable carbohydrates, especially sucrose, to the oral microflora. Whether or not these three key elements actually lead to dental caries is profoundly influenced by moderating factors including elapsed time, saliva composition (including buffering

J. R. Shaffer (✉) · M. L. Marazita
Department of Human Genetics, Graduate School of Public Health, University of Pittsburgh, Pittsburgh, PA, USA
e-mail: jrs51@pitt.edu

M. L. Marazita
Center for Craniofacial and Dental Genetics, Department of Oral Biology, School of Dental Medicine, University of Pittsburgh, Pittsburgh, PA, USA

Department of Psychiatry and Clinical and Translational Science Institute, School of Medicine: University of Pittsburgh, Pittsburgh, PA, USA

S. T. Sonis (ed.), *Genomics, Personalized Medicine and Oral Disease*,
DOI 10.1007/978-3-319-17942-1_6

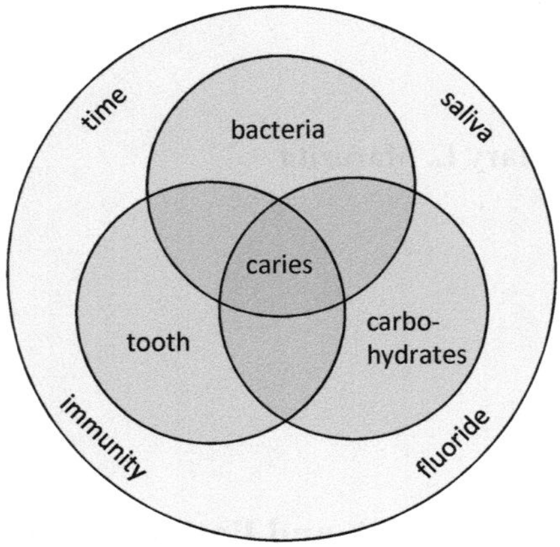

Fig. 1 Adaptation of the modified Keyes diagram from [85]. Development of dental caries requires three key elements: cariogenic micro-organisms, dietary carbohydrates, and a mineralized tooth surface. Cariogenesis is moderated by additional factors, including time, characteristics of the saliva, immune defense, and fluoride exposures. Genetic factors are speculated to influence a number of these key elements and moderating factors. These include genes influencing craniofacial, tooth, and enamel development, host immunity, dietary behaviors, oral hygienic behaviors, saliva composition and flow rate, predisposition toward tobacco use, etc

capacity) and flow, exposure to fluoride, and the host immune system. In a non-disease state, demineralization of the tooth caused by localized changes in pH by acid-producing bacteria is offset by natural processes of remineralization promoted by salivary factors and exogenous fluoride. However, an imbalance of these opposing processes causes loss of mineralization, and eventually, dental caries.

Many factors influence the balance between dissolution and re-precipitation of mineralized tissues, including diet, structure and morphology of the tooth, structure and composition of the enamel, regulators of the immune system, oral hygiene, tobacco use, medications (especially those affecting saliva flow and oral microflora), socio-economic status, access to oral health care, and behaviors reflecting cultural attitudes toward oral health. In turn, many of these factors are hypothesized or known to be influenced by host genetics. Despite detailed understanding of the fundamental mechanisms of cariogenesis, as well as some understanding of the risk factors affecting this process, prevention of dental caries remains extremely difficult. Normal preventative efforts such as exposure to fluoridated public water, daily brushing with fluoridated tooth paste, and routine professional dental cleaning is insufficient for some individuals.

The complex interplay among the numerous exogenous and host factors that affect dental caries may be amenable to personalized interventions seeking to shift the dissolution-remineralization balance away from decay. Therefore, the purpose of this chapter is to review the current literature regarding genetic and genomic studies of dental caries in order to glean insights into the potential for future applications of personalized medicine. We start by revisiting the original evidence that dental caries is heavily influenced by host genetics, and then highlight contemporary studies seeking to catalog these variants. Included in this review is a discussion of the importance and challenges related to dental caries phenotype definitions, and how recent advances in this area have influenced genetics and genomics studies. We conclude this chapter with a section discussing the possibilities for personalized medicine in the prevention and treatment of dental caries.

Family-Based Studies

Studies in families have provided valuable insight into the role of genetics in dental caries experience. Some of the earliest evidence for a genetic component to dental caries susceptibility dates back to twin studies that showed greater concordance between monozygotic than dizygotic twins for a variety of caries assessments [3, 9, 20, 24, 26, 27, 32, 53]. Likewise, surveys of unselected school children showed similarity in caries experience among siblings [39]. Altogether, these studies represented strong evidence for the familial nature of dental caries. However, familial concordance may be due to shared environmental factors in addition to genetics, and therefore familial correlations, alone, do not prove the genetic basis of dental caries. To eliminate this issue, study designs that disentangle the sharing of genetic and environment factors are needed, such as studies of twins reared apart. Toward this end, studies by Borass [8] and Conry [17] included twins raised in separate environments, and provided compelling evidence that susceptibility to dental caries is in fact genetic.

More recent studies in twins [10–12] and families [73, 74, 77, 90] further confirmed the genetic nature of caries susceptibility in larger cohorts than the early studies. Indeed, heritability estimates ranged from 30 to 70% depending on the specific caries assessment and population. Moreover, heritability studies have suggested that significant differences exist for the role of genetics across dentitions and across tooth surfaces. For example, heritability estimates are generally greater for the primary dentition than for the permanent dentition [90], and are greater for smooth tooth surfaces than for pit-and-fissure surfaces [74]. Similarly, agnostic computational approaches have suggested that both genetic and environmental contributors to dental caries may differ by tooth morphology and position in the mouth [73, 77, 78]. Overall, these studies indicate that the genetics of dental caries may be quite complex.

Other family-based studies have attempted to determine the mode of inheritance of dental caries, and what chromosomal regions may harbor caries-susceptibility genes. Through complex segregation analysis, Werneck et al. [96] determined that dental caries in Brazilian families was segregating under a major gene model—that is, the inheritance pattern of caries experience in families was consistent with a single dominant or co-dominant gene having an enormous effect on caries risk, or multiple genes with modest effects that cumulatively affect risk. Taking this hypothesis one step further, a genome-wide linkage analysis of Filipino families by Vieira et al. [89] sought to locate the chromosomal regions where caries loci may be harbored.

Both multipoint parametric (under dominant and recessive inheritance models) and non-parametric linkage analyses were performed. While no chromosomal region was implicated at the traditional level of linkage deemed genome-wide significant, multiple suggestive regions of linkage were identified, including chromosomes 5q13, 14q11, and X127 for low caries susceptibility and 13q31 and 14q24 for high caries susceptibility. These chromosomal regions may serve as high-priority targets for fine-mapping studies, including genetic association studies.

Conclusion The cumulative evidence from family studies amassed over nearly nine decades has overwhelmingly supported the notion that common variation in dental caries experience is familial and in large part due to genetic factors. And, while these studies pointed to the important role of genetics in susceptibility to dental caries, they did not determine the specific genes involved. Instead, they set the stage for genetic and genomic studies of dental caries, made possible through recent technological advances developed under the Human Genome Project. Such studies seek to identify, catalog, and understand the complete set of dental caries genes, and the mechanisms through which they influence disease.

Genetic Association and Bioinformatics Studies

Candidate Gene Studies The search for specific genetic polymorphisms that explain common variation in dental caries experience has followed two complementary study designs: candidate gene studies and genome-wide association studies (GWAS). Candidate gene studies require prior knowledge of which genes may be of interest, often based on the extensive literature of tooth development in animal models (not reviewed here). Candidate gene studies facilitate hypothesis-driven research into the role of genetic variation in susceptibility to dental caries. In contrast, GWAS and other agnostic genome-wide approaches including genome-wide meta-analysis and gene set-based analysis, are "hypothesis-generating" study designs, the results of which are used to nominate novel caries genes for follow-up in hypothesis-driven investigations. Both candidate gene and genome-wide approaches have been successful in identifying genetic variants associated with caries susceptibility. Here, we review the major findings of each of these approaches.

Several candidate gene studies have identified loci associated with dental caries. The most extensively-studied category of putative caries genes are the enamel matrix and related genes, which have been investigated in numerous candidate gene association studies. Overall, these studies suggest that genetic variation in genes involved in enamel formation influences dental caries experience (see Table 1), however, specific genes and/or variants have yielded inconsistent results across studies. The emerging picture is that the effect of genetic variation in enamel formation genes on caries experience may be quite complex, and may be influenced by phenotype definition, demography, and environment. Other categories of candidate genes include genes related to immunity, saliva, craniofacial and tooth development, and

Table 1 Summary of candidate gene association studies

Gene/locus	Gene name	Function(s) related to cariogenesis	Genetic association result(s)
Enamel matrix and related genes			
AMBN (4q13.3-q21)	Ameloblastin	Organization of enamel matrix and mineralization/development of tooth enamel	Associated with dental caries experience and enamel microhardness in some studies [65, 79], but not others [18, 35, 75 81]
AMELX (Xp22.31-p22.1)	Amelogenin	Organization of enamel matrix and mineralization/development of tooth enamel	Associated with dental caries experience and enamel microhardness in some studies [18, 36, 35 65, 79] but not others [18, 25, 63]
ENAM (4q13.3)	Enamelin	Organization of enamel matrix and mineralization/development of tooth enamel	Associated with dental caries experience and enamel microhardness in some studies [35, 65, 79] but not others [18, 63, 75, 81, 92]
KLK4 (19q13.41)	Kallikrein-related peptidase 4	Degradation of enamel matrix proteins	Associated with dental caries experience [92]
MMP20 (11q22.3)	Matrix metalloproteinase 20	Degradation of enamel matrix proteins	Associated with dental caries experience, and may interact with oral hygiene and ethnicity [84]
TFIP11 (22q12.1)	Tuftelin-interacting protein 11	Not fully understood; interacts with tuftelin during tooth development and may be involved in differentiation of ameloblasts or odontoblasts, or in organization or initiation of the enamel matrix	Associated with dental caries [35, 75] and with enamel microhardness after artificial formation of carious lesions [79]
TUFT1 (1q21)	Tuftelin	Not fully understood; expressed during tooth development and present in mineralized enamel	Associated with dental caries experience and enamel microhardness [18, 65, 79], and may interact with presence of *Streptococcus mutans* [81]
Immunity and host defense genes			
CD14 (5q31.1)	Cluster of differentiation 14	Detection of bacterial lipopolysaccharide and other pathogen-associated molecules	Salivary *CD14* is absent in individuals with active carious lesions [5]
DEFB1 (8p23.1)	Beta-defensin 1	Antimicrobial peptide made by neutrophils	Haplotypes are associated with dental caries experience [41, 64]
HLA-DR (6p21.3)	Human leukocyte antigen	Major histocompatibility complex genes related to immune system; involved in presenting extracellular antigens	not significantly associated with early childhood caries [1]; associated with dental caries experience in adolescents [86]; may be associated with presence of *Streptococcus mutans* [1]

Table 1 (continued)

Gene/locus	Gene name	Function(s) related to cariogenesis	Genetic association result(s)
HLA-DQ (6p21.3)	Human leukocyte antigen	Major histocompatibility complex genes related to immune system; involved in presenting extracellular antigens	Not significantly associated with early childhood caries [1]; associated with dental caries experience in adolescents [86]; may be associated with presence of *Streptococcus mutans* [1]
LPO (17q23.1)	lactoperoxidase	Salivary antibacterial peroxidase enzyme	Associated with dental caries experience in some samples, but not others [82]
LTF (3p21.31)	Lactotransferrin	Salivary antimicrobial enzyme	Associated with dental caries experience [2, 23] and *Streptococcus mutans* [23]
MASP2 (1p36.3–36.2)	Mannan-binding lectin serine peptidase 2	Bactericidal factor	Not associated with dental caries experience [63]
MBL2 (10q11.2)	Mannan-binding lectin (protein C) 2, soluble	Recognizes microorganisms and activates immune response	Associated with dental caries experience in some studies [63] but not others [98]
MUC7 (4q13.3)	Mucin 7	Salivary protein involved in clearing oral cavity of bacteria	Associated with dental caries experience in some studies [67] but not others [15]
TRAV4 (14q11)	T cell receptor alpha variable 4	Unknown	Associated with dental caries experience [13]
Saliva genes			
AQP5 (12.q13)	Aquaporin 5	Involved in saliva production	Associated with dental caries experience [92]
CA4 (17q23)	Carbonic anhydrase IV	Involved in bone resorption and saliva production	Not associated with dental caries experience [100]
PRH1 (12p13.2)	Proline-rich protein HaeIII subfamily 1	Saliva protein, protective for teeth	Associated with racial differences in dental caries experience and *Streptococcus mutans* [101]
Taste sense genes			
GNAT3 (17q21.11)	Guanine nucleotide binding protein, alpha transducing 3	May be involved in dietary preferences	Not associated with dental caries experience [95]
TAS1R2 (1p36.13)	Taste receptor, type 1, member 2	Sweet taste sense receptor	Associated with dental caries experience [44, 95]

Table 1 (continued)

Gene/locus	Gene name	Function(s) related to cariogenesis	Genetic association result(s)
TAS2R38 (7q34)	Taste receptor, type 2, member 38	Determines taste sense to glucosinolates	Associated with dental caries experience [66, 95]
Other genes			
ACTN2 (1q42-q43)	Actinin alpha 2	Cytoskeletal protein, organization of ameloblasts during tooth development	Associated with dental caries experience [82]
DSPP (4q21.3)	Dentin sialophosphoprotein	Mineralization of tooth dentin	Associated with dental caries experience [92]
EDARADD (1q42.3)	Ectodysplasin-A receptor-associated adapter protein	Death domain-containing protein involved in teeth and hair development	Associated with dental caries experience in some samples, but not others [82]
EPHA7 (6q16.1)	Ephrin type-A receptor 7	Involved in mediating development (including murine tooth development)	Associated with dental caries experience in some samples, but not others [82]
MMP2 (16q13-q21)	Matrix metalloproteinase 2	Degradation of collagen	Not associated with dental caries experience [84]
MMP9 (20q11.2-q13.1)	Matrix metalloproteinase 9	Degradation of collagen	Not associated with dental caries experience [84]
MMP13 (11q22.3)	Matrix metalloproteinase 13	Bone remodeling	Associated with dental caries experience [84]
MPPED2 (11p13)	Metallophosphoesterase domain containg 2	Unknown	Associated with dental caries experience [82]
MTR (1q43)	5-methyltetrahydro-folate-homocysteine methyltransferase	Unknown	Associated with dental caries experience in some samples, but not others [82]
SLC2A2 (3q26.1-q26.2)	Glucose transporter 2/solute carrier family 2, member 2	Glucose transporter	Associated with dental caries experience [44]
SPP1 (4q22.1)	Secreted phosphoprotein 1	Osteoclast attachment to mineralized bone matrix	Not associated with dental caries experience [92]
TIMP2 (17q25)	Metalloproteinase inhibitor 2	Involved in regulation of protease activity in extra-cellular matrix	Not associated with dental caries experience [84]

Table 1 (continued)

Gene/locus	Gene name	Function(s) related to cariogenesis	Genetic association result(s)
ZMPSTE24 (1p34.2)	Zinc metallopeptidase STE24	Unknown	Associated with dental caries experience in some samples, but not others [82]
5q12.1-q13.3	Four genes	Unknown	Follow-up of linkage analysis peak [89]; associated with dental caries experience [80]
13q31.1	Unknown evolutionarily conserved region	Unknown	Follow-up of linkage analysis peak [89]; associated with dental caries experience in some samples but not others [42]
Xq25.1–27.2	24 genes	Various/unknown	Follow-up of linkage analysis peak [89]; not associated with dental caries experience [43]

taste sense, which are also summarized in Table 1. Additional research is needed to replicate observed associations and characterize the effects of risk variants in the context of other environmental and genetic risk factors. Moreover, rigorous meta-analyses of existing candidate gene studies are needed, especially for enamel matrix genes, to statistically evaluate the accumulated evidence for or against the role of specific variants in cariogenesis.

Genome-Wide Studies Candidate gene studies are well-suited for investigating genes with known or hypothesized functions potentially related to the processes of cariogenesis. However, other *hypothesis-free* gene-mapping methods are needed to discover novel caries genes for which no *a priori* knowledge is available. Genome-wide association studies (GWAS), which typically interrogate more than a million anonymous polymorphisms across the genome for evidence of association with dental caries phenotypes, have been useful for this purpose. As a hypothesis-generating approach, the goal of GWAS usually is to identify a set of "top hits"—that is, polymorphisms showing the strongest evidence of association with the phenotype—for further investigation in follow-up studies. Given the high burden of multiple comparisons, strict control for type I error is needed, and therefore extremely small p-values are necessary to declare genome-wide significance, usually defined as $p < 5E\text{-}8$. This corresponds to the Bonferroni adjustment for 1 million tests. Note, however, that this threshold is very conservative given that fewer than 1 million *independent* tests are actually performed per GWAS scan due to the linkage-disequilibrium (LD, i.e., correlational) structure of the human genome. Under the test assumptions, any association observed at genome-wide significance is very unlikely to be a false positive. However, given the hypothesis-generating nature of GWAS, and the conservative nature of the typical Bonferroni threshold, hits falling below a specified threshold but still showing strong statistical evidence are also of great interest for developing hypotheses and designing follow-up studies. Such "suggestive" associations are assumed to include a mix of true positive and false positive hits.

The first GWAS of dental caries was performed by Shaffer et al. in 1305 white children 3–12 years of age [70]. This study focused on a dichotomous (affected vs. unaffected) caries phenotype for the primary dentition, and failed to identify any genome-wide significant associations [72]. However, several suggestive associations in or near novel genes with compelling biological relevance were observed, including a locus near three genes, *ACTN2*, *MTR*, and *EDARADD*, and additional loci near *MPPED2* and *LPO* (Table 2). These hits were tested in an independent sample of 1695 Danish children with some SNPs showing nominal statistical significance ($p < 0.05$), though no SNPs were significantly replicated after adjustment for multiple comparisons. Fine-mapping of GWAS top hits was performed in a follow-up candidate gene study of 3600 participants that re-examined the original cohort of white children, along with 11 additional independent samples of black and white children and adults [82]. For the follow-up study, continuous dental caries phenotypes were considered, and evidence of genetic association was interpreted against two benchmarks: (1) whether any samples other than the white children from the

Table 2 Summary of suggestive and significant (bolded) implicated genes from genome-wide studies

Study	Sample	Phenotype	Gene	Gene name	Function related to cariogenesis
[70]	1305 white children	binary	*ACTN2* (1q42-q43)	Actinin alpha 2	Cytoskeletal protein, organization of ameloblasts during tooth development [71]
			EDARADD (1q42.3)	Ectodysplasin-A receptor-associated adapter protein	Death domain-containing protein involved in teeth and hair development; mutations cause a disorder characterized by multiple dental, hair, and sweat gland defects [29]
			EPHA7 (6q16.1)	Ephrin type-A receptor 7	Involved in mediating development (including murine tooth development) [50]
			LPO (17q23.1)	Lactoperoxidase	Salivary antibacterial peroxidase enzyme
			MPPED2 (11p13)	Metallophosphoesterase domain containing 2	Unknown
			MTR (1q43)	5-Methyltetrahydrofolate-homocysteine methyltransferase	Unknown
			TFIP11 (22q12.1)	Tuftelin-interacting protein 11	Not fully understood; interacts with tuftelin during tooth development and may be involved in differentiation of ameloblasts or odontoblasts, or in organization or initiation of the enamel matrix
			ZMPSTE24 (1p34.2)	Zinc metallopeptidase STE24	Unknown
[91]	1483 white adults	DMFS	*CNIH1* (14q22.2)	Cornichon family AMPA receptor auxiliary protein 1	Unknown
			ISL1 (5q11.1)	ISL LIM homeobox 1	Expressed exclusively in developing tooth and regulator of tooth and jaw development in mouse [56]
			PTK2B (8p21.1)	Protein tyrosine kinase 2 beta	Mediates p38-dependent *MAPK* signaling, which is important for development and host defense [83]
			RPS6KA2 (6q27)	Ribosomal protein S6 kinase, 90 kDa, polypeptide 2	May be involved in *MAPK* signaling, which is important for development and host defense

Table 2 (continued)

Study	Sample	Phenotype	Gene	Gene name	Function related to cariogenesis
	5960 older adults	Proportion of affected surfaces, self-report	*FZD1* (7q21)	Frizzled family receptor 1	*WNT* signaling protein; the *WNT* pathway is involved in tooth development [49]
			RHOU (1q42.11-q42.3)	Ras homolog family member U	Mediator of *WNT* signaling; the *WNT* pathway is involved in tooth development
	7443	DMFS, proportion of affected surfaces, self-report	*ADAMTS3* (4q13.3)	ADAM metallopeptidase with thrombospondin type 1 motif 3	Highly expressed during tooth development in mouse [46]
			TLR2 (4q32)	Toll-like receptor 2	Immune response against cariogenesis; expressed on cell surface of odontoblasts [33, 58, 87]
[99]	1483 white adults	DMFS	*CNTN5* (11q22.1)	Contactin 5	Unknown
			COL4A2 (13q34)	Collagen, type IV, alpha 2	Unknown
[103]	1,006 white children	Smooth dfs	*AJAP1* (1p36.32)	Adherens junctions associated protein 1	Largely unknown; possible role in *MMP* activity during tooth development via interaction with basagin
			ITGAL (16p11.2)	Integrin, alpha L	Involved in leukocyte cellular adhesion and lymphocyte signaling; implicated in periodontitis [48]
			RPS6KA2 (6q27)	Ribosomal protein S6 kinase, 90 kDa, polypeptide 2	May be involved in MAPK signaling, which is important for host defense
			PLUNC family (20q11)	Palate, lung and nasal epithelium associated	*PLUNC* family genes (*BPIL1, BPIL3, BPIFA4P, BPIFA1*) are involved in respiratory, nasal, and oral defense [7, 19]

Table 2 (continued)

Study	Sample	Phenotype	Gene	Gene name	Function related to cariogenesis
	979 white children	Pit-and-fissure dfs	*KPNA4* *(3q25.33)*	Karyopherin alpha 4	Unknown
			MPPED2 (11p13)	Metallophosphoesterase domain containing 2	Unknown
[102]	1004 white adults	Smooth DMFS	*8q21.3*	Unknown	Unknown
			BCORL1 (Xq25-q26.1)	BCL6 corepressor-like 1	Unknown
			CXCR1 (2q35)	Chemokine (C-X-C motif) receptor 1	Receptor of *IL8*, involved in immune response to oral infection [34, 60]
			CXCR2 (2q35)	Chemokine (C-X-C motif) receptor 2	Receptor of *IL8*, involved in immune response to oral infection [34, 60, 88]
		Pit-and-fissure DMFS	*BCOR* (Xp11.4)	BCL6 corepressor	Mutation causes OFCD syndrome characterized by multiple dental anomalies among other developmental defects [59]
			INHBA (7p15-p13)	Inhibin, beta A	Subunit of activin and inhibin, members of *TGF-B* family, which play roles in development [54]; necessary for tooth development [14, 21, 22]
[76][a]	920 white adults	Cluster-based partial DMFS	*ABCG2* (4q22)	ATP-binding cassette, sub-family G (WHITE), member 2	Expressed in human dental pulp and developing tooth in mouse [31, 47]; increased expression in ameloblastic tumors ([45]
			AJAP1 (1p36.32)	Adherens junctions associated protein 1	Largely unknown; possible role in *MMP* activity during tooth development via interaction with basagin
			EDNRA (4q31.22)	Endothelin receptor type A	Cell signaling; knockout causes mouse craniofacial defects [69]
			IFT88 (13q12.1)	Intraflagellar transport 88 homolog	Mutations cause ectopic extra molars [62]

Table 2 (continued)

Study	Sample	Phenotype	Gene	Gene name	Function related to cariogenesis
			IL17D (13q11)	Interleukin 17D	Cytokine involved in host defense via *TLR4* [28]
			LYZL2 (10p11.23)	Lysozyme-like 2	Bacteriolytic factor
			NKX2–3 (10q24.2)	NK2 homeobox 3	Involved in tooth and salivary gland development [6]
			NR4A3 (9q22)	Nuclear receptor subfamily 4, group A, member 3	Up-regulated in dental follicle cells during osteogenic differentiation [57]
			PKD2 (4q22.1)	Polycystic kidney disease 2	Mutations cause dental loss and tooth fracture in mouse [38]
			SCPP family (4q21.3–22.1)	Secretory calcium-binding phospho-proteins	Dentin/bone sub-family of SCPPs (adjacent paralogous genes *SPP1, MEPE, IBSP, DMP1, DSPP*) are excellular matrix proteins of dentin/bone [37]
			SMAD7 (18q21.1)	SMAD family member 7	May be involved in development of tooth bud and enamel; expression induced by *TGFBR1* [97]
			TGFBR1 (9q22)	Transforming growth factor, beta receptor 1	Expressed in ameloblasts and regulates *MMP20* expression [40]
			TWSG1 (18p11.3)	Twisted gastrulation BMP signaling modulator 1	Mouse knock-out causes craniofacial and salivary gland defects [51, 55]

p-values thresholds for suggestive associations were specific to each publication. Genome-wide significance was defined as $p<5E\text{-}8$.

[a] Select suggestive genes from [76] are presented; other suggestive genes not shown in detail include: *ACAD9, ATXN7L1, CTNNA3, EXOC2, LPAR3, LPPR5, LZTS1, MYO16, NFASC, NT5DC1, PCDH9, PHTF2, SFTA1P, SPSB1, TRIB2, ZNF516,* and *ZNF827*, all of which have unknown functions related dental caries

original GWAS showed evidence of association, and (2) whether meta-analysis across all samples showed evidence of association. Significant evidence of genetic association was observed for *ACTN2* and *MPPED2*, but not for the other candidates. Moreover, potential differences between whites and blacks, and between primary and permanent dentitions, were observed. Overall, the follow-up study strengthened the hypothesis that genetic variation in *ACTN2* and *MPPED2* influences dental caries experience, but did not conclusively disprove the role of other genes nominated in the original GWAS. Additional work is needed to confirm these associations and determine the mechanisms through which they exert their effects on cariogenesis.

The first GWAS of dental caries in the permanent dentition included five independent cohorts totaling over 7000 adults [91]. Due to differences in data collection, demographics, and study design, GWAS scans of the five samples were performed separately, and results were combined via three meta-analyses: (1) comparatively younger Appalachian data sets with high-quality caries assessments, (2) comparatively older national data sets with inferior caries assessments, and (3) all five samples together. No SNP was associated with dental caries at the level of genome-wide significance, although eight suggestive loci were implicated. These included *RPS6KA2*, *PTK2B*, *CNIH*, and *ISL1* implicated in the younger Appalachian cohort, *RHOU* and *FZD1* implicated in the older national cohorts, and *ADAMTS3* and *TLR2* in all cohorts combined (Table 2). These hits included genes with plausible biological roles that may influence cariogenesis, for example, signaling cascades, tooth development, and immune response to oral pathogens. To date, no independent studies have attempted to replicate these putative caries genes.

Multi-Marker Tests and Bioinformatics Two of the five cohorts from Wang et al.'s GWAS of caries in the permanent dentition [91] were jointly re-analyzed using LD-based mapping methods that simultaneously incorporate information from multiple markers, including the LD between them, in a single test for genetic association [99]. Genome-wide analysis using these newly developed methods identified different associations than seen from the prior single-marker scans summarized above. The results included top hits in genes *CNTN5* (p-value = 2E-8 in LD-based mapping vs. p-value = 0.8 for single marker analysis), and *COL4A2* (p-value = 1E-7 in LD-based mapping vs. p-value = 0.06 for single marker analysis; Table 2). The fact that re-analysis of the GWAS data using different analytical approaches yielded different top hits is reasonable given that each type of test has different assumptions. In addition to nominating new caries genes for further investigation, this study illustrates the benefit of exploring GWAS data with multiple complementary approaches and with new methods as they become available.

Bioinformatics approaches that incorporate GWAS data along with information from public databases are another category of complementary approaches that may be extremely useful for gene discovery. Gene set-based (also known as gene enrichment) analysis examines the joint effects of multiple markers, for example, all markers within a gene, or within a group of related genes such as a gene family or known biological pathway. These methods draw information on gene boundaries and sets of related genes from genomic annotation databases. Such methods have been applied to dental caries GWAS studies. For example, Wang et al. re-analyzed data from the original GWAS of caries in the primary dentition of children [72] using four complementary gene set analysis methods [94]. They identified 13 significantly associated gene sets out of 1331 gene sets (annotated by Gene Ontology, www.geneontology.org) including those related to sphingoid metabolic processes, ubiquitin protein ligase activity, regulation of cytokine secretion, and ceramide metabolic processes. These groups of genes are involved in wide-ranging functions that may be involved in cariogenesis, though no "smoking guns" were observed. Nevertheless, these sets of genes may aid in interpreting future results, developing new hypotheses concerning the genetic contributors to dental caries, and expanding potential targets for clinical applications.

Conclusion Altogether, genetic association studies have been successful in identifying genetic polymorphisms that influence the risk of caries. Candidate gene studies have shown that genetic variation in genes with wide-ranging functions (i.e., tooth enamel, immunity, saliva, taste, etc.) influence caries, while GWAS studies have nominated new genes with plausible biological roles to add to the list of candidates. However, individual variants appear to explain only a small portion of variation in dental caries, and the cumulative effect across all implicated loci appears to explain only a fraction of the phenotypic heritability. This problem of "missing heritability", which is observed for nearly all heritable complex phenotypes in humans [52], suggests that additional pieces to the dental caries puzzle have yet to be discovered. In particular, previous genetic studies may have only detected a small fraction of the total set of polymorphisms influencing dental caries, many of which may have weak individual effects. Likewise, given the complex interplay between host, environment, and microbial flora that leads to dental caries, interactions among risk factors may be immensely important. Gene-environment interactions, gene-gene interactions, rare polymorphisms, and polymorphisms with small effects are all likely to be difficult to detect, and may benefit from strategies that seek to reduce sources of heterogeneity. One such area that may benefit from further consideration is the phenotypic definition of dental caries.

Dental Caries Phenotypes

Careful attention to disease definitions and phenotype characterization are essential for any application of personalized medicine. The most common dental caries phenotypes used in epidemiological and genetic studies are dichotomous characterizations of decay and the DMFS/T indices. Many studies, especially those of dental caries in children, dichotomize caries experience based on whether the participant has zero versus one or more teeth showing evidence of decay. Other definitions for dichotomizing caries experience, such as high and low caries experience also have been used. DMFS/T quantitative indices, which represent the count of the number **S**urfaces or **T**eeth that are **D**ecayed, **M**issing due to decay, or **F**illed (i.e., restored), are also widely used. Both dichotomous measures and DMFS/T indices are extremely useful tools for surveying disease and comparing prevalence and severity of disease among populations, and both of these phenotypes have been utilized for GWAS and candidate gene studies. However, a major flaw of both dichotomous phenotypes and DMFS/T indices for genetics and genomics studies is that these assessments are based on the observable consequence of disease rather than its pathogenesis. Moreover, dichotomous measures of caries over-simplify the variation observed in dental caries experience, whereas DMFS/T indices exhibit zero-inflated and skewed distributions that are problematic for statistical modeling. DMFS/T indices may also include measurement error because the inclusion of **M**issing due to decay and **F**illed surfaces/teeth are inherently less conclusive indicators of disease than is present **D**ecay. Additionally, these measures of dental caries fail to consider the locations of carious lesions (e.g., which specific surfaces are affected), which may be biologically informative given the non-uniform distribution of caries across the tooth surfaces of the complete dentition. Indeed, differences in the prevalence of disease across tooth surfaces has long been noted [39]. Various systems, both historical and contemporary, for defining categories of tooth surfaces with respect to cariogenesis are shown in Fig. 2.

Some tooth surfaces, such as the occlusal surfaces of the molars, exhibit much higher rates of decay than other surfaces, such as those of the mandibular incisors. Given that tooth surfaces show variation in caries prevalence, it follows that sur-

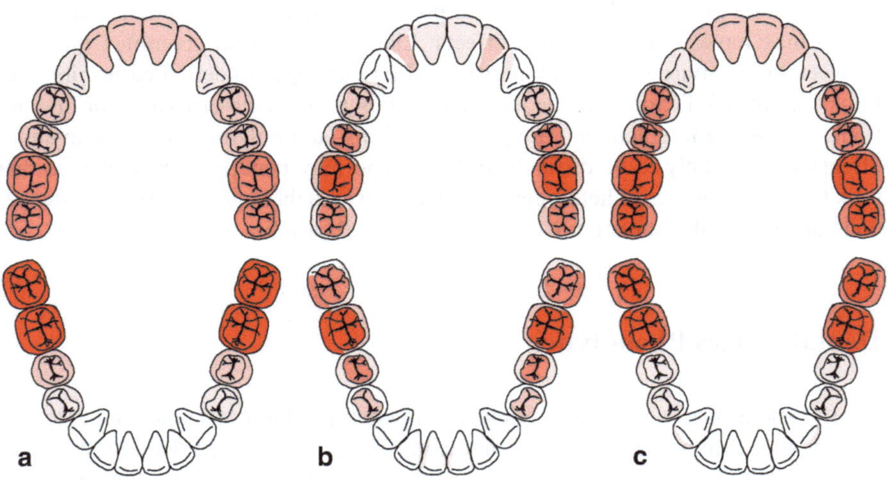

Fig. 2 Systems for categorizing teeth or tooth surfaces with respect to dental caries. **a** Klein et al. [39] mapped the trajectories of tooth-specific caries attack per 100 children across time in order to distinguish five categories of teeth based on how quickly caries develops. In order of increasing time-to-attack: 1. mandibular molars, 2. maxillary molars, 3. maxillary premolars + second mandibular molar + maxillary incisors, 4. maxillary canines + mandibular first molar, 5. mandibular incisors and canines. **b** Batchelor and Shieham (2004) [4] categorized 128 tooth surfaces into five bins based on surface-level caries prevalence. Saturation of shading indicates increasing prevalence. **c** Shaffer et al. [77] used hierarchical cluster analysis to categorized 128 tooth surfaces into five clusters based on co-occurrence of decay. Similarity was observed across all three approaches to categorizing teeth

faces may be differentially affected by caries risk factors. In this way, caries experience may be conceptualized as the cumulative result of multiple superimposed and in some cases overlapping patterns of decay, each due to a specific risk factor. Efforts to tease out the effects of risk factors may benefit from novel dental caries phenotypes that take into account the patterns of decay. For example, Shaffer et al. used principal components analysis and factor analysis—two agnostic methods of extracting signals from multidimensional data sets—to identify decay patterns across the dentition [73, 78]. Similarly, the application of cluster analysis methods to surface level caries data has been used in children [68] and adults [77] (Fig. 2c). These studies have yielded four major conclusions: (1) that total caries experience can be partitioned into stable and reproducible patterns; (2) that these patterns are associated with different risk factors (i.e., sex, race, and fluoride exposures); (3) that association with some risk factors would have been missed if only traditional caries phenotypes (i.e., DMFT/S) were considered; and (4) that heritability differs among the caries patterns. These features indicate that modeling, rather than ignoring, the spatial patterns of dental decay may be informative for genetics and genomics applications.

Another important issue related to caries phenotype definition is how to account for the progression of the carious lesion. Figure 3 illustrates this process. Demineralization of sound tooth enamel leads to the earliest detectable evidence of decay, the "white-spot" lesion (also called non-cavitated or precavitated lesion), in which the tooth surface remains intact. This type of early demineralization is sometimes reversible, at least initially. If not resolved, caries progression will continue irrevocably through the enamel, and in turn through the dentin and into the tooth pulp.

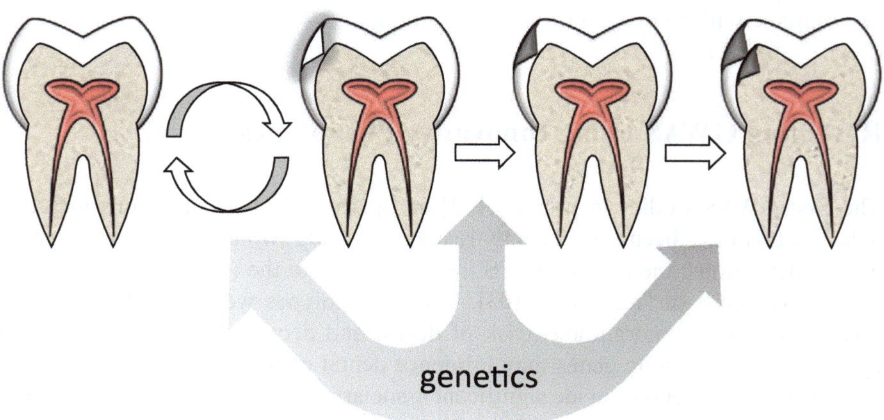

Fig. 3 From *left to right*, the progression of caries from sound enamel, to development of a repairable *white-spot* lesion, to decay of the enamel, to decay of the dentin (and eventually into the dental pulp; not shown). It is currently unknown to what extent genetic factors affect the progression of dental caries through each phase of decay. Likewise, it is unknown whether the same or different genes impact each phase. However, evidence reported by [90] suggests that genetics is especially important for the initial decay leading to *white-spot* lesions.

Whether phases of caries progression—that is, initial demineralization and/or remineralization of a white spot lesion, development of a cavitated lesion, and advancement of decay through the enamel, dentin, and pulp—are differentially affected by genetic factors is currently unknown. Wang et al. showed that the count of white-spot lesions was alone more heritable than other indictors of caries in the permanent dentition, including decayed, missing, or filled teeth [90]. Similarly, inclusion of white-spot lesions as a component of the DMFS index increased its heritability compared to the more widespread version of the DMFS that excludes white-spots [90]. This work suggests an important role of genetics for the processes of initial demineralization and/or remineralization. Therefore, white-spot lesions may be an important feature to include in caries assessments, although, unfortunately, many current genetic and genomic studies have not collected data on these. Additional work, especially longitudinal studies that assess evidence of initial demineralization, and of frank decay across multiple time points, are needed to determine the role of genetics at each stage of disease progression.

Conclusion Phenotype definition is critical for the success of genetic and genomic studies of complex traits, as well as the application of personalized medicine approaches. Ideally, the phenotype should capture the variation due to susceptibility genes, and minimize the variation due to noise and non-genetic risk factors. While relevant tissues are extremely accessible, challenges persist for designing dental caries assessments that yield biologically-informative phenotypes optimal for genetics and genomics studies. There is no consensus on the best dental caries phenotype, and each option has strengths and limitations. However, the increase in heritability obtained by including white-spot lesions, and by modeling the patterns of decay across the dentition, forms a compelling argument in favor of phenotypes that incorporate these features. In light of this concern, GWAS data sets have been re-examined with an eye to phenotype definition.

Revisiting GWAS Using Innovative Phenotypes

The first GWAS studies in primary [72] and permanent [91] dentitions included data sets that have been subsequently reanalyzed using two continuous caries phenotype definitions: the partial DMFS indices limited to the (1) pit-and-fissure and (2) smooth tooth surfaces [102, 103]. These phenotypes were intended to group surfaces of similar morphology, risk of decay, and decay progression, under the hypothesis that separate genes may influence dental caries of the two surface categories. Notably, genome-wide significant associations were observed for both dentitions.

GWAS of pit-and-fissure and smooth surface phenotypes in the primary dentition yielded a number of additional signals including the significant association with variants in *KPNA4* (and nearby genes; *p*-value = 2.0E-9) for the pit-and-fissure phenotype [103]. Suggestive associations were observed for *AJAP1*, *RPS6KA2*, *ITGAL*, and *PLUNC* family genes for the smooth surface phenotype (Table 2). These genes all have varying degrees of evidence suggesting plausible functions relevant to dental caries susceptibility. Notably, the *PLUNC* family includes adjacent genes on chromosome 20q11 that are involved in oral, nasal, and respiratory defense.

GWAS of the permanent dentition using pit-and-fissure and smooth surface caries phenotypes has also been reported [102]. One of the most exciting results for pit-and-fissure surfaces in the permanent dentition was the suggestive association (p-value = 1.8E-7) with markers near *BCOR* on chromosome Xp11. Various lines of evidence suggest that *BCOR* is involved in tooth development, and therefore may plausibly influence subsequent risk of caries. For example, mutations in *BCOR* cause oculofaciocardiodental (OFCD) syndrome [30, 59], characterized by dental anomalies including radiculomegaly, hypodontia, fusion and duplication of teeth, persistent primary teeth, delay in development and eruption of teeth, and defective enamel (in addition to other developmental defects) [61, 70]. Moreover, silencing *Bcor* expression in mouse causes dentiogenesis and delay of root development [16]. Interestingly, *BCORL1*, a gene sharing sequence similarity to *BCOR*, was among the top suggestive associations for the smooth surfaces phenotype. Other suggestive hits include *INHBA* for pit-and-fissure surfaces, and the locus harboring *CXCR1* and *CXCR2* for smooth surfaces (see Table 2).

A second re-analysis of the permanent dentition reported GWAS results of innovative phenotypes that further partitioned the dentition into five categories of surfaces—called clusters based on their derivation via cluster analysis—that covary with respect to caries status [76] (Fig. 2c). Surfaces within a given cluster were statistically more likely to share affection status, either sound or carious, with each other than with surfaces in other clusters. This analysis was performed under the hypothesis that dental caries of each cluster of teeth could be attributed to a different (but perhaps partially overlapping) set of risk factors, both genetic and non-genetic. For example, the cluster of teeth exhibiting the highest heritability of dental caries was the mandibular incisors, which have the lowest risk of caries. This observation is sensible under the hypothesis that participants exhibiting decay on these otherwise highly-resistant teeth may carry genetic predispositions. Consistent with this notion, the most significant result from GWAS scans of the five clusters was for the mandibular incisors, which showed significant association with *LYZL2* (p-value = 9E-9), a bacteriolytic factor thought to be involved in host defense. Another significant association was observed for *AJAP1* (p-value = 2E-8), a gene that was among the suggestive hits for smooth surfaces in the primary dentition in an independent sample of study participants. Several suggestive hits were also observed, some with biologically plausible roles in cariogenesis (Table 2).

Conclusion Dividing the dentition into categories of tooth surfaces, either based on *a priori* groupings (i.e., pit-and-fissure vs. smooth surfaces), or based on agnostic pattern extraction such as via cluster analysis, appears to benefit the GWAS approach. Heritability was increased compared to the dichotomous and DMFS phenotypes used in the original GWAS analyses. Moreover, results from re-analysis of data yielded genome-wide significant associations, which eluded the original studies. This suggests that carefully defining the dental caries phenotype, especially with regard to reducing heterogeneity in risk factors, may be fruitful for genetics and genomics approaches. Moreover, these phenotypes may serve as measures of disease for future applications of personalized medicine that tailor preventative and treatment options to the specific genetic and environmental liabilities of the individual.

Prospects and Challenges for Personalized Medicine

Personalized medicine is a medical model that makes greater use of patient genomic information (and potentially other 'omics data including microbiomics/meta-genomics, epigenomics, transcriptomics, proteomics, lipidomics, and metabolomics, etc.) to inform health care decisions. The goal is to utilize such large-scale molecular data sources to enhance customization of medical care to fit the patient's individual biological and environmental liabilities. Currently, major areas for personalized medicine are risk assessment and diagnostic testing, in which genomic information is used to determine the patient's risk of developing a disease, or to diagnose which disease or subtype of disease (particularly useful for cancers) a patient has developed. Such risk assessment and diagnostic testing may be extremely useful for preventative care, especially for late-onset diseases and complex diseases such as dental caries that are also influenced by environmental factors. Actions and decisions related to disease surveillance, environmental and/or behavioral modifications, and family planning may be informed by genomic-based risk assessments.

Another area of personalized medicine, pharmacogenomics, takes this medical model one step further to inform details of treatment options. Drug efficacy and dosage is affected by numerous factors, including patient biology; therefore, genomic and other 'omic information may be tremendously useful in tailoring drug prescriptions to a particular patient.

The underlying assumption of personal medicine, for both risk assessment/diagnostic testing and pharmacogenomics, is that genomic information will be available as part of a patients' medical record. Given the quickly declining cost of acquiring 'omics information, we speculate that this assumption will be met in the near future. If 'omics data becomes a routine part of the patient's medical history, the implications for oral health care are potentially huge.

Before personalized medicine for dental caries can gain a foothold, more work is needed to fully understand the genetic contributors to cariogenesis. Specifically, previously-identified risk variants will need to be validated and their effects will need to be rigorously characterized. This will require additional genetic and genomic studies of dental caries, including studies in populations of various racial backgrounds. Furthermore, comprehensive and systematic approaches that pull together results across genetic/genomic studies, and incorporate evidence from other sources such as gene expression studies and animal models, will be needed in order to derive global conclusions regarding the contributors to cariogenesis. One study, representing a first attempt at 'putting it all together', reported prioritization of dental caries candidate genes based on information from association studies, linkage analysis, gene expression studies, protein-protein interactions, and literature mining [93]. This work indicated that three categories of genes should be prioritized: genes related to cytokine networks (cell signaling), genes in the matrix metalloproteinases family (protein degradation), and genes related to *TFG-β* (development, cell proliferation, and immune system). More work in this area is absolutely necessary in order for a personalized medicine model of dental caries to be developed.

In addition to further genetic association studies, other study designs are needed to fully understand the biological contributors to dental caries, and to identify sources of the "missing heritability". For example, to date, no next-generation (i.e., high-throughput, big data-generating) sequencing studies of dental caries have been attempted. Such work may enable identification of rare variants or *de novo* mutations contributing to dental caries, which cannot easily be detected via classical genetic association studies. Likewise, epigenetic studies that interrogate the role of genomic modifiers such as DNA methylation sites, and studies of copy number variation that seek to determine the role of interpersonal differences in how many copies of a gene or genomic region are present, as opposed to the differences in the nucleotide sequence itself, may identify other types of variants that impact risk of dental caries. These areas are completely absent in the current dental caries literature.

Interactions between the host genome and exogenous environmental factors or oral microbes are also critically important in developing a complete picture of dental caries etiology. (Note, the impact of the microbiome on oral disease is an exciting and currently evolving area of research, which is discussed in a separate chapter 3) Gene-by-environment interactions, as they relate to dental caries, is currently under-studied, although a limited number of studies have addressed this issue. Slayton et al. identified a significant interaction between *TUFT1*, an enamel matrix gene, and *Streptococcus mutans* affecting dental caries in children [81] (Table 1). Similarly, in the original GWAS study of the primary dentition, Shaffer et al. investigated gene-by-fluoride interactions by stratifying children into two groups based on low vs. sufficient home water source fluoride concentration [72]. Three genes involved in gene-by-fluoride interactions were identified: *TFIP11*, an enamel matrix-related gene, which had larger effects in the low-fluoride group, and *EPHA7* and *ZMPSTE24*, plausible genes which had larger effects in the sufficient fluoride group (Table 2). Likewise, Shaffer et al. identified gene-by-fluoride interactions for variants in enamel matrix genes, TUFT1 and AMBN. In both cases, individuals with the risk allele experienced greater dental caries only if not exposed to fluoride [75]. Though very little work in this area has been reported, these results illustrate the potential role of modifying environments in how dental caries risk genes exert their effects.

Gene-by-environmental interactions are a promising area for personalized medicine, especially for environments that are amenable to modification, which can then be leveraged for preventative and/or treatment strategies. The oral environment, in particular, may be readily modified, perhaps through rinses, chewing gums, or other hygienic practices. A vision of personalized medicine of dental caries may involve manipulating aspects of the oral environment (see modified Keyes' model in Fig. 1) based on the patient's specific genetic liabilities. In addition, personalized medicine may involve customized therapies to prevent disease or its recurrence. A number of existing or theoretical treatments that have limited utility across the population as a whole, such as chemoprophylactic agents, antimicrobial peptides, vaccines, probiotics, microbial replacement therapy, and remineralization agents including fluorides and casein phosphopeptides, may nevertheless be particularly beneficial for select patients. Identification of such patients based on genomic information may enable implementation of personalized treatment and prevention strategies. A more distant vision for the future of dental caries prevention may even include gene therapy approaches, which alter the patient's genes or regulatory elements in order to minimize genetic susceptibility to dental caries. At present these therapies and strategies are all purely speculative, though they are certainly plausible given our current trajectories of caries gene-mapping efforts and 'omics data collection.

Challenges The personalized medicine model is an exciting vision, and may become feasible within the near future, even for diseases as complex as dental caries. Whether such a shift in paradigm will be possible for dental caries depends greatly on a number of developments that have yet to be fully realized. Foremost among these is the need for decreased cost and increased utilization of clinical-quality 'omics assessments, and the development of systems of data storage, sharing, and annotation to facilitate the use of these data in the clinical setting. Given the potential impact of personalized medicine, progress in this area appears all but certain. Applications related to dental caries will not be a major driving force, however, the eventual large-scale collection of patient 'omics data will likely happen, motivated by the cost-effectiveness of personalized medicine applications in other areas of healthcare. Subsequently, the data will become available for oral healthcare applications.

Other developments seem less certain. A personalized medicine model for dental caries will require extensive understanding of the biological contributors to cariogenesis, whereas our current understanding of these is limited. In fact, the clinical utility of genetic (or other 'omic-based) information may depend on the existence of inter-individual differences in risk factors of large effects; the existence of these variants is currently unknown. Perhaps in lieu of major risk variants, the accumulation of multiple risk variants of modest effect, but that operate through a common cariogenic mechanism (e.g., deficiencies in host immunity or saliva flow rate/buffering capacity) may be useful for identifying targets of personalized interventions. Currently, the utility of 'omics information for dental caries is purely speculative. Lastly, the advent of personalized medicine, in general, will require meeting the social and legal challenges stemming from patent issues, privacy rights, and systems for healthcare coverage and reimbursement. Personalized medicine, if fully realized, is certain to bring enormous changes to the healthcare enterprise. Therefore, some restructuring will be necessary to accommodate these changes. As long as personalized medicine applications are shown to be cost-effective and promote health, solutions for these social and legal issues will surely be met.

Conclusion In summary, a number of important and necessary investigations of the host genetic contribution to susceptibility to dental caries have yielded a foundation of knowledge that is crucial for personalized medicine. However, further advances in technology, science, and policy are all needed to meet the promise of personalized medicine applications for dental caries; some of these advances seem assured, whereas others are less certain. Given the potential gains for improved treatment or prevention of the most common chronic disease, worldwide, we advocate that these advances must continue to be major priorities in dental caries research.

References

1. Altun C, Guven G, Orkunoglu F, Cehreli ZC, Karaaslan A, Basak F, Akbulut E. Human leukocyte antigen class II alleles and dental caries in a child population. Pediatr Dent. 2008;30:154–9.
2. Azevedo LF, Pecharki GD, Brancher JA, Cordeiro CA Jr, Medeiros KG, Antunes AA, Arruda ES, Werneck RI, De Azevedo LR, Mazur RF, Moyses SJ, Moyses ST, Faucz FR, Trevilatto PC. Analysis of the association between lactotransferrin (LTF) gene polymorphism and dental caries. J Appl Oral Sci. 2010;18:166–70.
3. Bachrach F, Young M. A comparison of the degree of resemblance in dental characters shown in pairs of twins of identical and fraternal types. Br Dent J. 1927;48:1293–304.
4. Batchelor PA, Sheiham A. Grouping of tooth surfaces by susceptibility to caries: a study in 5–16 year-old children. BMC Oral Health. 2004;4:2.
5. Bergandi L, Defabianis P, Re F, Preti G, Aldieri E, Garetto S, Bosia A, Ghigo D. Absence of soluble CD14 in saliva of young patients with dental caries. Eur J Oral Sci. 2007;115:93–6.
6. Biben C, Wang CC, Harvey RP. NK-2 class homeobox genes and pharyngeal/oral patterning: Nkx2–3 is required for salivary gland and tooth morphogenesis. Int J Dev Biol. 2002;46:415–22.
7. Bingle CD, Craven CJ. PLUNC: a novel family of candidate host defence proteins expressed in the upper airways and nasopharynx. Hum Mol Genet. 2002;11:937–43.
8. Boraas JC, Messer LB, Till MJ. A genetic contribution to dental caries, occlusion, and morphology as demonstrated by twins reared apart. J Dent Res. 1988;67:1150–5.
9. Bordoni N, Dono R, Manfredi C, Allegrotti I. Prevalence of dental caries in twins. ASDC J Dent Child. 1973;40:440–3.
10. Bretz WA, Corby PM, Hart TC, Costa S, Coelho MQ, Weyant RJ, Robinson M, Schork NJ. Dental caries and microbial acid production in twins. Caries Res. 2005a;39:168–72.
11. Bretz WA, Corby PM, Schork NJ, Robinson MT, Coelho M, Costa S, Melo Filho MR, Weyant RJ, Hart TC. Longitudinal analysis of heritability for dental caries traits. J Dent Res. 2005b;84:1047–51.
12. Bretz WA, Corby PM, Melo MR, Coelho MQ, Costa SM, Robinson M, Schork NJ, Drewnowski A, Hart TC. Heritability estimates for dental caries and sucrose sweetness preference. Arch Oral Biol. 2006;51:1156–60.
13. Briseno-Ruiz J, Shimizu T, Deeley K, Dizak PM, Ruff TD, Faraco IM Jr, Poletta FA, Brancher JA, Pecharki GD, Kuchler EC, Tannure PN, Lips A, Vieira TC, Patir A, Koruyucu M, Mereb JC, Resick JM, Brandon CA, Letra A, Silva RM, Cooper ME, Seymen F, Costa MC, Granjeiro JM, Trevilatto PC, Orioli IM, Castilla EE, Marazita ML, Vieira AR. Role of TRAV locus in low caries experience. Hum Genet. 2013;132:1015–25.
14. Brown CW, Houston-Hawkins DE, Woodruff TK, Matzuk MM. Insertion of Inhbb into the Inhba locus rescues the Inhba-null phenotype and reveals new activin functions. Nat Genet. 2000;25:453–7.

15. Buczkowska-Radlinska J, Szyszka-Sommerfeld L, Wozniak K. Anterior tooth crowding and prevalence of dental caries in children in Szczecin, Poland. Community Dent Health. 2012;29:168–72.
16. Cai J, Kwak S, Lee JM, Kim EJ, Lee MJ, Park GH, Cho SW, Jung HS. Function analysis of mesenchymal Bcor in tooth development by using RNA interference. Cell Tissue Res. 2010;341:251–8.
17. Conry JP, Messer LB, Boraas JC, Aeppli DP, Bouchard TJ Jr. Dental caries and treatment characteristics in human twins reared apart. Arch Oral Biol. 1993;38:937–43.
18. Deeley K, Letra A, Rose EK, Brandon CA, Resick JM, Marazita ML, Vieira AR. Possible association of amelogenin to high caries experience in a Guatemalan-Mayan population. Caries Res. 2008;42:8–13.
19. Fabian TK, Hermann P, Beck A, Fejerdy P, Fabian G. Salivary defense proteins: their network and role in innate and acquired oral immunity. Int J Mol Sci. 2012;13:4295–320.
20. Fairpo CG. Total caries experience in monozygotic and like-sexed dizygotic twins of caucasoid origin aged 5 to 15 years. Arch Oral Biol. 1979;24:491–4.
21. Ferguson CA, Tucker AS, Christensen L, Lau AL, Matzuk MM, Sharpe PT. Activin is an essential early mesenchymal signal in tooth development that is required for patterning of the murine dentition. Genes Dev. 1998;12:2636–49.
22. Ferguson CA, Tucker AS, Heikinheimo K, Nomura M, Oh P, Li E, Sharpe PT. The role of effectors of the activin signalling pathway, activin receptors IIA and IIB, and Smad2, in patterning of tooth development. Development. 2001;128:4605–13.
23. Fine DH, Toruner GA, Velliyagounder K, Sampathkumar V, Godboley D, Furgang D. A lactotransferrin single nucleotide polymorphism demonstrates biological activity that can reduce susceptibility to caries. Infect Immun. 2013;81:1596–605.
24. Finn SB, Caldwell RC. Dental Caries in Twins–I. A comparison of the caries experience of monozygotic twins, dizygotic twins and unrelated children. Arch Oral Biol. 1963;8:571–85.
25. Gasse B, Grabar S, Lafont AG, Quinquis L, Opsahl Vital S, Davit-Beal T, Moulis E, Chabadel O, Hennequin M, Courson F, Droz D, Vaysse F, Laboux O, Tassery H, Al-Hashimi N, Boillot A, Carel JC, Treluyer JM, Jeanpierre M, Beldjord C, Sire JY, Chaussain C. Common SNPs of AmelogeninX (AMELX) and dental caries susceptibility. J Dent Res. 2013;92:418–24.
26. Goldberg S. The dental arches of identical twins. Dental Cosmos. 1930;72:869–81.
27. Goodman HO, Luke JE, Rosen S, Hackel E. Heritability in dental caries, certain oral microflora and salivary components. Am J Hum Genet. 1959;11:263–73.
28. Guzzo C, Ayer A, Basta S, Banfield BW, Gee K. IL-27 enhances LPS-induced proinflammatory cytokine production via upregulation of TLR4 expression and signaling in human monocytes. J Immunol. 2012;188:864–73.
29. Headon DJ, Emmal SA, Ferguson BM, Tucker AS, Justice MJ, Sharpe PT, Zonana J, Overbeek PA. Gene defect in ectodermal dysplasia implicates a death domain adapter in development. Nature. 2001;414:913–6.
30. Hilton E, Johnston J, Whalen S, Okamoto N, Hatsukawa Y, Nishio J, Kohara H, Hirano Y, Mizuno S, Torii C, Kosaki K, Manouvrier S, Boute O, Perveen R, Law C, Moore A, Fitzpatrick D, Lemke J, Fellmann F, Debray FG, Dastot-Le-Moal F, Gerard M, Martin J, Bitoun P, Goossens M, Verloes A, Schinzel A, Bartholdi D, Bardakjian T, Hay B, Jenny K, Johnston K, Lyons M, Belmont JW, Biesecker LG, Giurgea I, Black G. BCOR analysis in patients with OFCD and Lenz microphthalmia syndromes, mental retardation with ocular anomalies, and cardiac laterality defects. Eur J Hum Genet. 2009;17:1325–35.
31. Honda MJ, Nakashima F, Satomura K, Shinohara Y, Tsuchiya S, Watanabe N, Ueda M. Side population cells expressing ABCG2 in human adult dental pulp tissue. Int Endod J. 2007;40:949–58.
32. Horowitz SL, Osborne RH, Degeorge FV. Caries experience in twins. Science. 1958;128:300–1.
33. Horst OV, Horst JA, Samudrala R, Dale BA. Caries induced cytokine network in the odontoblast layer of human teeth. BMC Immunol. 2011;12:9.
34. Huang GT, Potente AP, Kim JW, Chugal N, Zhang X. Increased interleukin-8 expression in inflamed human dental pulps. Oral Surg Oral Med Oral Pathol Oral Radiol Endod. 1999;88:214–20.

35. Jeremias F et al. Genes expressed in dental enamel development are associated with molar-incisor hypomineralization. Arch Oral Biol. 2013;58(10):1434–42.
36. Kang SW, Yoon I, Lee HW, Cho J. Association between AMELX polymorphisms and dental caries in Koreans. Oral Dis. 2011;17:399–406.
37. Kawasaki K, Weiss KM. SCPP gene evolution and the dental mineralization continuum. J Dent Res. 2008;87:520–31.
38. Khonsari RH, Ohazama A, Raouf R, Kawasaki M, Kawasaki K, Porntaveetus T, Ghafoor S, Hammond P, Suttie M, Odri GA, Sandford RN, Wood JN, Sharpe PT. Multiple postnatal craniofacial anomalies are characterized by conditional loss of polycystic kidney disease 2 (Pkd2). Hum Mol Genet. 2013;22:1873–85.
39. Klein H, Palmer CE. Studies on dental caries: XII. Comparison of the caries susceptibilty of the various morphological types of permanent teeth. J Dent Res. 1941;20:203–16.
40. Klopcic B, Maass T, Meyer E, Lehr HA, Metzger D, Chambon P, Mann A, Blessing M. TGF-beta superfamily signaling is essential for tooth and hair morphogenesis and differentiation. Eur J Cell Biol. 2007;86:781–99.
41. Krasone K, Lace B, Akota I, Care R, Deeley K, Kuchler EC, Vieira AR. Genetic variation in the promoter region of beta-defensin 1 (DEFB 1) is associated with high caries experience in children born with cleft lip and palate. Acta Odontol Scand. 2014;72:235–40.
42. Kuchler EC, Deeley K, Ho B, Linkowski S, Meyer C, Noel J, Kouzbari MZ, Bezamat M, Granjeiro JM, Antunes LS, Antunes LA, De Abreu FV, Costa MC, Tannure PN, Seymen F, Koruyucu M, Patir A, Mereb JC, Poletta FA, Castilla EE, Orioli IM, Marazita ML, Vieira AR. Genetic mapping of high caries experience on human chromosome 13. BMC Med Genet. 2013;14:116.
43. Kuchler EC, Feng P, Deeley K, Fitzgerald CA, Meyer C, Gorbunov A, Bezamat M, Reis MF, Noel J, Kouzbari MZ, Granjeiro JM, Antunes LS, Antunes LA, De Abreu FV, Costa MC, Tannure PN, Seymen F, Koruyucu M, Patir A, Vieira AR. Fine mapping of locus Xq25.1–27-2 for a low caries experience phenotype. Arch Oral Biol. 2014;59:479–86.
44. Kulkarni GV, Chng T, Eny KM, Nielsen D, Wessman C, El-Sohemy A. Association of GLUT2 and TAS1R2 genotypes with risk for dental caries. Caries Res. 2013;47:219–25.
45. Kumamoto H, Ohki K. Detection of CD133, Bmi-1, and ABCG2 in ameloblastic tumors. J Oral Pathol Med. 2010;39:87–93.
46. Le Goff C, Somerville RP, Kesteloot F, Powell K, Birk DE, Colige AC, Apte SS. Regulation of procollagen amino-propeptide processing during mouse embryogenesis by specialization of homologous ADAMTS proteases: insights on collagen biosynthesis and dermatosparaxis. Development. 2006;133:1587–96.
47. Li L, Kwon HJ, Harada H, Ohshima H, Cho SW, Jung HS. Expression patterns of ABCG2, Bmi-1, Oct-3/4, and Yap in the developing mouse incisor. Gene Expr Patterns. 2011;11:163–70.
48. Lima PM, Souza PE, Costa JE, Gomez RS, Gollob KJ, Dutra WO. Aggressive and chronic periodontitis correlate with distinct cellular sources of key immunoregulatory cytokines. J Periodontol. 2011;82:86–95.
49. Liu F, Chu EY, Watt B, Zhang Y, Gallant NM, Andl T, Yang SH, Lu MM, Piccolo S, Schmidt-Ullrich R, Taketo MM, Morrisey EE, Atit R, Dlugosz AA, Millar SE. Wnt/beta-catenin signaling directs multiple stages of tooth morphogenesis. Dev Biol. 2008;313:210–24.
50. Luukko K, Loes S, Kvinnsland IH, Kettunen P. Expression of ephrin-A ligands and EphA receptors in the developing mouse tooth and its supporting tissues. Cell Tissue Res. 2005;319:143–52.
51. Mackenzie B, Wolff R, Lowe N, Billington CJ Jr, Peterson A, Schmidt B, Graf D, Mina M, Gopalakrishnan R, Petryk A. Twisted gastrulation limits apoptosis in the distal region of the mandibular arch in mice. Dev Biol. 2009;328:13–23.
52. Manolio TA, Collins FS, Cox NJ, Goldstein DB, Hindorff LA, Hunter DJ, Mccarthy MI, Ramos EM, Cardon LR, Chakravarti A, Cho JH, Guttmacher AE, Kong A, Kruglyak L, Mardis E, Rotimi CN, Slatkin M, Valle D, Whittemore AS, Boehnke M, Clark AG, Eichler EE, Gibson G, Haines JL, Mackay TF, Mccarroll SA, Visscher PM. Finding the missing heritability of complex diseases. Nature. 2009;461:747–53.
53. Mansbridge JN. Heredity and dental caries. J Dent Res. 1959;38:337–47.

54. Mather JP, Moore A, Li RH. Activins, inhibins, and follistatins: further thoughts on a growing family of regulators. Proc Soc Exp Biol Med. 1997;215:209–22.
55. Melnick M, Petryk A, Abichaker G, Witcher D, Person AD, Jaskoll T. Embryonic salivary gland dysmorphogenesis in Twisted gastrulation deficient mice. Arch Oral Biol. 2006;51:433–8.
56. Mitsiadis TA, Angeli I, James C, Lendahl U, Sharpe PT. Role of Islet1 in the patterning of murine dentition. Development. 2003;130:4451–60.
57. Morsczeck C, Schmalz G, Reichert TE, Vollner F, Saugspier M, Viale-Bouroncle S, Driemel O. Gene expression profiles of dental follicle cells before and after osteogenic differentiation in vitro. Clin Oral Investig. 2009;13:383–91.
58. Mutoh N, Tani-Ishii N, Tsukinoki K, Chieda K, Watanabe K. Expression of toll-like receptor 2 and 4 in dental pulp. J Endod. 2007;33:1183–6.
59. Ng D, Thakker N, Corcoran CM, Donnai D, Perveen R, Schneider A, Hadley DW, Tifft C, Zhang L, Wilkie AO, Van Der Smagt JJ, Gorlin RJ, Burgess SM, Bardwell VJ, Black GC, Biesecker LG. Oculofaciocardiodental and Lenz microphthalmia syndromes result from distinct classes of mutations in BCOR. Nat Genet. 2004;36:411–6.
60. Noda D, Hamachi T, Inoue K, Maeda K. Relationship between the presence of periodontopathic bacteria and the expression of chemokine receptor mRNA in inflamed gingival tissues. J Periodontal Res. 2007;42:566–71.
61. Oberoi S, Winder AE, Johnston J, Vargervik K, Slavotinek AM. Case reports of oculofaciocardiodental syndrome with unusual dental findings. Am J Med Genet A. 2005;136:275–7.
62. Ohazama A, Haycraft CJ, Seppala M, Blackburn J, Ghafoor S, Cobourne M, Martinelli DC, Fan CM, Peterkova R, Lesot H, Yoder BK, Sharpe PT. Primary cilia regulate Shh activity in the control of molar tooth number. Development. 2009;136:897–903.
63. Olszowski T, Adler G, Janiszewska-Olszowska J, Safranow K, Kaczmarczyk M. MBL2, MASP2, AMELX, and ENAM gene polymorphisms and dental caries in Polish children. Oral Dis. 2012;18:389–95.
64. Ozturk A, Famili P, Vieira AR. The antimicrobial peptide DEFB1 is associated with caries. J Dent Res. 2010;89:631–6.
65. Patir A, Seymen F, Yildirim M, Deeley K, Cooper ME, Marazita ML, Vieira AR. Enamel formation genes are associated with high caries experience in Turkish children. Caries Res. 2008;42:394–400.
66. Pidamale R, Sowmya B, Thomas A, Jose T. Genetic sensitivity to bitter taste of 6-n Propylthiouracil: A useful diagnostic aid to detect early childhood caries in pre-school children. Indian J Hum Genet. 2012;18:101–5.
67. Pol J. Association of the polymorphism of MUC7 gene encoding the low-molecular-weight mucin MG2 with susceptibility to caries. Ann Acad Med Stetin. 2011;57:85–91.
68. Psoter WJ, Pendrys DG, Morse DE, Zhang HP, Mayne ST. Caries patterns in the primary dentition: cluster analysis of a sample of 5169 Arizona children 5–59 months of age. Int J Oral Sci. 2009;1:189–95.
69. Ruest LB, Xiang X, Lim KC, Levi G, Clouthier DE. Endothelin-A receptor-dependent and -independent signaling pathways in establishing mandibular identity. Development. 2004;131:4413–23.
70. Schulze BR, Horn D, Kobelt A, Tariverdian G, Stellzig A. Rare dental abnormalities seen in oculo-facio-cardio-dental (OFCD) syndrome: three new cases and review of nine patients. Am J Med Genet. 1999;82:429–35.
71. Sehic A, Risnes S, Khan QE, Khuu C, Osmundsen H. Gene expression and dental enamel structure in developing mouse incisor. Eur J Oral Sci. 2010;118:118–30.
72. Shaffer JR, Wang X, Feingold E, Lee M, Begum F, Weeks DE, Cuenco KT, Barmada MM, Wendell SK, Crosslin DR, Laurie CC, Doheny KF, Pugh EW, Zhang Q, Feenstra B, Geller F, Boyd HA, Zhang H, Melbye M, Murray JC, Weyant RJ, Crout R, Mcneil DW, Levy SM, Slayton RL, Willing MC, Broffitt B, Vieira AR, Marazita ML. Genome-wide association scan for childhood caries implicates novel genes. J Dent Res. 2011;90:1457–62.

73. Shaffer JR, Feingold E, Wang X, Tcuenco KT, Weeks DE, Desensi RS, Polk DE, Wendell S, Weyant RJ, Crout R, Mcneil DW, Marazita ML. Heritable patterns of tooth decay in the permanent dentition: principal components and factor analyses. BMC Oral Health. 2012a;12:7.
74. Shaffer JR, Wang X, Desensi RS, Wendell S, Weyant RJ, Cuenco KT, Crout R, Mcneil DW, Marazita ML. Genetic susceptibility to dental caries on pit and fissure and smooth surfaces. Caries Res. 2012b;46:38–46.
75. Shaffer JR et al. Effects of enamel matrix genes on dental caries are moderated by fluoride exposures. Hum Genet. 2015;134(2):159–67.
76. Shaffer JR, Feingold E, Wang X, Lee M, Tcuenco K, Weeks DE, Weyant RJ, Crout R, Mcneil DW, Marazita ML. GWAS of dental caries patterns in the permanent dentition. J Dent Res. 2013a;92:38–44.
77. Shaffer JR, Feingold E, Wang X, Weeks DE, Weyant RJ, Crout R, Mcneil DW, Marazita ML. Clustering tooth surfaces into biologically informative caries outcomes. J Dent Res. 2013b;92:32–7.
78. Shaffer JR, Polk DE, Feingold E, Wang X, Cuenco KT, Weeks DE, Desensi RS, Weyant RJ, Crout R, Mcneil DW, Marazita ML. Demographic, socioeconomic, and behavioral factors affecting patterns of tooth decay in the permanent dentition: principal components and factor analyses. Community Dent Oral Epidemiol. 2013c;41:364–73.
79. Shimizu T, Ho B, Deeley K, Briseno-Ruiz J, Faraco IM Jr, Schupack BI, Brancher JA, Pecharki GD, Kuchler EC, Tannure PN, Lips A, Vieira TC, Patir A, Yildirim M, Poletta FA, Mereb JC, Resick JM, Brandon CA, Orioli IM, Castilla EE, Marazita ML, Seymen F, Costa MC, Granjeiro JM, Trevilatto PC, Vieira AR. Enamel formation genes influence enamel microhardness before and after cariogenic challenge. PLoS One. 2012;7:e45022.
80. Shimizu T, Deeley K, Briseno-Ruiz J, Faraco IM Jr, Poletta FA, Brancher JA, Pecharki GD, Kuchler EC, Tannure PN, Lips A, Vieira TC, Patir A, Yildirim M, Mereb JC, Resick JM, Brandon CA, Cooper ME, Seymen F, Costa MC, Granjeiro JM, Trevilatto PC, Orioli IM, Castilla EE, Marazita ML, Vieira AR. Fine-mapping of 5q12.1–13.3 unveils new genetic contributors to caries. Caries Res. 2013;47:273–83.
81. Slayton RL, Cooper ME, Marazita ML. Tuftelin, mutans streptococci, and dental caries susceptibility. J Dent Res. 2005;84:711–4.
82. Stanley BO, Feingold E, Cooper M, Vanyukov MM, Maher BS, Slayton RL, Willing MC, Reis SE, Mcneil DW, Crout RJ, Weyant RJ, Levy SM, Vieira AR, Marazita ML, Shaffer JR. Genetic Association of MPPED2 and ACTN2 with Dental Caries. J Dent Res. 2014;93:626–32.
83. Takaoka A, Tanaka N, Mitani Y, Miyazaki T, Fujii H, Sato M, Kovarik P, Decker T, Schlessinger J, Taniguchi T. Protein tyrosine kinase Pyk2 mediates the Jak-dependent activation of MAPK and Stat1 in IFN-gamma, but not IFN-alpha, signaling. EMBO J. 1999;18:2480–8.
84. Tannure PN, Kuchler EC, Lips A, Costa Mde C, Luiz RR, Granjeiro JM, Vieira AR. Genetic variation in MMP20 contributes to higher caries experience. J Dent. 2012;40:381–6.
85. Ten CJM. The need for antibacterial approaches to improve caries control. Adv Dent Res. 2009;21:8–12.
86. Valarini N, Maciel SM, Moura SK, Poli-Frederico RC. Association of dental caries with HLA Class II allele in Brazilian adolescents. Caries Res. 2012;46:530–5.
87. Veerayutthwilai O, Byers MR, Pham TT, Darveau RP, Dale BA. Differential regulation of immune responses by odontoblasts. Oral Microbiol Immunol. 2007;22:5–13.
88. Viana AC, Kim YJ, Curtis KM, Renzi R, Orrico SR, Cirelli JA, Scarel-Caminaga RM. Association of haplotypes in the CXCR2 gene with periodontitis in a Brazilian population. DNA Cell Biol. 2010;29:191–200.
89. Vieira AR, Marazita ML, Goldstein-Mchenry T. Genome-wide scan finds suggestive caries loci. J Dent Res. 2008;87:435–9.
90. Wang X, Shaffer JR, Weyant RJ, Cuenco KT, Desensi RS, Crout R, Mcneil DW, Marazita ML. Genes and Their Effects on Dental Caries May Differ between Primary and Permanent Dentitions. Caries Res. 2010;44:277–84.
91. Wang X, Shaffer JR, Zeng Z, Begum F, Vieira AR, Noel J, Anjomshoaa I, Cuenco KT, Lee MK, Beck J, Boerwinkle E, Cornelis MC, Hu FB, Crosslin DR, Laurie CC, Nelson SC, Doheny KF, Pugh EW, Polk DE, Weyant RJ, Crout R, Mcneil DW, Weeks DE, Feingold E,

Marazita ML. Genome-wide association scan of dental caries in the permanent dentition. BMC Oral Health. 2012a;12:57.

92. Wang X, Willing MC, Marazita ML, Wendell S, Warren JJ, Broffitt B, Smith B, Busch T, Lidral AC, Levy SM. Genetic and environmental factors associated with dental caries in children: the Iowa Fluoride Study. Caries Res. 2012b;46:177–84.
93. Wang Q, Jia P, Cuenco KT, Feingold E, Marazita ML, Wang L, Zhao Z. Multi-dimensional prioritization of dental caries candidate genes and its enriched dense network modules. PLoS One. 2013a;8:e76666.
94. Wang Q, Jia P, Cuenco KT, Zeng Z, Feingold E, Marazita ML, Wang L, Zhao Z. Association signals unveiled by a comprehensive gene set enrichment analysis of dental caries genome-wide association studies. PLoS One. 2013b;8:e72653.
95. Wendell S, Wang X, Brown M, Cooper ME, Desensi RS, Weyant RJ, Crout R, Mcneil DW, Marazita ML. Taste genes associated with dental caries. J Dent Res. 2010;89:1198–202.
96. Werneck RI, Lazaro FP, Cobat A, Grant AV, Xavier MB, Abel L, Alcais A, Trevilatto PC, Mira MT. A major gene effect controls resistance to caries. J Dent Res. 2011;90:735–9.
97. Xu X, Jeong L, Han J, Ito Y, Bringas P Jr, Chai Y. Developmental expression of Smad1–7 suggests critical function of TGF-beta/BMP signaling in regulating epithelial-mesenchymal interaction during tooth morphogenesis. Int J Dev Biol. 2003;47:31–9.
98. Yang Y, Wang W, Qin M. Mannose-binding lectin gene polymorphisms are not associated with susceptibility to severe early childhood caries. Hum Immunol. 2013;74:110–3.
99. Yang J, Zhu W, Chen J, Zhang Q, Wu S. Genome-wide two-marker linkage disequilibrium mapping of quantitative trait loci. BMC Genet. 2014;15:20.
100. Yarat A, Ozturk LK, Ulucan K, Akyuz S, Atala H, Isbir T. Carbonic anhydrase VI exon 2 genetic polymorphism in Turkish subjects with low caries experience (preliminary study). In Vivo. 2011;25:941–4.
101. Zakhary GM, Clark RM, Bidichandani SI, Owen WL, Slayton RL, Levine M. Acidic proline-rich protein Db and caries in young children. J Dent Res. 2007;86:1176–80.
102. Zeng Z, Shaffer JR, Wang X, Feingold E, Weeks DE, Lee M, Cuenco KT, Wendell SK, Weyant RJ, Crout R, Mcneil DW, Marazita ML. Genome-wide association studies of pit-and-fissure- and smooth-surface caries in permanent dentition. J Dent Res. 2013;92:432–7.
103. Zeng Z, Feingold E, Wang X, Weeks DE, Lee M, Cuenco DT, Broffitt B, Weyant RJ, Crout R, Mcneil DW, Levy SM, Marazita ML, Shaffer JR. Genome-wide association study of primary dentition pit-and-fissure and smooth surface caries. Caries Res. 2014;48:330–8.

Periodontal Disease

Arne S. Schäfer

Epidemiology

Periodontitis Is a Complex Disease

Periodontitis (PD) is an inflammatory disease of the oral cavity caused by bacteria forming a biofilm on the tooth root surface and the gingiva. The inflammatory reaction leads to gingival bleeding, pocket formation, destruction of attachment and of alveolar bone, and eventually to tooth loss [116]. PD affects human populations worldwide at prevalence rates of 11 % for the severe forms and is the major cause of tooth loss in adults above 40 years [34, 90].

Sub-Forms of Periodontitis

Chronic Periodontitis (CP) is a common disease with a prevalence of 72–95 % in adults of Caucasian ethnicity older than 45 years, depending on age and severity of attachment loss [34, 85]. It is mostly observed in adults and is characterized by a slow progress of the disease. Smoking, diabetes, stress and age are strong predisposing factors and, related to these, socioeconomic factors such as low education and low income are also clearly associated with the progression of CP.

Aggressive Periodontitis (AgP) is a rare disease phenotype with a prevalence of ~0.1 % in European Caucasians [147] and is diagnosed in adolescents and young adults (<35 years of age) based on rapid attachment loss and severe destruction of the alveolar bone. Because of the very young age of disease development and the absence of long-lasting predisposing lifestyle factors, like smoking or diabetes, it is considered that genetic risk variants have a fundamental role in onset and development of AgP.

A. S. Schäfer (✉)
Research Center ImmunoSciences (RCIS), CC03, Center of Dento-Maxillo-Facial Medicine, Dept. of Periodontology and Synoptic Dentistry, Charité - University Medicine Berlin, Aßmannshauser Str. 4–6, 14197 Berlin, Deutschland
e-mail: arne.schaefer@charite.de

S. T. Sonis (ed.), *Genomics, Personalized Medicine and Oral Disease*,
DOI 10.1007/978-3-319-17942-1_7

Etiopathogenesis

Role of the Oral Microflora in the Disease Etiology of Periodontitis

Oral microbes are traditionally regarded as the principal cause of periodontitis. This is mainly because all forms of periodontitis were largely shown to be associated with specific bacterial pathogens. In contrast to this view, recent advances in sequencing technologies that allowed simultaneous investigation of the entire spectrum of periodontal pocket communities pointed to the perspective, in which the transition from health to disease is attributed to a shift in the global balance of the microbial flora rather than to the specific appearance of individual pathogens. The concept of this hypothesis, also known as dysbiosis or microbial-shift, is that periodontitis is the result of a decrease in the number of commensal microbes and/or an increase in the number of pathogens, instead of the presence of specific bacterial species [25]. Likewise, studies that investigated how standard periodontal disease treatment altered the microbial diversity in periodontal pockets in individuals of the same ethnicity found no clear differences between samples collected prior to treatment with those collected post-treatment from the same individual [134]. Further, strong differences in the oral microbial communities between different ethnicities that share a common food, nutritional and lifestyle background were described [93]. This supported recent findings that the core oral microbiome in unrelated individuals tends to be minimal at lower taxonomic levels [37]. As a consequence, it is being considered that the individual background exerts a selection pressure on the oral microbiome by defining the environment for bacterial colonization. This selection pressure is genetic rather than environmental, since ethnicities as well as individuals demonstrate significant microbial divergence within the same living conditions. In this view, increased quantities of oral pathogens may simply be a symptom and consequence of the disrupted subgingival environment rather than the primary cause of PD. Accordingly, several epidemiological and longitudinal clinical studies showed that the presence of bacteria does not invariably induce periodontal attachment loss. In the opposite, these studies identified the existence of high-risk groups in PD by showing that relatively many teeth are lost in relatively few patients [150, 14] and that only a minority of the patient populations accounted for most of all lost teeth, independent of specific oral bacteria [52, 96, 154]. The concept has recently emerged that there is not a single microbial composition that represents a periodontal state that is associated with health or disease but that the host genotype chooses the composition and quantities of the individual's "normal oral microbiome" and shapes the antimicrobial response, which can vary between individuals of a population [104, 105].

This highlights the importance of understanding the patients' genomic differences as critical factors when moving from health to a diseased state. It also shows the relevance for the assessment of the individual's genetic disease susceptibility for the development of targeted therapies.

Heritability of PD

First evidence that PD has a genetic cause was provided by comparisons of the presence of PD in both members of a pair of twins, which is expressed by the concordance rate of monozygous (MZ) and dizygous (DZ) twins. A population-based twin study on early-onset PD showed a twofold-increased risk of genetically identical MZ twins for early-onset PD compared to DZ twins [20]. Twin studies that assessed the heritability of PD in adults also unanimously reported a heritable component [20, 99–101] and estimated that a large proportion (38–82 %) of the variance of periodontal disease parameters could be attributed to genetic factors. The heritability of CP was estimated to be ~50%, which was unaltered following covariate adjustments for smoking, dental hygiene, age and gender. No heritability was observed for gingivitis.

Adipositas as a Likely Susceptibility Factor of PD

The susceptibility to PD is determined by additional internal and external factors like age, smoking, stress, and obesity [73]. Related to this, diabetes and atherosclerosis are also associated with PD, the predisposition to which is again partly determined by the individual's genetic constitution, and can lead to changes in the penetrance of the genetic variants.

Adipositas and Inflammation Whereas the negative effects on the immune system are well known for age [95, 118], stress [8, 9], and smoking [122], accumulating evidence indicates a substantial role of adipose tissues with regard to chronic low-grade inflammation. Adipose tissue is not only involved in energy storage but also functions as an endocrine organ that secretes multiple bioactive substances such as the pro-inflammatory cytokines TNF-alpha and IL6 [110]. The continuously increased expression of these factors as a cause of adipocyte dysfunction affects the immune system and promotes inflammation. This potentially promotes a variety of obesity-linked chronic conditions and diseases such as diabetes and atherosclerosis [111] and, as we see it today, also PD. Coinciding with obesity, PD is associated with medical conditions that have obvious metabolic connections, such as type 2 diabetes and cardiovascular disease.

Type 2 diabetes (T2D) and PD T2D promotes the occurrence, progression and severity of PD [119]. It leads to a hyperinflammatory response to the periodontal microbiota and impairs the resolution of inflammation and tissue repair, which accelerates periodontal destruction [80]. Reciprocally, the release of pro-inflammatory mediators such as the cytokines TNF-alpha and interleukins promote insulin tolerance [143]. A bi-directional relationship between both diseases is supported [17].

Cardiovascular Disease (CVD) and PD Strong evidence of associations between the presence of CAD and PD is derived from multiple randomized clinical trials and shows that the association between both diseases is independent of the shared

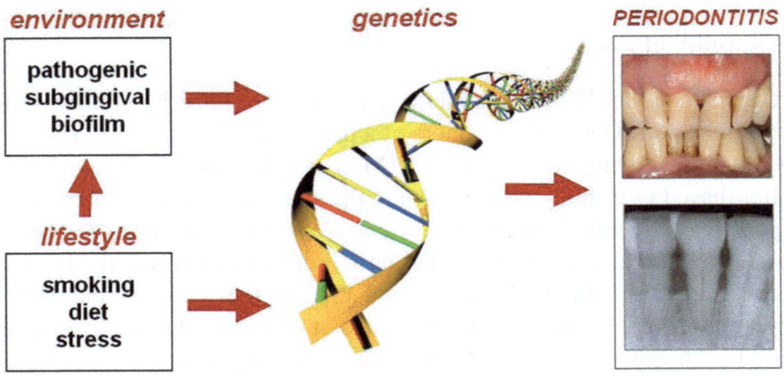

Fig. 1 Periodontitis is a complex disease

risk factor smoking [86]. However, other shared risk factors could confound the relationship between CVD and PD. Increasing age, obesity, T2D and the socio-economic status are all factors associated with both CVD and PD and the independence of the relationship of both diseases from these factors is not yet supported by evidence of longitudinal clinical studies. Yet, the impact of periodontitis on atherosclerotic CVD is biologically plausible, because translocated circulating oral microbiota may directly or indirectly induce inflammatory responses that have impacts on the pathogenesis of atherothrombogenesis [148] but at the time being, a causative relationship between CVD and PD is not being supported by clear experimental evidence [86].

The modifying disease genes play a central role and determine susceptibility to periodontitis. With this intrinsic characteristic, lifestyle factors pave the way to the disease, which is eventually initiated by an infection (environmental factors). Both have an individual as well as an additive effect on the (patho)physiological response of the patient, which is, in the final step, determined by the individual genetic predisposition. Therefore, in the complexity of periodontitis, gene–gene interactions as well as gene–environmental, gene–lifestyle and environmental–lifestyle interactions play a role in the development of the phenotype [132] (Fig. 1).

Current Strategies to Identify the Genetic Basis of PD

Candidate Gene Association Studies (CGAS)

Concepts of CGAS Genetic association studies of candidate genes that were selected **based on literature review** have been the most important strategy for the identification of disease genes in PD. **These studies required an *a priori* hypothesis on the involvement of the selected gene in the disease risk** [158]. In principle, there are two different selection strategies for a candidate gene. It can be asked whether specific loci within a specific biological pathway are involved in the increase of the genetic risk of the disease, or effects of a variant from other diseases are tested. Both approaches can give answers whether the selected genes carry genetic variants, which increase the risk of PD.

The hypotheses for the selection of the genes of interest in CGAS are depending on the current knowledge of the underlying molecular biological disease mechanisms and hundreds of genes that could have an influence are not selected because their functions are unknown or they lie within pathways, which have not been implicated with the disease. Because the knowledge on these genes is usually very **incomplete,** the selection of candidate genes is necessarily arbitrary and often does not reflect the situation in nature.

Limited Success of CGAS in PD Corresponding to the understanding of PD as an infectious disease, most studies focused on genes, which were selected for their roles in the immune system, but some also investigated genes that are involved in tissue destructive processes or metabolism [79, 88]. In addition to the limitations of the CGAS due to incomplete insight into the genetic and molecular biological disease mechanisms, most of the CGAS on PD additionally showed the following shortcomings, which raised questions about both type 1 errors (false positives) and type 2 errors (false negatives):

- Insufficient sample sizes
- Inappropriate case selection
- missing replications of the results
- incomplete genetic analysis of the selected candidate loci

As a consequence, most studies were unable to draw unambiguous conclusions from their findings, even when the outcome was negative.

Statistical Power

The statistical power (SP) is determined to a large extent by the size of the case-control samples of the initial explorative study and the subsequent replication sample. The SP increases with sample size and correlates with allele frequency and the genetic effect of the putative risk variant [71]. This is why variants with a high OR are more likely to be detected than rare variants with a small effect (Fig. 2). Most disease associated variants increase the susceptibility modestly, and to identify a common risk variant, often >1000 well-defined cases and at least the same number of controls are necessary to reach sufficient statistical power.

Genome-Wide Association Studies (GWAS)

A milestone in the conduction of genetic studies to identify the molecular biological basis of PD were technical advances in the middle of the last decade, that allowed a hypothesis-free approach with the simultaneous testing of up to several millions

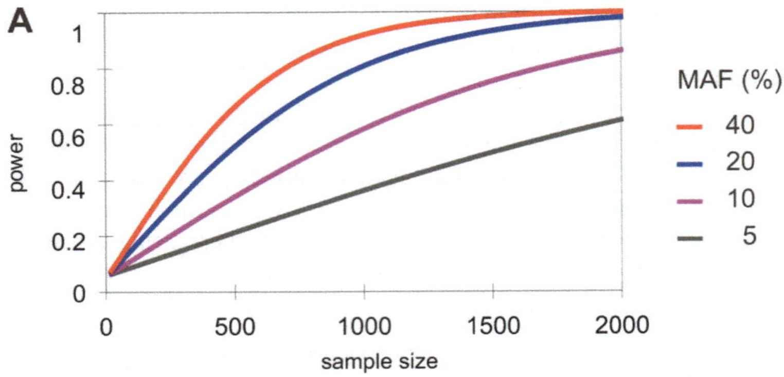

Fig. 2 Statistical power in relation to the sample size, allele frequency and odds ratio. To identify a genetic risk variant with a minor allele frequency (MAF) of e.g. 20 % in the general population, ≈ 1000 cases and 2000 controls are required to achieve the necessary statistical power of 0.8 (The statistical power was calculated as described in [32] for an average odds ratio (OR) of 1.3, and 2 times as many controls as cases were considered. A power of 0.8 is regarded as statistically significant). In CGAS of PD, case samples of more than hundred were very rare [132]

of genetic variants spread across the entire genome in thousands of selected cases and controls. These studies are termed genome-wide association studies (GWAS). A succession of GWAS for almost any complex human disease was performed since the year 2007 that particularly started with the milestone publication of the Wellcome Trust Case Control Consortium [159]. Since the begin of this era, which is also referred to as the genetic gold rush of human genetics, many if not most of the common genetic susceptibility variants were identified for almost any common complex human disease. A concrete example of the power of GWAS and the enormous progress made in recent medical genetic research is illustrated by inflammatory bowel disease (IBD). During the comparably long pre-GWAS era, only two risk genes *IBD5* and *NOD2* were identified [56, 107, 121]. During 5 years, from 2005–2010 several independent GWAS identified 37 additional genetic risk loci. This process has currently come to an end with a recent mega-analysis that encompassed almost all available case-controls samples, comprising > 75,000 individuals [67]. Including this study, a total of > 160 genetic risk loci of IBD are described.

Validated Genetic Risk Factors of PD

In PD, progress in the identification of genetic risk loci has been much slower compared with other complex human diseases. This was mainly because of difficulties to generate large case samples of homogenous ethnic background and severe disease phenotypes. As a consequence, despite an unclear number of CGAS that were performed for PD, evidence that is based on solid statistical associations, which were confirmed in large homogenous case-control samples with a clear replication in the same phenotype and/or additional molecular biological data that add to the proposed functional role of the associated region, is scarce. Thus, few genes can currently be considered as true genetic susceptibility factors for PD.

NPY (Neuropeptide Y)

A GWAS described an association with severe CP downstream of the coding region of *NPY* (chr.7) in a large sample of European American individuals [30]. In this study, the strongest association was observed for SNP rs2521634 in 958 severe CP cases and 1909 controls with $P=3.5\times10^{-7}$ and a genetic effect (odds ratio [OR]) = 1.5 (95 % confidence interval [CI] = 1.3–1.7). Another GWAS that systematically analyzed gene-sex interactions in German AgP cases and controls observed as the most significant association of the interaction term of sex and the genetic constitution, a sexually dimorphic role of genetic variants upstream *NPY* to be associated with AgP. SNP rs198712 showed the strongest association in interaction with sex ($P=4.03\times10^{-6}$; 721 AgP cases, 1472 matched controls) with odds ratios in males and females of 1.63 and 0.69, respectively.

Interestingly, sex-dependent effects of NPY had previously been described in mice. NPY loss-of-function mice showed different anxiogenic responses in behavioral tests in males and females, indicating a sexually dimorphic role of NPY in behavioral stress responses [112]. Also, gastrointestinal inflammation, known to enhance anxiety in a sex-dependent manner, produced different behavioral responses to stress challenges in female and male NPY knockout mice [112]. Further, NPY knockout mice showed sex-dependent responses in food intake, upper gastrointestinal transit and faecal pellet output induced by restrained and novel environment stresses [39]. NPY activates the hypothalamic–pituitary–adrenal (HPA) axis and modulates the visceral stress responses mediated through corticotrophin-releasing hormone (CRH) pathways [27]. Additionally, NPY is potently anxiolytic [69], acting through Y 1 receptors in the amygdala to inhibit CRH signaling and terminate the behavioral stress and anxiety responses. In accordance to the function in mice, NPY also influences many physiological processes in human, including stress response [169] and stress-induced obesity [76]. Often, stress-associated and eating-disorders, which showed sex-dependent effects in NPY knockout mice, also have a different prevalence among women and men in humans. Other studies showed that a nonsynonymous SNP (Leu/Pro transition) within *NPY* was associated with serum triglyceride concentrations and birth weight, high serum cholesterol and LDL cholesterol levels [70]. Furthermore, the Leu/Pro transition within *NPY* was reported to be associated with alcohol dependence in large population samples from the United States [81] and Finland [72]. In the disease etiology of PD, obesity, stress and alcohol consumption are strong risk indicators [2, 44]. In this context it is interesting, that both GWAS linked variants at the genetic region of *NPY* to severe phenotypes of periodontitis. Women often pay different attention to their diet, smoke and drink less alcohol and also perform preventive oral health measures more actively than men [4]. Likewise, women lose fewer teeth in longitudinal studies [12]. NPY is also the most abundant neuropeptide in bone [1] and has recently been shown to have a role in maintaining the balance between hard tissue formation and resorption, processes that are relevant to the definition of periodontitis [50]. The immunomodulatory effects of NPY are also thought to alter the pro-inflammatory T-helper type 1 (Th1):anti-inflammatory T-helper type 2 (Th2) balance and binding of NPY to Y1 receptors on a variety of immune cells is thought to be responsible for promoting the anti-inflammatory Th2 response [6]. NPY is therefore potentially important in the coordination of inflammation and bone metabolism, both of which are central to the pathogenesis of PD [89]. Accordingly, presence of NPY Y1 receptors were verified in the gingival tissue and in human gingival crevicular fluid (GCF), with significantly higher NPY levels in GCF in healthy compared with periodontitis-affected sites [89].

ANRIL (CDKN2BAS)

The long noncoding antisense RNA (lncRNA) *ANRIL* (*CDKN2BAS*, chr.9) is well-known as the best replicated genetic risk factor for CAD [26, 51, 97, 123, 133, 159]. To date, it is also the best replicated genetic risk factor of AgP [35, 129, 130, 125], which showed a genetic effect of OR = 1.4 (95 % CI = 1.2–1.7; $P = 1 \times 10^{-4}$; 283 AgP cases, 923 matched healthy controls) in the original study (Schaefer, Richter et al. 2009). Variants within ANRIL also showed significant association in two independent case-control samples of German and Dutch origin with CP [129]. In addition to the associations with CAD and PD, *ANRIL* is independently associated on a genome-wide level with a variety of other diseases such as type 2 diabetes, endometriosis, intracranial aneurysma, megakarypopoiesis [46, 124, 137, 152, 157, 159, 163, 166] (Schaefer, Richter et al. 2009), and several forms of cancer [139, 7, 150, 151].

ANRIL is regulated at least by STAT1 signaling [48], a pathway that mediates response to inflammation upon stimulation of the pro-inflammatory cytokine Interferon gamma. A number of CAD associated variants at this chromosomal locus were predicted to disrupt annotated transcription factor binding sites [138]. Various studies investigated the association of GWAS lead SNPs of CAD, which are shared with PD, and the expression of *ANRIL, CDKN2A*, and *CDKN2B* and showed correlations of the risk haplotype with the expression of these genes in different tissues or cell types, indicating cell type specific splicing and tissue specific effects of the risk haplotype [13, 19, 23, 38, 54, 64, 66, 102], with some consistency among the larger studies that the risk haplotype is associated with decreased levels of the short more complex isoform DQ485454, indicating a role of the 5′ exons of ANRIL in the disease etiology of CAD and PD [13, 19, 64, 66, 84, 102]. In accordance with observations that lncRNAs often play a role in the trans-regulation of gene expression [98, 113], it was shown that decreased expression of *ANRIL* transcripts containing exon 13 are correlated with decreased expression levels of the distant genes *ADIPOR1* (chr.1), *VAMP3* (chr.1) and *C11ORF10* (chr.11) [10].

ADIPOR1 is a high-affinity receptor for globular adiponectin, a hormone which e.g. mediates AMPK and PPAR-alpha ligand activities [161], key regulators of fatty acid oxidation and regulation of glucose levels. A negative correlation of body fat and adiponectin levels is well established [142, 162]. Apart of the function of globular adiponectin to increase fatty acid oxidation, adiponectin also strongly increases insulin's ability to stimulate glucose uptake through increased glucose transporter 4 (GLUT4) gene expression and increased GLUT4 recruitment to the plasma membrane [16, 42]. The hormone plays a role in T2D and CAD, where it may also attenuate the inflammatory response associated with e.g. atherogenesis [109] and PD [74]. In this context, adiponectin is discussed to act as an anti-inflammatory signal by selectively increasing tissue inhibitors of metalloproteinases (TIMPs) [75]. It was further established that adiponectin suppresses phagocytic activity and lipopolysaccharide (LPS)-induced TNF (tumour necrosis factor) expression [165].

VAMP3 belongs to the VAMP/synaptobrevin family and plays a role in phagocytosis, where VAMP3 mediates delivery of TNF-alpha to the cell surface of the phagocytic cup, resulting in a subsequent release of TNF-alpha [103]. Interestingly, VAMP3 is also involved in GLUT4 mediated glucose metabolism. Elevated insulin levels induce translocation of GLUT4 to the plasma membrane to aid glucose utilization, which is mediated by VAMP3 [135, 136].

Located 2 kb upstream of *VAMP3* is the extremely large chromosomal region of *CAMTA1* (calmodulin-binding transcription activator1), spanning >1 megabase (Mb) and being rich in enhancer elements, indicating complex regulation of this chromosomal region. In a GWAS on periodontal pathogen colonization, a large stretch of the *CAMTA1/VAMP3* region was reported to be associated with increased quantities of oral pathogenic bacteria [30]. A study finemapped the *CAMTA1/VAMP3* region to search for potential PD-associated disease variants [10]. Within *CAMTA1*, several significant SNP associations were observed, among which the rare variant rs17030881 (MAF controls = 0.04%, MAF cases = 0.27%), located directly upstream of the periodontal pathogen associated region within a conserved noncoding element, showed the best association signal with AgP (P=0.002, OR=4.36 [95% CI 1.5–12.5]; 864 North-West European AgP cases, 3664 geographically matched healthy controls) and interestingly, also with CAD (P=0.009, OR=6.09 [95% C.I.=1.3–28.7]; 1471 North-German non-obese CAD cases (BMI<30) and 2737 geographically matched healthy controls). The association with CAD was validated in a meta-analysis of 13 genome-wide association studies of CAD [133] comprising 21,033 individuals with CAD and 44,065 controls of European descent. In this study, the rare allele of rs2301462 within the AgP associated region showed the best association with $P = 0.0011$ upon adjustment for age and sex [10].

C11ORF10 is a transmembran protein of currently unknown function. It locates 7 kb distal to the *FADS* 1–3 (fatty acid desaturases 1–3) gene cluster. Various SNPs across *C11ORF9*, *C11ORF10*, *FADS1* and *FADS2*, all in strong linkage disequilibrium ($r^2 < 0.8$), were reported to show genome-wide significant associations with the metabolic syndrome [165], altered phospholipids concentrations [82], T2D [33], and CD [40], but not with CAD. These SNPs showed nominal significant association with AgP, with intronic SNP rs1535 of *FADS2* showing the smallest p value with the same effect directions. ($P_{allelic}$=0.016, OR=0.84, 95% CI 0.72–0.98) [10].

PLG (Plasminogen)

In addition to the shared risk variants of PD and CAD within the genes *ANRIL* and *CAMTA1*, the gene *PLG* (Plasminogen; chr.6) also showed association with AgP ($P_{pooled}=5.9\times10^{-5}$, OR = 1.27, 95 % CI = 1.3–1.4 [adjusted for smoking and sex]; 818 cases, 5309 controls) [131]. The association of the intronic variant rs4252120 of *PLG*, which is also genome-wide associated with CAD [26], was replicated in a German AgP sample and validated in a Dutch AgP sample [131]. Plasminogen is the inactive precursor of plasmin and occurs in the plasma, where it is converted to plasmin at the cell surface. Plasmin has the ability to facilitate cell migration in tissues and to trigger signals, which affect various functions of monocytes, macrophages, dentritic cells, and others. Plasmin also has the ability to stimulate the production of cytokines, thereby contributing to inflammation. Therefore, excessive activation of plasmin in chronic inflammation might exacerbate the activation of inflammatory cells and the pathogenesis of the disease [144]. The identification of plasminogen activator inhibitor 1 (PAI-1), an inhibitor of the PLG-plasmin system, as an adipokine that is strongly upregulated in visceral adipose depots in obesity suggested a mechanistic link between obesity and atherosclerosis [140], as reviewed in [94]. In the context of the shared association of CAD and PD, it is of interest that the PLG-plasmin system is highly important for degradation of tissue barriers and cell migration [78]. Various pathogenic bacteria were found to bind human PLG on bacterial PLG receptors [11, 77], which turns them into proteolytic organisms. E.g., the common periodontal pathogen *Porphyromonas gingivalis* (*P.G.*) is able to activate human PLG expression and to inactivate human plasmin inhibitors, causing uncontrolled plasmin activity [47].

GLT6D1 (Glycosyltransferase 6 Domain Containing 1)

GLT6D1 (chr.9) was identified in the first GWAS on AgP, where it was replicated and validated in independent case-control samples of German and Dutch descent (Schaefer, Richter et al. 2010). The GWAS lead SNP rs1537415, located at large intron 2, reached genome-wide significance $P=5.5\times10^{-9}$ and a genetic effect of OR = 1.59 (95 % CI 1.36–1.86; 447 AgP cases, 1340 controls). *GLT6D1* encodes a hitherto unknown protein with an annotated glycosyltransferase 6 domain. Glycosyltransferases form a ubiquitous group of enzymes that catalyze the transfer of a sugar moiety from donor onto acceptor molecules [21]. Glycosylation is the most common post-translational modification of proteins and it is estimated that more than 50 % of all mammalian cellular and membrane-bound proteins are glycosylated, implicating an essential role in protein and cell function [156]. In particular, glycosylation has an essential role for the immune system [155], where it has various functions for immune cell trafficking [141], T cell receptor signaling [120], B-cell receptor signaling [22], antibody function [3], immune cell differentiation [92], pathogen recognition [108], and immune homeostasis [43]. *GLT6D1* is predominantly expressed in T-cells and the gingiva (Schaefer, Richter et al. 2010). Prediction of the binding affinities of transcription factor and molecular biological characterization of the effects of the AgP associated rare allele of rs1537415 indicated the presence of an impaired GATA-3 transcription factor binding site as a potential causative effect for the increased disease risk (Schaefer, Richter et al. 2010). GATA-3 is considered as the master TF in transcriptional control of Th2 differentiation [167].

Other Genetic Risk Loci of PD

A large number of candidate gene or candidate SNP association studies have been published in the field of periodontitis but only few studies complied with the standards of genetic association studies [61, 132, 153]. Mostly, these studies lacked the statistical power to reject or accept the null-hypothesis of no association and, for the same reason, the findings were not replicated in independent and sufficiently sized case-control samples of the same ethnic origin and diagnostic criteria. Further, the data were often not corrected for multiple testing and not adjusted for major covariates such as sex and smoking. As a consequence, almost all studies were highly susceptible to type 1 and type 2 errors.

To increase the statistical power, meta-analyses on candidate SNPs have increasingly been performed during the last years. These studies, which pooled multiple smaller studies are susceptible to stratification by chance effects from the original studies as consequence of their limited sample size, but also because of the heterogeneous origins and diagnostic criteria of the individual samples, which were combined in the meta-analyses. Accordingly, the meta-analyses that found a significant association involving heterogeneous study populations did not find a significant association when the analyses were restricted to a subgroup of more homogenous studies [18, 28, 58, 59, 65, 68, 106, 117, 168].

As a consequence, only very few studies were published in genetic expert journals, which guarantee the necessary scientific standards.

In order to clarify the putative role of a comprehensive set of candidate genes that were in the focus of genetic research in periodontitis during the last decade, a large CGAS on PD finemapped 17 genes (*ABO, CCR5, FCGR2A, FCGR2C, FCGR3A, FCGR2B, FCGR3B, IL1B, IL2, IL6, IL10, LTA, MMP9, NOD2, TLR2, TLR4, VDR*), and five genes (*CD14, IL1A, IL1RN, TNFRSF11B (OPG), IFNGR1* and *L-selectin*) were genotyped with limited SNP density [130] in 600 German AgP cases and 1440 matched controls. This study gave no support that the tested loci are true susceptibility factors of PD except for variants at the chromosomal region upstream to the anti-inflammatory cytokine gene *IL10*.

Likewise, none of the three large GWAS that have been performed for PD provided support for genetic variants that had earlier been reported as significant susceptibility factors for PD [30, 146, 128]. One of the two GWAS on CP suggested association of three loci, which was apart from the gene *NPY* for severe CP, the genes *NCR2* and *EMR1* for moderate CP. Although the associations of these genes were not replicated in an independent sample, the effect size direction was concordant in the replication sample compared with the explorative GWAS sample for these genes.

Application of Genomics to Guide Diagnosis of PD

The recent progress of medical research in human genetics is expected to offer new opportunities for the prevention, early detection, and treatment of complex diseases, in part by allowing health care providers to use individualized preventive and therapeutic strategies based on patients' genomic profiles. At the time being, the complex interactions between genetic and environmental causes of CP is poorly understood but the little results of the GWAS on CP indicate a high heterogeneity of this disease phenotype and large contributions of related life-style, socio-economic, and age related factors. In this view, it cannot be expected that testing of the carriership of alleles in single genes will allow stratifying patients for diagnosis and treatment of CP, claims that are endorsed by professional and industrial stakeholders. In the current situation where positive results from various single gene-disease association studies were often inconsistently replicated in subsequent studies, selection of studies showing a statistically significant association is insufficient evidence of causality [60], especially if the associations were modest. Polymorphisms, which are selected on this basis for inclusion in genomic risk profiles have insufficient scientific evidence and clinical validity to conclude that genomic profiles are useful in measuring genetic risk for common diseases. In this context, large general concerns were raised by both the American and the European Society of Human Genetics [63, 114]. To be meaningful, a genetic risk profile should combine information about the disease risk associated with multiple validated risk variants and related environmental and life-style factors. Creating such a profile would require extensive knowledge of first, true genetic risk alleles, and secondly of gene-gene and gene-environment interactions, which both are currently even much less well understood than the disease risk associated with individual polymorphisms [62]. Yet tests, which are based in the laboratory that created them, often require only evidence of analytical validity before being allowed to market. In the case of a genetic test, the assay may be analytically valid, if it accurately identifies a particular polymorphism. However, if that genetic variant has nothing to do with the disease of interest, then the test has no clinical validity.

In conclusion, although genomic profiling may have potential to enhance the effectiveness and efficiency of preventive interventions in the future, to date, the scientific evidence for associations between most reported genetic risk variants and the risk of PD is insufficient to support useful applications.

Application of Genomics to Guide Prevention and Treatment of PD

In general, disease related genes may be functionally related and/or interact with one another in biological pathways that are assumed to have the potential to contribute to disease development [91, 115]. It is probable that only a limited number of biological pathways have the potential to specifically contribute to the aetiology of a disease phenotype such as PD [15]. Although the current knowledge of the molecular genetic etiology of PD is yet limited to a few loci, which gave evidence of disease relevance by statistically significant associations and replication in large homogenous study populations and/or additional independent molecular biological data, the knowledge of these genes points to mechanisms of disease pathogenic mechanisms.

Molecular biological data placed ANRIL as a major genetic susceptibility factor of PD into a regulatory network that integrates glucose and fatty acid metabolism with immune response, providing molecular evidence for a mechanistic link between PD, obesity and inflammation. Deficiency of the adiponectin receptor ADIPOR1, the expression of which was shown to be regulated by ANRIL activity, results in reduced adiponectin-induced AMP-activated protein kinase (AMPK) activation and increased glucose production in mice [161]. It is well-established that AMPK activation leads to an increase in fatty acid oxidation and glucose uptake. Adiponectin affects T helper cell immunity by suppressing the production of interferon-gamma and TNF-alpha and by inducing production of the anti-inflammatory cytokine IL-10 [146]. It affects B-cell development through the induction of prostaglandin synthesis by upregulation of Cox-1 and Cox-2. Interestingly, variants upstream the coding regions of IL-10 and COX-2 were reported for homogenous case-control samples of AgP [128, 53, 83, 87, 130, 160], although meta-analyses of COX-2 candidate SNPs that used heterogeneous pooled samples from different ethnicities could not validate these findings [65, 117].

Adiponectin expression is generally downregulated in dysfunctional adipocytes that are associated with obesity, and the production of adiponectin by adipocytes is inhibited by pro-inflammatory cytokines such as TNF-alpha and IL-6, the expression of which are both strongly increased in adipose tissues [41, 55]. In contrast, adiponectin stimulates the production of the anti-inflammatory cytokine IL-10 [75] and is discussed to suppress phagocytic activity and lipopolysaccharide (LPS)-induced TNF expression [164]. The chromosomal region that harbors VAMP3 has been independently shown to be associated with increased susceptibility of PD. VAMP3 is involved in glucose metabolism but also plays a role in phagocytosis, where it mediates delivery of TNF-alpha to the cell surface of the phagocytic cup, resulting in a subsequent release of TNF-alpha [103]. Adding to the relevance of interactions between the metabolic and the immune systems for the etiology of PD is the identification that the adipokine plasminogen activator inhibitor 1 (PAI-1) is strongly upregulated in adipose tissues. Plasminogen, which is used by several oral pathogens to become proteolytic, was shown to be strongly associated with AgP and CAD.

Further adding to this context is the nominal association of the FADS gene cluster with AgP and the metabolic syndrome, diabetes and IBD and the associations of variants upstream *IL10* with AgP that are both in strong LD to GWAS lead SNPs of diabetes and IBD. Interestingly, a transcription factor binding site (TFBS) of PPARG (peroxisome proliferator-activated receptor gamma), a transcription factor that functions in the regulation of fatty acid storage and glucose metabolism was predicted at the chromosomal position of this association [130]. However, the binding of PPARG at this site awaits molecular biological evidence. Likewise, the associations of the chromosomal region of *NPY*, which were observed in independent GWAS for severe CP and AgP, indicate the relevance of the interplay of metabolic and immunological processes for PD. NPY is associated with many physiological processes such as the regulation of serum triglyceride, cholesterol and LDL cholesterol levels, with obesity as a clinical outcome.

The increased production of pro-inflammatory cytokines in adipose tissues such as TNF-alpha, IL6 and others can have local and systemic effects on inflammatory responses, e.g. by affecting T helper cell and B-cell development, thereby contributing to the initiation and progression of obesity-induced metabolic complications that are mediated by chronic inflammation. The interactions of the metabolic and immune system suggest pathogenic mechanisms that underlie many of the downstream complications of PD and its associated comorbidities and offer substantial therapeutic promise [94]. But here, research is at an early stage and there are several important questions to be answered in the future, notably: What is the relative contribution of adipocyte dysfunction to the development of PD? By what pathways does obesity provoke PD? To what extent are obesity and inflammation triggered in parallel or in sequence. And what might be the common initiation signals, or what signals link the two processes [94]?

Outlook

For common inflammatory diseases, large numbers of different common susceptibility alleles were identified in the last years, with generally moderate effects. This observation together with highly variable disease phenotypes between close relatives despite high concordance rates and genetic heritability, shaped a current hypothesis, which proposes that the specific disease phenotype can be seen as part of a large range of similar conditions that are attributed to the effects of combinations of different genetic risk alleles and their interactions with internal and external factors [45]. The individual effects of the risk alleles are expected to be moderate and little support for a significant impact of single susceptibility variants is available [58]. As a consequence, individual combinations and the sum of the effects of all risk variants are regarded to be conditional on the individual's disease susceptibility and clinical manifestations.

The few large genetic studies that were performed to elucidate the genetic risk factors of PD identified genetic susceptibility genes with functions that integrate glucose and fatty acid metabolism with immune response. The impairment of these pathways by genetic and/or physiological factors is likely to have strong detrimental effects and may be a common pathogenic denominator of PD. A further consequence of the disseminating meta-inflammation is the development of homotolerance, the unresponsiveness to the same agonist as result to longterm exposure. E.g., the high cytokine levels that are released from adipose tissues over long time-periods in the disseminating meta-inflammation result in impaired capability of the periodontal tissues to control proliferation of oral pathogens. For the late-onset forms, this process is further promoted by consequences of immunosenescence, the gradual deterioration of the immune system brought on by natural age advancement, which is also partly determined by the individual's genetic constitution.

The identified risk alleles currently only explain a small proportion of the heritability of PD. This is in large parts due to statistical limitations of the published studies, which allowed only the detection of few common genetic risk variants. But the little success in the identification of risk alleles that predispose to CP is probably also due to high heterogeneity of this disease phenotype and the large contributions of related life-style and socio-economic factors. Where the missing heritability is likely to lie is currently unknown. For the more severe forms it is probable that it is also located in pathways of innate host defense, which are key to the maintenance of periodontal health. For example, individuals with specific congenital deficiencies invariably develop PD [49] and the role of antimicrobial peptides in oral health and disease has repeatedly been described [24, 31]. Likewise, associations with AgP and CP were described for the β-defensin gene *DEFB1* (Schaefer, Richter et al. 2009). Additionally, the missing heritability can also be located in pathways of the adaptive immune system and tissue homeostasis, because destructive host inflammatory reactions as well as impairment of the self-renewing capacities of the periodontium clearly contribute to the damage associated with PD. However, the fundamental prerequisite to identify the missing genetic susceptibility factors will be the creation of large research consortia that enable the sampling of sufficiently powered and well-designed case–control populations. This challenge has successfully been taken for most complex diseases, which is greatly illustrated by the progress made for IBD. Whereas >6800 CD cases were required to increase the knowledge of genetic risk alleles of CD from eleven to 32 [5], genetic analysis of >21,000 CD cases resulted in a total of 71 distinct loci with genome-wide significant evidence for association with CD [40], and a combination of more than 75,000 cases and controls resulted in a total of 163 IBD loci that meet genome-wide significance thresholds [67]. This most recent study marked the preliminary end of the systematic genome-wide studies for the identification of common susceptibility alleles of IBD. Similar progress is being made for e.g. CAD (>63,000 cases; 46 loci) [26] and rheumatoid arthritis (>14,000 cases; 49 risk loci) [36].

The recruitment of appropriate and well-defined case–control populations will be a considerable challenge. But only after this challenge will have been taken, a complete identification of the genetic basis of PD will be possible, with the translational effects on diagnosis, classification, prevention, and the development of personalized medicine for oral health.

References

1. Ahmed M, Srinivasan GR, et al. Extraction and quantitation of neuropeptides in bone by radioimmunoassay. Regul Pept. 1994;51(3):179–88.
2. Amaral Cda S, Vettore MV, et al. The relationship of alcohol dependence and alcohol consumption with periodontitis: a systematic review. J Dent. 2009;37(9):643–51.
3. Arnold JN, Wormald MR, et al. The impact of glycosylation on the biological function and structure of human immunoglobulins. Annu Rev Immunol. 2007;25:21–50.
4. Astrom AN, Rise J. Socio-economic differences in patterns of health and oral health behaviour in 25 year old Norwegians. Clin Oral Investig. 2001;5(2):122–8.
5. Barrett JC, Hansoul S, et al. Genome-wide association defines more than 30 distinct susceptibility loci for Crohn's disease. Nat Genet. 2008;40(8):955–62.
6. Bedoui S, Miyake S, et al. Neuropeptide Y (NPY) suppresses experimental autoimmune encephalomyelitis: NPY1 receptor-specific inhibition of autoreactive Th1 responses in vivo. J Immunol. 2003;171(7):3451–8.
7. Bei JX, Li Y, et al. A genome-wide association study of nasopharyngeal carcinoma identifies three new susceptibility loci. Nat Genet. 2010;42(7):599–603.
8. Black PH. The inflammatory response is an integral part of the stress response: implications for atherosclerosis, insulin resistance, type II diabetes and metabolic syndrome X. Brain Behav Immun. 2003;17(5):350–64.
9. Black PH. The inflammatory consequences of psychologic stress: relationship to insulin resistance, obesity, atherosclerosis and diabetes mellitus, type II. Med Hypotheses. 2006;67(4):879–91.
10. Bochenek G, Hasler R, et al. The large non-coding RNA ANRIL, which is associated with atherosclerosis, periodontitis and several forms of cancer, regulates ADIPOR1, VAMP3 and C11ORF10. Hum Mol Genet. 2013;22(22):4516–27.
11. Boyle MD, Lottenberg R. Plasminogen activation by invasive human pathogens. Thromb Haemost. 1997;77(1):1–10.
12. Buchwald S, Kocher T, et al. Tooth loss and periodontitis by socio-economic status and inflammation in a longitudinal population-based study. J Clin Periodontol. 2013;40(3):203–11.
13. Burd CE, Jeck WR, et al. Expression of linear and novel circular forms of an INK4/ARF-associated non-coding RNA correlates with atherosclerosis risk. PLoS Genet. 2010;6(12):e1001233.
14. Burt BA, Ismail AI, et al. Risk factors for tooth loss over a 28-year period. J Dent Res. 1990;69(5):1126–30.
15. Carlborg O, Haley CS. Epistasis: too often neglected in complex trait studies? Nat Rev Genet. 2004;5(8):618–25.
16. Ceddia RB, Somwar R, et al. Globular adiponectin increases GLUT4 translocation and glucose uptake but reduces glycogen synthesis in rat skeletal muscle cells. Diabetologia. 2005;48(1):132–9.
17. Chapple IL, Genco R. Diabetes and periodontal diseases: consensus report of the Joint EFP/AAP Workshop on Periodontitis and Systemic Diseases. J Periodontol. 2013;84(4 Suppl):S106–12.
18. Chen LL, Li H, et al. Association between vitamin D receptor polymorphisms and periodontitis: a meta-analysis. J Periodontol. 2012;83(9):1095–103.
19. Congrains A, Kamide K, et al. Genetic variants at the 9p21 locus contribute to atherosclerosis through modulation of ANRIL and CDKN2A/B. Atherosclerosis. 2012;220(2):449–55.
20. Corey LA, Nance WE, et al. Self-reported periodontal disease in a Virginia twin population. J Periodontol. 1993;64(12):1205–8.
21. Coutinho PM, Deleury E, et al. An evolving hierarchical family classification for glycosyltransferases. J Mol Biol. 2003;328(2):307–17.
22. Crocker PR, Paulson JC, et al. Siglecs and their roles in the immune system. Nat Rev Immunol. 2007;7(4):255–66.
23. Cunnington MS, Santibanez Koref M, et al. Chromosome 9p21 SNPs Associated with Multiple Disease Phenotypes Correlate with ANRIL Expression. PLoS Genet. 2010;6(4):e1000899.

24. Dale BA, Fredericks LP. Antimicrobial peptides in the oral environment: expression and function in health and disease. Curr Issues Mol Biol. 2005;7(2):119–33.
25. Darveau RP. Periodontitis: a polymicrobial disruption of host homeostasis. Nat Rev Microbiol. 2010;8(7):481–90.
26. Deloukas P, Kanoni S, et al. Large-scale association analysis identifies new risk loci for coronary artery disease. Nat Genet. 2013;45(1):25–33.
27. Dimitrov EL, DeJoseph MR, et al. Involvement of neuropeptide Y Y1 receptors in the regulation of neuroendocrine corticotropin-releasing hormone neuronal activity. Endocrinology. 2007;148(8):3666–73.
28. Dimou NL, Nikolopoulos GK, et al. Fcgamma receptor polymorphisms and their association with periodontal disease: a meta-analysis. J Clin Periodontol. 2010;37(3):255–65.
29. Divaris K, Monda KL, et al. Genome-wide association study of periodontal pathogen colonization. J Dent Res. 2012;91(7 Suppl):S21–8.
30. Divaris K, Monda KL, et al. Exploring the genetic basis of chronic periodontitis: a genome-wide association study. Hum Mol Genet. 2013;22(11):2312–24.
31. Dommisch H, Chung WO, et al. Protease-activated receptor 2 mediates human beta-defensin 2 and CC chemokine ligand 20 mRNA expression in response to proteases secreted by Porphyromonas gingivalis. Infect Immun. 2007;75(9):4326–33.
32. Dupont WD, Plummer WD Jr. Power and sample size calculations for studies involving linear regression. Control Clin Trials. 1998;19(6):589–601.
33. Dupuis J, Langenberg C, et al. New genetic loci implicated in fasting glucose homeostasis and their impact on type 2 diabetes risk. Nat Genet. 2010;42(2):105–16.
34. Eke PI, Dye BA, et al. Prevalence of periodontitis in adults in the United States: 2009 and 2010. J Dent Res. 2012;91(10):914–20.
35. Ernst FD, Uhr K, et al. Replication of the association of chromosomal region 9p21.3 with generalized aggressive periodontitis (gAgP) using an independent case-control cohort. BMC Med Genet. 2010;11:119.
36. Eyre S, Bowes J, et al. High-density genetic mapping identifies new susceptibility loci for rheumatoid arthritis. Nat Genet. 2012;44(12):1336–40.
37. Faith JJ, Guruge JL, et al. The long-term stability of the human gut microbiota. Science. 2013;341(6141):1237439.
38. Folkersen L, Kyriakou T, et al. Relationship between CAD risk genotype in the chromosome 9p21 locus and gene expression. Identification of eight new ANRIL splice variants. PLoS One. 2009;4(11):e7677.
39. Forbes S, Herzog H, et al. A role for neuropeptide Y in the gender-specific gastrointestinal, corticosterone and feeding responses to stress. Br J Pharmacol. 2012;166(8):2307–16.
40. Franke A, McGovern DP, et al. Genome-wide meta-analysis increases to 71 the number of confirmed Crohn's disease susceptibility loci. Nat Genet. 2010;42(12):1118–25.
41. Fried SK, Bunkin DA, et al. Omental and subcutaneous adipose tissues of obese subjects release interleukin-6: depot difference and regulation by glucocorticoid. J Clin Endocrinol Metab. 1998;83(3):847–50.
42. Fu Y, Luo N, et al. Adiponectin promotes adipocyte differentiation, insulin sensitivity, and lipid accumulation. J Lipid Res. 2005;46(7):1369–79.
43. Garcia-Vallejo JJ, van Kooyk Y. Endogenous ligands for C-type lectin receptors: the true regulators of immune homeostasis. Immunol Rev. 2009;230(1):22–37.
44. Genco RJ, Borgnakke WS. Risk factors for periodontal disease. Periodontol 2000. 2013;62(1):59–94.
45. Gibson G. Rare and common variants: twenty arguments. Nat Rev Genet. 2011;13(2):135–45.
46. Gieger C, Radhakrishnan A, et al. New gene functions in megakaryopoiesis and platelet formation. Nature. 2011;480(7376):201–8.
47. Grenier D. Degradation of host protease inhibitors and activation of plasminogen by proteolytic enzymes from Porphyromonas gingivalis and Treponema denticola. Microbiology. 1996;142(Pt 4):955–61.
48. Harismendy O, Notani D, et al. 9p21 DNA variants associated with coronary artery disease impair interferon-gamma signalling response. Nature. 2011;470(7333):264–8.

49. Hart TC, Shapira L, et al. Neutrophil defects as risk factors for periodontal diseases. J Periodontol. 1994;65(5 Suppl):521–9.
50. Haug SR, Heyeraas KJ. Modulation of dental inflammation by the sympathetic nervous system. J Dent Res. 2006;85(6):488–95.
51. Helgadottir A, Thorleifsson G, et al. A common variant on chromosome 9p21 affects the risk of myocardial infarction. Science. 2007;316(5830):1491–3.
52. Hirschfeld L, Wasserman B. A long-term survey of tooth loss in 600 treated periodontal patients. J Periodontol. 1978;49(5):225–37.
53. Ho YP, Lin YC, et al. Cyclooxygenase-2 Gene-765 single nucleotide polymorphism as a protective factor against periodontitis in Taiwanese. J Clin Periodontol. 2008;35(1):1–8.
54. Holdt LM, Beutner F, et al. ANRIL expression is associated with atherosclerosis risk at chromosome 9p21. Arterioscler Thromb Vasc Biol. 2010;30(3):620–7.
55. Hotamisligil GS, Shargill NS, et al. Adipose expression of tumor necrosis factor-alpha: direct role in obesity-linked insulin resistance. Science. 1993;259(5091):87–91.
56. Hugot JP, Chamaillard M, et al. Association of NOD2 leucine-rich repeat variants with susceptibility to Crohn's disease. Nature. 2001;411(6837):599–603.
57. Hunt KA, Mistry V, et al. Negligible impact of rare autoimmune-locus coding-region variants on missing heritability. Nature. 2013;498(7453):232–5.
58. Huynh-Ba G, Lang NP, et al. The association of the composite IL-1 genotype with periodontitis progression and/or treatment outcomes: a systematic review. J Clin Periodontol. 2007;34(4):305–17.
59. Huynh-Ba G, Lang NP, et al. Association of the composite IL-1 genotype with peri-implantitis: a systematic review. Clin Oral Implants Res. 2008;19(11):1154–62.
60. Ioannidis JP, Trikalinos TA. An exploratory test for an excess of significant findings. Clin Trials. 2007;4(3):245–53.
61. Ioannidis JP, Boffetta P, et al. Assessment of cumulative evidence on genetic associations: interim guidelines. Int J Epidemiol. 2008;37(1):120–32.
62. Janssens AC, van Duijn CM. An epidemiological perspective on the future of direct-to-consumer personal genome testing. Investig Genet. 2010;1(1):10.
63. Janssens AC, Gwinn M, et al. A critical appraisal of the scientific basis of commercial genomic profiles used to assess health risks and personalize health interventions. Am J Hum Genet. 2008;82(3):593–9.
64. Jarinova O, Stewart AF, et al. Functional analysis of the chromosome 9p21.3 coronary artery disease risk locus. Arterioscler Thromb Vasc Biol. 2009;29(10):1671–7.
65. Jiang L, Weng H, et al. Association between cyclooxygenase-2 gene polymorphisms and risk of periodontitis: a meta-analysis involving 5653 individuals. Mol Biol Rep. 2014;41:4795–801.
66. Johnson AD, Hwang SJ, et al. Resequencing and clinical associations of the 9p21.3 region: a comprehensive investigation in the Framingham heart study. Circulation. 2013;127(7):799–810.
67. Jostins L, Ripke S, et al. Host-microbe interactions have shaped the genetic architecture of inflammatory bowel disease. Nature. 2012;491(7422):119–24.
68. Karimbux NY, Saraiya VM, et al. Interleukin-1 gene polymorphisms and chronic periodontitis in adult whites: a systematic review and meta-analysis. J Periodontol. 2012;83(11):1407–19.
69. Karl T, Duffy L, et al. Behavioural profile of a new mouse model for NPY deficiency. Eur J Neurosci. 2008;28(1):173–80.
70. Karvonen MK, Pesonen U, et al. Association of a leucine(7)-to-proline(7) polymorphism in the signal peptide of neuropeptide Y with high serum cholesterol and LDL cholesterol levels. Nat Med. 1998;4(12):1434–7.
71. Kathiresan S, Newton-Cheh C, et al. On the interpretation of genetic association studies. Eur Heart J. 2004;25(16):1378–81.
72. Kauhanen J, Karvonen MK, et al. Neuropeptide Y polymorphism and alcohol consumption in middle-aged men. Am J Med Genet. 2000;93(2):117–21.

73. Kinane DF, Peterson M, et al. Environmental and other modifying factors of the periodontal diseases. Periodontol 2000. 2006;40:107–19.
74. Kraus D, Winter J, et al. Interactions of adiponectin and lipopolysaccharide from Porphyromonas gingivalis on human oral epithelial cells. PLoS One. 2012;7(2):e30716.
75. Kumada M, Kihara S, et al. Adiponectin specifically increased tissue inhibitor of metalloproteinase-1 through interleukin-10 expression in human macrophages. Circulation. 2004;109(17):2046–9.
76. Kuo LE, Kitlinska JB, et al. Neuropeptide Y acts directly in the periphery on fat tissue and mediates stress-induced obesity and metabolic syndrome. Nat Med. 2007;13(7):803–11.
77. Lahteenmaki K, Kuusela P, et al. Bacterial plasminogen activators and receptors. FEMS Microbiol Rev. 2001;25(5):531–52.
78. Lahteenmaki K, Edelman S, et al. Bacterial metastasis: the host plasminogen system in bacterial invasion. Trends Microbiol. 2005;13(2):79–85.
79. Laine ML, Loos BG, et al. Gene polymorphisms in chronic periodontitis. Int J Dent. 2010;2010:324719.
80. Lalla E, Papapanou PN. Diabetes mellitus and periodontitis: a tale of two common interrelated diseases. Nat Rev Endocrinol. 2011;7(12):738–48.
81. Lappalainen J, Kranzler HR, et al. A functional neuropeptide Y Leu7Pro polymorphism associated with alcohol dependence in a large population sample from the United States. Arch Gen Psychiatry. 2002;59(9):825–31.
82. Lemaitre RN, Tanaka T, et al. Genetic loci associated with plasma phospholipid n-3 fatty acids: a meta-analysis of genome-wide association studies from the CHARGE Consortium. PLoS Genet. 2011;7(7):e1002193.
83. Li G, Yue Y, et al. Association of matrix metalloproteinase (MMP)-1, 3, 9, interleukin (IL)-2, 8 and cyclooxygenase (COX)-2 gene polymorphisms with chronic periodontitis in a Chinese population. Cytokine. 2012;60(2):552–60.
84. Liu Y, Sanoff HK, et al. INK4/ARF transcript expression is associated with chromosome 9p21 variants linked to atherosclerosis. PLoS One. 2009;4(4):e5027.
85. Locker D, Slade GD, et al. Epidemiology of periodontal disease among older adults: a review. Periodontol 2000. 1998;16:16–33.
86. Lockhart PB, Bolger AF, et al. Periodontal disease and atherosclerotic vascular disease: does the evidence support an independent association?: a scientific statement from the American Heart Association. Circulation. 2012;125(20):2520–44.
87. Loo WT, Wang M, et al. Association of matrix metalloproteinase (MMP-1, MMP-3 and MMP-9) and cyclooxygenase-2 gene polymorphisms and their proteins with chronic periodontitis. Arch Oral Biol. 2011;56(10):1081–90.
88. Loos BG, John RP, et al. Identification of genetic risk factors for periodontitis and possible mechanisms of action.J Clin Periodontol. 2005;32(6 Suppl):159–79.
89. Lundy FT, El Karim IA, et al. Neuropeptide Y (NPY) and NPY Y1 receptor in periodontal health and disease. Arch Oral Biol. 2009;54(3):258–62.
90. Marcenes W, Kassebaum NJ, et al. Global burden of oral conditions in 1990–2010: a systematic analysis. J Dent Res. 2013;92(7):592–7.
91. Marchini J, Donnelly P, et al. Genome-wide strategies for detecting multiple loci that influence complex diseases. Nat Genet. 2005;37(4):413–7.
92. Marth JD, Grewal PK. Mammalian glycosylation in immunity. Nat Rev Immunol. 2008;8(11):874–87.
93. Mason MR, Nagaraja HN, et al. Deep sequencing identifies ethnicity-specific bacterial signatures in the oral microbiome. PLoS One. 2013;8(10):e77287.
94. Mathis D, Shoelson SE. Immunometabolism: an emerging frontier. Nat Rev Immunol. 2011;11(2):81.
95. McElhaney JE, Effros RB. Immunosenescence: what does it mean to health outcomes in older adults? Curr Opin Immunol. 2009;21(4):418–24.
96. McFall WT Jr. Tooth loss in 100 treated patients with periodontal disease. A long-term study. J Periodontol. 1982;53(9):539–49.

97. McPherson R, Pertsemlidis A, et al. A common allele on chromosome 9 associated with coronary heart disease. Science. 2007;316(5830):1488–91.
98. Mercer TR, Dinger ME, et al. Long non-coding RNAs: insights into functions. Nat Rev Genet. 2009;10(3):155–9.
99. Michalowicz BS, Aeppli D, et al. Periodontal findings in adult twins. J Periodontol. 1991;62(5):293–9.
100. Michalowicz BS, Aeppli DP, et al. A twin study of genetic variation in proportional radiographic alveolar bone height. J Dent Res. 1991;70(11):1431–5.
101. Michalowicz BS, Diehl SR, et al. Evidence of a substantial genetic basis for risk of adult periodontitis. J Periodontol. 2000;71(11):1699–707.
102. Motterle A, Pu X, et al. Functional analyses of coronary artery disease associated variation on chromosome 9p21 in vascular smooth muscle cells. Hum Mol Genet. 2012;21:4021–9.
103. Murray RZ, Kay JG, et al. A role for the phagosome in cytokine secretion. Science. 2005;310(5753):1492–5.
104. Nibali L, Donos N, et al. Periodontal infectogenomics. J Med Microbiol. 2009;58(10):1269–74.
105. Nibali L, Henderson B, et al. Genetic dysbiosis: the role of microbial insults in chronic inflammatory diseases. J Oral Microbiol. 2014;6.
106. Nikolopoulos GK, Dimou NL, et al. Cytokine gene polymorphisms in periodontal disease: a meta-analysis of 53 studies including 4178 cases and 4590 controls. J Clin Periodontol. 2008;35(9):754–67.
107. Ogura Y, Bonen DK, et al. A frameshift mutation in NOD2 associated with susceptibility to Crohn's disease. Nature. 2001;411(6837):603–6.
108. Osorio F, Reis e Sousa C. Myeloid C-type lectin receptors in pathogen recognition and host defense. Immunity. 2011;34(5):651–64.
109. Ouchi N, Kihara S, et al. Novel modulator for endothelial adhesion molecules: adipocyte-derived plasma protein adiponectin. Circulation. 1999;100(25):2473–6.
110. Ouchi N, Kihara S, et al. Obesity, adiponectin and vascular inflammatory disease. Curr Opin Lipidol. 2003;14(6):561–6.
111. Ouchi N, Parker JL, et al. Adipokines in inflammation and metabolic disease. Nat Rev Immunol. 2011;11(2):85–97.
112. Painsipp E, Herzog H, et al. Sex-dependent control of murine emotional-affective behaviour in health and colitis by peptide YY and neuropeptide Y. Br J Pharmacol. 2011;163(6):1302–14.
113. Pandey RR, Mondal T, et al. Kcnq1ot1 antisense noncoding RNA mediates lineage-specific transcriptional silencing through chromatin-level regulation. Mol Cell. 2008;32(2):232–46.
114. Patch C, Sequeiros J, et al. Genetic horoscopes: is it all in the genes? Points for regulatory control of direct-to-consumer genetic testing. Eur J Hum Genet. 2009;17(7):857–9.
115. Peng G, Luo L, et al. Gene and pathway-based second-wave analysis of genome-wide association studies. Eur J Hum Genet. 2010;18(1):111–7.
116. Pihlstrom BL, Michalowicz BS, et al. Periodontal diseases. The Lancet. 2005;366(9499):1809–20.
117. Prakash G, Umar M, et al. (2013). COX-2 gene polymorphisms and risk of chronic periodontitis: a case-control study and meta-analysis. Oral Dis. 2013;21:38–45.
118. Prelog M. Aging of the immune system: a risk factor for autoimmunity? Autoimmun Rev. 2006;5(2):136–9.
119. Preshaw PM, Alba AL, et al. Periodontitis and diabetes: a two-way relationship. Diabetologia. 2012;55(1):21–31.
120. Rabinovich GA, Croci DO. Regulatory circuits mediated by lectin-glycan interactions in autoimmunity and cancer. Immunity. 2012;36(3):322–35.
121. Rioux JD, Daly MJ, et al. Genetic variation in the 5q31 cytokine gene cluster confers susceptibility to Crohn disease. Nat Genet. 2001;29(2):223–8.
122. Rom O, Avezov K, et al. Cigarette smoking and inflammation revisited. Respir Physiol Neurobiol. 2013;187(1):5–10.

123. Samani NJ, Erdmann J, et al. Genomewide association analysis of coronary artery disease. N Engl J Med. 2007;357(5):443–53.
124. Saxena R, Voight BF, et al. Genome-wide association analysis identifies loci for type 2 diabetes and triglyceride levels. Science. 2007;316(5829):1331–6.
125. Schaefer AS, Richter GM, et al. Identification of a shared genetic susceptibility locus for coronary heart disease and periodontitis. PLoS Genet. 2009;5(2):e1000378.
126. Schaefer AS, Richter GM, et al. A 3′ UTR transition within DEFB1 is associated with chronic and aggressive periodontitis. Genes Immun. 2009;11(1):45–54.
127. Schaefer AS, Richter GM, et al. COX-2 is associated with periodontitis in Europeans. J Dent Res. 2010;89(4):384–8.
128. Schaefer AS, Richter GM, et al. A genome-wide association study identifies GLT6D1 as a susceptibility locus for periodontitis. Hum Mol Genet. 2010;19(3):553–62.
129. Schaefer AS, Richter GM, et al. CDKN2BAS is associated with periodontitis in different European populations and is activated by bacterial infection. J Med Genet. 2011;48(1):38–47.
130. Schaefer AS, Bochenek G, et al. Validation of reported genetic risk factors for periodontitis in a large-scale replication study. J Clin Periodontol. 2013;40(6):563–72.
131. Schaefer AS, Circ Cardiovasc Genet. 2015;8(1):159–67.
132. Schafer AS, Jepsen S, et al. Periodontal genetics: a decade of genetic association studies mandates better study designs. J Clin Periodontol. 2011;38(2):103–7.
133. Schunkert H, Konig IR, et al. Large-scale association analysis identifies 13 new susceptibility loci for coronary artery disease. Nat Genet. 2011;43(4):333–8.
134. Schwarzberg K, Le R, et al. The personal human oral microbiome obscures the effects of treatment on periodontal disease. PLoS One. 2014;9(1):e86708.
135. Schwenk RW, Luiken JJ, et al. Regulation of sarcolemmal glucose and fatty acid transporters in cardiac disease. Cardiovasc Res. 2008;79(2):249–58.
136. Schwenk RW, Angin Y, et al. Overexpression of vesicle-associated membrane protein (VAMP) 3, but not VAMP2, protects glucose transporter (GLUT) 4 protein translocation in an in vitro model of cardiac insulin resistance. J Biol Chem. 2012;287(44):37530–9.
137. Scott LJ, Mohlke KL, et al. A genome-wide association study of type 2 diabetes in Finns detects multiple susceptibility variants. Science. 2007;316(5829):1341–5.
138. Shea J, Agarwala V, et al. Comparing strategies to fine-map the association of common SNPs at chromosome 9p21 with type 2 diabetes and myocardial infarction. Nat Genet. 2011;43(8):801–5.
139. Shete S, Hosking FJ, et al. Genome-wide association study identifies five susceptibility loci for glioma. Nat Genet. 2009;41(8):899–904.
140. Shimomura I, Funahashi T, et al. Enhanced expression of PAI-1 in visceral fat: possible contributor to vascular disease in obesity. Nat Med. 1996;2(7):800–3.
141. Sperandio M, Gleissner CA, et al. Glycosylation in immune cell trafficking. Immunol Rev. 2009;230(1):97–113.
142. Stefan N, Bunt JC, et al. Plasma adiponectin concentrations in children: relationships with obesity and insulinemia. J Clin Endocrinol Metab. 2002;87(10):4652–6.
143. Stumvoll M, Goldstein BJ, et al. Type 2 diabetes: principles of pathogenesis and therapy. Lancet. 2005;365(9467):1333–46.
144. Syrovets T, Lunov O, et al. Plasmin as a proinflammatory cell activator. J Leukoc Biol. 2012;92(3):509–19.
145. Teumer A, Holtfreter B, et al. Genome-wide association study of chronic periodontitis in a general German population. J Clin Periodontol. 2013;40(11):977–85.
146. Tilg H, Moschen AR. Adipocytokines: mediators linking adipose tissue, inflammation and immunity. Nat Rev Immunol. 2006;6(10):772–83.
147. Tonetti M, Mombelli A. Aggressive periodontitis. In: Lindhe J, Karring T, Lang NP, Editors. Clinical periodontology and implant dentistry. Oxford, Blackwell Munksgaard; 2008. pp. 428–58.

148. Tonetti MS, Van Dyke TE. Periodontitis and atherosclerotic cardiovascular disease: consensus report of the Joint EFP/AAP Workshop on Periodontitis and Systemic Diseases. J Periodontol. 2013;84(4 Suppl):S24–9.
149. Trott JR, Cross HG. An analysis of the principle reasons for tooth extractions in 1813 patients in Manitoba. Dent Pract Dent Rec. 1966;17(1):20–7.
150. Turnbull C, Ahmed S, et al. Genome-wide association study identifies five new breast cancer susceptibility loci. Nat Genet. 2010;42(6):504–7.
151. Turnbull C, Rapley EA, et al. Variants near DMRT1, TERT and ATF7IP are associated with testicular germ cell cancer. Nat Genet. 2010;42(7):604–7.
152. Uno S, Zembutsu H, et al. A genome-wide association study identifies genetic variants in the CDKN2BAS locus associated with endometriosis in Japanese. Nat Genet. 2010;42(8):-707–10.
153. Vaithilingam RD, Safii SH, et al. Moving into a new era of periodontal genetic studies: relevance of large case-control samples using severe phenotypes for genome-wide association studies. J Periodontal Res. 2014;49:683–95.
154. Van der Velden U, Abbas F, et al. Java project on periodontal diseases. The natural development of periodontitis: risk factors, risk predictors and risk determinants. J Clin Periodontol. 2006;33(8):540–8.
155. van Kooyk Y, Rabinovich GA. Protein-glycan interactions in the control of innate and adaptive immune responses. Nat Immunol. 2008;9(6):593–601.
156. van Kooyk Y, Kalay H, et al. (2013). Analytical tools for the study of cellular glycosylation in the immune system. Front Immunol. 2013;4:451.
157. Voight BF, Scott LJ, et al. Twelve type 2 diabetes susceptibility loci identified through large-scale association analysis. Nat Genet. 2010;42(7):579–89.
158. Wilkening S, Chen B, et al. Is there still a need for candidate gene approaches in the era of genome-wide association studies? Genomics. 2009;93:415–9.
159. WTCCC. Genome-wide association study of 14,000 cases of seven common diseases and 3000 shared controls. Nature. 2007;447(7145):661–78.
160. Xie CJ, Xiao LM, et al. Common single nucleotide polymorphisms in cyclooxygenase-2 and risk of severe chronic periodontitis in a Chinese population. J Clin Periodontol. 2009;36(3):198–203.
161. Yamauchi T, Nio Y, et al. Targeted disruption of AdipoR1 and AdipoR2 causes abrogation of adiponectin binding and metabolic actions. Nat Med. 2007;13(3):332–9.
162. Yang WS, Lee WJ, et al. Weight reduction increases plasma levels of an adipose-derived anti-inflammatory protein, adiponectin. J Clin Endocrinol Metab. 2001;86(8):3815–9.
163. Yasuno K, Bilguvar K, et al. Genome-wide association study of intracranial aneurysm identifies three new risk loci. Nat Genet. 2010;42(5):420–5.
164. Yokota T, Oritani K, et al. Adiponectin, a new member of the family of soluble defense collagens, negatively regulates the growth of myelomonocytic progenitors and the functions of macrophages. Blood. 2000;96(5):1723–32.
165. Zabaneh D, Balding DJ. A genome-wide association study of the metabolic syndrome in Indian Asian men. PLoS One. 2010;5(8):e11961.
166. Zeggini E, Weedon MN, et al. Replication of genome-wide association signals in UK samples reveals risk loci for type 2 diabetes. Science. 2007;316(5829):1336–41.
167. Zheng W, Flavell RA. The transcription factor GATA-3 is necessary and sufficient for Th2 cytokine gene expression in CD4 T cells. Cell. 1997;89(4):587–96.
168. Zhong Q, Ding C, et al. Interleukin-10 gene polymorphisms and chronic/aggressive periodontitis susceptibility: a meta-analysis based on 14 case-control studies. Cytokine. 2012;60(1):47–54.
169. Zhou Z, Zhu G, et al. Genetic variation in human NPY expression affects stress response and emotion. Nature. 2008;452(7190):997–1001.

Aphthous Stomatitis

Ricardo Carneiro Borra and Priscila Maria de Andrade

Introduction and Concepts

The main cause of RAS is unknown; numerous studies have tried to associate the etiology of the disease with specific bacteria and viruses, but without success. Currently, RAS has been classified as an auto-inflammatory disease based on its relationship with a probable dysfunction of the innate immunological response without evidence of alterations in the adaptive immune reaction. Many of the common characteristics of this group of diseases, such as (1) genetic predisposition, (2) multifactorial origin, (3) local triggering factor, (4) primary dysfunction of the innate immune system related with aberrant responses to Pathogen-Associated Molecular Patterns (PAMPs) or Damage-Associated Molecular Patterns (DAMPs), and 5) prominent neutrophil response associated with intensification of the cascades of inflammatory cytokines (IL-1β and TNF-α) are present in patients with RAS [1].

Clinical Classification

RAS can manifest in different ways and depending on the morphology, the size, duration and the distribution of the lesions can be classified in minor, major, or herpetiform. However, independent of the type of disease, the lesions appear as non-specific oral ulcers, which heal themselves spontaneously and recur after variable periods of time [2]. Currently, some authors have categorized the RAS disease as simple or complex (idiopathic or secondary), based on the degree of damage and

R. C. Borra (✉)
Genetic and Evolution Department, Federal University of São Carlos (UFSCar), São Carlos, Brazil
e-mail: rcborra@gmail.com

P. M. de Andrade
Medicine Department, Federal University of São Carlos (UFSCar), São Carlos, Brazil
e-mail: pmborra@gmail.com

S. T. Sonis (ed.), *Genomics, Personalized Medicine and Oral Disease*,
DOI 10.1007/978-3-319-17942-1_8

the aggressiveness of the clinical situation. Simple RAS, considered the most prevalent, is characterized by lesions of diverse morphology which recur after variable periods of time with distinct intervals of remission. Complex RAS is characterized by rising of more than three lesions in the oropharynx and/or in the genital regions, which appear sequentially or in very short intervals of time and always in the absence of other characteristics, which could dismiss Behcet´s Disease. Complex aphthosis may be sub-classified as primary (idiopathic) when the manifestations of lesions surge independently, or as secondary in situations when the lesions are associated with other systemic conditions such as AIDS, cyclical neutropenia, Crohn´s Disease, ulcerative rectocolitis, celiac disease, vitamin deficiencies, hematological alterations, or PFAPA or MAGIC syndromes, and so forth [3].

Regardless of the classification, the diagnosis of RAS is performed based on the history and clinical manifestations of the disease. There are no laboratory exams to diagnose this disease. Histologically, RAS is characterized by an ulceration of the buccal mucosa covered by a fibropurulent exudate and a chronic non-specific inflammatory infiltrate confined to the lamina propria. In the pre-ulcerative stage, the suprabasal degeneration of the epithelium begins, accompanied by a lymphocyte infiltrate in the lamina propria, compatible with viral and immunological etiologies. As the lesion progresses, the epithelium suffers ulceration and the infiltrate will be composed predominantly by neutrophils and, in the more advanced stages, by monocytes or macrophages and occasionally by eosinophils as well. In the peripheral region of the ulcer, the quantity of lymphocytes and macrophages increases while that of the neutrophils decreases [4–6].

Studies have shown that RAS patients present aberrations in the proportions of T CD4+ and CD8+ lymphocytes, which are important for immunological regulation and vigilance. In patients with RAS, there is a decrease in the number of CD4+ lymphocytes, with a reduction in the ratio of CD4+/CD8+ [2, 7–9], which may favor the development of the cytotoxic immune response mediated by T CD8+ lymphocytes against the oral epithelium [10]. The target present in the epithelial cells capable of stimulating the reaction of the immune system is unknown. Heat Shock Proteins (HSPs) have been considered possible candidates. The lymphocytes from RAS patients present higher proliferation indeces than the lymphocytes of individuals without the disease when stimulated with peptides coming from *Microbacterium bovis* HSP or with peptides derived of the human homolog HSPs [11, 12]. The increase in the lymphocyte proliferation was also observed in cultures of lymphocytes exposed to the *S. mutans and S. sanguis* bacteria or to the D glycosyltransferase antigen [13]. The most probable explanation of why certain components of the microorganisms inhabiting the oral cavity induce an inflammatory response in some individuals and not in others involves the regulatory mechanisms of the immune response.

TH1 Polarization of Immune Response in RAS

The gastrointestinal tract is one of the parts of the body where the greatest contact exists between bacterial and food antigens and the immunological system. In this environment, it is especially important for the immune response to remain under strict control. The peripheral tolerance is the post-natal physiological mechanism responsible for the inhibition of humoral and cellular responses against auto- and harmless foreign antigens that penetrate through the mucous membranes. Loss of oral tolerance may explain the appearance of auto-inflammatory and hypersensitive reactions against food proteins and common bacterial components from local microbiota. The inducement of the peripheral tolerance is associated with the preferential activation of the Th2 (IL4), Tr1 (IL10), and Th3 (TGF-B) -type lymphocytes responses, and with the activity of the CD4+ CD25+ T regulatory cells [14].

Recently, the abnormal immune response of the cellular type has been considered an important factor in the development of oral lesions in RAS [15]. Many chronic inflammatory diseases that affect the gastrointestinal tract are characterized by the loss of peripheral immunological tolerance with a consequent polarization of the Th1-type immune response [16]. The analysis of the gene expression in RAS lesions and normal mucous membranes showed an increase of the transcripts of such genes as IL-2, INF-γ, and STAT1, among others, and hyper-expression of the IP10, MIG, MIP1α, and MIP1β chemokines, which are associated with the activation and attraction of cells responsible for the Th1-type response [17–20]. Furthermore, the defense cells present in the blood show the same pattern of immune response observed in the lesions. *In vitro* studies showed that peripheral blood mononuclear cells (PBMC) from RAS patients stimulated with phytohaemagglutinin are able to produce greater quantities of Th1-type cytokines (IFN-γ, TNF-α, IL-2, IL-6 and IL-8) in the active and remission phases, compared to PBMCs from individuals without the disease. In contrast, the production of the anti-inflammatory cytokines like IL-10 and TGF-B, or the number and the inhibitory activity of $CD4^{+}CD25^{+}$regulatory T cells are lower in RAS patients relative to the control counterparts [21, 22].

Although RAS is characterized by localized lesions, its causes seem to be systemic in character. The factors that positively influence the appearance of RAS, such as stress, medications (anti-inflammatory drugs, beta-blockers, IFN-α therapy, and imiquimod), hormonal alterations, and systemic diseases (Behcet´s disease, celiac disease, and Crohn´s disease), have been correlated with the stimulation of the Th1 immune response pattern. Factors that negatively influence the appearance of RAS such as pregnancy, use of tobacco, and some medications (tetracycline, dexamethasone, pentoxifylline, dapsone, colchicine, and thalidomide), have been related to the Th2 profile. The relation between these factors for worsening or improvement of RAS and the Th1/Th2 balance suggests the existence of a hyper-responsiveness Th1 immunological state in patients with this disease (Fig. 1) [23].

This increase in responsiveness can be confirmed in a number of patients by pathergy testing. The oblique introduction of a thick caliber needle (20 Gauge) into the skin of the forearm or in the labial submucosa induce, after 24 to 48 h, the appearance of an erythematous or pustular nodules in the skin, or ulcerations of the buccal mucosa in cases considered as positive, especially if the needle has been contaminated by the patient´s own saliva [24]. Pathergy is a clinical phenomenon related to the alteration of the innate or adaptive immunity associated with Th1 or Th2 reactivity triggered by the trauma [25, 26]. The positivity of the pathergy test has mainly been associated with RAS patients that present atopic diathesis [27].

Despite the Th1 character of RAS, a variable number of patients possess histories of allergic diseases and patterns of serological response associated with the Th2

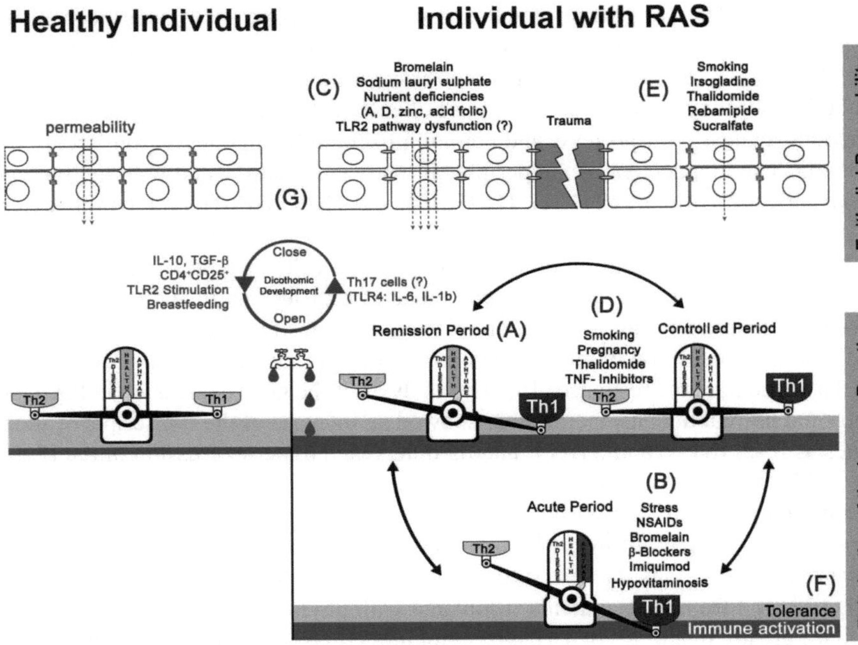

Fig. 1 The Th1-type hyper reactivity response model of the RAS—(*A*) Th1/Th2/Treg imbalance of RAS of polygenic character. (*B*) The factors associated with the Th1/Th2/Treg imbalance (*C*) and those related to the increase of the permeability of the epithelium, allow that buccal antigens induces a disproportionate immune response against epithelial cells and the onset of the disease. (*D*) The environmental factors that correct the Th1/Th2 imbalance and/or (*E*) reduce the permeability of the epithelium are able to prevent the triggering of the disease by a normal stimulus. (*F and G*) The factors that augment peripheral tolerance, such as IL-10, TGF-β, and CD4+ CD25+ can neutralize the Th1/Th2 imbalance and aiding both the prevention of RAS onset and the amelioration of its outcome. The production and activation of CD4+ CD25+ cells that is inversely proportional to Th17 cells, (*G*) can certainly influence the maintenance of the peripheral tolerance in the buccal cavity [24]

type immune response [20, 28, 29]. In a study quantifying the level of Th1 and Th2 cytokines in the blood, we found a high expression (above average) of INF-γ, IL-12 (Th1) and IL-4 (Th2) in half of the patients with RAS (unpublished data) showing that the pattern of the immunological disorder is heterogeneous. That the lesion is characterized by the Th1 immune response does not exclude the possibility of the patient showing, concomitantly, a hyper-reactive Th2 disorder, such as occurs in Behcet´s disease where the Th1 and Th17 responses are involved in the etiology of the disease and the Th2 is associated with the development and severity of the clinical outcomes [30]. In these cases, it is probable that the diverse alterations occur in common pathways that modulate the Th1 and Th2 immune responses such as in the dysfunction of the CD4+ Foxp3+ group of lymphocytes encountered in intestinal inflammatory diseases and in RAS [22, 31].

TNF-α as the Key Cytokine for RAS

Among the Cytokines associated with RAS, TNF-**α** plays a central part in the pathophysiology of the disease. Studies show that the gene expression of the transcripts related to the Th1–type immune response (TNF-**α**, IL-2, IFN-γ) increases in the canker lesion, while anti-inflammatory cytokine IL-10 decreases in comparison with traumatic oral ulcers. The leukocytes from peripheral blood of RAS patients produce elevated levels of TNF-**α** in comparison with the control group. Immunohistochemical studies show that the cells of the inflammatory infiltrate present in the lesions express a higher quantity of TNF-**α** in relation to traumatic oral ulcers [20, 21, 32–34]. The systemic use of *etanercept*, a synthetic protein inhibitor of TNF-**α**, facilitates healing and reduces the number of oral lesions considered as being recalcitrant in RAS patients [35]. TNF-**α** is one of the main inflammatory mediators secreted by macrophages, which among many other activities, is capable of inducing the production of collagenases, the apoptosis, the chemotaxis of neutrophils and monocytes, the increase of the cytotoxic activity of neutrophils and the necrosis in tumor cells.

The TNF-α production is stimulated mainly by the activation of cellular membrane proteins known as pattern recognition receptor (PRR) and whose one of the main families is represented by Toll-like Receptors (TLRs). The family of TLRs is composed of 11 members (TRL1—TRL11), each of which has a capacity to recognize different types of conserved molecular signatures from bacteria, viruses, fungi, and even of host proteins. TLR1 recognizes triacyl-lipopeptides from bacteria and micobacteria. TLR2 identifies lipoproteins from diverse pathogens, peptidoglycans, lipoteichoic acid from Gram + bacteria, zymosan from fungi, and the HSP70 protein from the host. TRL3 recognizes the double helix of viral RNA, while TLR4 identifies LPS of Gram- bacteria and the host components HSP60, HSP70, fibronectin, oligosaccharides of hyaluronic acid, polysaccharides of heparan sulfate, and fibrinogen. TLR5 recognizes flagellin, and TLR6, lipopeptides from mycoplasm. TLR7 and TLR8 identify the synthetic compounds imidazoquinoline (Imiquimod), loxoribine, and bropirimine. TLR9 identifies islands of CpG from bacterial DNA. The binders for TLR10 have not been identified yet, and that of TLR11 is Profilin (*Toxoplamagondii* and urogenic *Escherichia coli*) [36].

In view of these recognition functions, the TLRs perform essential roles connecting the innate and adaptive immune responses. The interaction between the TLRs and their respective ligand is able to induce the expression of co-stimulating molecules, the secretion of pro-inflammatory cytokines, phagocytosis, antigen processing, the migration of professional antigen-presenting cells to the lymphocyte forming centers, and influencing the polarization of the immune response [37, 38].

In the past, it was imagined that the immune response was activated every time the host's defense system found an unknown component. But, according to the new paradigm of immunology, the activation of the immune system requires not only the recognition of the unknown, but principally the establishing of the danger offered by this component. The pattern recognition receptors, such as the TLRs, which identify the molecular signatures associated with the pathogens (PAMPs) and the molecules liberated during tissue damage (DAMPs), are responsible for the classification of the potentially damaging effects of the situation [39]. When they are exposed to the signatures of pathogenic microorganisms, they supply a danger signal to activate the immune response, but when they encounter non-pathogenic microorganisms, the co-stimulatory signal is not produced, causing the inhibition of the immune response [38, 40]. Stimulation of TLR2, TLR2/1 or TLR2/6 by ligands encountered in commensal bacteria can yield a response characterized by little IL-12p70, vigorous IL-10, and a preference toward Th2 or T regulatory responses while, the stimulation of DCs by the others TLRs can result in the induction of Th1 type immune reaction [38].

Importance of the Professional Antigen-Presenting Cells to the Maintenance of Bucal Tolerance

The induction of the primary immunological response is initiated by the presentation of exogenous antigens processed by the dendritic cells (DC), which are considered the only professional Antigen-Presenting Cells (APCs) capable of activating the T helper lymphocytes and directing the nature of the immune response in the secondary lymphoid organs. These same types of cells may present the processing antigen directly to the CD4+ T cells or the CD8+ memories T cells in the peripheral tissue to activate the immune response [41]. The buccal cavity can be considered a privileged immunological site, where, despite the constant exposure to antigens coming from commensal microorganisms and diverse food substances, a homeostatic state prevails. For this to be possible, tolerance mechanisms must counterbalance the stimuli of activation of the immunological system. The dendritic cells of the oral mucosa probably have an active participation in this process. They are composed of a heterogeneous population of APC cells, where some are responsible for the induction of the pro-tolerogenic activity and others for the Th1 pro-inflammatory response [42, 43].

In the oral cavity, four principal types of professional antigen-presenting cells can be found. Langerhans cells (LCs), the first type, are observed in greater numbers in the inter-epithelial region of the buccal (vestibular), lingual, gum and sublingual mucous membranes where they spread its extensions in the direction of the neighboring cells or on the surface of the epithelium to form a network. LCs are responsible for monitoring of the external stimuli and for the tolerogenic activity. The expression of the TLR2 and TLR4 is much greater in these cells in relation to the epidermal LCs. When they are stimulated by lipopolysaccharide (LPS), the buccal LCs exhibit increased expression of co-inhibiting molecules B7.H1 and B7.H3, and diminished expression of co-stimulating molecules B7.2 (CD86), inducing the polarization of the T lymphocytes into regulatory type [44, 45]. A study of the LCs of the skin showed that this type of APC recognized, through the heterodimer TLR2/6, the signatures of commensal Gram-positive bacteria, and produce IL-10 [46]. The interstitial dendritic cells (CD11b+/CD11c+), which comprise the other subtype of DCs, are found in greater density in the lamina propria of the buccal, gum and sublingual mucous membrane regions respectively, and are responsible for the activation of the CD4+ and CD8+ cells in the secondary lymphoid organs. The interstitial dendritic cells CD11b+/CD11c−, the third subtype, reside in the lamina propria of all regions, and are the largest DCs of the sublingual area. In mice, this group of APCs also exercises an immunological tolerance induction activity. Finally, the plasmacytoid dendritic cells (pDCs) are found infrequently in the healthy mucous membrane, although they are observed in greater density in the submucosa, in the muscular region of the floor of the mouth and in all oral inflammated mucosa [42, 47]. Research into APC cells in RAS is scarce. There is only a single study which found an increase in the density of interstitial dendritic cells (Factor XIIIa+) in ulcerated RAS lesions in comparison with traumatic ulcers and clinically normal mucous membranes. The location of the Factor XIIIa+ DCs was restricted to the area of mononuclear cells and the perivascular region [48].

The precursors of the dendritic cells originating in the bone marrow are attracted to the supra-basal and basal regions of the epitheliums by chemokines such as RANTES, MIP-1, MIP-2, and MCP-1, and they are stimulated to differentiate into DCs by the factors produced by the cells of the mucous membrane micro-environment, such as, the TNF-α, GM-CSF, IL-3, and TGF-b1, among others [49]. The DCs in their immature form exercise vigilance activity, and when they encounter exogenous antigens capable of activating the Pattern Recognition Receptors (PRRs), they undergo a maturation process, diminishing their phagocytic capacity and enhancing the mechanisms involved in the processing of antigens. In parallel, there is an attenuation of the expression of chemokine receptors that keep them on the peripheral site and an increase of those that attract them (CCR7) to the secondary lymphoid organs for the activation of the type of immune response that is effective against the pathogen encountered. The Langerhans cells of the buccal cavity migrate more slowly than the interstitial DCs, and express a lesser quantity of co-stimulatory molecules [42].

The APCs of the oral mucosa may be influenced by environmental factors. For example, the nicotine from cigarette smoking modulates the capacity of the dendritic cells to respond to LPS, modifying its activation pattern of the Th1-type immune response to the Th2-type [42]. This may partially explain the association between the low prevalence of RAS in the group of individuals who smoke [50].

Importance of the Epithelial Cell in Maintaining Buccal Tolerance

The balance between the immunological activation and peripheral tolerance is associated with the manner in which the organism can discriminate between commensal and pathogenic microorganisms in the buccal cavity. The epithelial cells, keeping most of the time intimate contact with the agents and components of the external environment, have an enormous participation in the process of stimulating or modulating the immunological response. The keratinocytes, aside from their barrier protective activities, are essential for the differentiation of the Langerhans cells [51]. E-cadherin, an adhesion molecule found normally in keratinocytes, is present only in the APCs of the LC type. The E-cadherin of the LCs is responsible for the stable union of the immune cell to the keratinocytes, impeding the massive migration of the immature LCs to the secondary lymphoid organs. The loss of adhesion of the LCs due to infections, tissue disruption and inflammatory cytokines, for example, is sufficient to permit the induction of the maturation of the LCs through the activation of the channels related to the β-catenin. In certain dendritic cells, the disconnection of E-cadherin is capable of carrying them to their maturation without stimulating the liberation of cytokine IL-12, considered to be the third signal for the activation of the immune response, thus producing a tolerogenic activity. A similar mechanism probably acts on the buccal LCs, so that they can function by modulating the immunological response of the mucous membranes [51].

Importance of the Epithelial Celular Permeability to the Maintenance of Buccal Tolerance

The loss of integrity of the oral mucous membrane has a great importance for the etiopathogeny of RAS. The buccal mucosa is considered a relatively impermeable tissue when compared with the intestinal mucosa. The buccal stratified squamous epithelium, which is made up of various layers of non-keratinized or ortho-keratinized cells connected by desmosomes, forms a barrier against antigens of the oral cavity. The lingual submucosa is an exception; in function of its thickness and the absence of keratin, it is much more permeable than the rest of the buccal cavity. Thus, aphthae are more prevalent in non-keratinized regions of the oral mucosa, where micro traumas are probably sufficient to permit the penetration of antigens in the lamina propria. In the buccal cavity, the sites with higher concentrations of Langerhans cells are those covered by thinner non-keratinized mucosa, such as the floor of the mouth, the underside of the tongue, the oropharynx region, the labial mucosa, and the soft palate, which also correspond to the locations with a greater prevalence of ulcerated RAS lesions [52]. Since the epithelial protection is lesser in these locations, and the absorption of antigens is potentially greater [42], any local or systemic alteration which diminishes the tolerance or affects the epithelial barrier could favor the appearance of an immunological response at these locations. We can speculate that the increase of LCs and of their tolerogenic activity could counterbalance the greater antigenic stimulation in these areas.

In this sense, circumstances that increase the permeability of the mucosa, such as acute trauma or the reduction of the induced epithelial protection, including, for example, by bromelain, sodium lauryl sulfate, or diverse types of nutritional deficiency, factors considered as being precipitants of RAS, could possibly favor the contact of products of the oral microbiota with the cells of a hyper-reactive immune system. On the other hand, frequent exposure to products of cigarette combustion and the use of medications that increase the intracellular adhesion (irsogladine [53], sucralfate [54]) could actuate in the opposite direction, improving the efficiency of the epithelial barrier and preventing the interaction of exogenous antigens with the immune system cells.

Importance of the TLR in the Control of Immune Response and the Permeability of the Buccal Mucosa

The TLRs, besides being involved in the activation of immune system cells, are fundamental to the control of the permeability of the epithelial barrier. The diverse types of TLRs have already been found in different layers of the buccal mucosa epithelium. Studies in the gut have shown that stimulation of TLR2 decreases the mucosa permeability by increasing the function of the tight junctions [55]. Deficient TLR2 signaling may cause an imbalance in the commensal-dependent epithelial barrier defense, facilitate mucosal permeability, and lead to an increase in susceptibility to chronic mucosal inflammatory diseases. The deficiency of this receptor has also already been implicated in the development of colitis in an experimental animal model [56]. On the other side, the reduction in permeability mediated by TLR2 stimulation is capable of improving the outcomes of experimental colitis [55, 57]. The presence of the heterozygote polymorphism in the TLR2 genome sequence, commonly associated with the minor activity of the receptor, has already been related to the most severe phenotype of ulcerative recto-colitis [58].

RAS patients seem to present an alteration in the stimulation activity of the TLRs, which might be associated with the decrease of epithelial protection, loss of tolerance, or stimulation of the cytotoxic immune response against the epithelium. In a previous work, studying the activity of the PBMCs stimulated by diverse types of ligands to the TLR, we noted the existence of a deficiency in the response of the PBMCs of RAS patients when stimulated by the HKLM and LTA, inductors of the TLR2 homodimer receptors, and PamC3SK4, an activator of the TLR2/1 heterodimer [59]. Similar results were encountered in other research which quantified the stimulatory activity in the PBMCs of patients with Behcet´s disease and the macrophages of patients with Crohn´s disease exposed to TLR1/2 ligands [60, 61]. The alteration in the functioning of the TLR2 seems to be unrelated to the variability in the expression of the mRNA [62]; however, the possibility of modification of the transcript cannot be discarded, since alternative splices of the TLR2 associated with the aggressiveness of the disease were observed in patients with Behcet´s disease [60].

Diverse types of cells express TLRs, and the disorganization in the functioning of the TLR2 in the PBMCs may indicate the presence of a defect in the activation/regulation channels of this type of receptor in other cell groups. As with the PBMCs, the epithelial cells present TLR2, and the inhibition of their activity has already been related to enabling microorganisms to penetrate the underlying tissue [63]. Alterations in the expression pattern of TLR2 have been observed, as well as in the epithelial cells of RAS patients [64]. The 2, 5, 6, 7, and 8 TLRs are organized in a polarized fashion in normal epithelium, concentrating themselves principally in the cells near the basal layer. However, in cases of RAS, with the apparent loss of polarization, the TLRs extend themselves throughout the entire thickness of the epithelium [64]. Analyzing the data on genetic expressions of aphthous lesions, mucosa without ulceration from patients, and normal mucosa of controls individuals without the disease, deposited in the GeoDataSet (NHI, GSE37265), we find a greater expression of TLR1 up to 10 in the ulcerated lesions from RAS patients in comparison with the non-ulcerated tissue samples from patients and controls. The only exception was associated with the TLR5, which was greater in the non-ulcerated mucosa of RAS patients in relation to the genetic expressions of ulcerated lesions and normal tissues. This pattern has also been observed in lesions and PBMC cells of RAS patients [62]. On the other hand, when comparing the non-ulcerated tissue of RAS patients and of individuals without the disease, a greater expression of the TLR2, 9 and 10 was observed (Table 1). It is still too early to affirm that these results are related to the influence of the inflammatory process adjacent to the biopsied area, or if they represent an alteration in the mucous pattern of the RAS patient, since the area studied seems to exhibit a considerable subjacent inflammatory process [64].

The benefits of cigarettes or of nicotine in the decrease of lesions and the control of outbreaks in RAS patients are already well known [65]. It was always believed that this occurred because of the increase in the resistance of buccal mucosa from the stimulus of the keratinization. However, the immunomodulatory effects of

Table 1 Expression pattern or activity of the TLRs present in RAS patients

TLR	PBMC—(activity)	PBMC (mRNA)	Ulcerated Lesion (mRNA)	Ulcerated Lesion (mRNA)	Healthy Mucosa (mRNA)	Keratinocytes (Upper protein)
1	*Loss of response from TLR2 and TLR1*	NS	NS	↑	NS	NS
2		NS	↑	↑	↑	*Loss of polarization*
3	NS	NS	↓	↑	NS	NS
4	NS	NS	NS	↑	NS	NS
5	NS	↓	↓	↓	↑	*Loss of polarization*
6	NS	NS	NS	↑	NS	*Loss of polarization*
7	NS	NS	NS	↑	NS	*Loss of polarization*
8	NS	NS	NS	↑	NS	*Loss of polarization*
9	NS	NS	NS	↑	↑	NS
10	NS	NS	NS	↑	↑	NS
Article	[59]	[62]	[62]	GSE37265	GSE37265	[64]

ns considered non-significant

nicotine have been documented. The immune system cells possess cholinergic receptors, and their activation has already been connected to the increased functioning of TLR2 and TLR9. In patients with sarcoidosis, a type of chronic granulomatous Th1disease, the use of nicotine is able to restore the responsiveness of TLR2 and expand the T cell regulators in the group of patients with low TLR2 activity [66]. In epithelial cells of the lungs, the lack of TLR2 activity is associated with an increase in the expression of IL-8, induced by the dependent activity of the nicotine receptor [67]. In a study using epithelial cells of the gum tissue, nicotine did not exert the same effect on the IL-8 level; nonetheless, the results cannot be compared, since TNF-α was used in combination with nicotine to treat epithelial cells stimulated with a TLR2 ligand [68]. In PBMCs, nicotine was capable of inhibiting the liberation of TNF-α and IFN-γ (Th1 cytokines), but did not affect the secretion of IL-6, IL-1β and IL-10 [69]. Pulmonary macrophages, PBMCs, and monocytes from individuals that smoke presents a decrease in the pro-inflammatory activity of the TLRs (TNF-α, IL-1β, IL-6, IL-8, RANTES), but not in the anti-inflammatory activity (IL-10, IL-1RA) [70]. Generally, these studies indicate a modulator effect from cigarettes, mediated by the interference in the TLR activity, principally those related to TLR2.

Research into the importance of the TLRs in RAS is still in its initial stage, and many questions still remain open. Is the association between the deregulation of the TLRs and RAS of the cause-and-effect type, or are they concomitant manifestations of the pathogenicity of the disease? To answer this type of question, the emphasis of research must shift from case-control observations to experimental tests, so that the functioning of the cells involved in RAS is better analyzed.

Future Perspectives

Even though the etiopathogeny of RAS is still unknown, it is considered a multifactorial, complex disease where the deficient regulation of the immune system, the increase in the innate response associated with the unbalance of the local microbiota, and the dysfunction of the epithelial barrier are probably involved in the emergence and evolution of ulcerative lesions.

In this type of disease, affected individuals normally possess a combination of genetic and environmental factors, which propagate the development of the phenotype involved. Unlike the monogenic diseases, the genetic factors connected to this complex disease are of a polygenic character, and because of this they are difficult to identify. In such cases, the combination of genes of low frequency is responsible for the establishment of the conditions which lead to the diseased phenotypes. Therefore, studies that try to attribute the cause to specific genes or find a Mendelian inheritance type are inconclusive. Aside from this, since the genetic factors are heterogeneous, the origin of the disease for each individual may be different, despite the same final outcome. This hinders the precise identification of the causes and the establishment of a unique therapeutic protocol, which functions in all cases.

In this scenario, the application of personalized medicine, in which the treatment is prescribed for a very specifically determined group of patients, will be of fundamental importance. But for this, knowledge of the altered signaling pathways and the Pharmacogenomics associated with each individual will be indispensable.

Currently, it is possible to establish the association between complex diseases and genetic variations utilizing techniques of large scale genotyping. Unlike the traditional studies of polymorphisms, the Genome-Wise Association Studies (GWAS)—as this approach is known—permit thousands of Single Nucleotide Polymorphisms (SNPs) and other DNA alterations to be researched simultaneously. Despite the enormous benefits, huge barriers still limit their utilization. The first challenge is the number of individuals needed to obtain the desirable statistical power. As the quantity of data required and the individual variability of the population are very great, the chance of obtaining a positive association between SNPs and some other characteristics not related to the disease increase. To prevent this from occurring, this type of study normally analyzes the genome of hundreds, or even thousands, of individuals. Another great challenge is the interpretation of the relevance of the susceptible loci, since large parts of the identified segments are located in non-coding regions of the DNA, complicating the planning for testing hypotheses. This difficulty has been circumvented through the utilization of information from functional studies of genetic or protein expression [71], or through the comparison of sets of SNP candidates with known signaling pathways [72].

The high-throughput gene expression analysis is another area in expansion, complementary to the GWAS type studies. The quantification of gene expression in tissues or cells in large scale has permitted identification of candidate pathways involved in the pathogenesis of the diseases. In this type of approach, microarray platforms have currently been substituted by the sequencing of the transcriptome of target tissues. As advantages in relation to technology of the microarray, the next-generation sequencing (NGS) not only allows the precision quantification of the mRNA expression, but also permits the identification of the variety of SNPs in the sequenced exons. This type of analysis has produced an enormous quantity of information, accessible on the US National Center for Biotechnology Information (NCBI) site (http://www.ncbi.nlm.nih.gov/gds). As a function of greater availability of data, the next challenge is to unify the different types of information, and to use data mining methodologies in order to find answers to the clinical questions.

In the meantime, diseases with high interdependencies of factors, such as RAS, form a complex biological system that cannot be fully understood by using only reductionist approaches. In this case, the search for patterns and signatures related to the diseases is much more significant than the search for a specific target gene or protein. As the complex diseases are caused by multiple alterations, the identification of altered pathways becomes more important for recognizing the process that is triggering the disease. For this, the analysis of Biological Systems, based on co-expression networks, is an important mathematical tool, which has been widely used to discriminate the pathways related to the affected phenotype [73].

The networks of co-expression (matrix of co-expression), used in this type of analysis, can be constructed by measuring the level of association between the genes or proteins expressions obtained from large scale experiments. These networks describe the co-variation between pairs of transcribed genes or proteins. Each gene or protein of the matrix represents a node in the correlation network. The nodes (e.g.: genes or proteins) that possess correlation with a large number of other nodes are considered key to the regulation of biological phenomena. Any interference that occurs in these highly connected nodes, known as "hubs", has the potential to destabilize the system in which they participate.

In the analysis of expression commonly performed, the concern is focused on identifying the differentially expressed genes or proteins, without considering whether the target in question presents a high or low value of connectivity with the other elements. In this way, a gene or a protein considered differentially expressed and with a low connectivity often present a small influence in the biological context in which it appears. On the other hand, the genes or proteins that are considered regulators are normally part of the highly preserved signal transduction pathway, and are mostly responsible for the central control of the biological phenomena.

The nodes that present the same correlation pattern can be grouped in units called "modules", which are strongly enriched into specific functional categories or cellular markers. One of the advantages in the elucidation of the functional significance of the modules, in comparison with isolated genes or proteins, is the greater reproducibility [74]. The adoption of a strategy for data mining based on modules may simplify the identification of more stable biomarkers in relation to the methodologies centered on genes or proteins. Since the modules are composed of many elements (genes or proteins), the experimental noise that corrupts the expression signals of sporadic nodes will hardly affect the pattern of expression of the modules. Another advantage in the utilization of modules as the unit of comparison is their high reproducibility. This characteristic furnishes a natural structure for comparisons between the species, the tissues, and between different physiological or pathological conditions [74, 75].

Furthermore, the nodes with high connectivity among the modules—intra-modular hubs—and which usually are related to the disease under study are often of clinical importance. For example, intra-modular hubs in a module of cellular proliferation in studies of cancers present an association with the life expectancy of the patients [76, 77].

The evidence shows that this methodology may lead to important biological discoveries [74]. This type of approach has been successfully used in the study of cross regulations between the immune system, the microbiota, the epithelium, and intestinal metabolism [78]. The results from a previous study showed inter-relationships between two apparently unrelated networks, that of the lipid metabolism and that of the immune system regulation. The authors correlated the results with clinical findings and established the explanation for the affinity between functional defects of the immune system and lipid absorption deficiency [78]. This type of analysis was also applied with success to the study of systemic lupus erythematosus. The authors of this study mapped the changes in gene expression, using the modules to construct disease signatures, which permit the visualization and functional interpretation of microarray data in a more stable and reproductive way [79]. In the study, about polarization of immune response in RAS executed in 2004, we had success using these concepts to characterize the expression of the Th1 and Th2 modules in the ulcerated lesions [19].

Conclusion

Recurrent Aphthous Stomatitis is an entity, which represents a complex biological system. Unlike earlier methods, the identification of specific signatures formed by regulators nodes may be of great use in determining the main signaling pathways and in defining preferential therapeutic targets. With the evolution of technology, there currently exists the possibility of analyzing all the human transcripts in individual samples, in an attempt to identify the regulator nodes associated with RAS.

References

1. McGonagle D, McDermott MF. A proposed classification of the immunological diseases. PLoS Med. 2006;3(8):e297.
2. Scully C, Gorsky M, Lozada-Nur F. The diagnosis and management of recurrent aphthous stomatitis: a consensus approach. J Am Dent Assoc. 2003;134(2):200–7.
3. Letsinger JA, McCarty MA, Jorizzo JL. Complex aphthosis: a large case series with evaluation algorithm and therapeutic ladder from topicals to thalidomide. J Am Acad Dermatol. 2005;52(3 Pt 1):500–8.
4. Natah SS, et al. Increased density of lymphocytes bearing gamma/delta T-cell receptors in recurrent aphthous ulceration (RAU). Int J Oral Maxillofac Surg. 2000;29(5):375–80.
5. Mills MP, et al. Quantitative distribution of inflammatory cells in recurrent aphthous stomatitis. J Dent Res. 1980;59(3):562–6.
6. Stenman G, Heyden G. Premonitory stages of recurrent aphthous stomatitis. I. Histological and enzyme histochemical investigations. J Oral Pathol. 1980;9(3):155–62.
7. Pedersen A, Hougen HP, Kenrad B. T-lymphocyte subsets in oral mucosa of patients with recurrent aphthous ulceration. J Oral Pathol Med. 1992;21(4):176–80.
8. Savage NW, Seymour GJ, Kruger BJ. T-lymphocyte subset changes in recurrent aphthous stomatitis. Oral Surg Oral Med Oral Pathol. 1985;60(2):175–81.
9. Ship JA, et al. Recurrent aphthous stomatitis. Quintessence Int. 2000;31(2):95–112.
10. Thomas DW, Bagg J, Walker DM. The in vitro cytotoxic effect of leukocytes from patients with recurrent aphthous ulceration upon mouse 3T3 fibroblasts. J Oral Pathol. 1988;17(8):421–5.
11. Hasan A, et al. Recognition of a unique peptide epitope of the mycobacterial and human heat shock protein 65–60 antigen by T cells of patients with recurrent oral ulcers. Clin Exp Immunol. 1995;99(3):392–7.
12. Hasan A, et al. Defining a T-cell epitope within HSP 65 in recurrent aphthous stomatitis. Clin Exp Immunol. 2002;128(2):318–25.
13. Sun A, Chia JS, Chiang CP. Increased proliferative response of peripheral blood mononuclear cells and T cells to Streptococcus mutans and glucosyltransferase D antigens in the exacerbation stage of recurrent aphthous ulcerations. J Formos Med Assoc. 2002;101(8):560–6.
14. Toussirot EA. Oral tolerance in the treatment of rheumatoid arthritis. Curr Drug Targets Inflamm Allergy. 2002;1(1):45–52.
15. Savage NW, Seymour GJ. Specific lymphocytotoxic destruction of autologous epithelial cell targets in recurrent aphthous stomatitis. Aust Dent J. 1994;39(2):98–104.

16. Garside P, Mowat AM, Khoruts A. Oral tolerance in disease. Gut. 1999;44(1):137–42.
17. Lewkowicz N, et al. Expression of Thl/Th2/Th3/Th17-related genes in recurrent aphthous ulcers. Arch Immunol Ther Exp (Warsz). 2011;59(5):399–406.
18. Dalghous AM, Freysdottir J, Fortune F. Expression of cytokines, chemokines, and chemokine receptors in oral ulcers of patients with Behcet's disease (BD) and recurrent aphthous stomatitis is Th1-associated, although Th2-association is also observed in patients with BD. Scand J Rheumatol. 2006;35(6):472–5.
19. Borra RC, et al. The Thl/Th2 immune-type response of the recurrent aphthous ulceration analyzed by cDNA microarray. J Oral Pathol Med. 2004;33(3):140–6.
20. Buno IJ, et al. Elevated levels of interferon gamma, tumor necrosis factor alpha, interleukins 2, 4, and 5, but not interleukin 10, are present in recurrent aphthous stomatitis. Arch Dermatol. 1998;134(7):827–31.
21. Miyamoto NT Jr., et al. Immune-expression of HSP27 and IL-10 in recurrent aphthous ulceration. J Oral Pathol Med. 2008;37(8):462–7.
22. Lewkowicz N, et al. Predominance of Type 1 cytokines and decreased number of CD4(+) CD25(+ high) T regulatory cells in peripheral blood of patients with recurrent aphthous ulcerations. Immunol Lett. 2005;99(1):57–62.
23. Barros FM, et al. Possible Association between Thl Immune Polarization and Epithelial Permeability with Toll-Like Receptors 2 Dysfunction in the Pathogenesis of the Recurrent Aphthous Ulceration. Ulcers. 2010;2010:11.
24. Sequeira FF, Daryani D. The oral and skin pathergy test. Indian J Dermatol Venereol Leprol. 2011;77(4):526–30.
25. Ozdemir M, et al. Pathergy reaction in different body areas in Behcet's disease. Clin Exp Dermatol. 2007;32(1):85–7.
26. Togashi A, et al. Skin prick test with self-saliva in patients with oral aphthoses: a diagnostic pathergy for Behcet's disease and recurrent aphthosis. Inflamm Allergy Drug Targets. 2011;10(3):164–70.
27. Veller-Fornasa C, Gallina P. Recurrent aphthous stomatitis as an expression of pathergy in atopics. Acta Dermatovenerol Alp Panonica Adriat. 2006;15(3):144–7.
28. Sun A, Kwan HW. Serum IgD and IgE concentrations in recurrent aphthous ulcers and oral lichen planus. Zhonghua Ya Yi Xue Hui Za Zhi. 1986;5(1):7–11.
29. Scully C, Yap PL, Boyle P. IgE and IgD concentrations in patients with recurrent aphthous stomatitis. Arch Dermatol. 1983;119(1):31–4.
30. Vaccarino L, et al. Pathological implications of Thl/Th2 cytokine genetic variants in Behcet's disease: data from a pilot study in a Sicilian population. Biochem Genet. 2013;51(11–12):967–75.
31. Li L, Boussiotis VA. The role of IL-17-producing Foxp3+ CD4+ T cells in inflammatory bowel disease and colon cancer. Clin Immunol. 2013;148(2):246–53.
32. Taylor LJ, et al. Increased production of tumour necrosis factor by peripheral blood leukocytes in patients with recurrent oral aphthous ulceration. J Oral Pathol Med. 1992;21(1):21–5.
33. Sun A, et al. Expression of interleukin-2 receptor by activated peripheral blood lymphocytes upregulated by the plasma level of interleukin-2 in patients with recurrent aphthous ulcers. Proc Natl Sci Counc Repub China B. 2000;24(3):116–22.
34. Natah SS, et al. Immunolocalization of tumor necrosis factor-alpha expressing cells in recurrent aphthous ulcer lesions (RAU). J Oral Pathol Med. 2000;29(1):19–25.
35. Sand FL, Thomsen SF. Efficacy and safety of TNF-alpha inhibitors in refractory primary complex aphthosis: a patient series and overview of the literature. J Dermatolog Treat. 2013;24(6):444–6.
36. Parker LC, Prince LR, Sabroe I. Translational mini-review series on Toll-like receptors: networks regulated by Toll-like receptors mediate innate and adaptive immunity. Clin Exp Immunol. 2007;147(2):199–207.
37. Frazao JB, Errante PR, Condino-Neto A. Toll-like receptors' pathway disturbances are associated with increased susceptibility to infections in humans. Arch Immunol Ther Exp (Warsz). 2013;61(6):427–43.

38. Pulendran B, Ahmed R. Translating innate immunity into immunological memory: implications for vaccine development. Cell. 2006;124(4):849–63.
39. Pradeu T, Cooper EL. The danger theory: 20 years later. Front Immunol. 2012;3:287.
40. Li J, Lee DS, Madrenas J. Evolving Bacterial Envelopes and Plasticity of TLR2-Dependent Responses: basic Research and Translational Opportunities. Front Immunol. 2013;4:347.
41. Barrett AW, Cruchley AT, Williams DM. Oral mucosal Langerhans' cells. Crit Rev Oral Biol Med. 1996;7(1):36–58.
42. Hovav AH. Dendritic cells of the oral mucosa. Mucosal Immunol. 2014;7(1):27–37.
43. Novak N, et al. Human skin and oral mucosal dendritic cells as 'good guys' and 'bad guys' in allergic immune responses. Clin Exp Immunol. 2010;161(1):28–33.
44. Jaitley S, Saraswathi T. Pathophysiology of Langerhans cells. J Oral Maxillofac Pathol. 2012;16(2):239–44.
45. Allam JP, et al. Toll-like receptor 4 ligation enforces tolerogenic properties of oral mucosal Langerhans cells. J Allergy Clin Immunol. 2008;121(2):368–374e1.
46. Flacher V, et al. Human Langerhans cells express a specific TLR profile and differentially respond to viruses and Gram-positive bacteria. J Immunol. 2006;177(11):7959–67.
47. Cutler CW, Jotwani R. Dendritic cells at the oral mucosal interface. J Dent Res. 2006;85(8):678–89.
48. Natah SS, et al. Factor XIIIa-positive dendrocytes are increased in number and size in recurrent aphthous ulcers (RAU). J Oral Pathol Med. 1997;26(9):408–13.
49. Upadhyay J, et al. Langerhans Cells and Their Role in Oral Mucosal Diseases. N Am J Med Sci. 2013;5(9):505–514.
50. Sawair FA. Does smoking really protect from recurrent aphthous stomatitis? Ther Clin Risk Manag. 2010;6:573–7.
51. Van den Bossche J, Van Ginderachter JA. E-cadherin: from epithelial glue to immunological regulator. Eur J Immunol. 2013;43(1):34–7.
52. Daniels TE. Human mucosal Langerhans cells: postmortem identification of regional variations in oral mucosa. J Invest Dermatol. 1984;82(1):21–4.
53. Nanke Y, et al. Irsogladine is effective for recurrent oral ulcers in patients with Behcet's disease: an open-label, single-centre study. Drugs R D. 2008;9(6):455–9.
54. Rattan J, et al. Sucralfate suspension as a treatment of recurrent aphthous stomatitis. J Intern Med. 1994;236(3):341–3.
55. Cario E. Barrier-protective function of intestinal epithelial Toll-like receptor 2. Mucosal Immunol. 2008;1(Suppl 1):S62–6.
56. Ey B, et al. Loss of TLR2 worsens spontaneous colitis in MDR1A deficiency through commensally induced pyroptosis. J Immunol. 2013;190(11):5676–88.
57. Cario E, Gerken G, Podolsky DK. Toll-like receptor 2 enhances ZO-1-associated intestinal epithelial barrier integrity via protein kinase C. Gastroenterology. 2004;127(1):224–38.
58. Pierik M, et al. Toll-like receptor-1, -2, and -6 polymorphisms influence disease extension in inflammatory bowel diseases. Inflamm Bowel Dis. 2006;12(1):1–8.
59. Borra RC, et al. Toll-like receptor activity in recurrent aphthous ulceration. J Oral Pathol Med. 2009;38(3):289–98.
60. Seoudi N, et al. The role of TLR2 and 4 in Behcet's disease pathogenesis. Innate Immun. 2014;20(4):412–22.
61. Sewell GW, et al. Defective tumor necrosis factor release from Crohn's disease macrophages in response to Toll-like receptor activation: relationship to phenotype and genome-wide association susceptibility loci. Inflamm Bowel Dis. 2012;18(11):2120–7.
62. Gallo C, et al. Differential expression of toll-like receptor mRNAs in recurrent aphthous ulceration. J Oral Pathol Med. 2012;41(1):80–5.
63. Meng J, et al. Morphine induces bacterial translocation in mice by compromising intestinal barrier function in a TLR-dependent manner. PLoS One. 2013;8(1):e54040.
64. Hietanen J, et al. Recurrent aphthous ulcers–a Toll-like receptor-mediated disease? J Oral Pathol Med. 2012;41(2):158–64.

65. Hill SC, Stavrakoglou A, Coutts IR. Nicotine replacement therapy as a treatment for complex aphthosis. J Dermatolog Treat. 2010;21(5):317–8.
66. Julian MW, et al. Nicotine treatment improves Toll-like receptor 2 and Toll-like receptor 9 responsiveness in active pulmonary sarcoidosis. Chest. 2013;143(2):461–70.
67. Greene CM, et al. Inhibition of Toll-like receptor 2-mediated interleukin-8 production in Cystic Fibrosis airway epithelial cells via the alpha7-nicotinic acetylcholine eceptor. Mediators Inflamm. 2010;2010:423241.
68. Mahanonda R, et al. Cigarette smoke extract modulates human beta-defensin-2 and interleukin-8 expression in human gingival epithelial cells. J Periodontal Res. 2009;44(4):557–64.
69. Kox M, et al. GTS-21 inhibits pro-inflammatory cytokine release independent of the Toll-like receptor stimulated via a transcriptional mechanism involving JAK2 activation. Biochem Pharmacol. 2009;78(7):863–72.
70. Chen H, et al. Tobacco smoking inhibits expression of proinflammatory cytokines and activation of IL-1R-associated kinase, p38, and NF-kappaB in alveolar macrophages stimulated with TLR2 and TLR4 agonists. J Immunol. 2007;179(9):6097–106.
71. Hou L, Zhao H. A review of post-GWAS prioritization approaches. Front Genet. 2013;4:280.
72. Reilly D, et al. Use of systems biology approaches to analysis of genome-wide association studies of myocardial infarction and blood cholesterol in the nurses' health study and health professionals' follow-up study. PLoS One. 2013;8(12):e85369.
73. Chen B, Butte AJ. Network medicine in disease analysis and therapeutics. Clin Pharmacol Ther. 2013;94(6):627–9.
74. Langfelder P, Mischel PS, Horvath S. When is hub gene selection better than standard meta-analysis? PLoS One. 2013;8(4):e61505.
75. Oldham MC, et al. Functional organization of the transcriptome in human brain. Nat Neurosci. 2008;11(11):1271–82.
76. Ivliev AE, t Hoen PA, Sergeeva MG. Coexpression network analysis identifies transcriptional modules related to proastrocytic differentiation and sprouty signaling in glioma. Cancer Res. 2010;70(24):10060–70.
77. Horvath S, et al. Analysis of oncogenic signaling networks in glioblastoma identifies ASPM as a molecular target. Proc Natl Acad Sci USA. 2006;103(46):17402–7.
78. Shulzhenko N, et al. Crosstalk between B lymphocytes, microbiota and the intestinal epithelium governs immunity versus metabolism in the gut. Nat Med. 2011;17(12):1585–93.
79. Chaussabel D, et al. A modular analysis framework for blood genomics studies: application to systemic lupus erythematosus. Immunity. 2008;29(1):150–64.

Lichen Planus

Nicola Luigi Bragazzi and Claudio Nicolini

Introduction

Oral Lichen Planus (OLP) was first clinically identified by a British physician, Doctor Erasmus Wilson (1809–1884) in 1869 [191], in a group of 50 patients, even though he simply considered OLP as a variant of Lichen Ruber Planus described by the dermatologist Ferdinand Ritter von Hebra (1816–1880) in 1860 [48]. The first detailed description of the oral lesion dates back to 1895, when Doctor Thieberg established the diagnostic criteria of OLP [132]. In the same year, Louis Frédéric Wickam [190] identified white striations currently known as Wickam's *striae*. Some years later, François Henry Hallopeau reported the first case of OLP-related oral carcinoma in 1910 [71].

OLP is a common chronic inflammatory T-cell mediated oral mucocutaneous disease with an overall age-standardized prevalence of 1.27% (0.96% in men and 1.57% in women) [110].

The histopathological features of OLP are hydropic degeneration of the basal cell layer, hyperkeratosis, acanthosis, irregular ridges and a dense band-like infiltration of T lymphocytes mainly in the *lamina propria* [6, 7, 16, 21]. Although the aetiology of OLP is still unknown [134, 161], it has been widely accepted that immunological impairments are very critical among the multiple aetiological factors. Previous studies have suggested that it represents a cell-mediated immunologic response to an induced antigenic change in the oral mucosa [79, 116, 169]. The inflammatory response in OLP is characterized by the accumulation and expansion of T helper type 1 (Th1) lymphocytes. Cross-talk between $CD8^+$ T-cells secreting

N. L. Bragazzi (✉) · C. Nicolini
Department of Experimental Medicine (DIMES), Nanobiotechnology and Biophysics Laboratories (NBL), University of Genoa, Via Antonio Pastore 3, 16132 Genoa, Italy
e-mail: 2896313@studenti.unige.it

C. Nicolini
Nanoworld Institute Fondazione EL.B.A Nicolini (FEN), Pradalunga, Largo Redaelli 7, Bergamo, Italy
e-mail: info@fondazioneelba-nicolini.org

S. T. Sonis (ed.), *Genomics, Personalized Medicine and Oral Disease*,
DOI 10.1007/978-3-319-17942-1_9

TNF-α and CD4⁺ lymphocytes could foster further secretion of IL2 and IFN-γ, and thus contribute to maintaining the Th1 profile that can be found in OLP, sometimes leading to the chronic form. It is not known whether the CD4 or CD8 activation may trigger the process leading to OLP ("*one-cell hypothesis*") or if both are responsible at the same time of the OLP pathogenesis ("*two-cells hypothesis*") [170]. However, a "*cooperation*" between the helper/inducer $CD4^+$ and the suppressor/cytotoxic $C8^+$ lymphocytes seems to play a major role in the disease.

The normal oral mucosa may be an immune privileged site (similar to the eye, testis, and placenta), and breakdown of immune privilege could result into OLP and possibly other autoimmune oral mucosal diseases, which could be inter-related [170].

The specific antigen-mediated immunological reaction begins with the permeation of this unknown antigen from the oral cavity into the oral epithelium, where it meets the interdigitating dendritic Langerhans cells (LCs) activation [180], which are immunocompetent Antigen Presenting Cells (APCs) situated above all along the epithelial-stromal junction. CD1a+/Langerin+LCs form a plexus and capture, entrap and present this antigen to CD4+ lymphocytes. From the collected histopathological evidences, the number of LCs in epithelium and within the dermal lymphoid infiltration is found to be significantly increased [68, 180]; however the number of CD38+ LCs does not appear to be increased [68]. The nature of this antigen is still unknown, whether of epithelial origin or from damaged or apoptotic cells, as well as the detail of its presentation—whether major histocompatibility complex type 1 (MHC-I) or type II (MHC-II) mediated [170]. Speculations have given raise to two different hypotheses: the "chance encounter" and the "directed migration" argument [170]. CD4+ cells, through intercellular adhesion molecule 1 (ICAM-1) and lymphocyte function-associated type 1 (LFA-1) pathways, promote epithelial destruction. Afterwards, cytokine production (namely, IL1, IL8, IL10, and IL12), intense chemochine production also favoured by chemochine receptor upregulation (CXCR3, CXCL10) [39], ICAM-1 and VCAM-1 expression can activate CD8+ lymphocytes leading to the chronic form of the disease [5, 182, 183]. CD8+ accumulation in the superficial lamina propria leads to basement membrane disruption with subsequent formation of duplications, breaks, ruptures, and branches, to intra-epithelial T-cell migration, and keratinocyte apoptosis with the formation of the Civatte bodies (homogenous, cytoid and hyaline eosinophilic globules), as well as to the previously described histopathological features [170].

Also non-specific antigen-mediated reactions such as mastocite degranulation with subsequent release of histamine, chymase and tryptase, as well as of TNF and matrix metalloproteinases (MMPs) activation (in particular, the stromelysin MMP-3, and the gelatinase MMP-9) could play a role [170, 185, 202]. Besides producing factors that are able to cleave collagen type IV, mastocites induce T-cells to produce the chemokine RANTES, which in its turn stimulate mast-cells and lead to a chronic activation of lymphocytes.

In the last years, growing evidence about the role of autoantibodies against p63 [122] and E74-like factor type 3 (ELF-3) [35] has been increasing [39]. Deficient antigen-specific TGF- β1-mediated immunosuppression may contribute to OLP chronic forms [170]. B-cells and plasma cells involvement in OLP pathogenesis seems irrelevant since minimal serological changes or deposits of immunoglobulin (such as IgM) or complement (like C3, C4) are found in OLP lesions [170].

From a clinical point of view, OLP is characterized by 4 *P*s: purple, polygonal, pruritic, and papules [136]. Pathognomonic signs include the above-mentioned Wickam's white striations, grouped and coalescent pruritic lineal papules, plaques, erythema, blisters, ulcerations and erosions that usually affect buccal mucosa, the dorsum of the tongue and gingiva in a symmetrical and bilateral way. Gingival OLP is also termed as "*desquamative gingivitis*" by some scholars and clinicians. Clinical features and accompanying symptoms of OLP lesions range from asymptomatic reticular white lesions in atrophic mucosa to erosive-ulcerative areas accompanied with burning pain, bleeding and discomfort, complained by at least two third of the patients [169]. The histological lesion is characterized by a sub-epithelial inflammatory infiltrate, composed of different mononuclear cells, dominated by T lymphocytes, and limited to the basal keratinocyte layer [45].

OLP lesions are characterized by a higher degree of cell turnover than healthy tissue [155]. In fact, the proliferation, the maturation and the apoptosis of basal keratinocytes require a fine regulation at a genomic level [53]. Moreover, the World Health Organization (WHO) defined OLP as a pre-malignant condition making "*oral mucosa more sensitive to exogenous carcinogens and thus to develop oral carcinoma*"[137]. The rate of malignant transformation seems to vary in the range 0–5.3 % [39, 79]. However, the molecular mechanisms determining the possible development of cancer in OLP lesions are not well understood yet [97, 98].

There are different kinds of OLP from the clinical point of view: reticular, papular, plaque-like, erosive, atrophic and bullous [6]. Besides these clinical forms of idiopathic OLP, there are also familial variants of OLP: at least 200 cases described in the literature and associated with HLA-A3, HLA-B7, HLA-B18, HLA-Cw8 and HLA-DR1 [136].

OLP should be kept distinct from other oral lichenoid lesions or oral lichenoid reactions (OLL/OLR) or lichenoid stomatitis, which include: Oral Lichenoid Contact Lesions (OLCL), Oral Lichenoid Drug Eruptions or Oral Lichenoid Drug Reactions (OLDE/OLDR), Oral Lichenoid Lesions of Graft versus Host Disease (OLL-GVHD), lichenoid stomatitis, autoimmune-related OLRs and lichenoid dysplasia [8, 43].

OLCL is due to allergic contact stomatitis and seen in direct topographic relationship to implants and dental prostheses which can contain metals such as nickel, gold, palladium and dental restorative materials, most commonly mercury and other compounds used for amalgam fillings, dental acrylics, composites and resin-based materials. OLCL may be also due to the usage of cinnamate-flavored toothpaste or other flavourings [170].

OLDE/OLDR can be caused by drugs like oral hypoglycaemic drugs, sulfonylureas, anti-hypertensive agents and angiotensin converting enzyme (ACE) inhibitors, diuretics (such as benzothiadiazides), beta-blockers, non-steroidal anti-inflammatory agents (NSAIDs), statins, phenothiazines, anti-malarials, dapsone, gold salts and penicillamine for the management of rheumatoid arthritis, interferon and anti-retrovirals, and antiblastics like tumor necrosis factor α(TNF-α) inhibitors.

OLL-GVHD in the setting of patients is usually encountered acute, but predominantly, chronic graft versus host disease (cGVHD), usually following allogeneic bone marrow transplantation.

Other lichenoid reactions can be due to the Koebner phenomenon (mechanical traumas after calculus deposits, sharp teeth, rough surfaces, cheek or tongue biting and oral surgical procedures [169].

Other disorders of the "lichen disease" family include oral lichen nitidus, which affects usually children, oral lichen planus pigmentosus [91] and other rather rare and uncommon oral pathologies [79, 136].

OLRs have been associated also with tobacco [64] a betel quid chewing [145] and alcohol consumption [39, 140, 171, 181].

From these observations, OLP may be seen as a spectrum of disorders [8].

Also the role of bacteria like Helicobacter pylori [160] and viruses, such as herpes simplex, Epstein Barr virus [198], hepatitis C virus or HCV [22, 23], Human Papillomavirus or HPV [63, 175], has been discussed, even though it remains elusive [170].

OLP is indeed a complex multi-factorial disease [67, 75], such as Alzheimer's dementia, Crohn's enteritis and cardiovascular disorders. Usually, these diseases present a relatively mild phenotype and are slowly progressive. The physiopathology of complex pathologies is characterized by various biologic pathways, dependent upon the contribution of a large number of genes [75, 154]. Therefore, the knowledge of molecular mechanisms of complex multi-factorial diseases must deal with a large number of genes [176]. These genes form complex networks of interactions, which may be direct (physical interactions between the proteins, confirmed by experimental techniques, such as NMR or crystallography) or indirect (involvement in the same metabolic pathway or co-expression in different conditions) (Fig. 1).

At present, several studies have analysed the role of different genes in the pathogenesis and evolution of OLP [2, 31, 40, 42, 60, 74, 78, 86–88, 95, 107, 108, 114, 125–127, 130, 138, 149, 158, 174, 197, 204]. However these studies were in most

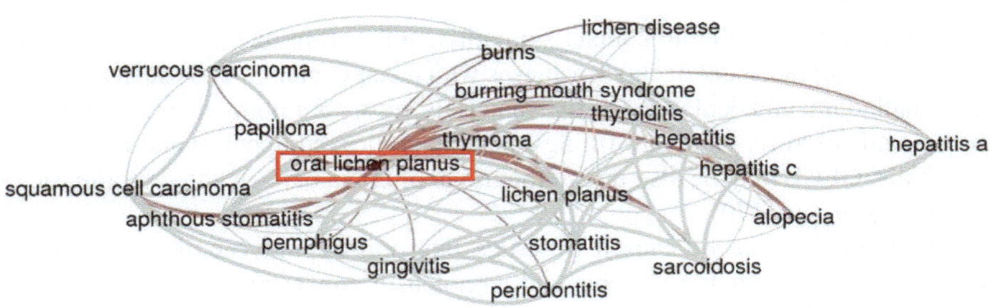

Fig. 1 Oral Lichen Planus (*OLP*) as complex multifactorial disorder, interconnected with other diseases. (Obtained with MalaCards software. Copyright © Weizmann Institute of Science—www.malacards.org)

cases dealing with one single gene or with a single gene family; to our knowledge, there are very few studies and analysis on the interactions among all the different genes involved in OLP, using also OMICS technologies [198].

The role of miRNAs in oral diseases is also emerging, since evidences are growing and accumulating in favor of this hypothesis. For example, miRNAs circulating in peripheral blood and in serum [102], as well tissutal and salivary miRNAs [52] can be used as helpful biomarkers of early-stage oral cancer [201]. Salivary miRNAs are transported in vesicles called exosomes, that preserve their stability [112, 124]. In particular, the interest in studying salivary microRNAome is increasing because of the non invasivity and effectiveness of the procedure, and also because of the advancement in salivary diagnostic technologies [14, 20, 26, 27 131, 195, 196]. Several scholars maintain the importance of integrating it with other diagnostic alphabets (namely transcriptomics, proteomics, metabolomics and oral microbiomics), that constitute the five pillars of a new, recently emerging approach to oral diseases called "salivaomics" [192] and more generally speaking "oralomics" [9, 19, 92, 151].

The importance of the OMICS studies in medicine and in dentistry is that they could enable a personalized treatment of the disorders [44, 55, 62, 121].

However, notwithstanding the importance of incorporating the oral microRNAome in the OMICS-based study of oral pathologies, so far only few miRNAs related to them have been discovered and described in the literature. Most published studies focus on oral cancer, while for example gingival and periodontal disorders are less studied.

The aim of the data-mining based approach that we introduce in this manuscript is to accelerate and facilitate discoveries in the field of oral diseases.

Bioinformatics indeed can play a central role in the analysis and interpretation of genomic and proteomic data [30, 193]. Recently, a bioinformatics method, defined as the "Leader Gene approach" has been proposed [17]. This search/statistics algorithm is based on the systematic search for the genes involved in a given process, on the calculation of an interaction map and on their ranking according to the number of all experimentally established interactions, as derived from free Web-available databases, such as STRING (Search Tool for the Retrieval of Interacting Genes, Heidelberg, Germany) [49, 184]. Genes belonging to the highest rank are defined as "leader genes" because they may be assumed to play an important role in the analysed processes. The "Leader Gene approach" can suggest a list of few genes potentially relevant in a given cellular process, according to the already available experimental data [18]. Moreover, the interaction map among all the genes involved in the same process may be useful in interpreting the experimental and clinical results and in planning new targeted experimentation. Interestingly, such experimentation may be simpler to be analysed than mass-scale molecular genomics. This method gave promising results when applied to the human T lymphocyte cell cycle, human kidney transplant, osteogenesis and periodontitis [59, 84, 106, 120, 129,141, 163–165]. These results were also integrated with a targeted experimental analysis, to draw an overall picture of these processes.

In this theoretical work, genes involved in human OLP are identified and ranked according to their number of interactions, to preliminarily obtain a broader view of molecular mechanisms of OLP and to plan targeted experimentation.

In the second part, we used previously predicted Class A and Class B genes in order to predict putatively associated miRNAs, exploiting the miRGene database. In order to show an application of this data mining-based bioinformatics approach, this book-chapter explores the *in silico* prediction of OLP-associated miRNAs.

Methods

The *ab-initio* leader gene approach has been already described in detail elsewhere [17, 59].

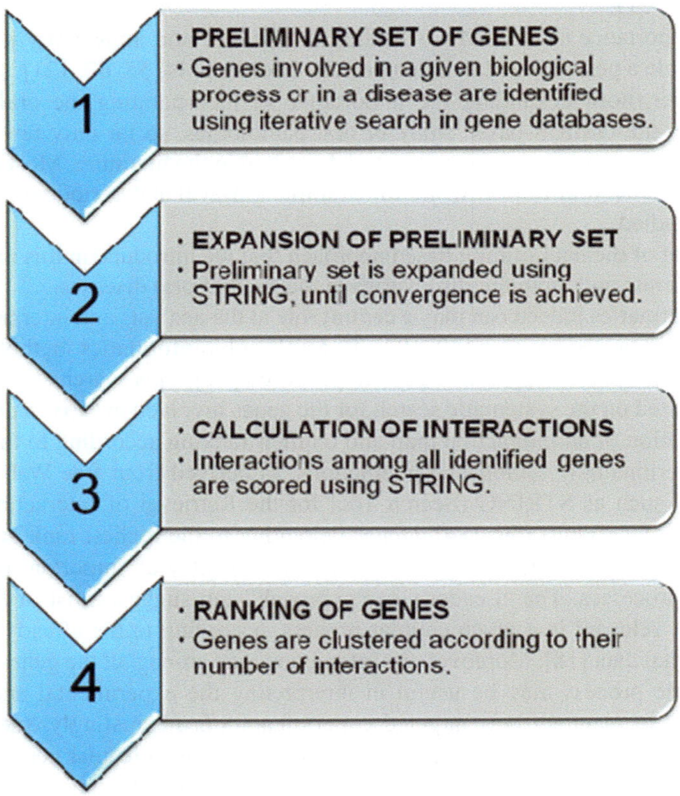

Briefly, at first, the key genes involved in OLP are identified by iterative search of large-scale gene databases. In particular, several search strategies were implemented and iteratively repeated until convergence. At first, a preliminary systematic query of inter-related databases—PubMed (accessible at http://www.ncbi.nlm.nih.gov/pubmed/), GeneBank (accessible at http://www.ncbi.nlm.nih.gov/gene/), GenAtlas (accessible at http://genatlas.medecine.univ-paris5.fr/), GeneCards [144; accessible at http://www.genecards.org/], The Disease and Gene Annotations (DGA) [135], Dis-GeNET (accessible at http://www.disgenet.org/web/DisGeNET/v2.1), GeneRIF [81, 82], MalaCards ([143] accessible at http://www.malacards.org/), Online Mendelian Inheritance in Man (OMIM) (accessible at http://www.ncbi.nlm.nih.gov/omim/), Genetic Testing Registry (GTR) (accessible at http://www.ncbi.nlm.nih.gov/gtr/), Genetic Association Database (GAD) (available at http://geneticassociationdb.nih.gov/), Human Gene Mutation Database (HGMD) (accessible at www.hgmd.cf.ac.uk/), OralCard [9, 151; accessible at http://bioinformatics.ua.pt/OralCard/], PheGenI [142; accessible at http://www.ncbi.nlm.nih.gov/gap/phegeni/], Indian Genetic Disease Database (IGDD) [139; accessible at http://www.igdd.iicb.res.in/], KEGG DISEASE (accessible at http://www.genome.jp/kegg/disease/), MedGen (accessible at http://www.ncbi.nlm.nih.gov/medgen/), GeneDis (accessible at http://life2.tau.ac.il/GeneDis/), DISEASES (accessible at http://diseases.jensenlab.org/Search), The Genome-wide Repository of Associations between SNPs and Phenotypes (GRASP) [93; accessible at http://apps.nhlbi.nih.gov/grasp], A Semantically Integrated Disease-associated Database (SIDD) ([28] accessible at http://mlg.hit.edu.cn/SIDD/), and SpliceDisease [188; accessible at http://202.38.126.151:8080/SDisease/) —was performed, using a proper string of pertinent keywords chosen by experts as well as MeSH (Medical Subject Headings) terms and all their possible Boolean logics-based combinations. In order to avoid possible bias due to different nomenclature systems, we used official Human Gene Organisation (HUGO) nomenclature (accessible at http://www.hugo-international.org/). Only human genes were considered. In this way, it was possible to identify a list of candidate genes potentially involved in OLP pathogenesis.

The preliminary set of genes was then expanded using the web-available software STRING (version 7.0), considering only direct interactions (i.e.: physical contact between encoded proteins, gene expression microarray data, or direct linkage in the same pathway), with a high degree of confidence (above 0.9—confidence value in STRING ranges between 0 and 0.99, with 0.99 being the highest confidence). In this way, it is possible to identify new genes directly linked to those with an already established role in OLP, and therefore potentially involved in this disease.

In order to discard false positives, results were then filtered using a further search in literature and gene databases. The process was repeated until no new gene potentially involved in OLP was identified.

Then, an interaction map among the identified genes was calculated using STRING. This software can give a combined association score to each interaction, representing the degree of confidence for each interaction. For every gene identified, we summed the different combined association scores with the other genes. The sum of all these scores is defined as the weighted number of links (WNL).

Genes were then clustered, using hierarchical or K-means algorithms (156, 178), according to their WNL. The genes belonging to the highest rank are defined as leader genes; these genes have a significant higher WNL if compared with the other ones. The other ranks are termed class B, class C, class D genes and so on, according to their WNL scores. Genes with no identified interactions (i.e. WNL =0) are defined as orphan genes.

Differences among various classes in terms of WNL were statistically evaluated using an ANOVA test, with a Tukey-Kramer *post-hoc* test. Statistical significance was set at a *p-value* <0.001, in order to ensure a high level of data reliability.

Moreover, interacting genes were classified as up-regulated, down-regulated or neutral in respect to OLP pathogenesis. For neutral genes, we mean that they do not exhibit fold expression changes in the disease versus health control condition or genes for which there is not a universal consensus in the literature and in the databases.

Topological analysis was carried out with Cytoscape [13, 37, 157] and FANMOD [189], while onthological analysis was performed with BinGO [103].

In the second part, in order to predict the potential miRNAs network related to OLP, we used our previously identified "leader genes" [128], namely both the Class A genes (JUN, EGFR, FOS, IL2, and ITGB4) and the Class B genes (CASP3, CD247, IL2RA, IFNG, MMP2, LAMC2). Then, we mined the miRGen database using the "Targets" option (available at http://www.microrna.gr/mirgen). This software relies on a relational database technology and enables a comprehensive research, integrating both the most widely used databases and repositories with the list of experimentally validated miRNAs and the bioinformatics tools for the microRNAs prediction (namely, DIANA-microT, miRanda, miRBase, PicTar, and TargetScanS) [3, 111].

We checked the biological significance of our obtained miRNAs networks mining the extant literature and using ad hoc bioinformatics tools (such as the miR2Disease Base, accessible at http://watson.compbio.iupui.edu:8080/miR2Disease/searchDiseasePre.jsp) [80]. In order to verify the statistical significance of the enrichment of our miRNA-related list, we randomly generated a list of 11 genes (five for Class 1 and six for Class 2). We used the RSA tool for this purpose (accessible at http://rsat.ulb.ac.be/rsat/random-genes_form.cgi), selecting "Homo sapiens" as source organism. Using the above-mentioned tools we generated a network of miRNAs associated to the obtained list of random genes. The two miRNAomes were compared using the statistical test for comparing two proportions/percentages. This computation was done with the commercial software MedCalc.

Topological properties of the obtained graphs portraying the OLP-related microRNAomes have been also studied. We investigated the clustering coefficient (a measure of degree to which nodes in a graph tend to cluster together), the network diameter and radius, the network centralization, the number of shortest paths (in percentage), the characteristic path length, the average number of neighbors, and the network density (that is to say, the proportion of all the possible ties that are actually present, a measure which is computed dividing the sum of the existing ties by the number of all possible ties). Network density reflects the speed at which signaling information diffuse among the nodes.

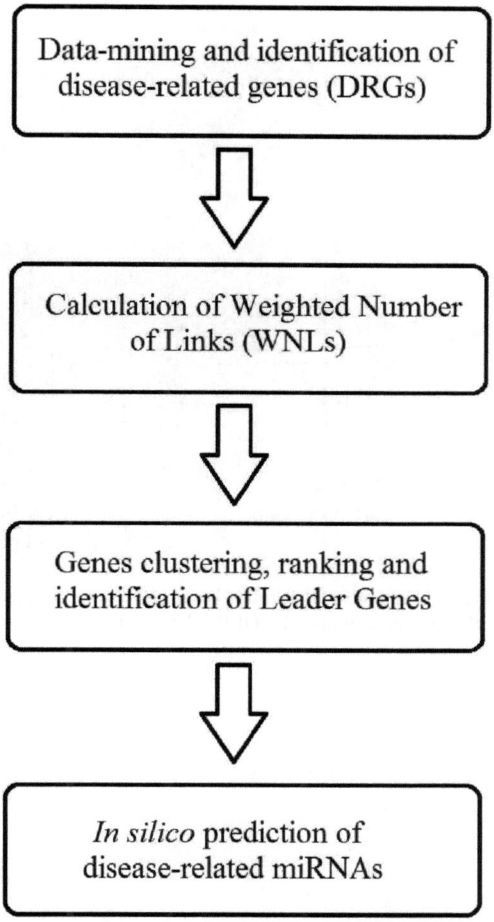

Fig. 2 A flow-chart of the biomolecular data mining strategy based on leader genes algorithm for identifying diseases-associated microRNAs

All the parameters were studied using Cytoscape software and reported in Table 2 and shown in Fig. 3. Networks are visualized using Medusa software [76] (Fig. 2).

Results

The preliminary set obtained by means of the first key words–based query in databases was expanded two times via STRING, until it reached convergence. Once convergence was reached, the expanded data set included 132 genes involved or potentially involved in human OLP. Figure 1 shows the final interaction map among this set of genes.

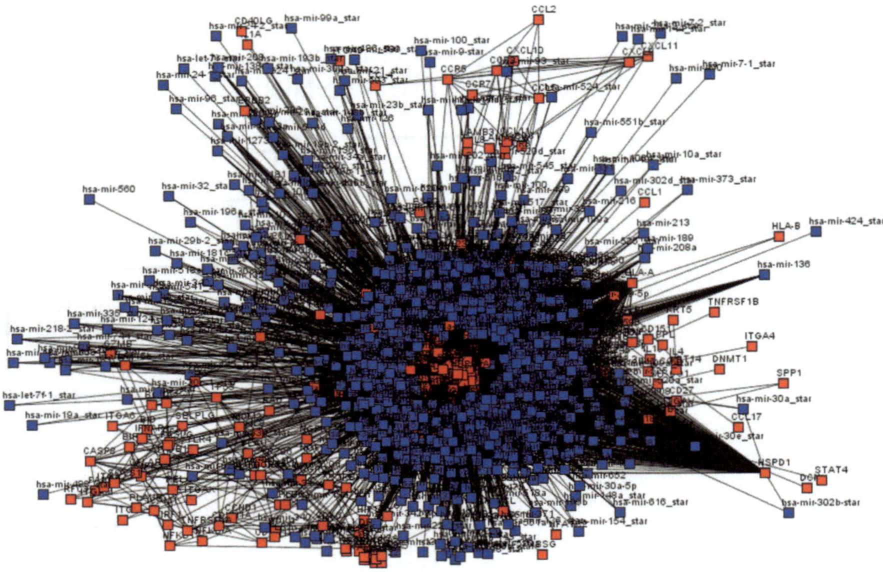

Fig. 3 The entire OLP-related microRNAome (nodes $n = 1124$, edges $n = 22281$): *in red* the OLP-related genes, *in blue* the OLP-related microRNAs

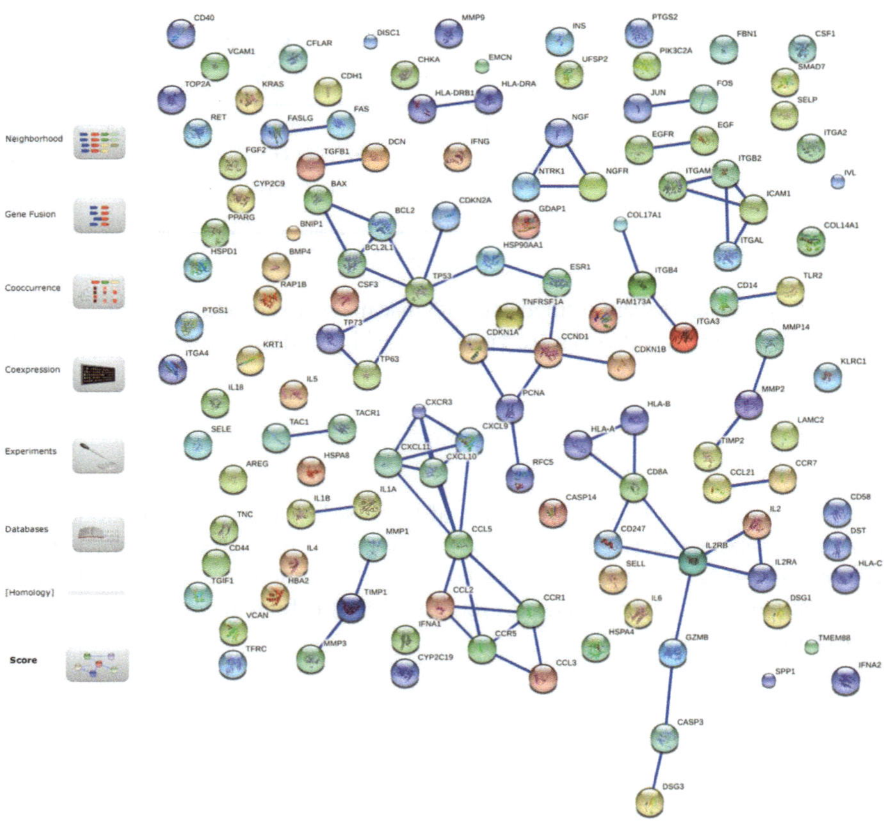

The WNL for each gene in this dataset is represented in Fig. 2.

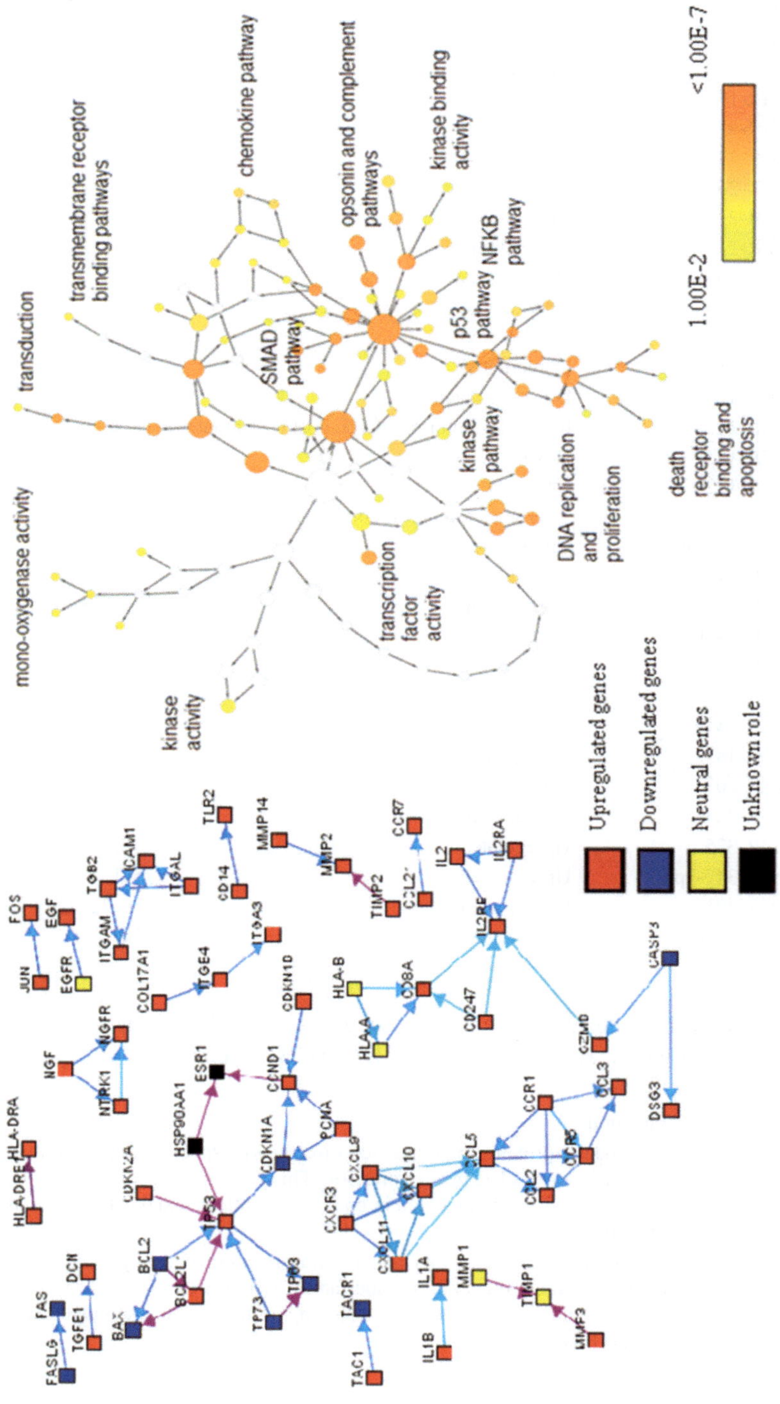

Table 1 Experimentally validated OLP-related miRNAs

Experimentally validated miRNA	Role and molecular mechanisms	References
let-7d*	Upregulated	[123]
let-7i	Upregulated	[56]
has-miR-15a	Upregulated	[56]
has-miR-21	Upregulated	[33, 56, 123]
has-miR-23b	Upregulated	[123]
has-miR-26b	Downregulated	[34, 56]
has-miR-27a	Upregulated	[123]
has-miR-27b	Downregulated	[203]
has-miR-30a	Downregulated	[56]
has-miR-30b	Upregulated	[123]
has-miR-30c	Upregulated	[123]
has-miR-31	Upregulated	[56]
has-miR-103	Downregulated	[123]
has-miR-125a	It is an inhibitor of CCL5, is downregulated	[77]
has-miR-125b	Downregulated	[33]
has-miR-132	Upregulated	[56]
has-miR-137	Aberrant promoter methylation	[32]
has-miR-140-5p	Upregulated	[123]
has-miR-143	Upregulated	[56, 123]
has-miR-146a	Upregulated	[11, 57]
has-miR-146b-5p	Upregulated	[56]
has-miR-151-3p	Downregulated	[123]
has-miR-155	Upregulated	[11, 57]
has-miR-181a	Downregulated	[123]
has-miR-183	Upregulated	[56]
has-miR-203	Upregulated	[33]
has-miR-223	Upregulated	[123]
has-miR-335	Upregulated	[56]
has-miR-342-3p	Upregulated	[56]
has-miR-425	Upregulated	[56, 123]
Has-miR-923	Downregulated	[56]

Cluster analysis of the WNL identified five genes belonging to the highest cluster, i.e., the leader genes: JUN, EGFR, FOS, IL2, ITGB4 (Table 1).

Gene	Function
JUN	It encodes a protein which interacts directly with specific target DNA sequences to regulate gene expression. This gene is mapped to 1p32-p31, a chromosomal region involved in both translocations and deletions in human malignancies
EGFR	The protein encoded by this gene is a transmembrane glycoprotein that is a member of the protein kinase superfamily. EGFR is a cell surface protein that binds to epidermal growth factor. Binding of the protein to a ligand leads to cell proliferation

Gene	Function
FOS	This gene encodes leucine zipper proteins that can dimerize with proteins of the JUN family, thereby forming the transcription factor complex AP-1. As such, the FOS proteins have been implicated as regulators of cell proliferation, differentiation, and transformation. In some cases, expression of the FOS gene has also been associated with apoptotic cell death
IL2	The protein encoded by this gene is a secreted cytokine that is important for the proliferation of T and B lymphocytes
ITGB4	Integrins mediate cell-matrix or cell-cell adhesion, and transduced signals that regulate gene expression and cell growth. This gene encodes the integrin beta 4 subunit, a receptor for the laminins. This subunit is likely to play a pivotal role in the biology of invasive carcinoma

Gene symbol	WNL	Gene symbol	WNL	Gene symbol	WNL	Gene symbol	WNL
JUN	19,868	OPN	6155	SELP	3320	CD14	1506
EGFR	18,932	VCAN	5964	FAS	3176	TIL4	1498
FOS	15,284	ITGAL	5555	MMP3	2867	IL1B	1433
IL2	14,851	CDKN1A	5460	NTRK1	2794	KET	1330
ITGB4	13,881	TNC	5433	MIG	2793	TP73	1327
CASP3	12,262	CCR5	5315	b-R1	2793	UND	1266
CD247	12,054	IL4	5295	SELL	2788	CD40	1266
IL2RA	11,619	MMP9	5277	COL17A1	2763	BNIP1	1266
IFNG	11,588	BCL2	5225	CCR1	2735	WS	1256
MMP2	11,524	FASLG	5156	CYP2C9	2699	PIK3C2A	1254
LAMC2	10,827	IFNA1	5128	CYP2C19	2699	RFC5	999
CTNNB	10,796	PTGS2	5029	PPARG	2697	CD58	999
CDH1	9671	DCN	5004	BAX	2632	HLA-DRA	989
P53	9653	MMP1	4689	PTGS1	2598	TACR1	986
NEU	9460	SISd	4462	TIMP2	2504	TAC1	986
KRAS	9118	PCNA	4421	HSPD1	2501	BLR2	986
ITGB2	8942	NGFR	4421	CCL3	2494	Erb	960
HLA-A	8876	TIMP1	4315	FGF2	2432	CD31	930
CD49B	8564	HLA-B	4287	IL18	2431	VCAM1	899
CD8A	8362	CXCR3	4175	INS	2400	NKG2	899
ITGA3	8328	TNFR1	4143	DSG1	2329	BMP4	899
CR3A	8239	BB2	4084	SELE	2165	RET	800
H2	7797	CDKN2A	4016	HSPA4	2027	IL1	800
IL2RB	7391	K60	3963	NGFB	1994	HSF	800
CED	7370	MCAF	3952	FBN	1899	CSF3	800
EGF	7254	HLA-C	3738	TOP2A	1889	TFRC	633
HSP90AA1	6981	KIP1	3695	IFNA2	1798	EMCN	633
ESR1	6920	HLA-DRB1	3686	MMP14	1792	CSF1	633
CD44	6715	CSPB	3657	DSG3	1765	TGIF1	627
BCL1	6701	CFLAR	3624	DST	1742	ENSG00000109321	627
CD49D	6375	HSPA8	3572	CD2	1632	DSC1	625
SMAD7	6254	IL5	3398	CCL21	1563	gfg	623
BCL2L1	6172	crg-2	3370	EHK1	1530	IVL	532

The analysis of variance (ANOVA) revealed a statistically significant difference in the WNL. In particular, the post-hoc test revealed that leader genes had a significantly higher WNL when compared to class B genes (*p-value* <0.001), and that class B genes differed significantly from other classes (*p-value* <0.001 *versus* class C). The established or putative role of leader genes in OLP is summarized in Table 2.

Figure 3 shows up-regulated and down-regulated genes, with data obtained by means of data-mining. Interestingly all leader genes but EGFR were up-regulated, while EGFR appeared neutral in respect to OLP pathogenesis. Evidence concerning EGFR regulation is indeed controversial and not conclusive, while some authors report a down-regulated expression, other scholars report opposite findings. These contrasting findings may be due to heterogeneity of the studied samples and with or to the study design, among others.

Topological analysis is summarized in Table 3. Interestingly all leader genes were widely distributed in the network (in term of topological parameters, such as

Table 2 In silico predicted Class A leader genes-related miRNAs (n=192). Underlined in light blue the miRNAs directly associated to OLP and experimentally validated; in yellow the miRNAs linked with oral diseases but not directly with OLP and that could play a putative role in OLP pathogenesis

miR-7	miR-146	miR-320	miR-522
let-7a	miR-146a	miR-328	miR-524*
let-7b	miR-146b	miR-331	miR-527
let-7c	miR-148	miR-337	miR-532
let-7d	miR-148a	miR-338	miR-539
let-7d*	miR-148b	miR-362	miR-543
let-7e	miR-149	miR-367	miR-545
let-7f	miR-152	miR-368	miR-548c
let-7g	miR-153	miR-369-3p	miR-551a
let-7i	miR-154*	miR-369-5p	miR-551b
miR-9	miR-155	miR-370	miR-554
miR-15a	miR-181a	miR-372	miR-561
miR-17-3p	miR-181b	miR-373	miR-562
miR-17-5p	miR-181c	miR-376a	miR-567
miR-20	miR-181d	miR-376b	miR-574
miR-25	miR-182	miR-382	miR-576
miR-27a	miR-182*	miR-383	miR-577
miR-27b	miR-184	miR-409-3p	miR-578
miR-29a	miR-185	miR-409-5p	miR-580
miR-29b	miR-186	miR-411	miR-581
miR-29c	miR-187	miR-449b	miR-584
miR-30a	miR-190	miR-451	miR-586
miR-30a-3p	miR-196a	miR-452	miR-587
miR-30e-3p	miR-196b	miR-485-5p	miR-589
miR-32	miR-199a*	miR-487a	miR-591
miR-34a	miR-200a	miR-487b	miR-597
miR-34c	miR-200b	miR-490	miR-600
miR-92	miR-200c	miR-491	miR-603
miR-93	miR-204	miR-493-3p	miR-607
miR-95	miR-205	miR-493-5p	miR-608
miR-98	miR-206	miR-495	miR-611
miR-99a	miR-208	miR-499	miR-612
miR-99b	miR-210	miR-502	miR-614
miR-101	miR-211	miR-509	miR-622
miR-105	miR-219	miR-517b	miR-624
miR-125a	miR-221	miR-517c	miR-625
miR-125b	miR-222	miR-518a	miR-631
miR-126*	miR-296	miR-518e	miR-633
miR-128a	miR-299-3p	miR-518f	miR-642
miR-128b	miR-299-5p	miR-519a	miR-647
miR-130b	miR-301	miR-519b	miR-650
miR-136	miR-302	miR-519c	miR-651
miR-139	miR-302a	miR-520a	miR-654
miR-140	miR-302b	miR-520b	miR-661
miR-141	miR-302b*	miR-520c	miR-663
miR-142-3p	miR-302c	miR-520d	miR-765
miR-142-5p	miR-302c*	miR-520d*	miR-770-5p
miR-144	miR-302d	miR-520e	miR-802

stress, eccentricity and radiality) and showed higher topological coefficients at the topological analysis.

Table 3 In silico predicted Class B leader genes-related miRNAs ($n=126$). Underlined in light blue the miRNAs directly associated to OLP and experimentally validated; in yellow the miRNAs linked with oral diseases but not directly with OLP and that could play a putative role in OLP pathogenesis

miR-1	miR-181a	miR-508
let-7a	miR-181b	miR-509
let7-b	miR-181c	miR-512-5p
let7c	miR-181d	miR-513
let-7d	miR-182	miR-518a
let-7e	miR-187	miR-518b
let-7f	miR-191	miR-518c
let-7g	miR-193a	miR-518d
let-7i	miR-193b	miR-518e
miR-23a	miR-196a	miR-519e
miR-23b	miR-205	miR-520d
miR-24	miR-206	miR-548b
miR-26a	miR-214	miR-551b
miR-26b	miR-302b*	miR-559
miR-27a	miR-324-3p	miR-564
miR-27b	miR-325	miR-565
miR-29a	miR-328	miR-566
miR-29b	miR-331	miR-567
miR-29c	miR-338	miR-571
miR-30a-5p	miR-346	miR-577
miR-30b	miR-363	miR-579
miR-30c	miR-367	miR-582
miR-30d	miR-369-3p	miR-588
miR-30e-5p	miR-369-5p	miR-590
miR-95	miR-370	miR-592
miR-98	miR-382	miR-593
miR-106b	miR-383	miR-598
miR-107	miR-409-5p	miR-599
miR-122a	miR-410	miR-610
miR-125a	miR-411	miR-615
miR-125b	miR-421	miR-621
miR-132	miR-425-5p	miR-622
miR-135a	miR-431	miR-624
miR-136	miR-452	miR-627
miR-137	miR-454-3p	miR-639
miR-138	miR-455	miR-643
miR-140	miR-489	miR-655
miR-143	miR-490	miR-660
miR-144	miR-500	miR-767-5p
miR-151	miR-503	miR-768-3p
miR-153	miR-505	miR-769-3p
miR-154	miR-507	miR-801

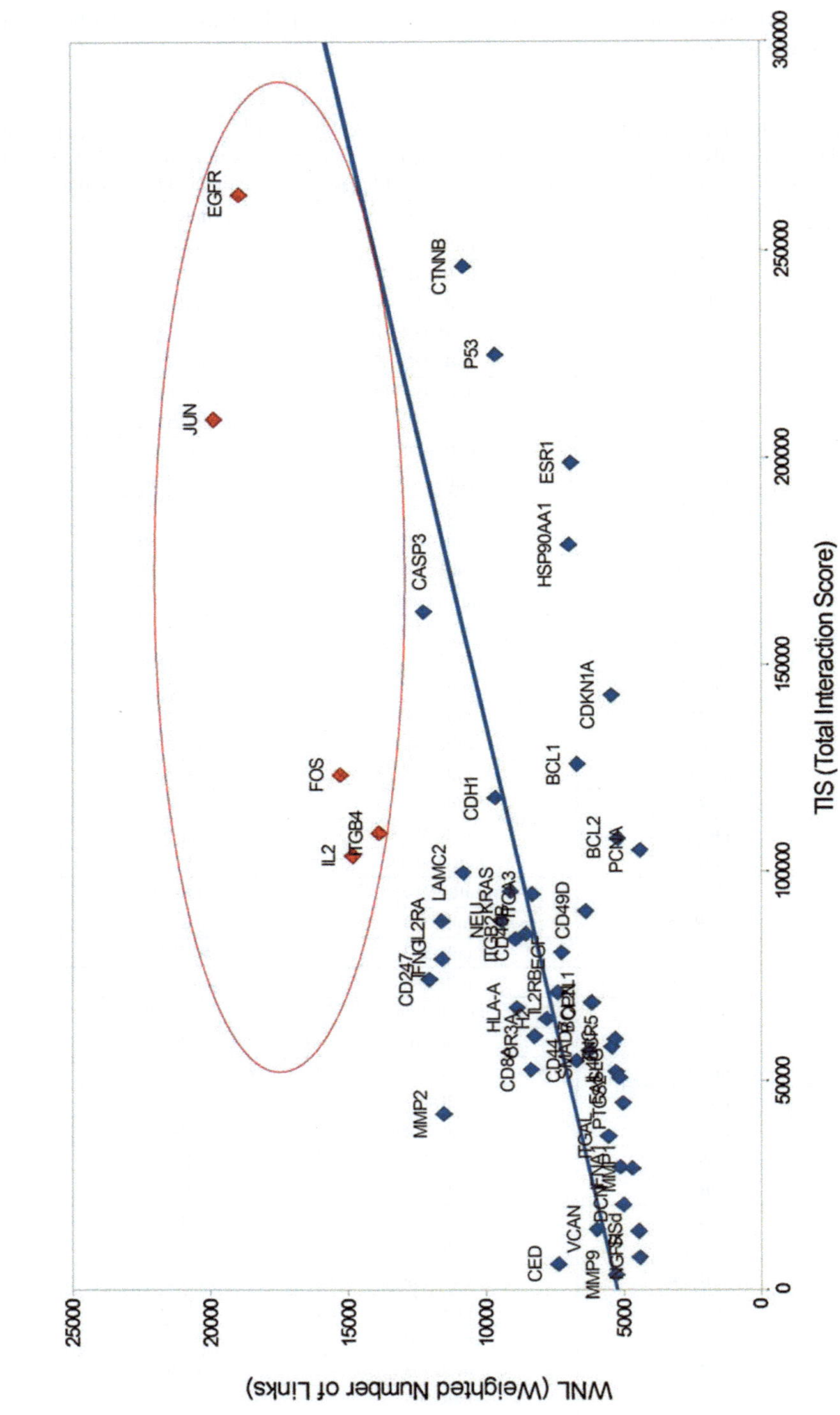
TIS (Total Interaction Score)
WNL (Weighted Number of Links)
0
50000
100000
150000
200000
250000
300000
0
5000
10000
15000
20000
25000
EGFR
JUN
FOS
IL2
ITGB4
CTNNB
P53
ESR1
HSP90AA1
CASP3
CDKN1A
BCL1
CDH1
BCL2
PCNA
LAMC2
IL2RA
CD247
KRAS
NEU
CD49D
HLA-A
MMP2
CD44
CED
VCAN
MMP9

OLP network parameter	Value
Clustering coefficient	0.33
Network centralization	0.07
Average number of neighbours	3.22
Network density	0.02
Shortest paths (%)	52.00
Average path length	4.66
FFL[a] (%)	17.83

[a] feed-forward loops

In the second part of our analysis, we focused on the OLP-related miRNAs. As mentioned before, only few studies in the literature report OLP-related miRNAs. These articles [32–34, 56, 77, 203] have been reviewed and summarized in Table 1.

All the *in silico* predicted miRNAs are reported in Table 2 for Class A leader genes-related molecules and in Table 3 for the Class B leader genes-related miRNAs.

In table 2, miRNAs that have been already experimentally validated are underlined in light blue (they represent the 35.48% (11/31) of the entire list). Class A leader genes-related miRNAs predict the 54.17% (104/192) of the OLP+oral diseases-associated miRNAs validated in the literature.

Compared to the related randomly generated microRNAome, the numbers of experimentally validated miRNAs did not statistically differ (9/31 versus 11/31), but the numbers of oral disease-associated miRNAs yielded a statistical significance (p-value <0.05).

In Table 3, miRNAs that have been already found to be associated with OLP are underlined in light blue (they are the 32.26% (12/31) of the entire list) against 9/31 for the randomly generated miRNAs (not statistically significant).

Class B leader genes-related miRNAs predict the 46.03% (58/126) of the OLP+oral diseases-associated miRNAs validated in the literature.

Taken together the Class A and Class B leader genes-related miRNAs, they can predict the 48.39% (15/31) of the OLP and the of OLP+ oral diseases-associated miRNAs validated in the literature.

The entire OLP-related microRNAome is too complex to be properly studied (shown in Fig. 2), since the number of nodes and edges of the graph is overwhelming. For this reason, after using a holistic and highly integrated approach, a reductionist methodology should be exploited in order to underpin a more essential panel of miRNAs, that could be better investigated and elucidated via *ad hoc* targeted experiments (Fig. 3).

From the topological analysis (summarized in Table 4), the two graphs representing the Class A and Class B-related microRNAomes (shown in Fig. 4) appear to be the most significant part of the entire microRNAome graph, being the bulk of it in term of network diameter and radius, network centralization, topological density and clustering coefficients, while preserving other topological parameters (such as the scale-free behavior, the characteristic path length and the percentage of shortest paths) [12, 104, 159, 167].

Table 4 Comparison of the topological properties among the entire OLP-related microRNAome, and the ones associated to Class A and B leader genes

Topological parameter	Entire OLP-related microRNAome	Class A leader genes-related microRNAome	Class B leader genes-related microRNAome
Clustering coefficient	0.19	0.96	0.73
Network diameter	3	2	2
Network radius	2	1	1
Network centralization	0.94	0.99	0.99
Shortest paths (%)	99	99	99
Characteristic path length	1.99	1.99	1.99
Average number of neighbors	41.49	5.78	7.7
Network density	0.04	0.01	0.01

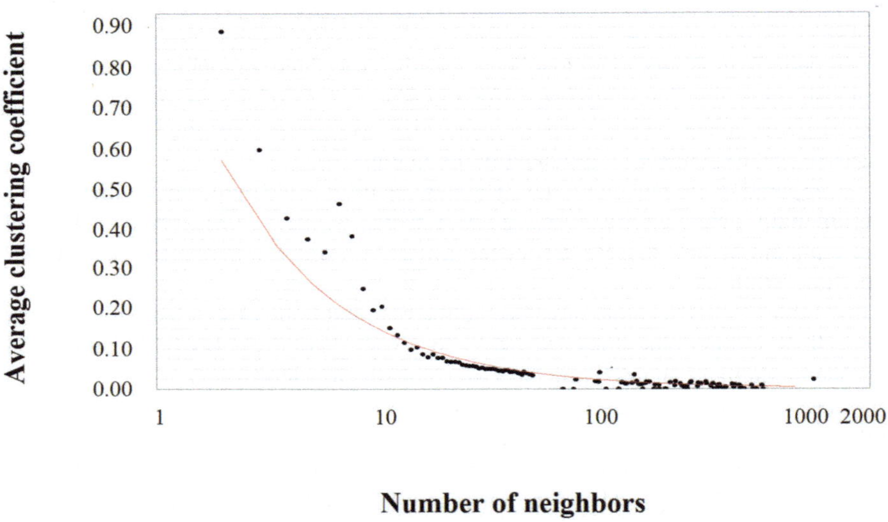

Fig. 4 OLP-related microRNAome topological properties

Discussion

Genetic and genomics research is rapidly increasing our understanding of the molecular basis of some diseases and may also suggest new diagnostic and treatment strategies. Many oral diseases have a genetic basis. Studies of these pathological conditions suggest that multiple gene interactions are important determinants of susceptibility. In this study, genes involved or potentially involved in human Oral Lichen Planus (132 genes in total) are identified with a data-mining approach in order to obtain a broader view of molecular mechanisms of this condition. A map of interactions among these genes is also drawn, as well as a map of the involved pathways. Genes are ranked according to the number and confidence of the interactions in the whole gene set. In particular, it is possible to identify a small set of five genes with a higher number of interactions weighted for the confidence of these links (WNL) than the other ones included in the gene set: namely, JUN, EGFR, FOS, IL2, ITGB4. These genes are defined as leader genes, according to previous studies conducted on different cellular and pathological processes, including periodontitis and may be supposed to play a major role in the pathogenesis of OLP because their WNL was the highest in the whole gene set (Fig. 4).

Experimental Evidence of Leader Genes Involvement in OLP

This analysis was conducted completely in blind: we did not look at scientific literature when identifying leader genes. This choice could be considered as a further proof of the validity of the method: after *ab-initio* identification of leader genes, scientific literature was searched to see if there is an established evidence (epidemiological, clinical, or biochemical) for the involvement of leader genes in OLP. However, if no evidence is found for a given gene, it might be important to verify if there are known direct links to some other leader gene playing an established role in OLP. In this case, a possible involvement in the disease may be preliminarily supposed and could be verified with a targeted experimentation.

The bibliographic research revealed that among the five genes identified as leader genes, only three were specifically associated with OLP. This search confirmed that every gene identified as a leader gene can be supposed to play a major role in OLP at a molecular level. The analysis of the interaction map allowed the identification of different groups of genes corresponding to the typical aspects of OLP lesion: changes of the epithelial basement, regulation of cell cycle regulation and interleukine/chemokine signalling (see Fig. 3). Interestingly, there is at least one leader gene for each area (Fig. 5).

Epithelial basement membrane changes are common in OLP and include cellular damage as well as the degeneration of the basal keratinocyte anchoring elements, which may result in a weakened epithelial-connective tissue interface and histological cleft formation (the so-called Max-Joseph space).

Noteworthy, it was suggested that dysregulation in integrin pathways and inhibition of the TGFB1/Smad pathway could play a major role in the pathogenesis of OLP (38).

The b4 integrin encoded by the gene ITGB4 has different functions, ranging from epithelial cell adhesion (by interacting with the basement membrane component laminin) to cell motility (interacting with F-actin) and also to immunity regulation. Reflecting this wide array of functions, ITGB4 has been involved in a variety of oral diseases (being a potential biomarker of the Warthin's tumour of the parotid gland, the tongue squamous cell carcinoma, the ameloblastoma and other oral cavity tumours, and playing as well a role in the Sjogren disease and in most pre-malignant lesions, from oral leukoplakia to oral pemphigoid). Moreover, ITGB4 expression may reflect the response to a dental implant, determining the success of it together

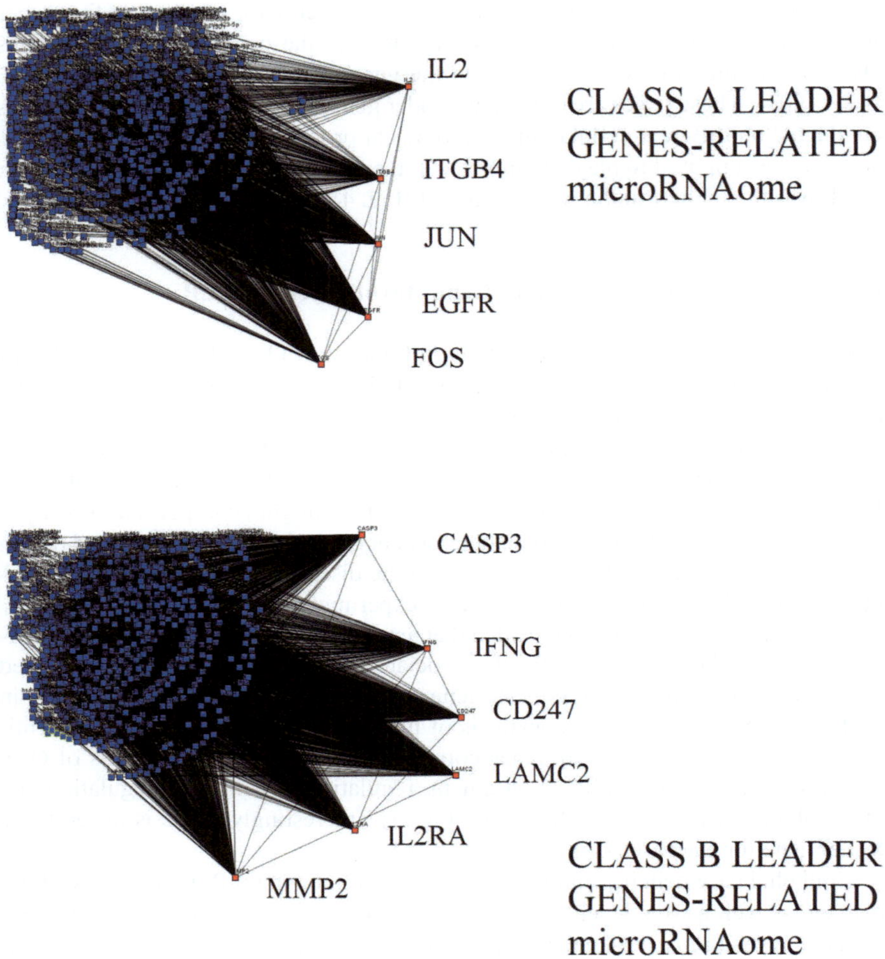

Fig. 5 Class A leader genes-related microRNAome (*upper*), class B leader genes-related microRNAome (*below*). *In red* the OLP-related genes, *in blue* the OLP-related microRNAs

with osseointegration, a well-known parameter of the clinical outcome (39–46). Together with a4 integrin, b4 integrin selectively accumulates in the oral mucosa in patients with OLP but does not seem to play a role in cutaneous lichen planus [187].

Matrix proteins (such as collagen types I, III, V, VI and undulin) and integrins are altered in OLP [61, 70, 72, 83, 85, 90].

EGFR too plays a key role in OLP even if its precise function is still not understood [24, 41, 46, 47, 50, 65, 73, 89, 117, 150, 186, 196, 199, 205]. EGFR has been linked to other many oral pathologies and has been found to have a role in oral infections and to be as well a good biomarker for both the diagnosis and the prognosis of oropharyngeal and oral cavity squamous cell carcinomas, salivary gland cancer and other oral tumours. EGFR may play a key role in the malignant transformation of oral pre-malignant lesions and the contradictory experimental findings about its expression and regulation in OLP may reflect different clusters of OLP in term of prognosis, considering that a fraction of OLP lesion transforms into a cancer [54, 58, 105, 119, 173]. Some scholars have proposed EGFR-targeted therapeutics as an emerging treatment for oral tumours (56), while other authors have suggested to use anti-EGFR monoclonal antibodies for monitoring oral cancers in vivo [46, 47].

FOS and JUN control cell cycle regulation but their role in OLP is still very little known. Also these genes have been related to a variety of oral disorders, from oral infections to oral cancers, and interestingly they may play an important role in the transition from pre-malignant to malignant lesions [69, 113, 152, 166, 179].

The third area in the interaction map is represented by interleukine and chemokine signalling and T lymphocytes/macrophages infiltration. There is a large evidence on a role for immune misregulation, specifically involving the cellular immune system [29, 31, 57, 62, 96, 146, 147, 148].

IL2 has been found to play a major role in oral cancers, thus becoming an important drug target (64; Gaur; Ribeiro). It has been also linked with oral infections, periodontitis, oral autoimmune diseases and other pathologies (65–66). Also IL6 as proinflammatory interleukin has been found to be over expressed in patients with OLP [1, 66, 148]. Pro-inflammatory IL1-α, IL8 [172, 148], as well as anti-inflammatory IL4 and IL10 [162, 177] play a role in the pathogenesis of OLP. Other interleukins such IL17 and IL23 proinflammatory signalling axis [100] may play a role in the pathogenesis of OLP, increasing the percentage of Th17 cells and IL-17 production in the CD4+ T cells from reticular OLP patients, enhancing the expression of β-defensin-2, -3, CCL-20, IL-8, and TNF-α in human oral keratinocytes.

Noteworthy, at present no direct link at a genomic level is identified between this area and the basal membrane alteration and cell cycle control areas. This finding may suggest other possible targeted experimentations.

Topological Analysis

We recently complemented the leader gene approach with a systems biology and topological analysis of the obtained graphs and networks. This is preliminary for further bioinformatics analysis and disease simulations using ad-hoc software. Topological analysis, in fact, can shed light on how molecular pathways work and how a disease develops and evolves.

Our analysis showed that our network exhibits a power law behaviour in agreement with the Scale-free theory of bio-networks and has more FFL (feed-forward loops) than one would expect to find in a random graph. The topological properties of leader genes and their role in controlling each pathway emerged from onthological analysis confirm our results.

OLP-Related miRNAs

microRNAs (miRNAs) are a family of small and short (usually 19–25 nucleotides long), single-stranded, endogenous, non-coding RNA molecules. They play a key role both in physiology and in pathology, and their role in oral diseases is emerging. Here we introduce a biomolecular strategy for predicting diseases-related miRNAs, based upon our previous published "Leader Genes algorithm". We managed to find most of the already established OLP-related miRNAs. Moreover, we found also other miRNAs that have not been directly linked with OLP yet but have a role in other oral diseases. For the remaining miRNAs their role is still unknown. This biomolecular strategy can foster further research in the field of the biology of oral diseases, suggesting to the researchers and molecular biologists some targets to focus on and to explore with *ad hoc* targeted experiments.

Limitations

Even if bioinformatics and data mining are supposed to play a major role in the analysis of genomics and proteomics data, the results of this study are to be considered more as well-supported hypotheses than as proven statements. This theoretical analysis used data mining, i.e., sorting thorough large amounts of data and picking up relevant information to potentially discover new knowledge. Therefore, because this approach is completely based on previous information, it is only able to generate new hypotheses. A targeted experimentation, e.g. with microarrays or RT-PCR, must be conducted to verify the hypotheses.

Noteworthy, only direct interactions, i.e., those based directly on experimental observations described in the public domain and available in specific databases, such as STRING were considered in the calculation of interactions. Direct interactions include physical interactions between encoded proteins (e.g., ligand–receptor contact), gene expression data derived from microarray experiments, and proved involvement in the same metabolic pathways. Only interactions with a high degree of confidence in the STRING database, i.e., those with a stronger experimental evidence, were considered. In this way, it is possible to limit, at least partially, a possible bias related to database mining.

On these basis, a limited circular reasoning–related bias might not represent a problem, because the results will be confirmed by experimentation. Moreover, these theoretical results are well supported by literature findings on the contribution of single genes to OLP as follows from the above described multiple experimental evidence. Noteworthy if we plot WNL for each genes in OLP network against global connectivity we can see (as in Fig. 5) that leader genes are above the regression line, confirming the hypothesis of a central and specific role of these genes in the OLP pathogenesis.

Conclusions

These data could further confirm that an approach based on bioinformatics and data-mining of existing databases could be a starting point to improve our knowledge about cellular processes and molecular mechanism of diseases and to plan targeted experimentation. In particular, the detailed analysis of gene interaction maps and the ranking of genes according to their number and confidence of interactions as well as the prediction of OLP-related miRNAs might have great value in the identification of new targets for a focused experimental analysis, which may confirm each hypothesis and suggest potential risk factors and therapy targets [99, 118, 133, 153, 168]. Noteworthy, a proper combination of experimental and theoretical results is necessary to draw a significant picture of a complex phenomenon, such as gene expression in a particular biologic system.

In this study, some genes with a potential major role in OLP were identified and are preliminarily divided into three different groups according to their function. Subsequently, they were used to predict OLP-related miRNAs. Even with the limitations of any theoretical study, these preliminarily results might suggest targeted DNA or protein microarray as well as RT-PCR experiments, focused on significant genes and simpler to be analysed than mass scale molecular genomics.

References

1. Abdel-Haq A, Kusnierz-Cabala B, Darczuk D, Sobuta E, Dumnicka P, Wojas-Pelc A, Chomyszyn-Gajewska M. Interleukin-6 and neopterin levels in the serum and saliva of patients with lichen planus and oral lichen planus. J Oral Pathol Med. 2014. doi:10.1111/jop.12199 (Epub ahead of print).
2. Agha-Hosseini F, Mirzaii-Dizgah I. p53 as a neoplastic biomarker in patients with erosive and plaque like forms of oral lichen planus. J Contemp Dent Pract. 2013;14(1):1–3.
3. Alexiou P, Vergoulis T, Gleditzsch M, Prekas G, Dalamagas T, Megraw M, Grosse I, Sellis T, Hatzigeorgiou AG. miRGen 2.0: a database of microRNA genomic information and regulation. Nucleic Acids Res. 2010;38(Database issue):D137–41.
4. An N, Rausch-Fan X, Wieland M, Matejka M, Andrukhov O, Schedle A. Initial attachment, cell proliferation/viability and gene expression of epithelial cells related to attachment and wound healing in response to different titanium surfaces. Dent Mater. 2012;28(12):1207–14.
5. Andreadis D, Epivatianos A, Poulopoulos A, Nomikos A, Christidis K, Papazoglou G, Antoniades D, Barbatis C. Immunohistochemical detection of the expression of the cell adhesion molecules E-cadherin, desmoglein-2, beta4-integrin, ICAM-1 and HCAM (CD44s) in Warthin's tumour of the parotid gland. Oral Oncol. 2005;41(8):799–805.
6. Andreasen JO. Oral lichen planus: a clinical evaluation of 115 cases. Oral Surg Oral Med Oral Pathol. 1968;25:31–42.
7. Anuradha CH, Reddy BV, Nandan SR, Kumar SR. Oral lichen planus. A review. N Y State Dent J. 2008;74:66–68.
8. Aguirre Urizar JM. Letter to the editor: oral lichenoid disease. A new classification proposal. Med Oral Patol Oral Cir Bucal. 2008;13(4):E224.
9. Arrais JP, Rosa N, Melo J, Coelho ED, Amaral D, Correia MJ, Barros M, Oliveira JL. OralCard: a bioinformatic tool for the study of oral proteome. Arch Oral Biol. 2013;58(7):762–72.

10. Arão TC1, Guimarães AL, de Paula AM, Gomes CC, Gomez RS. Increased miRNA-146a and miRNA-155 expressions in oral lichen planus. Arch Dermatol Res. 2012;304(5):371–5. doi:10.1007/s00403-011-1197-x.
11. Au J, Patel D, Campbell JH. Oral lichen planus. Oral Maxillofac Surg Clin North Am. 2013;25(1):93–100, vii.
12. Barabasi AL, Ravasz E, Vicsek T. Deterministic scale-free networks. Phys A Stat Mech Appl. 2001;299(3–4):559–564.
13. Bauer-Mehren A. Integration of genomic information with biological networks using Cytoscape. Methods Mol Biol. 2013;1021:37–61.
14. Baum BJ, Yates JR 3rd, Srivastava S, Wong DT, Melvin JE. Scientific frontiers: emerging technologies for salivary diagnostics. Adv Dent Res. 2011;23(4):360–8.
15. Becker J, Schuppan D. Altered expression of extracellular matrix proteins and integrins in oral lichen planus (OLP). J Oral Pathol Med. 1995;24(4):159–64.
16. Boorghani M, Gholizadeh N, Taghavi Zenouz A, Vatankhah M, Mehdipour M. Oral lichen planus: clinical features, etiology, treatment and management; a review of literature. J Dent Res Dent Clin Dent Prospects. 2010;4(1):3–9.
17. Bragazzi NL, Nicolini C. a leader genes approach-based tool for molecular genomics: from gene-ranking to gene-network systems biology and biotargets predictions. J Comput Sci Syst Biol. 2013;6:165–176.
18. Bragazzi NL, Sivozhelezov V, Nicolini C. Leader gene: a fast data-mining tool for molecular genomics. J Proteomics Bioinform. 2011;4(4):83–86.
19. Bragazzi NL, Pechkova E, Nicolini C. Proteomics and proteogenomics approaches for oral diseases. Adv Protein Chem Struct Biol. 2014;95:125–62.
20. Burbelo PD, Bayat A, Lebovitz EE, Iadarola MJ. New technologies for studying the complexity of oral diseases. Oral Dis. 2012;18(2):121–6.
21. Carbone M, Arduino PG, Carrozzo M, Gandolfo S, Argiolas MR, Bertolusso G, Conrotto D, Pentenero M, Broccoletti R. Course of oral lichen planus: a retrospective study of 808 northern Italian patients. Oral Dis. 2009;15(3):235–43.
22. Carrozzo M. Oral diseases associated with hepatitis C virus infection. Part 2: lichen planus and other diseases. Oral Dis. 2008;14(3):217–28.
23. Carrozzo M, Gandolfo S. Oral diseases possibly associated with hepatitis C virus. Crit Rev Oral Biol Med. 2003;14(2):115–27.
24. Chandarana SP, Lee JS, Chanowski EJ, Sacco AG, Bradford CR, Wolf GT, Prince ME, Moyer JS, Eisbruch A, Worden FP, Giordano TJ, Kumar B, Cordell KG, Carey TE, Chepeha DB. Prevalence and predictive role of p16 and epidermal growth factor receptor in surgically treated oropharyngeal and oral cavity cancer. Head Neck. 2012. doi:10.1002/hed.23087 (Epub ahead of print).
25. Chen Y, Zhang W, Geng N, Tian K, Jack Windsor L. MMPs, TIMP-2, and TGF-beta1 in the cancerization of oral lichen planus. Head Neck. 2008;30(9):1237–45.
26. Cheng YS, Rees T, Jordan L, Oxford L, O'Brien J, Chen HS, Wong D. Salivary endothelin-1 potential for detecting oral cancer in patients with oral lichen planus or oral cancer in remission. Oral Oncol. 2011;47(12):1122–6.
27. Cheng YS, Jordan L, Rees T, Chen HS, Oxford L, Brinkmann O, Wong D. Levels of potential oral cancer salivary mRNA biomarkers in oral cancer patients in remission and oral lichen planus patients. Clin Oral Investig. 2014;18(3):985–93. doi:10.1007/s00784-013-1041-0.
28. Cheng L, Wang G, Li J, Zhang T, Xu P, Wang Y. SIDD: a semantically integrated database towards a global view of human disease. PLoS ONE. 2013;8(10):e75504.
29. Cheriyan VT, Thomas C, Balaram P. Augmentation of T-cell immune responses and signal transduction proteins in oral cancer patients: potential for IL-2-mediated immunotherapy. J Cancer Res Clin Oncol. 2011;137(10):1435–44.
30. Chiappelli F, Covani U, Giacomelli L. Proteomics as it pertains to oral pathologies and dental research. Bioinformation. 2011;5(7):277.
31. Covani U, Marconcini S, Giacomelli L, Sivozhelevov V, Barone A, Nicolini C. Bioinformatic prediction of leader genes in human periodontitis. J Periodontol. 2008;79(10):1974–83.

32. Dang J, Bian YQ, Sun JY, Chen F, Dong GY, Liu Q, Wang XW, Kjems J, Gao S, Wang QT. MicroRNA-137 promoter methylation in oral lichen planus and oral squamous cell carcinoma. J Oral Pathol Med. 2013;42(4):315–21.
33. Danielsson K, Ebrahimi M, Wahlin YB, Nylander K, Boldrup L. Increased levels of COX-2 in oral lichen planus supports an autoimmune cause of the disease. J Eur Acad Dermatol Venereol. 2012;26(11):1415–9.
34. Danielsson K, Wahlin YB, Gu X, Boldrup L, Nylander K. Altered expression of miR-21, miR-125b, and miR-203 indicates a role for these microRNAs in oral lichen planus. J Oral Pathol Med. 2012;41(1):90–5.
35. Danielsson K, Boldrup L, Rentoft M, Coates PJ, Ebrahimi M, Nylander E, Wahlin YB, Nylander K. Autoantibodies and decreased expression of the transcription factor ELF-3 together with increased chemokine pathways support an autoimmune phenotype and altered differentiation in lichen planus located in oral mucosa. J Eur Acad Dermatol Venereol. 2013;27(11):1410–6.
36. De Cecco L, Dugo M, Canevari S, Daidone MG, Callari M. Measuring microRNA expression levels in oncology: from samples to data analysis. Crit Rev Oncog. 2013;18(4):273–87.
37. Demchak B, Hull T, Reich M, Liefeld T, Smoot M, Ideker T, Mesirov JP. Cytoscape: the network visualization tool for GenomeSpace workflows. F1000Res. 2014;3:151.
38. de Sousa FA, Paradella TC, Carvalho YR, Rosa LE. Comparative analysis of the expression of proliferating cell nuclear antigen, p53, bax, and bcl-2 in oral lichen planus and oral squamous cell carcinoma. Ann Diagn Pathol. 2009;13(5):308–12.
39. Di Stasio D, Guida A, Salerno C, Contaldo M, Esposito V, Laino L, Serpico R, Lucchese AE. Oral lichen planus: a narrative review. Front Biosci (Elite Ed). 2014;6:370–6.
40. Ebrahimi M, Boldrup L, Coates PJ, Wahlin YB, Bourdon JC, Nylander K. Expression of novel p53 isoforms in oral lichen planus. Oral Oncol. 2008;44(2):156–61.
41. Ebrahimi M, Boldrup L, Wahlin YB, Coates PJ, Nylander K. Decreased expression of the p63 related proteins beta-catenin, E-cadherin and EGFR in oral lichen planus. Oral Oncol 2008;44(7):634–8.
42. Ebrahimi M, Nylander K, van der Waal I. Oral lichen planus and the p53 family: what do we know? J Oral Pathol Med. 2011;40(4):281–5.
43. Eisenberg E, Krutchkoff DJ. Lichenoid lesions of oral mucosa. Diagnostic criteria and their importance in the alleged relationship to oral cancer. Oral Surg Oral Med Oral Pathol. 1992;73(6):699–704.
44. Eng G, Chen A, Vess T, Ginsburg GS. Genome technologies and personalized dental medicine. Oral Dis. 2012;18(3):223–35.
45. Epstein JB, Wan LS, Gorsky M, Zhang L. Oral lichen planus: progress in understanding its malignant potential and the implications for clinical management. Oral Surg Oral Med Oral Pathol Oral Radiol Endod. 2003;96(1):32–7.
46. Ettl T, Baader K, Stiegler C, Müller M, Agaimy A, Zenk J, Kühnel T, Gosau M, Zeitler K, Schwarz S, Brockhoff G. Loss of PTEN is associated with elevated EGFR and HER2 expression and worse prognosis in salivary gland cancer. Br J Cancer. 2012;106(4):719–26.
47. Ettl T, Stiegler C, Zeitler K, Agaimy A, Zenk J, Reichert TE, Gosau M, Kühnel T, Brockhoff G, Schwarz S. EGFR, HER2, survivin, and loss of pSTAT3 characterize high-grade malignancy in salivary gland cancer with impact on prognosis. Hum Pathol. 2012;43(6):921–31.
48. Fox T. Clinical lecture on lichen ruber of hebra. Br Med J. 1871;1(537):392–4.
49. Franceschini A, Szklarczyk D, Frankild S, Kuhn M, Simonovic M, Roth A, Lin J, Minguez P, Bork P, von Mering C, Jensen LJ. STRING v9.1: protein-protein interaction networks, with increased coverage and integration. Nucleic Acids Res. 2013;41(Database issue):D808–15.
50. Fu J, Chen W, Sun Z. Gene expression of epidermal growth factor and epidermal growth factor receptor in oral lichen planus. Zhonghua Kou Qiang Yi Xue Za Zhi. 2005;40(6):455–8.
51. Gaffen SL, Liu KD. Overview of interleukin-2 function, production and clinical applications. Cytokine. 2004;28(3):109–23.
52. Gallo A, Alevizos I. Isolation of circulating microRNA in saliva. Methods Mol Biol. 2013;1024:183–90.

53. Gandarillas A. Epidermal differentiation, apoptosis, and senescence: common pathways? Exp Gerontol. 2000;35(1):53–62.
54. Gandolfo S, Richiardi L, Carrozzo M, Broccoletti R, Carbone M, Pagano M, Vestita C, Rosso S, Merletti F. Risk of oral squamous cell carcinoma in 402 patients with oral lichen planus: a follow-up study in an Italian population. Oral Oncol. 2004;40(1):77–83.
55. Garcia I, Kuska R, Somerman MJ. Expanding the foundation for personalized medicine: implications and challenges for dentistry. J Dent Res. 2013;92(7 Suppl):3S-10S.
56. Gassling V, Hampe J, Açil Y, Braesen JH, Wiltfang J, Häsler R. Disease-associated miRNA-mRNA networks in oral lichen planus. PLoS One. 2013;8(5):e63015.
57. Gaur P, Singh AK, Shukla NK, Das SN. Inter-relation of Th1, Th2, Th17 and Treg cytokines in oral cancer patients and their clinical significance. Hum Immunol. 2014;75(4):330–7.
58. Georgakopoulou EA, Troupis TG, Troupis G, Gorgoulis VG. Update of the cancer-associated molecular mechanisms in oral lichen planus, a disease with possible premalignant nature. J BUON. 2011;16(4):613–6.
59. Giacomelli L, Nicolini C. Gene expression of human T lymphocytes cell cycle: experimental and bioinformatic analysis. J Cell Biochem. 2006;99(5):1326–33.
60. Giacomelli L, Oluwadara O, Chiappe G, Barone A, Chiappelli F, Covani U. Relationship between human oral lichen planus and oral squamous cell carcinoma at a genomic level: a datamining study. Bioinformation. 2009;4(6):258–62.
61. Giannelli G, Brassard J, Foti C, Stetler-Stevenson WG, Falk-Marzillier J, Zambonin-Zallone A, Schiraldi O, Quaranta V. Altered expression of basement membrane proteins and their integrin receptors in lichen planus: possible pathogenetic role of gelatinases A and B. Lab Invest. 1996;74(6):1091–104.
62. Glick M. Personalized oral health care: providing '-omic' answers to oral health care queries. J Am Dent Assoc. 2012;143(2):102–4.
63. Gorsky M, Epstein JB. Oral lichen planus: malignant transformation and human papilloma virus: a review of potential clinical implications. Oral Surg Oral Med Oral Pathol Oral Radiol Endod. 2011;111(4):461–4.
64. Gorsky M, Epstein JB, Hasson-Kanfi H, Kaufman E. Smoking habits among patients diagnosed with oral lichen planus. Tob Induc Dis. 2004;2(2):103–8.
65. Grimm M, Iftner T, Altaki H, Iftner A, Peters JP, Munz A, Reinert S. Detection of mutation-specific epidermal growth factor receptor (E746-A750del) and lack of detection of human papillomavirus in oral squamous cell carcinoma. Int J Oral Maxillofac Surg. 2014. pii: S0901-5027(14)00162–3. doi:10.1016/j.ijom.2014.04.006 (Epub ahead of print).
66. Gu GM, Martin MD, Darveau RP, Truelove E, Epstein J. Oral and serum IL-6 levels in oral lichen planus patients. Oral Surg Oral Med Oral Pathol Oral Radiol Endod. 2004;98(6):673–8.
67. Gurkan C, Lapp H, Hogenesch JB, Balch WE. Exploring trafficking GTPase function by mRNA expression profiling: use of the SymAtlas web-application and the Membrome datasets. Methods Enzymol. 2005;403:1–10.
68. Gustafson J, Eklund C, Wallström M, Zellin G, Magnusson B, Hasséus B. Langerin-expressing and CD83-expressing cells in oral lichen planus lesions. Acta Odontol Scand. 2007;65(3):156–61.
69. Gutiérrez-Venegas G, Castillo-Alemán R. Characterization of the transduction pathway involved in c-fos and c-jun expression induced by Aggregatibacter actinomycetemcomitans lipopolysaccharides in human gingival fibroblasts. Int Immunopharmacol. 2008;8(11):-1513–23.
70. Häkkinen L, Kainulainen T, Salo T, Grenman R, Larjava H. Expression of integrin alpha9 subunit and tenascin in oral leukoplakia, lichen planus, and squamous cell carcinoma. Oral Dis. 1999;5(3):210–7.
71. Hallopeau H. Sur un cas de lichen de Wilson gingival avec neoplasia voisine dans la région maxillaire. Bull Soc Fr Dermatol Syphiligr 1910;17:32.
72. Hamidi S, Salo T, Kainulainen T, Epstein J, Lerner K, Larjava H. Expression of alpha(v)beta6 integrin in oral leukoplakia. Br J Cancer. 2000;82(8):1433–40.

73. Hartmann S, Seher A, Brands RC, Linz C, Lessner G, Böhm H, Kübler AC, Müller-Richter UD. Influence of epidermal growth factor receptor expression on the cetuximab and panitumumab response rates of head and neck carcinoma cells. J Craniomaxillofac Surg. 2014. pii: S1010-5182(14)00098-5. doi:10.1016/j.jcms.2014.03.018 (Epub ahead of print).
74. Hirota M, Ito T, Okudela K, Kawabe R, Yazawa T, Hayashi H, Nakatani Y, Fujita K, Kitamura H. Cell proliferation activity and the expression of cell cycle regulatory proteins in oral lichen planus. J Oral Pathol Med. 2002;31(4):204–12.
75. Hirschhorn JN. Genetic approaches to studying common diseases and complex traits. Pediatr Res. 2005;57(5 Pt 2):74R-77R.ù.
76. Hooper SD, Bork P. Medusa: a simple tool for interaction graph analysis. Bioinformatics. 2005;21(24):4432–3.
77. Hu JY, Zhang J, Cui JL, Liang XY, Lu R, Du GF, Xu XY, Zhou G. Increasing CCL5/CCR5 on CD4+ T cells in peripheral blood of oral lichen planus. Cytokine. 2013;62(1):141–5.
78. Ichimura M, Hiratsuka K, Ogura N, Utsunomiya T, Sakamaki H, Kondoh T, Abiko Y, Otake S, Yamamoto M. Expression profile of chemokines and chemokine receptors in epithelial cell layers of oral lichen planus. J Oral Pathol Med. 2006;35(3):167–74.
79. Ismail SB, Kumar SK, Zain RB. Oral lichen planus and lichenoid reactions: etiopathogenesis, diagnosis, management and malignant transformation. J Oral Sci. 2007;49(2):89–106.
80. Jiang Q, Wang Y, Hao Y, Juan L, Teng M, Zhang X, Li M, Wang G, Liu Y. miR2Disease: a manually curated database for microRNA deregulation in human disease. Nucleic Acids Res. 2009;37(Database issue):D98–104.
81. Jiang L, Edwards SM, Thomsen B, Workman CT, Guldbrandtsen B, Sørensen P. A random set scoring model for prioritization of disease candidate genes using protein complexes and data-mining of GeneRIF, OMIM and PubMed records. BMC Bioinformatics. 2014;15(1):315.
82. Jimeno-Yepes AJ, Sticco JC, Mork JG, Aronson AR. GeneRIF indexing: sentence selection based on machine learning. BMC Bioinformatics. 2013;14:171.
83. Jones J, Downer CS, Speight PM. Changes in the expression of integrins and basement membrane proteins in benign mucous membrane pemphigoid. Oral Dis. 1995;1(3):159–65.
84. Jovanovic V, Giacomelli L, Sivozhelezov V, Degauque N, Lair D, Soulillou JP, Pechkova E, Nicolini C, Brouard S. AKT1 leader gene and downstream targets are involved in a rat model of kidney allograft tolerance. J Cell Biochem. 2010;111(3):709–19.
85. Kainulainen T, Autio-Harmainen H, Oikarinen A, Salo S, Tryggvason K, Salo T. Altered distribution and synthesis of laminin-5 (kalinin) in oral lichen planus, epithelial dysplasias and squamous cell carcinomas. Br J Dermatol. 1997;136(3):331–6.
86. Karatsaidis A, Schreurs O, Axéll T, Helgeland K, Schenck K. Inhibition of the transforming growth factor-beta/Smad signaling pathway in the epithelium of oral lichen. J Invest Dermatol. 2003;121(6):1283–90.
87. Khan A, Farah CS, Savage NW, Walsh LJ, Harbrow DJ, Sugerman PB. Th1 cytokines in oral lichen planus. J Oral Pathol Med. 2003;32(2):77–83.
88. Kilpi A, Rich AM, Konttinen YT, Reade PC. Expression of c-erbB-2 protein in keratinocytes of oral mucosal lichen planus and subsequent squamous cell carcinoma. Eur J Oral Sci. 1996;104(3):278–84.
89. Kumagai K, Horikawa T, Gotoh A, Yamane S, Yamada H, Kobayashi H, Hamada Y, Suzuki S, Suzuki R. Up-regulation of EGF receptor and its ligands, AREG, EREG, and HB-EGF in oral lichen planus. Oral Surg Oral Med Oral Pathol Oral Radiol Endod. 2010;110(6):748–54.
90. Kurokawa A, Nagata M, Kitamura N, Noman AA, Ohnishi M, Ohyama T, Kobayashi T, Shingaki S, Takagi R; Oral, Maxillofacial Pathology, and Surgery Group. Diagnostic value of integrin alpha3, beta4, and beta5 gene expression levels for the clinical outcome of tongue squamous cell carcinoma. Cancer. 2008;112(6):1272–81.
91. Laskaris GC, Papavasiliou SS, Bovopoulou OD, Nicolis GD. Lichen planus pigmentosus of the oral mucosa: a rare clinical variety. Dermatologica. 1981;162(1):61–3.
92. Lee HJ. Exceptional stories of microRNAs. Exp Biol Med (Maywood). 2013;238(4):339–43.

93. Leslie R, O'Donnell CJ, Johnson AD. GRASP: analysis of genotype-phenotype results from 1390 genome-wide association studies and corresponding open access database. Bioinformatics. 2014;30(12):i185–94.
94. Li TJ, Cui J. COX-2, MMP-7 expression in oral lichen planus and oral squamous cell carcinoma. Asian Pac J Trop Med. 2013;6(8):640–3.
95. Leyva-Huerta ER, Ledesma-Montes C, Rojo-Botello RE, Vega-Memije E. P53 and bcl-2 immunoexpression in patients with oral lichen planus and oral squamous cell carcinoma. Med Oral Patol Oral Cir Bucal. 2012;17(5):e745–50.
96. Lisa Cheng YS, Jordan L, Gorugantula LM, Schneiderman E, Chen HS, Rees T. Salivary interleukins 6 and 8 in oral cancer patients and in patients with chronic oral inflammatory diseases. J Periodontol. 2014;85(7):956–65. doi:10.1902/jop.2013.130320.
97. Lodi G, Scully C, Carrozzo M, Griffiths M, Sugerman PB, Thongprasom K. Current controversies in oral lichen planus: report of an international consensus meeting. Part 1. Viral infections and etiopathogenesis. Oral Surg Oral Med Oral Pathol Oral Radiol Endod. 2005;100(1):40–51.
98. Lodi G, Scully C, Carrozzo M, Griffiths M, Sugerman PB, Thongprasom K. Current controversies in oral lichen planus: report of an international consensus meeting. Part 2. Clinical management and malignant transformation. Oral Surg Oral Med Oral Pathol Oral Radiol Endod. 2005;100(2):164–78.
99. Lodi G, Carrozzo M, Furness S, Thongprasom K. Interventions for treating oral lichen planus: a systematic review. Br J Dermatol. 2012;166(5):938–47.
100. Lu R, Zeng X, Han Q, Lin M, Long L, Dan H, Zhou G, Chen Q. Overexpression and selectively regulatory roles of IL-23/IL-17 axis in the lesions of Oral Lichen Planus. Mediators Inflamm. 2014;2014:701094.
101. Ma L, Wang H, Yao H, Zhu L, Liu W, Zhou Z. Bmi1 expression in oral lichen planus and the risk of progression to oral squamous cell carcinoma. Ann Diagn Pathol. 2013;17(4):-327–30.
102. Maclellan SA, Lawson J, Baik J, Guillaud M, Poh CF, Garnis C. Differential expression of miRNAs in the serum of patients with high-risk oral lesions. Cancer Med. 2012;1(2):-268–74.
103. Maere S, Heymans K, Kuiper M. BiNGO: a Cytoscape plugin to assess overrepresentation of gene ontology categories in biological networks. Bioinformatics. 2005;21(16):3448–9.
104. Mangan S, Alon U. Structure and function of the feed-forward loop network motif. Proc Natl Acad Sci U S A. 2003;100(21):11980–11985.
105. Manuel Gándara Rey J, Diniz Freitas M. High rate of malignant transformation in atypical oral lichen planus lesions. Med Oral. 2003;8(5):309.
106. Marconcini S, Covani U, Barone A, Vittorio O, Curcio M, Barbuti S, Scatena F, Felli L, Nicolini C. Real-time quantitative polymerase chain reaction analysis of patients with refractory chronic periodontitis. J Periodontol. 2011;82(7):1018–24.
107. Martínez-Lara I, González-Moles MA, Ruiz-Avila I, Bravo M, Ramos MC, Fernández-Martínez JA. Proliferating cell nuclear antigen (PCNA) as a marker of dysplasia in oral mucosa. Acta Stomatol Belg. 1996;93(1):29–32.
108. Martín-Ezquerra G, Salgado R, Toll A, Gilaberte M, Baró T, Alameda Quitllet F, Yébenes M, Solé F, Garcia-Muret M, Espinet B, Pujol RM. Multiple genetic copy number alterations in oral squamous cell carcinoma: study of MYC, TP53, CCDN1, EGFR and ERBB2 status in primary and metastatic tumours. Br J Dermatol. 2010;163(5):1028–35.
109. Mattila R, Alanen K, Syrjänen S. Desmocollin expression in oral atrophic lichen planus correlates with clinical behavior and DNA content. J Cutan Pathol. 2008;35(9):832–8.
110. McCartan BE, Healy CM. The reported prevalence of oral lichen planus: a review and critique. J Oral Pathol Med. 2008;37(8):447–53.
111. Megraw M, Sethupathy P, Corda B, Hatzigeorgiou AG. miRGen: a database for the study of animal microRNA genomic organization and function. Nucleic Acids Res. 2007;35(Database issue):D149–55.

112. Michael A, Bajracharya SD, Yuen PS, Zhou H, Star RA, Illei GG, Alevizos I. Exosomes from human saliva as a source of microRNA biomarkers. Oral Dis. 2010;16(1):34–8.
113. Mishra A, Bharti AC, Saluja D, Das BC. Transactivation and expression patterns of Jun and Fos/AP-1 super-family proteins in human oral cancer. Int J Cancer. 2010;126(4):819–29.
114. Mithani SK, Mydlarz WK, Grumbine FL, Smith IM, Califano JA. Molecular genetics of premalignant oral lesions. Oral Dis. 2007;13(2):126–33.
115. Modolo F, Martins MT, Loducca SV, de Araújo VC. Expression of integrin subunits alpha2, alpha3, alpha5, alphav, beta1, beta3 and beta4 in different histological types of ameloblastoma compared with dental germ, dental lamina and adult lining epithelium. Oral Dis. 2004;10(5):277–82.
116. Mollaoglu N. Oral lichen planus: a review. Br J Oral Maxillofac Surg. 2000;38(4):370–7.
117. Monteiro LS, Diniz-Freitas M, Garcia-Caballero T, Warnakulasuriya S, Forteza J, Fraga M. Combined cytoplasmic and membranous EGFR and p53 overexpression is a poor prognostic marker in early stage oral squamous cell carcinoma. J Oral Pathol Med. 2012;41(7):-559–67.
118. Nagao T, Warnakulasuriya S. Annual screening for oral cancer detection. Cancer Detect Prev. 2003;27(5):333–7.
119. Nagao T, Warnakulasuriya S, Gelbier S, Yuasa H, Tsuboi S, Nakagaki H. Oral pre-cancer and the associated risk factors among industrial workers in Japan's overseas enterprises in the UK. J Oral Pathol Med. 2003;32(5):257–64.
120. Nicolini C, Spera R, Stura E, Fiordoro S, Giacomelli L. Gene expression in the cell cycle of human T-lymphocytes: II. Experimental determination by DNASER technology. J Cell Biochem. 2006;97(5):1151–9.
121. Nicolini C, Bragazzi N, Pechkova E. Nanoproteomics enabling personalized nanomedicine. Adv Drug Deliv Rev. 2012;64(13):1522–31.
122. Nylander K, Coates PJ, Hall PA. Characterization of the expression pattern of p63 alpha and delta Np63 alpha in benign and malignant oral epithelial lesions. Int J Cancer. 2000;87(3):368–72.
123. Nylander E, Ebrahimi M, Wahlin YB, Boldrup L, Nylander K. Changes in miRNA expression in sera and correlation to duration of disease in patients with multifocal mucosal lichen planus. J Oral Pathol Med. 2012;41(1):86–9.
124. Ogawa Y, Taketomi Y, Murakami M, Tsujimoto M, Yanoshita R. Small RNA transcriptomes of two types of exosomes in human whole saliva determined by next generation sequencing. Biol Pharm Bull. 2013;36(1):66–75.
125. Ogmundsdóttir HM, Björnsson J, Holbrook WP. Role of TP53 in the progression of premalignant and malignant oral mucosal lesions. A follow-up study of 144 patients. J Oral Pathol Med. 2009;38(7):565–71.
126. Ogmundsdóttir HM, Hilmarsdóttir H, Björnsson J, Holbrook WP. Longitudinal study of TP53 mutations in eight patients with potentially malignant oral mucosal disorders. J Oral Pathol Med. 2009;38(9):716–21.
127. Oluwadara O, Giacomelli L, Christensen R, Kossan G, Avezova R, Chiappelli F. LCK, survivin and PI-3K in the molecular biomarker profiling of oral lichen planus and oral squamous cell carcinoma. Bioinformation. 2009;4(6):249–57.
128. Orlando B, Bragazzi N, Nicolini C. Bioinformatics and systems biology analysis of genes network involved in OLP (Oral Lichen Planus) pathogenesis. Arch Oral Biol. 2013;58(6):664–73.
129. Orlando B, Giacomelli L, Ricci M, Barone A, Covani U. Leader genes in osteogenesis: a theoretical study. Arch Oral Biol. 2013;58(1):42–9.
130. Pandey M, Prakash O, Santhi WS, Soumithran CS, Pillai RM. Overexpression of COX-2 gene in oral cancer is independent of stage of disease and degree of differentiation. Int J Oral Maxillofac Surg. 2008;37(4):379–83.
131. Patel RS, Jakymiw A, Yao B, Pauley BA, Carcamo WC, Katz J, Cheng JQ, Chan EK. High resolution of microRNA signatures in human whole saliva. Arch Oral Biol. 2011;56(12):1506–13.

132. Patil A, Prasad S, Ashok L, Sujatha GP. Oral bullous lichen planus: case report and review of management. Contemp Clin Dent. 2012;3(3):344–8.
133. Pavlic V, Vujic-Aleksic V. Phototherapy approaches in treatment of oral lichen planus. Photodermatol Photoimmunol Photomed. 2013. doi:10.1111/phpp.12074 (Epub ahead of print).
134. Payeras MR, Cherubini K, Figueiredo MA, Salum FG. Oral lichen planus: Focus on etiopathogenesis. Arch Oral Biol. 2013;58(9):1057–69. doi:10.1016/j.archoralbio.2013.04.004.
135. Peng K, Xu W, Zheng J, Huang K, Wang H, Tong J, Lin Z, Liu J, Cheng W, Fu D, Du P, Kibbe WA, Lin SM, Xia T. The Disease and Gene Annotations (DGA): an annotation resource for human disease. Nucleic Acids Res. 2013;41(Database issue):D553–60.
136. Pickert A. Concise review of lichen planus and lichenoid dermatoses. Cutis. 2012;90(3):-E1–3.
137. Pindborg JJRP, Smith CJ, van der Waal I. Histological typing of cancer and precancer of the oral mucosa. World Health Organization international histological classification of tumours. 2nd ed. Berlin: Springer; 1997.
138. Poomsawat S, Buajeeb W, Khovidhunkit SO, Punyasingh J. Overexpression of cdk4 and p16 in oral lichen planus supports the concept of premalignancy. J Oral Pathol Med. 2011;40(4):294–9.
139. Pradhan S, Sengupta M, Dutta A, Bhattacharyya K, Bag SK, Dutta C, Ray K. Indian genetic disease database. Nucleic Acids Res. 2011;39(Database issue):D933–8.
140. Prolo P, Chiappelli F, Cajulis E, Bauer J, Spackman S, Romeo H, Carrozzo M, Gandolfo S, Christensen R. Psychoneuroimmunology in oral biology and medicine: the model of oral lichen planus. Ann N Y Acad Sci. 2002;966:429–40.
141. Racapé M, Bragazzi N, Sivozhelezov V, Danger R, Pechkova E, Duong Van Huyen JP, Soulillou JP, Brouard S, Nicolini C. SMILE silencing and PMA activation gene networks in HeLa cells: comparison with kidney transplantation gene networks. J Cell Biochem. 2012;113(6):1820–32.
142. Ramos EM, Hoffman D, Junkins HA, Maglott D, Phan L, Sherry ST, Feolo M, Hindorff LA. Phenotype-Genotype Integrator (PheGenI): synthesizing genome-wide association study (GWAS) data with existing genomic resources. Eur J Hum Genet. 2014;22(1):144–7.
143. Rappaport N, Nativ N, Stelzer G, Twik M, Guan-Golan Y, Stein TI, Bahir I, Belinky F, Morrey CP, Safran M, Lancet D. MalaCards: an integrated compendium for diseases and their annotation. Database (Oxford). 2013;2013:bat018.
144. Rebhan M, Chalifa-Caspi V, Prilusky J, Lancet D. GeneCards: integrating information about genes, proteins and diseases. Trends Genet. 1997;13(4):163.
145. Reichart PA, Schmidtberg W, Scheifele C. Betel chewer's mucosa in elderly Cambodian women. J Oral Pathol Med. 1996;25(7):367–70.
146. Reichert S, Machulla HK, Klapproth J, Zimmermann U, Reichert Y, Gläser C, Schaller HG, Schulz S. Interleukin-2 -330 and 166 gene polymorphisms in relation to aggressive or chronic periodontitis and the presence of periodontopathic bacteria. J Periodontal Res. 2009;44(5):628–35.
147. Rhodus NL, Cheng B, Myers S, Bowles W, Ho V, Ondrey F. A comparison of the pro-inflammatory, NF-kappaB-dependent cytokines: TNF-alpha, IL-1-alpha, IL-6, and IL-8 in different oral fluids from oral lichen planus patients. Clin Immunol. 2005;114(3):278–83.
148. Rhodus NL, Cheng B, Myers S, Miller L, Ho V, Ondrey F. The feasibility of monitoring NF-kappaB associated cytokines: TNF-alpha, IL-1alpha, IL-6, and IL-8 in whole saliva for the malignant transformation of oral lichen planus. Mol Carcinog. 2005;44(2):77–82.
149. Ribeiro IP, Marques F, Caramelo F, Pereira J, Patrício M, Prazeres H, Ferrão J, Julião MJ, Castelo-Branco M, de Melo JB, Baptista IP, Carreira IM. Genetic gains and losses in oral squamous cell carcinoma: impact on clinical management. Cell Oncol (Dordr). 2014;37(1):29–39.
150. Ribeiro FA, Noguti J, Oshima CT, Ribeiro DA. Effective targeting of the epidermal growth factor receptor (EGFR) for treating oral cancer: a promising approach. Anticancer Res. 2014;34(4):1547–52.

151. Rosa N, Correia MJ, Arrais JP, Lopes P, Melo J, Oliveira JL, Barros M. From the salivary proteome to the OralOme: comprehensive molecular oral biology. Arch Oral Biol. 2012;57(7):853–64.
152. Sachdev R, Mandal AK, Singh I, Agarwal AK. Progressive rise of c fos expression from premalignant to malignant lesions of oral cavity. Med Oral Patol Oral Cir Bucal. 2008;13(11):E683–6.
153. Sami N, Bhol KC, Ahmed AR. Treatment of oral pemphigoid with intravenous immunoglobulin as monotherapy. Long-term follow-up: influence of treatment on antibody titres to human alpha6 integrin. Clin Exp Immunol. 2002;129(3):533–40.
154. Schrodi SJ, Mukherjee S, Shan Y, Tromp G, Sninsky JJ, Callear AP, Carter TC, Ye Z, Haines JL, Brilliant MH, Crane PK, Smelser DT, Elston RC, Weeks DE. Genetic-based prediction of disease traits: prediction is very difficult, especially about the future. Front Genet. 2014;5:162.
155. Scully C, El-Kom M. Lichen planus: review and update on pathogenesis. J Oral Pathol 1985;14:431–58.
156. Shannon W, Culverhouse R, Duncan J. Analyzing microarray data using cluster analysis. Pharmacogenomics. 2003;4(1):41–52.
157. Shannon P, Markiel A, Ozier O, Baliga NS, Wang JT, Ramage D, Amin N, Schwikowski B, Ideker T. Cytoscape: a software environment for integrated models of biomolecular interaction networks. Genome Res. 2003;13(11):2498–504.
158. Shi P, Liu W, Zhou ZT, He QB, Jiang WW. Podoplanin and ABCG2: malignant transformation risk markers for oral lichen planus. Cancer Epidemiol Biomarkers Prev. 2010;19(3):844–9.
159. Shimbel A. Structural parameters of communication networks. Bull Math Biophys. 1953;15:501–507.
160. Shimoyama T, Horie N, Kato T, Kaneko T, Komiyama K. Helicobacter pylori in oral ulcerations. J Oral Sci. 2000;42(4):225–9.
161. Silverman S, Jr, Gorsky M, Lozada-Nur F. A prospective follow-up study of 570 patients with oral lichen planus: persistence, remission, and malignant association. Oral Surg Oral Med Oral Pathol. 1985;60:30–4.
162. Simark-Mattsson C, Bergenholtz G, Jontell M, Eklund C, Seymour GJ, Sugerman PB, Savage NW, Dahlgren UI. Distribution of interleukin-2, -4, -10, tumour necrosis factor-alpha and transforming growth factor-beta mRNAs in oral lichen planus. Arch Oral Biol. 1999;44(6):499–507.
163. Sivozhelezov V, Giacomelli L, Tripathi S, Nicolini C. Gene expression in the cell cycle of human T lymphocytes: I. Predicted gene and protein networks. J Cell Biochem. 2006;97(5):1137–50.
164. Sivozhelezov V, Braud C, Giacomelli L, Pechkova E, Giral M, Soulillou JP, Brouard S, Nicolini C. Immunosuppressive drug-free operational immune tolerance in human kidney transplants recipients. Part II. Non-statistical gene microarray analysis. J Cell Biochem. 2008;103(6):1693–706.
165. Sivozhelezov V, Spera R, Giacomelli L, Hainsworth E, LaBaer J, Bragazzi NL, Nicolini C. Bioinformatics and fluorescence DNASER for NAPPA studies on cell transformation and cell cycle. In: LaBaer J, editor. Functional proteomics and nanotechnology-based microarays. Pan sotanford publishing. 2010. pp. 31–59.
166. Soejima K, Nakamura H, Tamai M, Kawakami A, Eguchi K. Activation of MKK4 (SEK1), JNK, and c-Jun in labial salivary infiltrating T cells in patients with Sjögren's syndrome. Rheumatol Int. 2007;27(4):329–33.
167. Stelzl U, Worm U, Lalowski M, Haenig C, Brembeck FH, Goehler H, Stroedicke M, Zenkner M, Schoenherr A, Koeppen S, Timm J, Mintzlaff S, Abraham C, Bock N, Kietzmann S, Goedde A, Toksöz E, Droege A, Krobitsch S, Korn B, Birchmeier W, Lehrach H, Wanker EE. A human protein-protein interaction network: a resource for annotating the proteome. Cell. 2005;122(6):957–68.

168. Stone SJ, McCracken GI, Heasman PA, Staines KS, Pennington M. Cost-effectiveness of personalized plaque control for managing the gingival manifestations of oral lichen planus: a randomized controlled study. J Clin Periodontol. 2013;40(9):859–67.
169. Sugerman PB, Savage NW. Oral lichen planus: causes, diagnosis and management. Aust Dent J. 2002;47(4):290–7.
170. Sugerman PB, Savage NW, Walsh LJ, Zhao ZZ, Zhou XJ, Khan A, Seymour GJ, Bigby M. The pathogenesis of oral lichen planus. Crit Rev Oral Biol Med. 2002;13(4):350–65.
171. Sujatha D, Hebbar PB, Pai A. Prevalence and correlation of oral lesions among tobacco smokers, tobacco chewers, areca nut and alcohol users. Asian Pac J Cancer Prev. 2012;13(4):1633–7.
172. Sun A, Wang JT, Chia JS, Chiang CP. Serum interleukin-8 level is a more sensitive marker than serum interleukin-6 level in monitoring the disease activity of oral lichen planus. Br J Dermatol. 2005;152(6):1187–92.
173. Sun L, Feng J, Ma L, Liu W, Zhou Z. CD133 expression in oral lichen planus correlated with the risk for progression to oral squamous cell carcinoma. Ann Diagn Pathol. 2013. pii:S1092-9134(13)00078-6. doi:10.1016/j.anndiagpath.2013.06.004 (Epub ahead of print).
174. Sutinen M, Kainulainen T, Hurskainen T, Vesterlund E, Alexander JP, Overall CM, Sorsa T, Salo T. Expression of matrix metalloproteinases (MMP-1 and -2) and their inhibitors (TIMP-1, -2 and -3) in oral lichen planus, dysplasia, squamous cell carcinoma and lymph node metastasis. Br J Cancer. 1998;77(12):2239–45.
175. Syrjänen S, Lodi G, von Bültzingslöwen I, Aliko A, Arduino P, Campisi G, Challacombe S, Ficarra G, Flaitz C, Zhou HM, Maeda H, Miller C, Jontell M. Human papillomaviruses in oral carcinoma and oral potentially malignant disorders: a systematic review. Oral Dis. 2011;17(Suppl 1):58–72.
176. Tabor HK, Risch NJ, Myers RM. Candidate-gene approaches for studying complex genetic traits: practical considerations. Nat Rev Genet. 2002;3(5):391–7.
177. Tao XA, Li CY, Rhodus NL, Xia J, Yang XP, Cheng B. Simultaneous detection of IFN-gamma and IL-4 in lesional tissues and whole unstimulated saliva from patients with oral lichen planus. J Oral Pathol Med. 2008;37(2):83–7.
178. Tsai CA, Lee TC, Ho IC, Yang UC, Chen CH, Chen JJ. Multi-class clustering and prediction in the analysis of microarray data. Math Biosci. 2005;193(1):79–100.
179. Turatti E, da Costa Neves A, de Magalhães MH, de Sousa SO. Assessment of c-Jun, c-Fos and cyclin D1 in premalignant and malignant oral lesions. J Oral Sci. 2005;47(2):71–6.
180. Upadhyay J, Upadhyay RB, Agrawal P, Jaitley S, Shekhar R. Langerhans cells and their role in oral mucosal diseases. N Am J Med Sci. 2013;5(9):505–14.
181. Valter K, Boras VV, Buljan D, Juras DV, Susić M, Pandurić DG, Verzak Z. The influence of psychological state on oral lichen planus. Acta Clin Croat. 2013;52(2):145–9.
182. Velozo J, Aguilera S, Alliende C, Ewert P, Molina C, Pérez P, Leyton L, Quest A, Brito M, González S, Leyton C, Hermoso M, Romo R, González MJ. Severe alterations in expression and localisation of {alpha}6{beta}4 integrin in salivary gland acini from patients with Sjogren syndrome. Ann Rheum Dis. 2009;68(6):991–6 (Epub 2008 Jul 14).
183. Villarroel Dorrego M, Correnti M, Delgado R, Tapia FJ. Oral lichen planus: immunohistology of mucosal lesions. J Oral Pathol Med. 2002;31(7):410–4.
184. von Mering C, Jensen LJ, Snel B, Hooper SD, Krupp M, Foglierini M, Jouffre N, Huynen MA, Bork P. STRING: known and predicted protein-protein associations, integrated and transferred across organisms. Nucleic Acids Res. 2005;33(Database issue):D433–7.
185. Walsh LJ. Mast cells and oral inflammation. Crit Rev Oral Biol Med. 2003;14(3):188–98.
186. Wali RK, Kunte DP, De La Cruz M, Tiwari AK, Brasky J, Weber CR, Gibson TP, Patel A, Savkovic SD, Brockstein BE, Roy HK Topical polyethylene glycol as a novel chemopreventive agent for oral cancer via targeting of epidermal growth factor response. PLoS One. 2012;7(6):e38047.

187. Walton LJ, Thornhill MH, Macey MG, Farthing PM. Cutaneous lymphocyte associated antigen (CLA) and alpha e beta 7 integrins are expressed by mononuclear cells in skin and oral lichen planus. J Oral Pathol Med. 1997;26(9):402–7.
188. Wang J, Zhang J, Li K, Zhao W, Cui Q. SpliceDisease database: linking RNA splicing and disease. Nucleic Acids Res. 2012;40(Database issue):D1055–9.
189. Wernicke S, Rasche F. FANMOD: a tool for fast network motif detection. Bioinformatics. 2006;22(9):1152–3.
190. Wickham LF. Sur un signe pathognomonique delichen du Wilson (lichen plan) stries et punctuations grisatres. Ann Dermatol Syph 1895;6:17–20.
191. Wilson E. On lichen planus. J Cutan Med Dis Skin 1869;3:117–132.
192. Wong DT. Salivaomics. J Am Dent Assoc. 2012;143(10 Suppl):19S-24S.
193. Wong DT. Salivary diagnostics: scientific and clinical frontiers. Adv Dent Res. 2011;23(4):350–2.
194. Wong DT. Salivary extracellular noncoding RNA: emerging biomarkers for molecular diagnostics. Clin Ther. 2015;37(3):540–51.
195. Wright JT, Hart TC. The genome projects: implications for dental practice and education. J Dent Educ. 2002;66(5):659–71.
196. Xia W, Lau YK, Zhang HZ, et al. Combination of EGFR, HER-2/neu, and HER-3 is a stronger predictor for the outcome of oral squamous cell carcinoma than any individual family members. Clin Cancer Res 1999;5:4164–74.
197. Xu Z, Shen Z, Shi L, Sun H, Liu W, Zhou Z. Aldehyde dehydrogenase 1 expression correlated with malignant potential of oral lichen planus. Ann Diagn Pathol. 2013;17(5):408–11.
198. Yan SK, Wei BJ, Lin ZY, Yang Y, Zhou ZT, Zhang WD. A metabonomic approach to the diagnosis of oral squamous cell carcinoma, oral lichen planus and oral leukoplakia. Oral Oncol. 2008;44(5):477–83.
199. Yang K, Zhang FJ, Tang H, Zhao C, Cao YA, Lv XQ, Chen D, Li YD. In-vivo imaging of oral squamous cell carcinoma by EGFR monoclonal antibody conjugated near-infrared quantum dots in mice. Int J Nanomedicine. 2011;6:1739–45 (Epub 2011 Aug 19).
200. Yildirim B, Sengüven B, Demir C. Prevalence of herpes simplex, Epstein Barr and human papilloma viruses in oral lichen planus. Med Oral Patol Oral Cir Bucal. 2011;16(2):e170–4.
201. Yoshizawa JM, Wong DT. Salivary microRNAs and oral cancer detection. Methods Mol Biol. 2013;936:313–24.
202. Younes F, Quartey EL, Kiguwa S, Partridge M. Expression of TNF and the 55-kDa TNF receptor in epidermis, oral mucosa, lichen planus and squamous cell carcinoma. Oral Dis. 1996;2(1):25–31.
203. Zhang WY, Liu W, Zhou YM, Shen XM, Wang YF, Tang GY. Altered microRNA expression profile with miR-27b down-regulation correlated with disease activity of oral lichen planus. Oral Dis. 2012;18(3):265–70.
204. Zhou XJ, Sugerman PB, Savage NW, Walsh LJ. Matrix metalloproteinases and their inhibitors in oral lichen planus. J Cutan Pathol. 2001;28(2):72–82.
205. Zhu W, Phan QT, Boontheung P, Solis NV, Loo JA, Filler SG. EGFR and HER2 receptor kinase signaling mediate epithelial cell invasion by *Candida albicans* during oropharyngeal infection. Proc Natl Acad Sci U S A. 2012;109(35):14194–9.

Blistering Diseases – Pemphigoid

Raman Mehrzad, Shawn Shetty and A. Razzaque Ahmed

Abbreviations

BP	Bullous pemphigous
MMP	Mucous membrane pemphigoid
LABD	Linear IgA bullous disease
EBA	Epidermolysis bullosa aquista
PG	Pemphigoid gestasionis
PV	Pemphigus vulgaris
PF	Pemphigus foliaceus
BMZ	basement membrane zone
AECP	anti-epiligrin cicatricial pemphigoid
ELISA	enzyme-linked immunosorbent assay

Introduction

This chapter contains current information on autoimmune blistering diseases that involve the skin and mucous membranes. It does not contain diseases that do not have a proven or presumed autoimmune basis. It also does not contain diseases that are genetic or genetically based.

Pemphigoid diseases are a group of autoimmune disorders characterized by the presences of autoantibodies against structural components of the dermal-epidermal junction. The cytoskeleton of the basal keratinocytes is bound to the extracellular matrix of the dermis via junctional proteins. When the pemphigoid autoantibodies bind to these proteins, it results in a separation of the epidermis and dermis. Clinically, these conditions share similar characteristics with tense blisters and ero-

A. R. Ahmed (✉) · R. Mehrzad · S. Shetty
Center for Blistering Diseases, 697 Cambridge Street, Suite 302,
Boston, MA 02135, USA
e-mail: arahmedmd@msn.com

S. T. Sonis (ed.), *Genomics, Personalized Medicine and Oral Disease*,
DOI 10.1007/978-3-319-17942-1_10

sions on the skin or mucous membranes and, in opposed to pemphigus, a negative Nikolsky sign. There may, however, be a significant difference in prognosis and treatment, making the exact diagnosis essential. However, due to similar symptomatology, establishing the exact diagnosis could be difficult based on clinical grounds only. Hence, diagnostic work up for detection autoantibodies is needed for confirmation.

Bullous Pemphigoid (BP)

Incidence

In Europe, BP is considered the most common autoimmune blistering disease. The vast majority of cases occur in patients over >60 years of age, and thus this is primarily a disease which effects the older population. A retrospective study of 869 patients in the United Kingdom showed that the median age of presentation was 80 years [1]. Most of the data on the incidence are derived from European reports, in which studies report an incidence of 4 to 22 cases per million individuals [2–4]. However, studies suggest that the incidence is increasing. A fairly recent retrospective analysis showed that the incidence rate of BP had tripled from 2000–2005 than between 1986–1992 [4], a finding that might be related to the availability of more sensitive and specific assay systems, and/or the increasing age of the general population.

BP is mainly considered a disease of the elderly, with onset usually in the late 70s. The incidence rises significantly to 150–300/million/year in patients above 80 years of age [5–10]. Hence, it appears that the incidence of BP linearly increases with advancing age. The disease is rarely seen in patients younger than 50 years (incidence <0.5/million/year) [5, 6, 9, 10]. Although the reason is unknown, multiple studies have reported at least a slight female predominance [2, 11].

Pathogenesis

The pathophysiology is not fully understood, however, it involves an autoantibody-mediated damage to the epithelial basement membrane zone. A destructive inflammatory cascade initiates as the autoantibodies bind to the epithelial basement membrane zone, and as a result, separation of the epidermis from the dermis in skin and epithelium from subepithelial tissue in mucous membranes with the formation of typical cutaneous and mucosal blisters and erosions [12, 13]. Interleukin 17, which is produced by innate immune cells, may play an important role in the maintenance and persistence of the disease [14]. Two hemidesmosomal proteins, BP180 and BP230, have been identified in BP [15–17]. The BP180 is a 180 kDa transmembrane glycoprotein that extends from the hemidesmosome plaque in basal keratinocytes into the lamina densa of the basement membrane zone, and is present in most patients with BP [18]. In the majority of patients with, antibodies against BP180, a transmembrane protein that extends from the hemidesmosomal dense plaque in basal keratinocytes into the lamina densa of the basement membrane zone [19], are present. The non-collagenous extracellular domain of BP180, known as the NC16A, are detectable by enzyme-linked immunosorbent assay (ELISA) in 80–90 % of affected patients [20–26]. The BP230 is a 230 kDa intracellular consistent of the hemidesmosomal plaque and belongs in the plakin family that is found in the basal keratinocytes. BP230 links keratin intermediate filaments to the hemidesmosomes [19]. Approximately 60–70 % of patients with BP have antibodies against BP230 [19]. The antibodies bind to the primary site of the BP230 at the globular C-terminal domain [12, 27]. The production of antibodies to other epitopes of BP180 also occurs in bullous pemphigoid, and may have clinical significance. As an example, antibodies directed against epitopes in the C-terminal end of BP180 have been associated with the presence of mucosal disease [28, 29].

IgG4 is the the predominant subclass of antibodies that react with the basement membrane zone in IgG4 antibodies also are present in the prodromal phase of BP that often precedes cutaneous blistering. Furthermore, IgG1 and IgG2 antibodies may be present, but are less frequently detected than IgG4 antibodies. Typically, there are no IgG3 antibodies [30]

Although the relevance is uncertain, IgA and IgE BP180 antibodies also may be present in BP [31, 32]. Furthermore, there is evidence for a pathogenic role for IgE basement membrane zone antibodies, in which exiciting levels may be association with severe disease [31, 32]

Antibodies to BP230 are detected in approximately 60–70 % of patients with BP. BP230 is an intracellular hemidesmosomal protein in the plakin family that is found in basal keratinocytes. BP230 links keratin intermediate filaments to hemidesmosomes and the primary site of antibody binding on BP230 is the globular C-terminal domain [12, 27]. However, it is not clear whether BP230 antibodies have a pathogenic role in BP. They may occur as a secondary event related to keratinocyte injury, or epitope spreading since serum levels do not reveal consistent correlation with disease activity [33, 34]

Clinical Presentation

Typically, BP presents with clear, tense blisters and erythema, with urticarial plaques. Classical lesions are a 1–3 cm tense bulla on an erythematous, urticarial, or non-inflammatory base, and blisters may be numerous and widespread [35]. The most common sites of blister formation are the abdomen and flexural aspects of the limbs. Crust and erosions are usual rest symptoms. Although mucosal lesions are rare, oral lesions affect between 10–20 % [36]. Nail changes also may develop [37]. Prior to blister formations, patients typically experience a prodromal state with isolated pruritus, or in combination with eczematous, papular and/or urticarial lesions [36].

Diagnosis

A combination of clinical symptoms, serology, and direct immunostaining is used to confirm the diagnosis of BP. Clinical features that support the diagnosis are formation of tense blisters and erosion occurring without any other identifiable cause, desquamitive gingivitis or mucousal lesions, and unexplained pruritic eczematous eruptions or urticarial plaques. Linear deposits of IgG or complement 3 is seen at the dermal-epidermal junction with direct immunofluorescence microscopy in greater than 90 % of cases [38] Skin that has been induced by 1 mol/L NaCl solution will provide the most sensitive immunofluorescence substrate for screening of serum antibodies. It will also differentiate between different autoantibody specificities [39]. ELISA is used to identify circulating antibodies against BP180, NC16A and BP230 [40]. The diagnostic sensitivity of these tests combined reaches up to 90 % [41].

Treatment

The goals of treatment are to decrease blister formation and pruritus, promote healing and improve quality of life. Treatment chosen should depend upon the extend and severity of disease, the impact of the disease on the quality of life of the person, and the associated comorbidities present in the patient at that time. First line treatment is with high potency topical corticosteroids, and a recommended choice is clobetasol 0.05 % cream. Systemic therapy, usually with prednisone, could be used as first line therapy when topical corticosteroids have not produced a clinical remission [42]. The use of topical corticosteroids as monotherapy for bullous pemphigoid (BP) is supported by a multicenter randomized trial which found that patients with extensive bullous pemphigoid who were treated with topical corticosteroids had better clinical outcomes than patients with extensive bullous pemphigoid who were treated with systemic glucocorticoid therapy [42].The response to prednisone is usually rapid, and >90 % with moderate or extensive BP achieves disease control by 21 days.

Azathrioprine is commonly used as a glucocorticoid-sparing agent with doses between 0.5 and 2.5 mg/kg per day. However, there are limited data on its efficacy [43]. A major side effect of this medication is myelosuppression and close monitoring is advised.

Mycophenolate Mofetil is another drug in which case reports [44] have shown a potential value in the treatment of BP, although the only randomized, unblended trial showed similar effect between methylprednisolone plus mycophenolate mofetil, compared to methylprednisone plus azathioprine. The main difference that mycophenolate mofetil was better tolerated, although the mean day of remission was shortened in the azathioprine group (42 ± 55 versus 24 ± 19 days, respectively). In adults, the dose is typically 1.5 to 2 g per day, with a maximum dose of 3 g per day.

In case reports and uncontrolled studies, Methotrexate has shown to be beneficial when used alone or in combination with topical or systemic steroids [45]. The typical dose is between 5 and 20 mg per week.

Tetracyclines and dapsone, or other anti-inflammatory agents administered with nicotinamide may also be of benefit, although data is limited. Usually, these agents should be considered in patients with mild BP and when a glucocorticoid-sparing regimen is required. In these patients, the risks of immunosuppressants often outweigh the benefits.

Rituximab, a humanized chimeric monoclonal antibody that targets and destroys CD20+B and pre-B cells, has in case reports and case series shown to be effective for refractory BP [46].

It has been reported that intravenous immunoglobin (IVIG) is effective in 86% of adult patients and improvement occurred within an average of three months [47].

The level of BP180 antibodies in serum usually correlates with the clinical activity of bullous pemphigoid (BP) [48]. Thus, many physicians use measurement of these antibodies in conjunction with the clinical evaluation to assess the response to therapy. Marked decreases in BP180 antibody levels and lesser decreases in BP230 antibody levels have been detected soon after the start of topical corticosteroid treatment [48].

Differential Diagnosis

The differential diagnosis for BP are: pemphigus, pemphigoid gestationis, laminin γ1 pemohigoid, Brunsting-Perry pemphigoid, mucous membrane pemphigoid, dermatitis herpetiformis, bullous lupus erythematosus, linear IgA bullous dermatosis, and epidermolysis bullosa acquisita.

Prognosis

The prognosis varies. Generally, BP follows a chronic, relapsing course [49]. Long-term remission may occur after months to years. BP is potentially a fatal disease with an estimated one-year overall mortality between 11–48% [50, 51]. The variation in the incidence of mortality in BP may entirely be dictated by the health care system of the country in which the study was done. In Europe, where many countries have state supported health care system, every death is recorded and reported. In the US, there is no requirement for reporting and therefore not every death in a patient with BP is recorded. The cause of death is multifactorial. In the majority of patients, the cause of death is opportunistic infection due to prolonged immunosuppression. Nevertheless, one study showed that the most common cause of death was heart disease, infection or neurological disease [50].

Mucous Membrane Pemphigoid

Introduction

Mucous membrane pemphigoid (MMP) is a rare, heterogeneous, autoimmune blistering disease that can affect multiple mucous membranes and the skin. Sera of these patients contain autoantibodies to various components of the basement membrane zone (BMZ). The disease may be divided into subsets based on the antigen to which the autoantibodies bind. MMP typically presents as relapsing and remitting mucosal inflammation and erosions. The oral cavity is the most common site of involvement, and cutaneous involvement also may be present

Incidence

The incidence of MMP has been estimated to 1.3–2.0 per million people [52]. In general, the diseases emerge earlier than does in BP, with a mean age onset between 60–65 years [53]. Although the reason is unknown, multiple studies have reported at least a slight female predominance.

Pathogenesis

MMP is an autoimmune disease with unknown etiology. Presently, six target antigens have been identified with the clinical phenotype of MMP; BP180 (75% of patients, BP230 (25% of patients), laminin-332 (25% of patients), both subunits of α6β4 integrin, and type VII collagen [54]. Laminin-332 is a major component of the basement membrane of skin and other epithelial tissues. It is composed of three subunits: α3, β3, and γ2 chains. The α3 chain is targeted in most of the cases with one of the subtypes of the disease anti-epiligrin cicatricial pemphigoid (AECP) [55]. Autoantibodies against laminin-322 serves as a specific marker for AECP as these are not found in any other skin or mucosal disease.

The main mechanism for lesion formation is antibody-mediated activation of the complement cascade by antibodies bound to specific antigens in the basement membrane zone. The activation of complement may stimulate the local recruitment of inflammatory cells that release proinflammatory mediators as well as proteases that directly damage the basement membrane zone [56].

Although not established, genetic factors, environmental exposures, and the phenomenon of epitope spreading are considered potential contributory factors. Human leukocyte antigen (HLA) alleles may play a role in bullous pemphigoid and MMP. The HLA-DQB1*0301 allele has an increased prevalence is reported in multiple studies [57]. The proposed mechanism through which certain HLA alleles might contribute to disease development involves the facilitation of antigen presentation of basement membrane zone antigens to T cells [57].

Infections or drugs that causes autoimmune reactions may play a role. As a result of cross-reactivity of antibodies that target infectious agents or drugs with antigens in the basement membrane zone, these disorders could potentially occur. Antibodies against hepatitis B, hepatitis C, Helicobacter pylori, Toxoplasma gondii, and cytomegalovirus were more prevalent among patients with bullous disease in a small case-control study [58]

Epitope spreading has been proposed as an explanation for cases in which pemphigoid occurred in the setting of other diseases Epitope spreading is the induction of an autoimmune response against normally tolerated host antigens and epitopes as a consequence of the exposure of these antigens and epitopes during immune-mediated tissue inflammation [59]. For instance, following conjunctival inflammation due to Stevens-Johnson syndrome, ocular MMP has developed.

Clinical Presentation

MMP is a chronic and progressive disease that presents with blisters formation or erosive disease in the mucosal surfaces of the mouth, eyes, nose, nasopharynx, hypopharynx, larynx, esophagus, genitals and/or anus. Skin is involved in approximately 25 % of cases [53]. The severity of the disease varies from single oral lesions in one or many mucous membranes, to widespread, extremely painful mucosal involvement. As a result of scarring, devastating sequels can appear on the affected site; one classic example is ocular involvement, which can progress to scar formation and lead to blindness. Moreover, laryngeal involvement can result in severe laryngeal stenosis, requiring tracheostomy and can occasionally result in death due to asphyxiation [53]

Diagnosis

MMP is diagnosed based on clinical symptoms of predominant mucosal lesions, and direct immunofluorescence microscopy showing deposition of IgG, C3, and in some patients, IgA along the dermal-epidermal junction. However, in approximately 20 % of cases with ocular disease, autoantibodies are not detected. Thus, indirect immunofluorescence microscopy on salt-split skin, epidermal or dermal staining can be seen [60].

Treatment

The goal of treatment of MMP is to decrease the progression of the disease, improve symptoms, and prevent adverse sequelae of chronic tissue inflammation and scarring. To date, there are few high quality clinical trials that have evaluated the optimal management for MMP. Thus, treatment of MMP can be challenging and should preferably be managed by a dermatologist specialized in blistering diseases. The primary goal is to slow the progression of the disease, improve symptoms and prevent adverse reactions of the chronic tissue inflammation and scaring [61]. For mild disease, characterized by patchy, localized oral lesions and the skin, local therapy are usually effective. Topical corticosteroids are the first-line treatment.

Topical tacrolimus, a calcineurin inhibitor, has shown to be effective in some patients who failed to respond well to topical corticosteroids [62]. Topical tacrolimus 0.1 % ointment initially is applied two to three times per day and is tapered as tolerated. Intralesional corticosteroids injections is an option for disease in the mouth that do not respond sufficiently to other local measures, although support for the efficacy are limited [60, 62]. Triamcinolone acetonide (10 mg/mL) is usually used, 0.1–0.5 mL, per injection site.

For moderate to severe disease, most often in patients who fails to respond to local therapy, or present with wide-spread oral disease, systematic glucocorticoids and dapsone are frequently beneficial. Prednisone 0.25–0.5 mg/kg per day as a single dose is a typical initial treatment. Dapsone 50–200 mg/day may be used for oral MMP.

For severe refractory disease, immunosuppressants such as azathioprine, mycophenolate mofetil, and cyclophosphamide are used in combination with systemic glucocorticoids, although support for the use of these agents primarily stems from uncontrolled studies [63]. Typically, the dose of prednisone is increased in refractory disease. A dose of 1 mg/kg per day of prednisone in addition to one of the following immunosuppressive agents has been suggested: Azathioprine (2–2.5 mg/kg per day; dose adjustments may be indicated based upon thiopurine methyltransferase activity), Mycophenolate mofetil (1–2.5 g per day), and cyclophosphamide (1–2 mg/kg per day). Moreover, (IVIG) may also be beneficial for patients with severe, refractory MMP, although the response to treatment is variable [64]. The typical dose is 1–2 g/kg body weight administered over 2–3 days every 2–6 weeks for 4–6 months.

Differential Diagnosis

The major differential diagnosis includes pemphigus, pemphigoid gestationis, anti-laminin gamma-1 pemphigoid, and BP.

Prognosis

Due to the rarity of the disease and the lack of clinical trials, the prognosis is not established.

Pemphigoid Gestationis

Introduction

Pemphigoid gestationis (PG), also known as herpes gestationis, is a blistering disease related with pregnancy with antibodies against BP180 NC16A. Women with this disorder have an increased fetal risk.

Incidence

The incidence of PG is 1/1700–1/50000 pregnancies. The disease occurs strictly in pregnant women, usually in the second or third trimester and in those with trophoblastic tumors. Approximately 10 % of patients are affected up to 4 weeks postpartum [65].

Pathogenesis

PG is characterized by IgG autoantibodies that bind to BP180 in the BMZ of the skin (34,180 lancet). These antibodies can potentially cross-react to the skin and cause maternal, and in some cases newborn, disease [66]. perivascular lymphocytic and eosinophilic infiltrate is noted in histology exam in the vesicles. Eosinophils can sometimes be noted in the Dermo-Epidermal Junction (DEJ). Basal cell necrosis and edema of the dermal papillae are usually noted.

Clinical Presentations

Typically, pruritic erythramatous papules and plaques, erythema multiforme-like changes, eczematous lesions or papulovesicles arise around the umbilicus, later spreading over the abdomen and thighs [67] in the 2nd or 3rd trimester or postpartum. Vesicles may also be present. Although lesions are not found in the face or mucous membranes, they may be seen on the palms and soles [67]. Less than 20 % have mucosal lesions

Diagnosis

Direct IF of a skin biopsy is used for the diagnosis by showing IgG and C3 deposit along the dermal-epidermal junction. Indirect IF may reveal the “herpes gestationis factor”, an IgG1 antibody that can fix complement at the BMZ. Nevertheless, only up to 25 % of cases have positive indirect IF studies [68]. Importantly, the degree of peripheral eosinophilia may correlate with disease severity; antibodies measured by indirect IF do not [68]. Two skin biopsies should be obtained to make the diagnosis of pemphigoid gestationis.

Treatment

First line therapy is highly potent topical steroids for the lesions, in combination with antihistamines to relief the pruritus. Oral antihistamines that are nonsedating are an acceptable alternative. If more severe disease, or if symptoms are resistance to local therapy, oral prednisone can be added at an initial dose of 0.25–0.5 mg/kg, and slowly tapered. Oral prednisone had no effect on the pregnancy in a study with 61 patients [69].

Differential Diagnosis

In early stages of disease, PG could resemble, papular urticarial papules and plaques of pregnancy (PUPPP). The two could be distinguished by the localization of the initial symptoms; PG is typically localized umbilicaly while PUPPP often begins in the striae. Other differentials are Dermatitis Herpetiformis (DH) and erythema multiforme.

Prognosis

PG is a benign disease in most cases and typically last for 4–6 months before complete resolution. It may remit prior to delivery, however, 3/4 of patients flare postpartum and at least 25 % subsequently flare with use of oral contraceptive pills or during menses [70]. Interestingly, in $<5\%$ of cases, the disease persist and converts to BP [70]. The disease relapses in 90 % of patients with their next pregnancy [70]. The disease is associated with mild placental insufficiency and in the newborn there is an increased risk of growth restriction and prematurity [71]. Occasionally, still births may occur.

Linear IgA Bullous Disease

Introduction

Linear IgA Bullous disease (LABD), also known as linear IgA disease, is a rare idiopathic or drug-induced autoimmune blistering disease, which name comes from its primary pathological feature; linear deposition of IgA at the dermal-epidermal junction. The clinical features can many times be difficult to distinguish from dermatitis herpetiformis, however, the diagnosis is usually confirmed with the specific immunopathological findings in LABD as well the absence of an associated gluten-sensitive enteropathy

Incidence

LABD is a rare disease. Reported incidence varies from country to country but range from less than 0.5–2.3 cases per million individuals/year [72]. Both children and adults are affected. In children, the presentation is usually between ages of 6 months and 10 years. Most cases in adults occur after the age of 60 [72].

Pathogenesis

Although the major target antigen is IgG antibodies against BP180, and the presence of IgA antibodies bound to the basement membranes zone (BMZ) is the primary immunopathology, the mechanism of lesion formation is not fully understood [73]. However, both humoral and cellular immune responses are most likely involved in the pathogenesis. An antibody-induced local inflammatory response results in tissue injury from the release of proteolytic enzymes by neutrophils and other inflammatory cells. This may contribute to the development of skin and mucosal lesions [73]. Most patients have IgA1 antibodies that target a 97 kDa antigen and a 120 kDa antigen with the BMZ, both which are antigen fragments of the extracellular portion of BP180 [73]. Many drugs are potential inciting factor to LABD, including common prescribed drugs such as vancomycin, NSAIDs, lithium, captopril, amiodarone and furosemide [73].

Genetic factors also may contribute. HLA B8, HLA Cw7, HLA DR3, HLA DQ2, and the tumor necrosis factor-2 allele have all been reported to be associated with the disease [74].

Clinical Presentation

LABD presents with lesions on the skin, mucous membrane or both areas. The skin lesions usually develop abruptly. The vesicles or bullae are tense. The trunk, extensor extremities, buttocks and face are common affected areas. Affected patients may be asymptomatic at first, but pruritus is common and may be severe. Importantly, the symptom distribution can mimic DH, which can make it difficult to distinguish both diseases.

Eighty-percent of patients have mucosal involvement, which typically presents as erosions or ulcers. Although, the oral and ocular mucosa are the most commonly affected mucosal sites, any musosal site can be affected, including, oral cavity, nose, pharynx, larynx conjunctiva, genitalia anus, and esophagus. Patients can develop mucosal scarring and develop symblepharon and ectropion [60]. LABD has been associated with ulcerative colitis, hematologic or solid malignancies [75].

Diagnosis

The diagnosis is made upon the combination of four different factors: (1) clinical symptoms, (2) linear deposition of IgA at the dermal-epidermal junction by direct immunofluorescence microscopy of a skin lesion or mucous membrane biopsy, (3) detection of IgA serum antibodies by indirect immunofluorescence microscopy on human salt-split skin, and (4) IgA reactivity against BP180 [76].

Differential Diagnosis

Dermatitis herpetiformis is a differential diagnosis that could sometimes be difficult to distinguish, mostly due to similar distribution of the symptoms. However, the lack of an associated gluten-sensitivity enteropathy confirms the diagnosis of LABD. Moreover, some overlap is seen with bullous pemphigoid (patients with dual IgG and IgA deposition along the junction), with mucous membrane pemphigoid (patients with predominant mucosal involvement), and IgA epidermolysis bullosa acquisita. Other differential diagnosis are bullous pemphigoid in which histopathologic examination usually reveals a higher proportion of eosinophils than are present in LABD; pemphigoid gestationis which is present in pregnant women; bullous lupus erythematosus which presents with patients in systemic lupus erythematosus; Epidermolysis bullosa acquisita which has deposits of IgG on direct immunofluorescence, at the basement membrane; and finally, MMP (discussed above).

Treatment

To date, there are no randomized controlled clinical trials that have evaluated the optimal treatment for therapy of LABD. Dapsone is typically the first line treatment, either with or without the combination of topical glucocorticosteroids [77]. It is well-tolerated and effective in these patients. Treatment is usually started with a low dose (<0.5 mg/kg per day in children or 25 or 50 mg per day in adults) and is subsequently titrated upward over several weeks as tolerated and in accordance with treatment response [60, 77]. The response is typically very dramatic with improvement noticed within the first few days of treatment. Hemolysis, methemoglobinemia, agranulocytosis, hypersensitivity syndrome, and peripheral motor neuropathy are some serious side effects with Dapsone and complete blood count (CBC) with differential, liver function tests and glucose-6-phosphate dehydrogenase (G6PD) level should be obtained prior to the starting therapy.

Topical corticosteroids can be used for patients with mild disease, however, it is usually used as adjunctive therapy. Alternatively to Dapson, Sulfapyridine can be used [77]. 1000 to 1500 mg per day of sulfapyridine should be the treatment dose for adults [77].

In case reports, successful treatment of LABD with colchicine has been documented [78].

For severe disease, some patients may need concomitant low-dose systemic glucocorticoids, usually 0.5–1 mg/kg per day, to suppress blister formation [77]. Treatment should usually be continued for several weeks after complete remission is achieved.

Prognosis

Patients usually respond well to treatment. Typically, the disease persists for months to several years prior to spontaneous resolution. Drug-induced LABD, however, resolves within a few weeks after removal of the contributing drug.

Epidermolysis Bullosa Acquisita

Introduction

Epidermolysis bullosa acquisita (EBA) is a rare, chronic subepidermal disease characterized by autoantibodies against type VII collagen. Skin fragility, noninflammatory tense bullae, milia, and scarring are the classical descriptions of EBA, but can in fact presents as an inflammatory bullous eruption, similar to BP.

Incidence

EBA is a rare disorder with an uncertain incidence. Reported incidence range between 0.2 and 0.5 new cases per 1 million [79].

Pathogenesis

The immunopathology is characterized by autoantibodies against type VII collagen. Type VII collagen is a major component of anchoring fibrils in the DEJ. Neutrophils are the major effector cells and are activated after interaction with skin-bound autoantibodies [80]. Glycolisation of type VII collagen-specific antibodies is essential for this interaction (60).

Genetics may play a role in EBA as HLA-DR2 was increased in these patients when compared with healthy controls [81].

Clinical Presentation

There are two forms of EBA: the classic or an inflammatory variant. Tense blisters and skin fragility, typically located in the extensor skin surfaces at trauma-prone areas, characterize the classic variant. These lesions usually heal with scarring and milia formation, and hyper and/or hypopigmenation is common rest sequels [82]. Esophageal stenosis or nail loss can occur. In contrast, the inflammatory subtype can resemble BP or MMP [82]. In fact, the disease is often referred as BP-EBA, MMP-EBA etc. due to the similarities of symptoms with other subepithelial blistering disease. Approximately 50 % of patients have mucous membrane involvement in both subcategories. Interestingly, concomitant inflammatory bowel disease is present in 20 % of patients [82].

Diagnosis

The diagnosis of EBA is based on linear deposit of IgG, IgA, and C3 at the dermal-epidermal junction by direct immunofluorescence microscopy [82]. Deposition of IgA, IgM, Factor B, and properdin may also be detected. However, because linear deposition of IgG and complement at the DEJ may be seen in multiple autoimmune subepithelial blistering disorders, the DIF findings do not provide a definitive diagnosis. ELISA could be used to detect anti-type VII collagen antibodies-different tissue extracts or the recombinant NC1 domain [82]. The diagnosis of EBA can be challenging, mainly due to similarities in clinical, pathological and immunohistological features with other subepithelial blistering disorders. Therefore, it is recommended to included the following in the initial assessment: full clinical history and skin examination, tissue biopsy for routine histopathology and direct immunofluresence microscopy and immunofluorescence on basement membrane zone-split skin.

Treatment

Treatment of EBA is challenging. As with LABD, there are no randomized controlled trials, and the disease is difficult to control, treat and refractory disease to many interventions is common. Generally, patients should be advised to avoid incidental and iatronic trauma to the skin, and gentle cleansing of skin during bathing. Systemic corticosteroids, 0.5–2.0 mg/kg in is tried initially at times. However, there is less response to classical EBA than inflammatory subtypes of EBA, and that overall, systemic glucocorticoids are less effective for this disease than for other subepithelial blistering diseases [82]. Therefore, in many countries, including the US, Colchicine, 0.5–3 mg per day, or Dapsone, to 2 mg/kg per day is first line treatment.

For severe or refractory cases, cylcosporin, azathrioprine, mycophenolate mofetil, plasmapheresis, IVIG and rituximab can be added [60].

Differential Diagnosis

The major differential diagnoses include BP and LABD. However, porphyria cutanea tarda, bullous systemic lupus erythematosus and recessive dystrophic epidermolysis bullosa

Prognosis

EBA is a chronic inflammatory disease that has periods of partial remissions and exacerbations. Treatment is very challenging, however, with therapy, patients have no increased risk of mortality.

Pemphigus

Introduction

Pemphigus is an autoimmune mucocutanaeous blistering disease of the skin. IgG autoantibodies target proteins of keratinocyte adhesion, desmoglein 1 (Dsg 1) and or desmogelin 3 (Dsg 3), thereby causing acantholysis [83, 84]. Autoantibodies affect cell adhesion molecules of the desmosome, which causes a loss of cell adhesion. Lesions are characterized by cutaneous and mucous blisters and erosions [84, 85]. There are different types of pemphigus: pemphigus vulgaris(PV), pemphigus foliaceus(PF), pemphigus erythematosus(PE), pemphigus herpetiformis(PH), pemphigus vegetans and IgA pemphigus [86, 87]. Paraneoplastic pemphigus(PNP) is an autoimmune syndrome that affects multiple organs. Autoantibodies in PNP also affect proteins in the plakin family (plectin, desmoplakin I, desmoplakin II, bullous pemphigoid antigen I, envoplakin, and periplakin). These plakin proteins are also involved in cell-cell adhesion of keratinocytes [88].

Pemphigus Vulgaris

Introduction

PV is a rare and life threatening chronic blistering disease of the skin and mucosae. PV patients develop IgG autoantibodies to Dsg 3 and about half the patients also having Dsg1 autoantibodies [89]. The start of the disease is most frequently seen in the oral mucosa and most times precedes cutaneous blisters. Patients with untreated pemphigus are prone to infections, loss of body fluids and proteins and to weight loss due to painful oral and esophageal erosions [85, 90].

Incidence

PV is the most common occurring form that frequently affects the oral cavity and skin or both. Other mucous membranes of eye, nose, pharynx, larynx, esophagus, trachea, vagina, penis, anal canal, and the nails can be affected [90]. Men and women are equally affected. While the mean age of onset is 50 to 60 years, PV has been observed in children and in the elderly. The incidence ranges from 0.76 to 5 new cases per million per year. The incidence is higher with 16–32 cases per million in those with Jewish Ancestry. The incidence of pemphigus in Central Europe is one to two cases per million persons per year [91, 92].

Pathogenesis

The pathomechanism of PV is based on autoantibodies damaging cell-cell cohesion and leading to cell-cell detachment of the epidermis and mucosae. The onset and course of PV depend on a variable interaction between predisposing and inducing factors. The precipitating factors are many and various. Factors may include the environment (eg, drug intake, viral infections, physical agents, contact allergens, diet), hormonal disorders or emotional stress. Invariably, they are somehow linked with the patient's lifestyle.

Drugs have to the potential of provoking acantholysis by interfering with the keratinocyte membrane chemistry and/or with the immune balance. Viral infections may trigger the outbreak of PV or simply complicate its clinical course. The precipitating effect might be due to the patient's immune system as a consequence of the viral attack, which over activates the immune response. PV may also be induced by physical agents like radiation, burns, or surgery, contact allergens like organophosphate pesticides, and dietary factors like garlic, onions, black pepper, red chili pepper, red wine, or tea. Induction by emotional stress is rare, but documented [91–96].

Clinical Presentation

Clinically, oral lesions are common and early manifestations and typically run a chronic course, causing blisters, erosions, and ulcers. Superficial ulcers can appear either on the labial and/or buccal mucosa. However, any mucosal surface, including the oropharynx and esophagus can be involved. Mouth lesions may be tender, preventing adequate food intake that leads to weight loss.

Many of these patients may have cutaneous findings of flaccid blisters and erosions with a tendency for the trunk, groin, axillae, scalp and face. Initially, there can be erythematous macules and patches that evolve into flaccid bullae. Fluids within the bullae are usually clear, but can become hemorrhagic, turbid, or pustulent. Lesions can be pruritic but are usually painful and accompanied by a burning sensation. The lesions are round to oval in shape, and range from skin-colored to erythematous. Over time, erosions crust over and the healed lesions leave hyperpigmented patches without scarring. The lesions may be accompanied by weakness and malaise.

A common finding is Nikolsky's sign. The direct Nikolsky sign refers to direct application of pressure on a blister, causing the extension of the blister. The indirect Nikolsky sign is when the application of friction on clinically normal skin induces a blister. Other clinical findings include nail dystrophy, paronychia, and subungual hematomas [86, 87, 90, 91, 94, 97].

Diagnosis

On biopsy of a lesion, PV is characterized by an intraepidermal suprabasilar cleft, within vivo deposition of lgG on perilesional tissue and a circulating autoantibody to keratinocyte cell surface molecules. Since cleavage occurs within the epidermis, blisters have a relatively thin roof and are loose and fragile; thus, skin erosions, rather than blistering, tends to be the predominant finding [86, 87].

Differential Diagnosis

Oral ulcers found in PV patients resemble many other conditions. This is why early diagnosis of PV is difficult with only oral lesions. PV may resemble herpetic stomatitis, apthous ulcers, erytheme multiforme, or bullous lichen planus. Cutaneous PV may resemble bullous pemphigoid, linear IgA dermatosis [86, 87, 90].

Pemphigus Vegetans

Pemphigus vegetans is a rare variant of PV. The lesions that form are fungoid or papillosquamous proliferation, usually along intertriginous areas or face. There are two sub types: a severe form called Neumann and 2) a mild type called Hallopeau. Lesions can first present as pustules and then later form vegetative plaques [98].

Pemphigus Foliaceus

Introduction

There are 2 predominant types of PF: idiopathic PF, which is found universally and occurs sporadically, and fogo selvage, an endemic variety linked exclusively to multiple distinct geographical areas. Rare variants of PF have also been described, including pemphigus erythematosus (PE, Senear-Usher syndrome) and drug-induced PF. IgA pemphigus and pemphigus herpetiformis (PH) have previously been described in the literature as variants of PF, but appear to be distinct subtypes of the general pemphigus category, both clinically and histopathologically [99, 100].

Pathogenesis

In PF, the body's immune system produces immunoglobulin G autoantibodies that target the intercellular adhesion glycoprotein desmoglein-1. The binding of these autoantibodies is principally expressed in the granular layer of the epidermis, results in the loss of intercellular connections between keratinocytes and the formation of subcorneal blisters within the epidermis [99].

Incidence

The worldwide incidence and prevalence of PF is very low. Due to the presence of endemic areas, these figures may vary considerably based on the specific region being studied. For instance, the incidence of PF in Tunisia has been found to be as high as 6.7 new cases per million per year. In Brazil, which has multiple foci of endemic PF, there is a region located in the state of Maso Grosso do Sul that has a prevalence equal to approximately 3 % of its population. The average age of non-endemic PF symptom onset ranges from 40 to 60 years of age. FS affects a larger number of children and young adults as symptoms usually begin during the second or third decade of life. Both sporadic and endemic PF are typically seen equally in men and women and affect those of all races and ethnicities [99].

Clinical Presentation

PF lesions are usually first seen on the trunk, but may also originate as localized lesions on the face or scalp. Unlike PV, there is typically no history of oral or other mucosal lesions. The patient may be unaware of the blisters because they rupture very easily. In these cases, there may only be a history of superficial sores or areas of crusting.

Most lesions appear on the chest, back, and shoulders. The lesions may become widespread. Patients with the mildest form of PF may only report a history of a small, solitary, recurrent scaly and crusty lesion of the face. Scaling represents the detachment of the overlying stratum corneum from the area of intra-epidermal acantholysis, and the stratum granulosum. A common clinical finding in PF is a positive Nikolsky's sign.

The primary lesions are flaccid, superficial vesicles and bullae of the skin. These lesions may not be seen on examination because of their fragile and subsequent transient nature. More often, only secondary lesions, such as shallow erosions, are seen. On certain areas of the body such as the face and scalp, the exudate from the erosive lesions dries quickly, leaving areas of crusting over an erythematous base. It may be years before the patient is correctly diagnosed.

In its most severe form, PF can produce an exfoliative erythroderma characterized by generalized erythema and diffuse scaling of the cutaneous surface. In these cases, it may also lead to alopecia. These patients require prompt hospitalization to prevent serious and sometimes fatal complications from metabolic instability [99, 97].

Differential Diagnosis

PF may resemble other pemphigus diseases, bullous impetigo, linear IgA dermatosis, subcorneal pustular dermatosis, or seborrhic dermatitis. Multiple drugs have been found to be associated with the development of PF. It is important to thoroughly review the patient's current medications. Drug-associated cases may persist or quickly clear after the offending agent is withdrawn. The most commonly implicated drugs are penicillamine, angiotensin-converting enzyme (ACE) inhibitors, and rifampicin. The demonstration of IgG autoantibodies identifies the lesion in the pemphigus group [99].

Pemphigus Erythematosus

Introduction

PE, also known as Senear–Usher syndrome, is a rare, benign, localized variant of PF. Serologically, PE patients have autoantibodies, which is similar to individuals with PF and cutaneous lupus erythematosus [99, 100].

Pathogenesis

There is a presence of immunoglobulin and complement at the dermo-epidermal junction resembling the lupus band test. Patients also exhibit antinuclear antibodies. In pemphigus erythematosus after UV irradiation, the anti-Dsg1 antibodies were deposited along dermo-epidermal junction mimicking the lupus band in ANA negative patients [99, 100].

Clinical Presentation

The clinical hallmarks of PE are seborrheic lesions in the nose, nasolabial folds, and malar areas resembling the "butterfly" distribution of lupus. Lesions may also affect the preauricular region. Hyperkeratotic scars with erythema and superficial blisters can be present on the chest. The oral mucosa, pharynx, and vulva are not involved [99, 100].

Diagnosis

The skin immunopathology of PE is characterised by acantholysis with immunoglobulin deposition in desmosomes and at the dermal-epidermal junction (lupus band test). Histology and serological marker antidesmoglein 1 of PF and PE are the same [100].

IgA Pemphigus

Introduction

IgA pemphigus represents a group of autoimmune intraepidermal blistering diseases presenting with a vesiculopustular eruption, neutrophil infiltration, acantholysis.

It is characterized by tissue-bound and circulating IgA antibodies targeting desmosomal or nondesmosomal cell surface components in the epidermis. The reaction between IgA and the keratinocyte cell surfaces is thought to be the leading pathogenic factor [101].

Incidence

IgA pemphigus is mostly seen in middle aged and older people. The onset of IgA pemphigus is reported to be subacute. There are two distinct types of IgA pemphigus: the subcorneal pustular dermatosis (SPD) type and the intraepidermal neutrophilic (IEN) type [101].

Pathogenesis

There is no clear explanation for the mechanism by which IgA autoantibodies produce characteristic skin lesions in IgA pemphigus. There are several hypotheses. IgA autoantibodies might bind to the Fc receptor CD89 on monocytes and granulocytes, resulting in accumulation of neutrophils and subsequent proteolytic cleavage of the keratinocyte cell-cell junction. The other issue to be considered is the possible epitopespreading phenomenon, in which an inflammatory event releases new target antigens, exposes them to the immune system, and then induces subsequent autoimmunity to new related antigens [101, 102].

Clinical Features

Patients with both types of IgA pemphigus clinically present with flaccid vesicles or pustules on erythematous or normal skin. The pustules tend to coalesce to form an annular or circinate pattern with crusts in the central area. The sites of predilection are the axillary and groin areas, but the trunk and proximal extremities are commonly involved. About half of IgA pemphigus patients suffer from pruritus, and mucous membrane involvement [101, 102, 103].

Diagnosis

Histopathologic examination of IgA pemphigus shows slight acantholysis and neutrophilic infiltration in the epidermis. Acantholysis in IgA pemphigus is much milder than that seen in classic pemphigus. In the SPD type of IgA pemphigus, pustules are located subcorneally in the upper epidermis, whereas in the IEN type, suprabasilar pustules in the lower or entire epidermis are present [101, 102].

Paraneoplastic Pemphigus

Introduction

PNP is defined as new mucocutaneous acantholytic disease characterized by the presence of autoantibodies (therefore named as pemphigus), in patients with neoplasia. It is believed that an immunological deregulation in antitumor antibodies leads to the production of autoantibodies that bind to epidermal proteins (plakin family) present in desmosomes and hemidesmosomes responsible for cell adhesion, thereby causing skin displacement [101, 104, 105].

Incidence

PNP is rare and the exact incidence of PNP is not known. It predominates in men of 45–70 years of age. However, case reports of the disease in children exist, and in them PNP has a predilection for those of Hispanic origin [101, 105, 106].

Clinical Features

The symptoms include the following: (I) pemphigus-like: superficial vesicules, flaccid blisters, erosions and crusts, occasional and limited erythema; (II) bullous pemphigoidlike: scaling erythematous papules that may be associated or not wiht tense blisters; (III) erythema multiforme-like: polymorphic lesions, mainly scaling erythematous papules with erosions or occasionally ulcers with difficult healing; (IV) graft versus host disease-like: disseminated dusky red scaly papules; (V) lichen planus-like: small squamous flat-topped violaceus papules and intense involvement of mucosal membranes

PNP lesions affect not only the oral mucosa, but also esophagus, stomach, duodenum, and colon. Oral involvement with painful stomatitis is seen in almost all cases and can often be the first symptom. Lesions may also involve the conjunctival and anorectal mucosa. Cutaneous manifestations range from papules and plaques similar to erythema multiforme, vesicles and blisters that resemble pemphigus vulgaris or even pruritic plaques similar to lichen planus [101, 104–106].

Diagnosis

The major histopathological feature of PNP is vacuolar or lichenoid interface dermatitis pattern. There may be intraepidermal cleft and acantholysis, or more rarely, subepidermal blisters. The clinical variants also have their respective histological features: (I) pemphigus-like: intra-epidermal cleft surrounded by mononuclear cells; (II) bullous pemphigoid-like: subepidermal cleft with or without basal cellular vacuolization, and moderate mononuclear infiltrate in dermo-epidermal junctions; (III) erythema multiforme-like: dyskeratosis without cleft or with areas of epidermal separation, due to basal cell disintegration, and distinct perivascular infiltrate; (IV) graft versus host disease-like: absence of epidermal separation, hyperkeratosis or hyperparakeratosis and dyskeratosis with or without vacuolar degeneration of basal cell layers and intense mononuclear interface dermatitis; (V) lichen planus-like: hypergranulosis, dyskeratosis and lichenoid mononuclear infiltrate [101, 104–108]

Differential Diagnosis

PNP may resemble pemphigus vulgaris, cicatricial pemphigoid, erythema multiforme, lichen planus, or persistent herpes simplex virus. In approximately, one third of patients, there is an underlying, undiagnosed malignancy at the first appearance of clinical features in PNP. Most associated malignancies develop in patients who are between 45 and 70 years old. Approximately 80 % are of hematological origin (B-cell lymphoproliferative disorders), such as non-Hodgkin lymphoma, chronic lymphocytic leukemia, Castleman's disease, thymoma, Waldernstrom's macroglobulinemia and follicular dendritic cell sarcoma. PNP is associated with high mortality rate secondary to sepsis, bleeding and respiratory failure. The search for malignancy should be conducted through a comprehensive physical examination [101, 105–108].

Pemphigus Herpetiformis

Introduction

PH is also known as acantholytic herpetiform dermatitis, herpetiform pemphigus, or mixed bullous disease. It is a clinical variant of PV or PF, that combines the clinical features of dermatitis herpetiformis with the immunopathologic features of pemphigus [101, 109].

Incidence

PH equally affects men and women, from 30 to 80 years of age, with rare case reports during childhood [101, 109]

Pathogenesis

PH resembles classic forms of pemphigus with intercellular deposits of IgG and C3 in the epidermis. The difference from PV or PF is that PH autoantibodies may recognize functionally less important epitopes of Dsg-1 or 3 and therefore do not lead directly to acantholysis. Autoantibodies in PH may induce signaling pathway of cytokines production by keratinocytes that attract inflammatory cells to the tissue, with focal intercellular edema and eosinophilic spongiosis [101, 110, 111].

Clinical Features

Patients presenting with PH are difficult to diagnosis when they first seek medical care. Clinical presentation is usually atypical. Patients usually show erythematous, gyrate, annular and edematous lesions, with clusters of small or abortive vesicles and/or pustules, frequently in herpetiform pattern. These features are not generally seen in PF and PV. Mucous lesions are not a frequent issue, but can be present in some patients. Pruritus is frequently associated and might be severe. PH can sometimes evolve into the classical forms of PV or PF. More than one biopsy may therefore be necessary for diagnosis of PH [101, 109–111].

Diagnosis

The histological findings can vary among patients and one patient can present different histological features at different times or biopsies. More than one biopsy may therefore be necessary for diagnosis of PH. On histology, there may be subcorneal pustules and/or intraepidermal vesicles filled with neutrophils and/or eosinophils and neutrophilic and/or eosinophilic spongiosis. Acantholysis may be minimal or absent. These characteristic findings differs PH histologically from PF and PV [101, 109–111].

Differential Diagnosis

Some diseases have been described together with PH, such as psoriasis, thyroid diseases, systemic lupus erythematosus, HIV infection and malignancies like lung cancer, esophageal carcinoma, and prostatic cancer.

Investigations for Pemphigus Diseases

To diagnose pemphigus, histology, immunofluorescence, and serological testing is required. The site of biopsy and the age of the lesion are important for both histology and immunofluorescence.

A tissue biopsy is an important modality in the diagnosis of the pemphigus. A punch biopsy preferably taken at the transitional edge of the blister and inflamed skin is needed. By including the edge of the blister, the site of blister formation can be better visualized. In an absence of blisters, a biopsy including an erosion and adjacent skin might be helpful; however, it often shows only non-specific inflammation [87–88].

The early histologic findings in PF are the formation of vacuoles within the intercellular spaces of the granular and/or upper spinous layers of the epidermis. The vacuoles become larger and eventually lead to subcorneal blister formation within the upper epidermis. There are variable amounts of acantholytic keratinocytes, neutrophils, and fibrin within the blisters [87–88].

The current gold standard of diagnostic testing is direct immunofluorescence microscopy to demonstrate tissue-bound autoantibodies and/or of C3 in the patient's skin or mucous membranes. Pemphigus will have autoantibodies attached to the cells. The fluorescent antibodies will bind to the autoantibodies and fluoresce under microscopy. Direct immunofluorescence will show IgG deposition on epithelial surface with suprabasilar blister cavity. IgG deposition is present in 100 % of all pemphigus patients. Demonstration of IgG deposition is diagnostic of a pemphigus disease, except in IgA pemphigus, where IgA is demonstrated [87–88].

Serology through indirect immunofluorescence microscopy of the patient's serum can be used as a screening test for circulating antibodies. Indirect immunofluorescence microscopy on monkey or guinea pig esophagus has become an established mode of testing for serum antibody in pemphigus. In patients with dermatitis herpetiformis, IgA reactivity against the endomysium can be visualized on monkey esophagus. Indirect immunofluorescence may directly correlate with clinical disease activity and might be useful to follow disease progression and response to therapy [87–88]

Definitive diagnostic testing follows, with the aid of various ELISA or Western blot studies involving the relevant target antigens. Definitive diagnostic testing follows, with the aid of various ELISA or Western blot studies involving the relevant target antigens. Some of these ELISAs are commercially available. These tests usually suffice to establish the diagnosis by serology in conjunction with a compatible clinical picture. Autoantibodies against desmoglein 1 in PF and desmoglein 3 in PV are correlated with disease activity. The corresponding ELISAs are, therefore, suitable tests for monitoring disease activity over time and can be a useful aid in setting the optimal dose of the immunosuppressive medication(s) used to treat the disease [87–88].

Treatment and Prognosis of Pemphigus Diseases

Goals of therapy include minimizing disease burden and improving quality of life. In most patients, this clinically translates into the absence of blistering or occasional blistering.

Non pharmalogical treatment includes, cleansing cutaneous erosions with antibacterial soap twice a day, followed by bandaging with non-stick gauzes. Patients should avoid aggressive oral hygiene practices, including flossing, when symptomatic lesions are present, due to increased risk of pain and/or bleeding. Gentle oral hygiene measures may consist of saline rinses [85].

In PV, systemic corticosteroids are employed as first-line treatment with lower doses used as maintenance. Higher doses of 120 mg/day result in a more rapid control of disease than lower doses, there is no evidence that the higher doses are beneficial in the long term. Therefore, it is recommended that 1 mg/kg per day be the initial dose for managing pemphigus. Once disease progression has been halted, a slow standardized tapering of corticosteroid is commenced over about a 4-month period. Monotherapy of pemphigus with oral corticosteroids causes frequent side effects, including systemic infections, diabetes mellitus, osteoporosis, thromboses , and gastrointestinal ulcers [85, 88, 91, 101, 112–118].

Steroid-sparing agents are employed to reduce the cumulative exposure and side effects associated with long-term steroid use. Such agents include azathioprine, mycophenolate mofetil, cyclophosphamide, and methotrexate. Newer agents included biologics, like intravenous immunoglobulin (IVIg) and rituximab. Rituximab is an anti-CD20 monoclonal antibody, associated with a reduction in autoantibodies and B cell depletion. Rituximab is typically prescribed for patients who are unable to taper steroids without flare of their disease or in patients who are still flaring despite combination therapy (steroids + steroid-sparing agent) [85, 88, 91, 101, 112–118].

The management of pemphigus in pediatric patients is divided into childhood (patient ≤12 years) pemphigus and juvenile (patients 13–18 years) pemphigus. In both groups the majority of patients have mucocutaneous disease. The mainstay of therapy is oral corticosteroids. About half to two thirds of the patients develop systemic side effects. The most concerning is growth retardation present in 50% of the patients. Others include infection, obesity, psychological, and social distress. Immunosuppressive agents are used in many patients for their steroid-sparing effects. The treatment lasts between 2 and 3 years. The prognosis in most cases reported was good. Intravenous immunoglobulin (IVIg) shows promise in early studies. Rituximab was effective in recalcitrant cases [117].

The treatment of PF, PE, and pemphigus vegetans are similar to PV. Disease is controlled is maintained with steroids and immunosuppressive agents. PF has a more favorable prognosis compared to PV [98, 99, 100].

The mainstays for treatment of IgA pemphigus are oral and topical corticosteroids. In addition, dapsone may be very useful in treating IgA pemphigus. Recently, mycophenolate mofetil and adalimumab, which are known to be effective in classic pemphigus, are also reported to be useful in treating IgA pemphigus [101–103]

The most widely suggested specific treatment in PNP combines prednisone with cyclosporine. However, the disease is generally resistant to therapy. The mortality of patients with PNP is 75–90 %. Respiratory failure due to bronchiolitis obliterans is one of the most important causes of death in patients with PNP. However, the disease course is highly variable, not only in severe cases, but also in indolent disease. The prognosis is worst in the presence of erythema multiforme-like lesions. Rituximab may be indicated, especially because of association with non-Hodgkin's lymphoma, though there are reports of complications and low therapeutic response. In general, treatment of neoplasia is not associated with improvement of PNP, except in cases associated to Castleman's disease. Complete response to neoplasia treatment may heal mucocutaneous lesions [104–108].

PH is considered to be less life-threatening than other types of pemphigus and generally to have a good prognosis, although some cases may progress into classic pemphigus and therefore require more intense treatment. PH usually responds well to monotherapy with dapsone which is considered the drug of first choice, or to a combination treatment with low doses of systemic corticosteroids. It has been reported that patients with PH and no circulating autoantibodies appear more likely to respond to therapy with dapsone [109–111].

Even with the use of corticosteroids and other immunosuppressive agents, there is still significant morbidity and mortality associated with pemphigus diseases. A common cause of death in PV is infection secondary to the immunosuppression required to treat the disease. Unfortunately, many of the drugs used to treat this disease have serious side effects, and patients must be monitored closely for infection, renal and liver function abnormalities, electrolyte disturbances, hypertension, diabetes, anemia, and gastrointestinal bleeding [85, 88, 91, 101, 112–118].

Conflict of interest None

References

1. Langan SM, Smeeth L, Hubbard R, Fleming KM, Smith CJ, West J. Bullous pemphigoid and pemphigus vulgaris—incidence and mortality in the UK: population based cohort study. BMJ. 2008;337:a180.
2. Marazza G, Pham HC, Schärer L, Pedrazzetti PP, Hunziker T, Trüeb RM, Hohl D, Itin P, Lautenschlager S, Naldi L, Borradori L. Autoimmune bullous disease Swiss study group. Incidence of bullous pemphigoid and pemphigus in Switzerland: a 2-year prospective study. Br J Dermatol. 2009;161(4):861–8.
3. Schmidt E, della Torre R, Borradori L. Clinical features and practical diagnosis of bullous pemphigoid. Dermatol Clin. 2011;29(3):427–38.
4. Joly P, Baricault S, Sparsa A, Bernard P, Bédane C, Duvert-Lehembre S, Courville P, Bravard P, Rémond B, Doffoel-Hantz V, Bénichou J. Incidence andmortality of bullous pemphigoid in France. J Invest Dermatol. 2012;132(8):1998–2004.
5. Marazza G, Pham HC, Scharer L, et al. Incidence of bullous pemphigoid and pemphigus in Switzerland: a 2-year prospective study. Br J Dermatol. 2009;161:861–8.

6. Bertram F, Brocker EB, Zillikens D, Schmidt E. Prospective analysis of the incidence of autoimmune bullous disorders in Lower Franconia, Germany. J Dtsch Dermatol Ges. 2009;7:434–40.
7. Gudi VS, White MI, Cruickshank N, et al. Annual incidence and mortality of bullous pemphigoid in the Grampian region of north-east Scotland. Br J Dermatol. 2005;153:424–7.
8. Joly P, Baricault S, Sparsa A, et al. Incidence and mortality of bullous pemphigoid in France. J Invest Dermatol. 2012;132:1998–2004.
9. Langan SM, Smeeth L, Hubbard R, Fleming KM, Smith CJ, West J. Bullous pemphigoid and pemphigus vulgaris—incidence and mortality in the UK: population based cohort study. BMJ. 2008;337:160–3.
10. Jung M, Kippes W, Messer G, Zillikens D, Rzany B. Increased risk of bullous pemphigoid in male and very old patients: a population-based study on incidence. J Am Acad Dermatol. 1999;41:266–8.
11. Kulthanan K, Chularojanamontri L, Tuchinda P, Sirikudta W, Pinkaew S. Prevalence and clinical features of Thai patients with bullous pemphigoid. Asian Pac J Allergy Immunol. 2011;29(1):66–72.
12. Kasperkiewicz M, Zillikens D, Schmidt E. Pemphigoid diseases: pathogenesis, diagnosis, and treatment. Autoimmunity. 2012;45(1):55–70.
13. Sagi L, Baum S, Agmon-Levin N, Sherer Y, Katz BS, Barzilai O, Ram M, Bizzaro N, SanMarco M, Trau H, Shoenfeld Y. Autoimmune bullous diseases the spectrum of infectious agent antibodies and review of the literature. Autoimmun Rev. 2011;10(9):527–35.
14. Le Jan S, Plée J, Vallerand D, Dupont A, Delanez E, Durlach A, Jackson PL, Blalock JE, Bernard P, Antonicelli F. Innate Immune Cell-Produced IL-17 Sustains Inflammation in Bullous Pemphigoid. J Invest Dermatol. 2014;134(12):2908–17.
15. Diaz LA, Ratrie H 3rd, Saunders WS, et al. Isolation of a human epidermal cDNA corresponding to the 180-kD autoantigen recognized by bullous pemphigoid and herpes gestationis sera. Immunolocalization of this protein to the hemidesmosome. J Clin Invest. 1990;86:1088–94.
16. Labib RS, Anhalt GJ, Patel HP, Mutasim DF, Diaz LA. Molecular heterogeneity of the bullous pemphigoid antigens as detected by immunoblotting. J Immunol. 1986;136:1231–5.
17. Stanley JR, Hawley-Nelson P, Yuspa SH, Shevach EM, Katz SI. Characterization of bullous pemphigoid antigen: a unique basement membrane protein of stratified squamous epithelia. Cell. 1981;24:897–903.
18. Knudson RM, Kalaaji AN, Bruce AJ. The management of mucous membrane pemphigoid and pemphigus. Dermatol Ther. 2010;23(3):268–80.
19. Fuchs E, Raghavan S. Getting under the skin of epidermal morphogenesis. Nat Rev Genet. 2002;3(3):199–209. (Review. PubMed PMID: 11972157).
20. Rzany B, Partscht K, Jung M, et al. Risk factors for lethal outcome in patients with bullous pemphigoid: low serum albumin level, high dosage of glucocorticosteroids, and old age. Arch Dermatol. 2002;138:903–8.
21. Chen YJ, Wu CY, Lin MW, et al. Comorbidity profi les among patients with bullous pemphigoid: a nationwide population-based study. Br J Dermatol. 2011;165:593–9.
22. Langan SM, Groves RW, West J. The relationship between neurological disease and bullous pemphigoid: a population-based case-control study. J Invest Dermatol. 2010;131:631–6.
23. Bastuji-Garin S, Joly P, Lemordant P, et al. Risk factors for bullous pemphigoid in the elderly: a prospective case-control study. J Invest Dermatol. 2011;131:637–43.
24. Seppanen A, Miettinen R, Alafuzoff I. Neuronal collagen XVII is localized to lipofuscin granules. Neuroreport. 2010;21:1090–4.
25. Claudepierre T, Manglapus MK, Marengi N, et al. Collagen XVII and BPAG1 expression in the retina: evidence for an anchoring complex in the central nervous system. J Comp Neurol. 2005;487:190–203.
26. Guo L, Degenstein L, Dowling J, Yu QC, Wollmann R, Perman B, Fuchs E. Gene targeting of BPAG1: abnormalities in mechanical strength and cell migration in stratified epithelia and neurologic degeneration. Cell. 1995;81:233–43.
27. Schmidt E, Zillikens D. Modern diagnosis of autoimmune blistering skin diseases. Autoimmun Rev. 2010;10(2):84–9.

28. Hofmann S, Thoma-Uszynski S, Hunziker T, Bernard P, Koebnick C, Stauber A, Schuler G, Borradori L, Hertl M. Severity and phenotype of bullous pemphigoid relate to autoantibody profile against the NH2- and COOH-terminal regions of the BP180 ectodomain. J Invest Dermatol. 2002;119:1065–73.
29. Di Zenzo G, Grosso F, Terracina M, Mariotti F, De PitÃ O, Owaribe K, Mastrogiacomo A, Sera F, Borradori L, Zambruno G. Characterization of the anti-BP180 autoantibody reactivity profile and epitope mapping in bullous pemphigoid patients. J Invest Dermatol. 2004;122(1):103–10.
30. Shirakata Y, Shiraishi S, Sayama K, Miki Y. Subclass characteristics of IgG autoantibodies in bullous pemphigoid and pemphigus. J Dermatol. 1990;17(11):661–6.
31. Döpp R, Schmidt E, Chimanovitch I, Leverkus M, Bröcker EB, Zillikens D. IgG4 and IgE are the major immunoglobulins targeting the NC16A domain of BP180 in Bullous pemphigoid: serum levels of these immunoglobulins reflect disease activity. J Am Acad Dermatol. 2000;42(4):577–83
32. Woodley DT. The role of IgE anti-basement membrane zone autoantibodies in bullous pemphigoid. Arch Dermatol. 2007;143(2):249–50.
33. Di Zenzo G, Thoma-Uszynski S, Fontao L, Calabresi V, Hofmann SC, Hellmark T, Sebbag N, Pedicelli C, Sera F, Lacour JP, Wieslander J, Bruckner-Tuderman L, Borradori L, Zambruno G, Hertl M. Multicenter prospective study of the humoral autoimmune response in bullous pemphigoid. Clin Immunol. 2008;128(3):415–26.
34. Roussel A, Benichou J, Randriamanantany ZA, Gilbert D, Drenovska K, Houivet E, Tron F, Joly P. Enzyme-linked immunosorbent assay for the combination of bullous pemphigoid antigens 1 and 2 in the diagnosis of bullous pemphigoid. Arch Dermatol. 2011;147(3):293–8.
35. Di Zenzo G, Marazza G, Borradori L. Bullous pemphigoid: physiopathology, clinical features and management. Adv Dermatol. 2007;23:257–88.
36. Venning VA, Wojnarowska F. Induced bullous pemphigoid. Br J Dermatol. 1995;132:831–2.
37. Tosti A, André M, Murrell DF. Nail involvement in autoimmune bullous disorders. Dermatol Clin. 2011;29(3):511–338.
38. Fleming TE, Korman NJ. Cicatricial pemphigoid. J Am Acad Dermatol. 2000;43(4):571–91. (quiz 591–4)
39. Gammon WR, Briggaman RA, Inman AO 3rd, Queen LL, Wheeler CE. Differentiating anti-lamina lucida and anti-sublamina densa anti-BMZ antibodies by indirect immunofluorescence on 1.0 M sodium chloride-separated skin. J Invest Dermatol. 1984;82:139–44.
40. Yoshida M, Hamada T, Amagai M, et al. Enzyme-linked immunosorbent assay using bacterial recombinant proteins of human BP230 as a diagnostic tool for bullous pemphigoid. J Dermatol Sci. 2006;41:21–30.
41. Charneux J, Lorin J, Vitry F, et al. Usefulness of BP230 and BP180-NC16a enzyme-linked immunosorbent assays in the initial diagnosis of bullous pemphigoid: a retrospective study of 138 patients. Arch Dermatol. 2011;147:286–91.
42. Joly P, Roujeau JC, Benichou J, Picard C, Dreno B, Delaporte E, Vaillant L, D'Incan M, Plantin P, Bedane C, Young P, Bernard P, Bullous Diseases French Study Group. A comparison of oral and topical corticosteroids in patients with bullous pemphigoid. N Engl J Med. 2002;346(5):321–7
43. Guillaume JC, Vaillant L, Bernard P, Picard C, Prost C, Labeille B, Guillot B, Foldès-Pauwels C, Prigent F, Joly P, et al. Controlled trial of azathioprine and plasma exchange in addition to prednisolone in the treatment of bullous pemphigoid. Arch Dermatol. 1993;129(1):49–53
44. Fox JC, Kenkare S, Petronic-Rosic V, Soltani K, Sethi A. Bullous pemphigoid in late childhood successfully treated with mycophenolate mofetil as an adjuvant therapy. Pediatr Dermatol. 2010;27(5):537–9.
45. Du-Thanh A, Merlet S, Maillard H, Bernard P, Joly P, Estève E, Richard MA, Pauwels C, Ingen-Housz-Oro S, Guillot B, Dereure O. Combined treatment with low-dose methotrexate and initial short-term superpotent topical steroids in bullous pemphigoid: an open, multicentre, retrospective study. Br J Dermatol. 2011;165(6):1337–43.

46. Kasperkiewicz M, Shimanovich I, Ludwig RJ, Rose C, Zillikens D, Schmidt E. Rituximab for treatment-refractory pemphigus and pemphigoid: a case series of 17 patients. J Am Acad Dermatol. 2011;65(3):552–8.
47. Czernik A, Toosi S, Bystryn JC, Grando SA. Intravenous immunoglobulin in the treatment of autoimmune bullous dermatoses: an update. Autoimmunity. 2012;45(1):111–8.
48. Schmidt E, Obe K, Bröcker EB, Zillikens D. Serum levels of autoantibodies to BP180 correlate with disease activity in patients with bullous pemphigoid. Arch Dermatol. 2000;136(2):174–8.
49. Bernard P, Reguiai Z, Tancrède-Bohin E, Cordel N, Plantin P, Pauwels C, Vaillant L, Grange F, Richard-Lallemand MA, Sassolas B, Roujeau JC, Lok C, Picard-Dahan C, Chosidow O, Vitry F, Joly P. Risk factors for relapse in patients with bullous pemphigoid in clinical remission: a multicenter, prospective, cohort study. Arch Dermatol. 2009;145(5):537–42.
50. Cortés B, Marazza G, Naldi L, Combescure C, Borradori L. Autoimmune Bullous Disease Swiss Study Group. Mortality of bullous pemphigoid in Switzerland: a prospective study. Br J Dermatol. 2011;165(2):368–74.
51. Joly P, Baricault S, Sparsa A, Bernard P, Bédane C, Duvert-Lehembre S, Courville P, Bravard P, Rémond B, Doffoel-Hantz V, Bénichou J. Incidence and mortality of bullous pemphigoid in France. J Invest Dermatol. 2012;132(8):1998–2004.
52. Bernard P, Vaillant L, Labeille B, et al. Incidence and distribution of subepidermal autoimmune bullous skin diseases in three French regions. Bullous Diseases French Study Group. Arch Dermatol. 1995;131:48–52.
53. Ahmed AR, Kurgis BS, Rogers RS 3rd. Cicatricial pemphigoid. J Am Acad Dermatol. 1991;24:987–1001.
54. Oyama N, Setterfield JF, Powell AM, Sakuma-Oyama Y, Albert S, Bhogal BS, Vaughan RW, Kaneko F, Challacombe SJ, Black MM. Bullous pemphigoid antigen II (BP180) and its soluble extracellular domains are major autoantigens in mucous membrane pemphigoid: the pathogenic relevance to HLA class II alleles and disease severity. Br J Dermatol. 2006;154(1):90–8
55. Kirtschig G, Marinkovich MP, Burgeson RE, Yancey KB. Anti-basement membrane autoantibodies in patients with anti-epiligrin cicatricial pemphigoid bind the alpha subunit of laminin 5. J Invest Dermatol. 1995t;105(4):543–8
56. Kasperkiewicz M, Zillikens D, Schmidt E. Pemphigoid diseases: pathogenesis, diagnosis, and treatment. Autoimmunity. 2012;45(1):55–70.
57. Yancey KB, Egan CA. Pemphigoid: clinical, histologic, immunopathologic, and therapeutic considerations. JAMA. 2000;284(3):350–6.
58. Sagi L, Baum S, Agmon-Levin N, Sherer Y, Katz BS, Barzilai O, Ram M, Bizzaro N, SanMarco M, Trau H, Shoenfeld Y. Autoimmune bullous diseases the spectrum of infectious agent antibodies and review of the literature. Autoimmun Rev. 2011;10(9):527–35.
59. Di Zenzo G, Thoma-Uszynski S, Calabresi V, Fontao L, Hofmann SC, Lacour JP, Sera F, Bruckner-Tuderman L, Zambruno G, Borradori L, Hertl M. Demonstration of epitope-spreading phenomena in bullous pemphigoid: results of a prospective multicenter study. J Invest Dermatol. 2011;131(11):2271–80.
60. Schmidt E, Zillikens D. Pemphigoid disease. Lancet. 2013;381:320–32.
61. Lazarova Z, Yancey KB. Cicatricial pemphigoid: immunopathogenesis and treatment Prognosis. Dermatologic Therapy. 2002;15:382.
62. Assmann T, Becker J, Ruzicka T, Megahed M. Topical tacrolimus for oral cicatricial pemphigoid. Clin Exp Dermatol. 2004;29(6):674–6.
63. Kourosh AS, Yancey KB. Therapeutic approaches to patients with mucous membrane pemphigoid. Dermatol Clin. 2011;29(4):637–41.
64. Segura S, Iranzo P, Martínez-de Pablo I, Mascaró JM Jr, Alsina M, Herrero J, Herrero C. High-dose intravenous immunoglobulins for the treatment of autoimmune mucocutaneous blistering diseases: evaluation of its use in 19 cases. J Am Acad Dermatol. 2007;56(6):960–7.

65. Ambros-Rudolph CM, Mullegger RR, Vaughan-Jones SA, Kerl H, Black MM. The specific dermatoses of pregnancy revisited and reclassifi ed: results of a retrospective two-center study on 505 pregnant patients. J Am Acad Dermatol. 2006;54:395–404.
66. Shimanovich I, Bröcker EB, Zillikens D. Pemphigoid gestationis: new insights into the pathogenesis lead to novel diagnostic tools. BJOG. 2002;109(9):970–6.
67. Roger D, Vaillant L, Fignon A, Pierre F, Bacq Y, Brechot JF, Grangeponte MC, Lorette G. Specific pruritic diseases of pregnancy. A prospective study of 3192 pregnant women. Arch Dermatol. 1994;130(6):734–9
68. Chimanovitch I, Schmidt E, Messer G, et al. IgG1 and IgG3 are the major immunoglobulin subclasses targeting epitopes within the NC16A domain of BP180 in pemphigoid gestationis. J Invest Dermatol. 1999;113:140–2.
69. Chi CC, Wang SH, Charles-Holmes R, et al. Pemphigoid gestationis: early onset and blister formation are associated with adverse pregnancy outcomes. Br J Dermatol. 2009;160:1222–8.
70. Jenkins RE, Hern S, Black MM. Clinical features and management of 87 patients with pemphigoid gestationis. Clin Exp Dermatol. 1999;24:255–9.
71. Jenkins RE, Shornick JK, Black BL. Pemphigoid gestationis. J Eur Acad Dermatol Venereol. 1993;2:163.
72. Fortuna G, Marinkovich MP. Linear immunoglobulin a bullous dermatosis. Clin Dermatol. 2012;30(1):38–50.
73. Marinkovich MP, Taylor TB, Keene DR, Burgeson RE, Zone JJ. LAD-1, the linear IgA bullous dermatosis autoantigen, is a novel 120-kDa anchoring fi lament protein synthesized by epidermal cells. J Invest Dermatol. 1996;106:734–38.
74. Collier PM, Wojnarowska F, Welsh K, McGuire W, Black MM. Adult linear IgA disease and chronic bullous disease of childhood: the association with human lymphocyte antigens Cw7, B8, DR3 and tumour necrosis factor influences disease expression. Br J Dermatol. 1999;141(5):867–75.
75. Paige DG, Leonard JN, Wojnarowska F, Fry L. Linear IgA disease and ulcerative colitis. Br J Dermatol. 1997;136(5):779–82.
76. Guide SV, Marinkovich MP. Linear IgA bullous dermatosis. Clin Dermatol. 2001;19(6):719–27.
77. Ng SY, Venning VV. Management of linear IgA disease. Dermatol Clin. 2011;29:629–30.
78. Aram H. Linear IgA bullous dermatosis. Successful treatment with colchicine. Arch Dermatol. 1984;120(7):960–1.
79. Wong SN, Chua SH. Spectrum of subepidermal immunobullous disorders seen at the National Skin Centre, Singapore: a 2-year review. Br J Dermatol. 2002;147:476–80.
80. Hirose M, Vafia K, Kalies K, et al. Enzymatic autoantibody glycan hydrolysis alleviates autoimmunity against type VII collagen. J Autoimmun. 2012;39:304–14
81. Gammon WR, Heise ER, Burke WA, Fine JD, Woodley DT, Briggaman RA. Increased frequency of HLA-DR2 in patients with autoantibodies to epidermolysis bullosa acquisita antigen: evidence that the expression of autoimmunity to type VII collagen is HLA class II allele associated. J Invest Dermatol. 1988;91(3):228–32.
82. Chen M, Kim GH, Prakash L, Woodley DT. Epidermolysis bullosa acquisita: autoimmunity to anchoring fi bril collagen. Autoimmunity. 2012;45:91–101.
83. Thiers BH. Pemphigus. J Am Acad Dermatol. 1981;4:603–5.
84. Ruocco V, Ruocco E, Lo Schiavo A, Brunetti G, Guerrera LP, Wolf R. Pemphigus: etiology, pathogenesis, and inducing or triggering factors: facts and controversies. Clin Dermatol. 2013;31:374–81.
85. Scully C, Challacombe SJ. Pemphigus vulgaris: update on etiopathogenesis, oral manifestations, and management. Crit Rev Oral Biol Med. 2002;13:397–408.
86. Baum S, Sakka N, Artsi O, Trau H, Barzilai A. Diagnosis and classification of autoimmune blistering diseases. Autoimmun Rev. 2014;13:482–9.
87. Kershenovich R, Hodak E, Mimouni D. Diagnosis and classification of pemphigus and bullous pemphigoid. Autoimmun Rev. 2014;13:477–81.
88. Santoro FA, Stoopler ET, Werth VP. Pemphigus. Dent Clin North Am. 2013;57:597–610.

89. Hertl M, Veldman C. Pemphigus—paradigm of autoantibody-mediated autoimmunity. Skin Pharmacol Appl Skin Physiol. 2001;14:408–18.
90. Bickle K, Roark TR, Hsu S. Autoimmune bullous dermatoses: a review. Am Fam Physician. 2002;65:1861–70.
91. Ruocco E, Baroni A, Wolf R, Ruocco V. Life-threatening bullous dermatoses: pemphigus vulgaris. Clin Dermatol. 2005;23:223–6.
92. Tron F, Gilbert D, Mouquet H, Joly P, Drouot L, Makni S, Masmoudi H, Charron D, Zitouni M, Loiseau P, Ben Ayed M. Genetic factors in pemphigus. J Autoimmun. 2005;24:319–28.
93. Gazit E, Loewenthal R. The immunogenetics of pemphigus vulgaris. Autoimmun Rev. 2005;4:16–20.
94. Bystryn JC, Rudolph JL. Pemphigus. Lancet. 2005;366:61–73.
95. Grando SA. Pemphigus autoimmunity: hypotheses and realities. Autoimmunity. 2012;45:7–35.
96. Tron F, Gilbert D, Joly P, Mouquet H, Drouot L, Ayed MB, Sellami M, Masmoudi H, Makni S. Immunogenetics of pemphigus: an update. Autoimmunity. 2006;39:531–9.
97. Uzun S, Durdu M. The specificity and sensitivity of Nikolskiy sign in the diagnosis of pemphigus. J Am Acad Dermatol. 2006;54:411–5.
98. Becker BA, Gaspari AA. Pemphigus vulgaris and vegetans. Dermatol Clin. 1993;11:429–52.
99. James KA, Culton DA, Diaz LA. Diagnosis & clinical features of pemphigus foliaceus. Dermatol Clin 2011;29:405–12.
100. Pérez-Pérez ME, Avalos-Díaz E, Herrera-Esparza R. Autoantibodies in senear-usher syndrome: cross-reactivity or multiple autoimmunity? Autoimmune Dis. 2012;2012:296214.
101. Porro AM, Caetano Lde V, Maehara Lde S, Enokihara MM. Non-classical forms of pemphigus: pemphigus herpetiformis, IgA pemphigus, paraneoplastic pemphigus and IgG/IgA pemphigus. An Bras Dermatol. 2014;89:96–106.
102. Hashimoto T. Immunopathology of IgA pemphigus. Clin Dermatol. 2001;19:683–9.
103. Beutner EH, Chorzelski TP, Wilson RM, Kumar V, Michel B, Helm F, Jablonska S. IgA pemphigus foliaceus. Report of two cases and a review of the literature. J Am Acad Dermatol. 1989;20:89–97.
104. Anhalt GJ, Kim SC, Stanley JR, Korman NJ, Jabs DA, Kory M, Izumi H, Ratrie H 3rd, Mutasim D, Ariss-Abdo L. Paraneoplastic pemphigus. An autoimmune mucocutaneous disease associated with neoplasia. N Engl J Med. 1990;323:1729–35.
105. Frew JW, Murrell DF. Paraneoplastic pemphigus (paraneoplastic autoimmune multiorgan syndrome): clinical presentations and pathogenesis. Dermatol Clin. 2011;29:419–25.
106. Zimmermann J, Bahmer F, Rose C, Zillikens D, Schmidt E. Clinical and immunopathological spectrum of paraneoplastic pemphigus. J Dtsch Dermatol Ges. 2010;8:598–606.
107. Nguyen VT, Ndoye A, Bassler KD, Shultz LD, Shields MC, Ruben BS, Webber RJ, Pittelkow MR, Lynch PJ, Grando SA. Classification, clinical manifestations, and immunopathological mechanisms of the epithelial variant of paraneoplastic autoimmune multiorgan syndrome: a reappraisal of paraneoplastic pemphigus. Arch Dermatol. 2001;137:193–206.
108. Horn TD, Anhalt GJ. Histologic features of paraneoplastic pemphigus. Arch Dermatol. 1992;128:1091–5.
109. Robinson ND, Hashimoto T, Amagai M, Chan LS. The new pemphigus variants. J Am Acad Dermatol. 1999;40:649–71.
110. Santi CG, Maruta CW, Aoki V, Sotto MN, Rivitti EA, Diaz LA. Pemphigus herpetiformis is a rare clinical expression of nonendemic pemphigus foliaceus, fogo selvagem, and pemphigus vulgaris. Cooperative Group on Fogo Selvagem Research. J Am Acad Dermatol. 1996;34:40–6.
111. Prado R, Brice SL, Fukuda S, Hashimoto T, Fujita M. Paraneoplastic pemphigus herpetiformis with IgG antibodies to desmoglein 3 and without mucosal lesions. Arch Dermatol. 2011;147:67–71.
112. Schmidt E, Zillikens D. The diagnosis and treatment of autoimmune blistering skin diseases. Dtsch Ärztebl Int 2011;108:399–405.
113. Daniel BS, Murrell DF. Management of pemphigus. F1000Prime Rep. 2014;6:32.

114. Hooten JN, Hall RP, Cardones AR. Updates on the management of autoimmune blistering diseases. Skin Therapy Lett. 2014;19:1–6.
115. Atzmony L, Hodak E, Gdalevich M, Rosenbaum O, Mimouni D. Treatment of pemphigus vulgaris and pemphigus foliaceus: a systematic review and meta-analysis. Am J Clin Dermatol. 2014;15:503–15.
116. Singh S. Evidence-based treatments for pemphigus vulgaris, pemphigus foliaceus, and bullous pemphigoid: a systematic review. Indian J Dermatol Venereol Leprol. 2011;77:456–69.
117. Gürcan H, Mabrouk D, Razzaque AA. Management of pemphigus in pediatric patients. Minerva Pediatr. 2011;63:279–91.
118. Leventhal JS, Sanchez MR. Is it time to re-evaluate the treatment of pemphigus? J Drugs Dermatol. 2012;11:1200–6.

Salivary Gland Diseases

Astrid Rasmussen, Christopher J. Lessard and Kathy L. Sivils

Introduction

Salivary gland diseases can be classified based on their underlying causes, which may be congenital, neoplastic, autoimmune, infectious, environmental or multifactorial. A common thread in all these ailments is salivary gland dysfunction, which is clinically manifested as xerostomia. A general outline of these disorders is shown in Table 1. The current chapter will be centered on the Sjögren's and sicca syndromes, other autoimmune and granulomatous diseases, monogenic and congenital malformations, as well as a brief overview of other miscellaneous causes. Benign and malignant salivary gland tumors are discussed extensively in Chaps. 12 (premalignant lesions) and 13 (oral cancer), while chapter 3 and 4 (host genomics and response to infectious agents) addresses the role of infectious diseases.

A. Rasmussen (✉)
Arthritis and Clinical Immunology Research Program, Oklahoma Medical Research Foundation, 825 NE 13th Street, MS#5, Oklahoma City, OK 73104, USA
e-mail: astrid-rasmussen@omrf.org

C. J. Lessard
Arthritis and Clinical Immunology Research Program, Oklahoma Medical Research Foundation, 825 NE 13th Street, MS#24, Oklahoma City, OK 73104, USA
e-mail: christopher-lessard@omrf.org

Department of Pathology, University of Oklahoma Health Sciences Center, 825 NE 13th Street, MS#24, 73104 Oklahoma City, OK, USA

K. L. Sivils
Arthritis and Clinical Immunology Research Program, Oklahoma Medical Research Foundation, 825 NE 13th Street, MS#57, Oklahoma City, OK 73104, USA
e-mail: kathy-sivils@omrf.org

S. T. Sonis (ed.), *Genomics, Personalized Medicine and Oral Disease*,
DOI 10.1007/978-3-319-17942-1_11

Autoimmune Diseases

Sjögren's Syndrome. OMIM 270150
Sjögren's Syndrome (SS) is a complex autoimmune disease characterized by lacrimal and salivary gland dysfunction, which is manifested as xerophthalmia and xerostomia [1–4]. The glandular damage is mediated by autoantibodies (antiRo/SSA and antiLa/SSB) and by lymphocytic infiltration of the glands [5, 6]. Approximately one third of patients with SS will develop extraglandular manifestations that may include fatigue, arthritis/arthralgias, peripheral neuropathy, pulmonary and renal disease, Raynaud's phenomenon, vasculitis, and a significantly increased risk of non-Hodgkin's lymphoma [7, 8]. Furthermore, significant proportions of patients with SS have overlapping features of other autoimmune diseases, most commonly autoimmune thyroid disease, rheumatoid arthritis, systemic lupus erythematosus and scleroderma. In spite of current population estimates that suggest that SS is the second most common autoimmune disease, preceded only by rheumatoid arthritis [2], SS is frequently misdiagnosed, underdiagnosed or diagnosed at late stages of the disease. Clinical diagnosis often takes 6–10 years, [9, 10] leading to a lag in potential preventive and therapeutic strategies and likely contributing to damage accrual.

The genetics of SS is largely understudied. Most genetic studies reported thus far are largely focused on candidate genes such as the HLA loci or genes demonstrating association in other autoimmune diseases; it is only within the last 18 months that the first genome-wide scans have been completed. The initial evidence supporting an important role of inherited factors in SS came from reports of increased concordance rates amongst monozygotic twins, reports of familial aggregation, and increased prevalence of other autoimmune diseases amongst relatives of SS patients. [2–4, 8, 11–17]. Female sibling or dizygotic twin rates of 2–4 % and estimated odds of female sibling concordance (λs) between 8 and 30 could be reasonable estimates for SS.

The HLA genes found to be associated with SS vary in different ethnic groups. In general, studies have primarily focused on alleles at the Class II HLA DR and DQ loci. The most consistent associations to date have been with DR2 and DR3 alleles at the DRB1 locus in Caucasian populations. In 2003, Gottenberg et al. confirmed the association of SS with HLA alleles DRB1*03 and DQB1* which was restricted to patients with anti-SSA and/or anti-SSB antibodies [18]. Particularly strong associations with these antibody responses were identified in patients heterozygous for DQw1 and DQw2 [19, 20]. Other genes that may also be involved in autoantibody production in SS include the *TAP2* gene (Transporter 2, ATP-binding cassette, sub-family B) and *TGF-β1* (Transforming Growth Factor-β1). The TAP genes, which are mapped to the MHC region, are important in peptide loading and cell surface expression of HLA Class I molecules. Gottenberg et al identified an allele at codon 10 of TGF-β1 with an elevated allele frequency in SS patients who had the HLA-DRB1*3 haplotype and elevated levels of anti-La/SSB autoantibodies [20]. They hypothesized that both the TGF-β1 polymorphism and the HLA-DRB1*3 haplotype act in combination to promote the production of anti-La/SSB

Table 1 Causes of salivary gland dysfunction

1.	*Neoplasia*
	a. Benign: Pleomorphic adenoma
	b. Malignant: Carcinomas, primary lymphoproliferative tumors, metastasis
2.	*Autoimmune*
	a. Sjögren's syndrome
	b. Granulomatosis with polyangiitis (Wegener's granulomatosis)
	c. IgG4-related disease
3.	*Granulomatous*
	a. Tuberculosis
	b. Sarcoidosis
4.	*Infectious*
	a. Mumps
	b. Bacterial parotitis (acute and chronic)
	c. HIV-related
	d. Other infectious parotitis: Tb, toxoplasmosis, cat-scratch disease
5.	*Environmental*
	a. Pharmaceuticals
	b. External beam radiation
	c. Internal radionuclides: ^{131}I
6.	*Miscellaneous (Most commonly sialolithiasis and sialoadenitis)*
	a. Idiopathic sialolithiasis
	b. Graft-versus-host disease
	c. Thyroid disease
	d. Parkinson's disease
	e. Cirrhosis
	f. Diabetes
	g. Malnutrition: anorexia, bulimia, ethanol related, other causes
7.	*Congenital malformations*
	a. Oculo-auriculo-vertebral spectrum (Goldenhar syndrome)
	b. Agenesis of salivary and lacrimal glands (ALSG syndrome)
	c. Lacrimo-auriculo-dento-digital syndrome (LADD syndrome, Levy-Hollister syndrome)
	d. Polycystic dysgenetic disease of parotid salivary glands
	e. Congenital deafness with labyrinthine aplasia, microtia and microdontia (LAMM syndrome)
	f. Hypohidrotic ectodermal dysplasia
	g. Brooke-Spiegler syndrome
8.	*Other monogenic disorders*
	a. Cystic Fibrosis

autoantibodies. A recent meta-analysis of 1166 cases and 6470 controls of diverse ethnic backgrounds derived from 23 studies [21] confirmed that significant risk was associated with pSS and HLA Class II alleles DRB1*03:01, DQA1*05:01, DQB1*02:01, while HLA DQA1*03:01, DQA1*05:01 and DQB1*05:01 alleles were protective factors. The genome wide association study (GWAS) by Lessard et al. replicated previously identified associations in the MHC, which represent the strongest genetic risk factors in their study (OR ~3.5). The most significant associations were HLA-DQB1 in the HLA class II region at rs115575857 with $P_{meta} = 7.65 \times 10^{-114}$ and a second set of variants not in linkage disequilibrium (LD) with rs115575857 that peaked at rs116232857 ($P_{meta} = 1.33 \times 10^{-96}$) [22]. Li et al., in their GWAS of Han Chinese SS patients, showed the strongest associations with *HLA-DRB1/HLA-DQA1* (rs9271588; $P_{combined} = 8.52 \times 10^{-37}$) and *HLA-DPB1* (rs4282438; $P_{combined} = 8.77 \times 10^{-25}$) [23].

Beyond the associations with the MHC, the majority of the early studies of SS genetics focused on candidate genes previously know to be important in immune function or other autoimmune diseases, including genes involved in interferon (IFN) pathways. Evidence for association with some of these genes has been reported in SS and notably includes interferon regulatory factor 5 (*IRF5*) and signal transducer and activator of transcription 4 (*STAT4*) [24–28]. IRF5 is a transcription factor that acts downstream of the toll-like receptors and type I interferons to promote the expression of numerous anti-viral and pro-inflammatory proteins [29–30]. Associations with several independent genetic effects within the IRF5 locus have been documented for SLE and are found in Asian, Caucasian, Hispanic, and African American populations [30–36]. In SS patients, the GT or TT genotype at an *IRF5* SNP (rs2004640) was found to be more prevalent when compared to controls. Moreover, an association in the 'G' risk allele of *TNPO3* at the SNP rs13246321, which is in linkage disequilibrium with rs10488631 within the *IRF5-TNPO3* locus previously demonstrated to have association with SS, was observed in a combined cohort of Norwegian and Swedish SS patients [27]. Finally, a CGGGG insertion/deletion (indel) polymorphism has yielded the most statistically significant result and appears to be responsible for the association in the promoter region [25–26, 37–38].

In 2008, a case-control study of SS patients found a weak association with a SNP (rs7574865) in *STAT4* that had already been associated with SLE and RA [24]. A second *STAT4* SNP shows a moderate association (rs7582694) with SS [26]. In addition to association analyses of SS with single variants, evaluation for more complex genetic models supports additive effects between the major risk alleles in the *IRF5* and *STAT4* [26, 39–41].

Several genes that function in adaptive immune responses, particularly in T and B cells, have been implicated in SS. A genetic association of SS has been recently identified by Nordmark et al. in Swedish and Norwegian SS patients, with tumor necrosis factor super family member 4 (*TNFSF4/OX40L*), a gene relevant to T cell functions.17 Genetic associations with potential relevance to B cell function have recently been identified in the region comprising two genes, *FAM167A* and the B lymphoid tyrosine kinase (*BLK*) locus [27]. *FAM167A and BLK* are transcribed in opposite directions, possibly from common promoter elements, and expression levels are inversely correlated. While the function of *FAM167A* remains unknown, *BLK* is expressed in B cells and is involved in cell signaling that results in activation of multiple nuclear transcription factors. Reduced expression of BLK is hypothesized to lead to a breakdown in tolerance by allowing autoreactive cells to escape deletion [27]. SS has also been associated to two SNPs (rs3843489 and rs869593) in *EBF1* (Early B cell factor 1). EBF1 is another vital transcription factor involved in the enhancement of transcriptional activity during B cell development; decreased or altered expression of EBF1 can lead to impaired B cell development [27].

Significantly increased levels of B-cell activating factor (BAFF, BLyS), a member of the tumor necrosis factor family, have been identified in the serum of SS patients compared to healthy controls and directly correlate to the degree of clinical activity and titer of circulating autoantibodies [42–46]. Thus, related genes have been studied as candidates for association with SS. Disease susceptibility for anti-Ro/SSA- and anti-La/SSB-positive SS has been associated with the CTAT haplotype of 4 SNPs located in the 5′ regulatory region of the BAFF gene, while the TTTT haplotype has been associated with elevated BAFF levels in SS [47].

Based on some recent insights into the potential role of NK cells in animal models of sialadenitis, a candidate gene case-control study of *NCR3/NKp30* was performed by Rusakiewicz et al. They found that the minor allele in the promoter SNP rs11575837 was protective from SS and resulted in reduced gene transcription and function [48, 49]

The last year has seen the completion of two GWA studies of SS, one in patients of European ancestry [22] and the other one in Han Chinese population [23]. The study by Lessard et al. included a total of 2685 SS cases and 9825 controls in the discovery and replication cohorts. Genotyping of the discovery cohort was performed for >1.1 million variants using the Illumina OMNI1-Quad array, while replication studies were conducted using the ImmunoChip and a custom array [22]. The Asian study by Li et al. included 542 cases and 1050 controls in the discovery stage genotyped using the Affymetrix Axiom Genome-Wide CHB 1 Array; the two replication experiments comprised 1303 cases and 2727 controls that were genotyped using the iPLEX MassARRAY platform (Sequenom) [23].

Beyond the previous associations within the MHC region, the study in European subjects confirmed associations to *IRF5-TNPO3, STAT4*, and *TNIP1* as risk loci for SS while identifying three new associated genes: *IL12A, DDX6-CXCR5*, and *FAM167A-BLK* [22]. Previous studies implicating *IRF5* as a risk locus had tested ~15 variants [26–29], the current study identified a total of 67 SNPs that exceeded the significance threshold for genome-wide association; further analysis suggested that three independent effects are likely but additional studies will be necessary to determine the precise causal variant in the promoter region. The associated variant identified in *STAT4* (rs10553577) is highly correlated with variants reported in previous studies. The variant in *TNIP1* identified in this study, rs6579837, has also been described within the H2 risk haplotype for lupus defined by Adrianto et al [50]; the exact function of *TNIP1* has not yet been defined, but it binds *TNFAIP3*, which in turn suppresses TLR-induced apoptosis by negatively regulating NF-κB51 [50].

Of the novel associations, *IL12A* encodes the p35 subunit that forms the IL12 heterodimer with the p40 subunit encoded by *IL12B* [51]. IL12 is an immunomodulatory cytokine primarily secreted by monocytes and dendritic cells, which plays a critical role in the differentiation of T-helper 1 cells and the production of interferon-γ by T cells and NK cells [52]. Variants in the 3′ end of *IL12A* have been reported for PBC, while 5′ effects have been described for celiac disease [53–55]; however, the effect in the region of *IL12A* in SS appears to be distinct from those previously reported. In addition, the IL12B subunit may also be affected by genetic effects in SS: *IRF5* can initiate transcription of *IL12B*, and once the IL12 heterodimer has been secreted from dendritic or NK cells, it binds receptors on T cells, thereby initiating a signaling cascade through *STAT4* phosphorylation [51]. In summary, *IL12A, IRF5* and *STAT4*, all associated with SS, interact and are involved in type I interferon signaling and overexpression of genes induced by type I interferon signaling correlates with titers of the classic SS autoantibodies anti-Ro and anti-La8 [22, 56, 57].

Associations to variants in the first intron of *BLK* and the promoter region it shares with *FAM167A* were identified in the European GWAS. Previous studies in SS had implicated a nearby region within *FAM167A* at rs1254979634 [27] and additional variants had been reported with other related diseases, such as SLE [57, 58]. *CXCR5* variants had been reported in association with multiple sclerosis and PBC [59, 60] and several SS studies have found *CXCR5* to be dysregulated in B cells in both the periphery and in salivary gland tissues [55, 61]. Lessard et al attributed the SS association with *CXCR5* to rs7119038, which is located ~16 kb 5′ of the coding region of the gene [22]. Finally, 29 additional regions were suggestive of association to SS but did not exceed the genome-wide significance threshold and warrant further study [22].

The GWAS of Han Chinese SS patients identified association with MHC class II genes and three non-MHC genes: *IRF5, TNFAIP3* and the novel *GTF2I*. The strongest association was with a SNP within *GTF2I* (rs117026326) and associated SNPs extending from *GTF2I* to *GTF2IRD1-GTF2I*. Interestingly, it was only a risk factor in the Chinese cohort and not in the cohort of European descent; the converse was true for *TNIP1*, which was only identified amongst the Europeans [23].

Further delineation of the complex etiology and diseases mechanisms in SS will require additional genome-wide association scans and large-scale replication studies in multiple ethnic backgrounds, but these typically identify only the common variants. Alternative approaches to delineate the causal and rare variants, will include whole genome sequencing, transcriptomics, and proteomics [62–70]. The ultimate goal will be to integrate these datasets with detailed clinical information, to achieve a global picture of Sjögren's pathogenesis.

Granulomatosis with polyangiitis (Wegener's Granulomatosis). OMIM 142857
Granulomatosis with polyangiitis (GPA), formerly known as Wegener's granulomatosis, is a systemic disease characterized by necrotizing granulomatous inflammation of the upper and lower respiratory tract, glomerulonephritis, vasculitis, and antineutrophil cytoplasmic autoantibodies (ANCA). It belongs to the small and medium-vessel ANCA-vasculitides, together with the Churg-Strauss syndrome (CSS), and

microscopic polyangiitis (MPA) [71]. A rare, but sometimes initial, presentation of GPA is by uni- or bilateral infiltration of the major salivary glands, particularly the parotid [72]. As is the case for many other autoimmune diseases, the best-studied genetic associations have been with markers across the HLA region, particularly in class II genes including HLA–DPB1 and HLA–DPA1, the strongest association being with the HLA–DPB1*04 allele [71, 73, 74]. Additional susceptibility loci have been confirmed near the gene for semaphorin 6A (SEMA6), *CTLA4, PTPN22, FCGR3B, FCAR,* and *PRTN3* [74–76].

Interestingly, more recent studies of the three ANCA-vasculitides have shown that the genetic associations are more closely associated with the autoantibody specificity than with the clinical phenotype. Anti–proteinase 3 ANCA is associated with HLA-DP and the genes encoding α_1-antitrypsin (*SERPINA1*) and proteinase 3 (*PRTN3*); while anti–myeloperoxidase ANCA is associated with HLA-DQ, *which is also the most common association with MPA* [77]. The *PTPN22* 620W allele confers susceptibility to the development of the GPA phenotype (but not of MPA or CSS), but particularly to the ANCA-positive subset of GPA [78]. The authors of these studies suggest that the results support for the concept that proteinase 3 ANCA–associated vasculitis and myeloperoxidase ANCA–associated vasculitis are distinct autoimmune syndromes [75, 79].

Further subphenotyping by renal phenotype or neutrophil response has been associated to specific FcR alleles: IgA ANCA responses, present in ~30% of patients and less common in subjects with severe renal disease, are driven by *FCAR* alelles, while IgG ANCA induced neutrophil activation is modulated by *FCGFR3B* polymorphisms [80]. Thus, it is tempting to speculate that further understanding of these variants may lead to prognostic and therapeutic targets for granulomatosis with polyangiitis.

IgG4-related Disease

IgG4-related disease (IgG4-RD) is a multi-organ systemic disease that unifies a number of clinical diagnoses previously thought to be independent conditions [81, 82]. Central features are the tumorous swelling of affected organs and elevated concentrations of IgG4 in serum in ~60–80% of the patients [83]. The affected tissues show diffuse lymphoplasmacytic infiltration, occasional eosinophils, characteristic "storiform" fibrosis, obliterative phlebitis, and infiltration by IgG4-bearing plasma cells and T lymphocytes [84, 85]. Two major presentations of this condition are type 1 autoimmune pancreatitis (AIP) "IgG4-related pancreatitis" and salivary gland disease, which may present as salivary gland enlargement (formerly "Mikulicz disease") or sclerosing sialadenitis (formerly "Küttner's tumor") [82, 86]. An additional characteristic, which has been suggested by some authors as a diagnostic criterion, is glucocorticoid responsiveness, particularly in earlier stages of the disease when tissue fibrosis is limited [85].

IgG4-RD may mimic many other conditions, particularly those with swollen gland lesions, such as malignancies, benign glandular diseases and Sjögren's syndrome [87–90]. Traditional descriptions of Mikulicz syndrome included patients, most commonly middle-aged to elderly males, with idiopathic bilateral, painless and symmetric swelling of the lacrimal, parotid and submandibular glands, accompanied by relatively mild xerophthalmia and xerostomia. A clear separation between Mikulicz syndrome from Sjögren's syndrome did not occur until 2005 [91, 92]. To clearly distinguish IgG4-RD manifestations from Sjögren's syndrome or other differential diagnoses, it is recommended that the terms "IgG4-related dacryoadenitis", "IgG4-related parotitis", and "IgG4-related sialadenitis" be used when the characteristic histopathologic features are confirmed in the glands [83, 93].

The etiology of IgG4-RD is still unclear, with evidence pointing to both an autoimmune disorder and allergic factors. While circulating autoantibodies are detected, a specific autoantigenic target has not been identified and it is not clear whether the autoantibodies are pathogenic [92]. Some studies have demonstrated molecular mimicry between *H. pylorii* and pancreatic epithelial cells, suggesting that the gastric infection triggers autoimmunity in genetically susceptible individuals [94].

Few and limited genetic studies have been undertaken, mostly in patients with AIP or IgG4-RD of Asian ancestry, therefore, larger studies including patients of multiple ethnic backgrounds are needed before conclusive associations can be established. So far, potential genetic links have been identified to HLA genes and to *FcRL3, CTLA4* and *KCNA3* [92, 95–98]. A study of AIP Japanese patients showed association with human leukocyte antigen (HLA) molecules DRB1*0405 and DQB*0401 which was not reproduced later in Korean subjects [94, 98]. Multiple SNPs at *CTLA4* are associated with susceptibility to various autoimmune diseases and several variants have been linked to Chinese AIP patients with increased risk for relapse. Association of AIP to *CTLA4* was also observed in Japanese patients but the risk alleles were different from those observed in Chinese studies [96, 99]. Studies focusing on genetic factors shared with known Th2 disorders or other autoimmune and chronic inflammatory conditions have been suggested as meritorious future directions [92, 100].

Granulomatous Diseases

Sarcoidosis. OMIM 181000
Sarcoidosis is a systemic granulomatous disorder characterized by the presence of non-caseating granulomas that may affect virtually any organ. While a well-defined cause remains unknown, sarcoidosis is likely to be the result of an exaggerated granulomatous reaction after exposure to unidentified antigens in individuals with genetic predisposition [101, 102]. Clinically, it is a very heterogeneous disease whose most common manifestations are persistent cough, the triad of skin/eye/lymph node involvement, erythema nodosum, fatigue, and abnormal chest radiographs. The diagnosis requires three criteria: clinical and radiological presentation, evidence of non-caseating granulomas, and no evidence of alternative explanations. Parotitis, manifested by bilateral parotid gland swelling, can be present in up to 4–6 % of patients with sarcoidosis either in isolation or as part of the Heerfordt syndrome [103]. This syndrome, also known as Heerfordt-Waldenström syndrome or uveoparotid syndrome, is characterized by a pathognomonic combination of parotid

gland enlargement, facial palsy, anterior uveitis, and fever; added manifestations may be cutaneous sarcoidosis and lacrimal gland enlargement [102, 103].

Sarcoidosis is generally a sporadic disease, but 5–16% of cases are familial [104–106]. Further evidence to support a strong genetic component includes a 80-fold increase in risk in monozygotic twins and striking differences in prevalence and clinical manifestation in different geographic areas and racial groups [107, 108,109]. A number of susceptibility loci for sarcoidosis have been identified; with the HLA class II alleles—particularly *DRB1* alleles—representing the main contributor across ethnicities. The most relevant protective alleles are *HLA-DRB1*01* and *HLA-DRB1*04*, while the main risk alleles are *HLA-DRB1*03, HLA-DRB1*11, HLA-DRB1*12, HLA-DRB1*14*, and *HLA-DRB1*15*. Furthermore, *HLA-DRB1*03* predisposes to disease with spontaneous resolution while *HLA-DRB1*14* and *HLA-DRB1*15* predispose to chronic disease. African-American patients have shown stronger associations with *HLA-DBQ1* than with *DRB1* alleles. A less prominent but consistent association has been described with HLA Class I alleles, in particular *HLA-B7* and *-B8* [101, 109]. Paradoxically, the *HLA-DRB1*04* allele, which protects against overall sarcoidosis, is considered a risk allele for ocular sarcoidosis and Heerfordt syndrome [110].

Genome-wide association studies have mainly identified two non-MHC loci that confer susceptibility to sarcoidosis: *BTNL2* and *ANXA11. BNTL2* has been associated with sarcoidosis in German and white American populations but not in African American patients; it maps within the MHC region and much discussion exists as to whether it represents an effect independent of class II alleles. The associations of this gene with several other diseases, including ulcerative colitis, multiple sclerosis, type 1 diabetes, rheumatoid arthritis and lupus appear to be driven by linkage disequilibrium with several HLA –DRB1 alleles [111, 112]. *ANXA11* has also been associated with numerous autoimmune disorders, including rheumatoid arthritis, lupus and Sjögren's syndrome; its association with sarcoidosis has been replicated in independent European, African and European-American populations [109, 112]. Studies of African-Americans, who have a higher prevalence of sarcoidosis and more severe disease, suggest that additional genes of African origin remain to be identified [113]. Genome wide scans in other populations have also identified new suggestive loci that remain unconfirmed.

Tuberculosis

Tuberculosis (Tb) is a granulomatous disease caused by infection by Mycobacterium tuberculosis. The primary target organ is the lung but almost any organ, including the salivary glands, may be affected. The tubercle bacillus is transmitted from person to person by aerosol, but only a proportion of those in contact with the infectious particles will become infected. Of those infected, less than 10% will develop clinical signs of Tb, while the majority will develop latent infection [114]. Observations of "familial predisposition" to Tb date to the eighteenth century, but it was not until 1933 that rigorous genetic epidemiological studies provided evidence of higher concordance rates in monozygotic than in dizygotic twin pairs [115, 116]. Hereditary susceptibility to multiple infections, including tuberculosis, is a feature of several primary immunodeficiency syndromes; the most common being severe combined immunodeficiency (SCID). Specific susceptibility of otherwise healthy subjects to poorly virulent mycobacteria, such as BCG vaccines and environmental mycobacteria, is associated with the syndrome of Mendelian susceptibility to mycobacterial diseases (MSMDs). Twelve distinct genetic disorders are responsible for MSMD and germline mutations have been found in the *IFNGR1, IFNGR2, STAT1, IL12B,* and *IL12RB1* genes; complete and partial interferon-γ receptor 1 (IFN-γR1) deficiency, complete interleukin (IL)-12 p40 deficiencies, and complete IL-12 receptor β1 (IL-12Rβ1) deficiency are genetic etiologies of MSMD. Some of the patients with MSMD who live in areas endemic for pathogenic Tb have also developed bonafide tuberculosis during childhood. It has been postulated that a significant proportion of children that develop Tb are carriers of one of the MSMD-related mutations [115–117].

Adult tuberculosis has not been associated with specific monogenic syndromes and no major susceptibility or protective locus has been identified, but multiple associations have been described including to genes involved in innate immunity pathways (v.gr. *SCL11A1/NRAMP, TNF, IL10*, IL-1 receptor antagonist, *ALOX5, LTA4H, TLR8,* and *VDR*) [118, 119]. Additional susceptibility loci include: *NRAMP1, SP110, CISH, TLR2, IFNGR1, CCL2, CD209*, and *MCP1*, [120, 121] while protection against Tb has been linked to: *MC3R, IFNGR1, TIRAP, IFNG*, and *IRGM*. [115, 116]. Polymorphisms in the Toll-like receptor (TLR) genes have variably been associated with protection or susceptibility to Tb in different populations. A recent meta-analysis indicated that *TLR2* G2258A is associated with increased Tb risk in Asians and Europeans, *TLR1* G1805T with increased Tb in Africans and American Hispanics, while *TLR6* C745T is associated with decreased TB risk [122]. Finally, recent studies support a significant role for the host genome in determining whether an exposed patient will develop clinical Tb or latent Tb infection [118].

Congenital Malformations

Oculo-auriculo-vertebral spectrum (Goldenhar Syndrome, Hemifacial Microsomia). OMIM164210

The Oculo-auriculo-vertebral (OAV) spectrum is a complex malformative process that may affect aural, oral and mandibular development as part of a morphogenesis and development of first and second branchial arch derivates. More severe cases may be accompanied by epibulbar dermoids (Goldenhar syndrome), vertebral, cardiac, and renal malformations [123, 124]. Earlobe malformations are an obligated feature and are accompanied by hemifacial microsomia in 65% of cases; the more severe cases may involve hypogenesis or agenesis of the mandibular ramus with agenesis of the ipsilateral parotid gland, displaced salivary gland tissue or salivary fistulas [125]. While most cases are sporadic, the evidence favors autosomal dominant inheritance with significant variability of expression [126, 127]. The underlying genetic defect is unknown, with linkage established in some families to a region of approximately 10.7 cM on chromosome 14q32, with a maximum multipoint lod score of 3.00 between microsatellite markers D14S987 and D14S65. However, other families did not show linkage to this region and genetic heterogeneity is likely [128].

Agenesis of salivary and lacrimal glands (ALSG syndrome; Aplasia of lacrimal and salivary glands; parotid aplasia or hypoplasia included). OMIM180920.

This rare autosomal dominant condition shows variable ranges of hypoplasia or agenesis of the lacrimal, parotid, submandibular, and sublingual glands accompanied by absence of the lacrimal puncta [129, 130]. From a clinical perspective, it results in xerophthalmia sometimes paradoxically accompanied by epiphora (constant tearing), and xerostomia, which in turn predisposes to dental erosion, caries, periodontal disease and oral infections. Thus, it should be part of the differential diagnosis of Sjögren's syndrome. The severity of the manifestations depends on the degree of hypoplasia and the number of glands affected and there is both inter and intra-familial variability of expression [129]. Mutations in the *FGF10* gene that lead to haploinsufficiency are the underlying cause, and the disorder is allelic with LADD syndrome [131–133].

Lacrimo-auriculo-dento-digital syndrome (LADD Syndrome, Levy-Hollister Syndrome). OMIM149730.

Initially described by Hollister in 1973, this syndrome encompasses variable aplasia or hypoplasia of the lacrimal puncta with obstruction of the naso-lacrimal ducts, earlobe malformations, aplasia or hypoplasia of the salivary glands, dental abnormalities (such as peg-shaped incisors, hypodontia, enamel defects), and upper limb malformations [134–136]. The agenesis of salivary glands has been noted in 50% of the cases and is manifest by xerostomia and/or absence of Stensen's papillae or ducts [137]. There is wide variability of expression with the more severe cases including genitourinary anomalies (hypospadias, nephrosclerosis, renal agenesis), and complex diaphragmatic and pulmonary malformations [138]. The underlying mutations are genetically heterogeneous: LADD syndrome is caused by heterozygous mutations in the *FGFR2* and *FGFR3* genes [139], which encode for fibroblast growth factor receptors 2 and 3 respectively, or in the *FGF10* gene, which encodes an FGFR ligand [139, 140]. The mutations in the *FGFR2* and *FGFR3* gene are concentrated in conserved intracellular tyrosine kinase domains, within loops that have a regulatory function [139].

Polycystic dysgenetic disease of parotid salivary glands. OMIM600343.
This is a rare benign disease characterized by recurrent and fluctuating bilateral nontender swelling of the parotid gland without decreasing salivary function [141, 142]. Age of onset is often in childhood or early adulthood but may occur at any time in life. Histologically, the functional acinar parenchyma of the glands is substituted by cystic changes, which have resulted a developmental abnormality of the intercalated duct system of the gland [143]. Several instances of transmission compatible with autosomal dominant inheritance have been reported, although the causal gene defects have not been identified [141, 144]. This disorder should be considered in the differential diagnosis of sialectasias of the parotid gland, salivary duct cysts, and lymphoepithelial cysts.

Brooke-Spiegler syndrome (BRSS) OMIM605041.
The Brooke-Spiegler syndrome (BRSS) is one of a group of syndromes characterized by the appearance of multiple skin appendage tumors including cylindromas, trichoepitheliomas, and spiradenomas [145–147]. Other syndromes with similar tumors and overlapping clinical features are Familial Cylindromatosis (FC; OMIM132700) [148] and Multiple Familial Trichoepithelioma-1 (MFT1; OMIM601606) [149]. BRSS patients progressively accumulate these tumors in the head and neck but they may also appear in other localizations such as the salivary glands, chest, breast, and back. It is important to note that the salivary gland lymphoepithelial lesions may be malignant and are more common in Chinese and Eskimo populations of Canada and Greenland [150]. Other malignancies such as basal cell carcinoma have also been described [151]. BRSS is an autosomal dominant disorder caused by heterozygous mutations in the *CYLD* gene [151, 152] and is allelic with FC and MFT1. The *CYLD* mutation frequencies were 85 % for BRSS, 100 % for FC, and 44 % for MFT1, the majority of them resulting in truncated proteins and without clear genotype-phenotype correlations [153, 154].

Hypohidrotic ectodermal dysplasia (HED) OMIM305100.
The hypohidrotic or anhidrotic ectodermal dysplasias are characterized by the triad of hypotrichosis, anodontia or hypodontia and anhidrosis or hypohidrosis [155–157]. In addition to very limited sweating, patients have reduced salivary flow and subjective dry mouth, which compound their oral symptoms by elevating the risk of caries and dental loss [158]. Their dryness extends to the skin, eyes, airways and mucous membranes due to defective development of exocrine glands; a subgroup may have dysmorphic features such as frontal bossing, depressed nasal bridge, prominent lips, fine linear wrinkles, and pigmentation about the eyes [157]. The cumulative prevalence of HED is ~1 in 5000–10,000 newborns, the most frequent form (~55–60 % of cases) being the X-linked ECTD1 which is due to mutations in the *EDA* gene [160, 161]. It is of note that EDA-related HED females show incomplete penetrance of the disease, with 60–80 % of them displaying some degree of hypodontia and reduced salivary flow [156, 159, 162]. The remaining ~40 % of cases are autosomal recessive (ECTD10B and ECTD11, caused by mutations in the *EDAR* and *EDARADD* genes, respectively) [163, 164, 165] or autosomal dominant (ECTD10A and ECTD11A, also caused by mutations in the *EDAR* and *EDARADD* genes, respectively) [165–167].

References

1. Besana C, Salmaggi C, Pellegrino C, et al. Chronic bilateral dacryo-adenitis in identical twins: a possible incomplete form of Sjogren syndrome. Eur J Pediatr. 1991;150:652–5.
2. Bolstad AI, Haga HJ, Wassmuth R, et al. Monozygotic twins with primary Sjogren's syndrome. J Rheumatol. 2000;27:2264–6.
3. Houghton KM, Cabral DA, Petty RE, et al. Primary Sjogren's syndrome in dizygotic adolescent twins: one case with lymphocytic interstitial pneumonia. J Rheumatol. 2005;32:1603–6.
4. Scofield RH, Kurien BT, Reichlin M. Immunologically restricted and inhibitory anti-Ro/SSA in monozygotic twins. Lupus. 1997;6:395–8.

5. Helmick CG, Felson DT, Lawrence RC, et al. National Arthritis Data Workgroup. Estimates of the prevalence of arthritis and other rheumatic conditions in the United States. Part I. Arthritis Rheum. 2008;58:15–25.
6. Gøransson LG, Haldorsen K, Brun JG, et al. The point prevalence of clinically relevant primary Sjögren's syndrome in two Norwegian counties. Scand J Rheumatol. 2011;40:221–4.
7. Fox RI. Sjögren's syndrome. Lancet. 2005;366:321–31.
8. Amador-Patarroyo MJ, Arbelaez JG, Mantilla RD, et al. Sjögren's syndrome at the crossroad of polyautoimmunity. J Autoimmun. 2012;39:199–205.
9. Daniels TE. Labial salivary gland biopsy in Sjögren's syndrome. Assessment as a diagnostic criterion in 362 suspected cases. Arthritis Rheum. 1984;27:147–56.
10. Reichlin, M, Scofield, RH. Ro (SS-A) antibodies. In: Shoenfeld Y, Gershwin ME, Meroni PL, editors. Textbook of Autoantibodies, 2nd ed. Amsterdam: Elsevier; 2006. p. 783–8.
11. Ramos-Casals M, Solans R, Rosas J, et al. Primary Sjögren syndrome in Spain: clinical and immunologic expression in 1010 patients. Medicine. 2008;87:210–9.
12. Anaya JM, Delgado-Vega AM, Castiblanco J. Genetic basis of Sjögren's syndrome. How strong is the evidence? Clin Dev Immunol. 2006;13:209–22.
13. Pavlidis NA, Karsh J, Moutsopoulos HM. The clinical picture of primary Sjögren's syndrome: a retrospective study. J Rheumatol. 1982:9:685–690.
14. Kassan SS, Moutsoupoulos HM. Clinical manifestations and early diagnosis of Sjögren Syndrome. Arch Intern Med. 2004;164:1275–1284.
15. Vitali C, Bombardieri S, Jonsson R, et al. Classification criteria for Sjögren's syndrome: a revised version of the European criteria proposed by the American-European Consensus Group. Ann Rheum Dis. 2002;61:554–8.
16. Rasmussen A, Ice JA, Li H, et al. Comparison of the American-European Consensus Group Sjögren's syndrome classification criteria to newly proposed American College of Rheumatology criteria in a large, carefully characterised SICCA cohort. Ann Rheum Dis. 2013. doi:10.1136/annrheumdis-2013-203845. [Epub ahead of print].
17. Cobb BL, Lessard CJ, Harley JB, et al. Genes and Sjögren's syndrome. Rheum Dis Clin North Am. 2008;34:847–68 (vii).
18. Gottenberg, J-E, Busson, M, Loiseau, P, et al. In primary Sjogren's syndrome, HLA class II is associated exclusively with autoantibody production and spreading of the autoimmune response. Arthritis Rheum.2003;48:2240–5.
19. Harley JB, Reichlin M, Arnett FC, et al. Gene interaction at HLA-DQ enhances autoantibody production in primary Sjögren's syndrome. Science. 1986;232(4754):1145–7.
20. Gottenberg JE, Busson M, Loiseau P, et al. Association of transforming growth factor beta1 and tumor necrosis factor alpha polymorphisms with anti-SSB/La antibody secretion in patients with primary Sjögren's syndrome. Arthritis Rheum. 2004;50:570–80.
21. Cruz-Tapias P, Rojas-Villarraga A, Maier-Moore S, Anaya JM. HLA and Sjögren's syndrome susceptibility. A meta-analysis of worldwide studies. Autoimmun Rev. 2012;11(4):281–7. doi:10.1016/j.autrev.2011.10.002.
22. Lessard CJ, Li H, Adrianto I, et al. Variants at multiple loci implicated in both innate and adaptive immune responses are associated with Sjögren's syndrome. Nat Genet. 2013;45(11):1284–92. doi:10.1038/ng.2792.
23. Li Y, Zhang K, Chen H, et al. A genome-wide association study in Han Chinese identifies a susceptibility locus for primary Sjögren's syndrome at 7q11.23. Nat Genet. 2013;45(11):1361–5.
24. Korman BD, Alba MI, Le JM, et al. Variant form of STAT4 is associated with primary Sjogren's syndrome. Genes Immun. 2008;9:267–70.
25. Miceli-Richard C, Comets E, Loiseau P, et al. Association of an IRF5 gene functional polymorphism with Sjogren's syndrome. Arthritis Rheum. 2007;56:3989–94.
26. Nordmark G, Kristjansdottir G, Theander E, et al. Additive effects of the major risk alleles of IRF5 and STAT4 in primary Sjogren's syndrome. Genes Immun. 2009;10:68–76.
27. Nordmark G, Kristjansdottir G, Theander E, et al. Association of EBF1, FAM167A(C8orf13)-BLK and TNFSF4 gene variants with primary Sjogren's syndrome. Genes Immun. 2011;12:100–9.

28. Takaoka A, Yanai H, Kondo S, et al. Integral role of IRF-5 in the gene induction programme activated by Toll-like receptors. Nature. 2005;434:243–9.
29. Taniguchi T, Ogasawara K, Takaoka A, et al. IRF family of transcription factors as regulators of host defense. Annu Rev Immunol. 2001;19:623–55.
30. Demirci FY, Manzi S, Ramsey-Goldman R, et al. Association of a common interferon regulatory factor 5 (IRF5) variant with increased risk of systemic lupus erythematosus (SLE). Ann Hum Genet. 2007;71:308–11.
31. Graham RR, Kozyrev SV, Baechler EC, et al. A common haplotype of interferon regulatory factor 5 (IRF5) regulates splicing and expression and is associated with increased risk of systemic lupus erythematosus. Nat Genet. 2006;38:550–5.
32. Kelly JA, Kelley JM, Kaufman KM, et al. Interferon regulatory factor-5 is genetically associated with systemic lupus erythematosus in African Americans. Genes Immun. 2008;9:187–94.
33. Kozyrev SV, Lewen S, Reddy PM, et al. Structural insertion/deletion variation in IRF5 is associated with a risk haplotype and defines the precise IRF5 isoforms expressed in systemic lupus erythematosus. Arthritis Rheum. 2007;56:1234–41.
34. Reddy MV, Velazquez-Cruz R, Baca V, et al. Genetic association of IRF5 with SLE in Mexicans: higher frequency of the risk haplotype and its homozygozity than Europeans. Hum Genet. 2007;121:721–7.
35. Shin HD, Sung YK, Choi CB, et al. Replication of the genetic effects of IFN regulatory factor 5 (IRF5) on systemic lupus erythematosus in a Korean population. Arthritis Res Ther. 2007;9:R32.
36. Sigurdsson S, Nordmark G, Goring HH, et al. Polymorphisms in the tyrosine kinase 2 and interferon regulatory factor 5 genes are associated with systemic lupus erythematosus. Am J Hum Genet. 2005;76:528–37.
37. Miceli-Richard C, Gestermann N, Ittah M, et al. The CGGGG insertion/deletion polymorphism of the IRF5 promoter is a strong risk factor for primary Sjögren's syndrome. Arthritis Rheum. 2009;60(7):1991–7.
38. Sigurdsson S, Göring HH, Kristjansdottir G, et al. Comprehensive evaluation of the genetic variants of interferon regulatory factor 5 (IRF5) reveals a novel 5 bp length polymorphism as strong risk factor for systemic lupus erythematosus. Hum Mol Genet. 2008;17(6):872–81.
39. Remmers EF, Plenge RM, Lee AT, et al. STAT4 and the risk of rheumatoid arthritis and systemic lupus erythematosus. N Engl J Med. 2007;357:977–86.
40. Morinobu A, Gadina M, Strober W, et al. STAT4 serine phosphorylation is critical for IL-12-induced IFN-gamma production but not for cell proliferation. Proc Natl Acad Sci U S A. 2002;99:12281–6.
41. Nishikomori R, Usui T, Wu CY, et al. Activated STAT4 has an essential role in Th1 differentiation and proliferation that is independent of its role in the maintenance of IL-12R beta 2 chain expression and signaling. J Immunol. 2002;169:4388–98.
42. Groom J, Kalled SL, Cutler AH, et al. Association of BAFF/BLyS overexpression and altered B cell differentiation with Sjögren's syndrome. J Clin Invest. 2002;109:59–68.
43. Mariette X, Roux S, Zhang J, et al. The level of BLyS (BAFF) correlates with the titre of autoantibodies in human Sjögren's syndrome. Ann Rheum Dis. 2003;62:168–71.
44. Pers JO, Daridon C, Devauchelle V, et al. BAFF overexpression is associated with autoantibody production in autoimmune diseases. Ann N Y Acad Sci. 2005;1050:34–9.
45. Schneider P, MacKay F, Steiner V, et al. BAFF, a novel ligand of the tumor necrosis factor family, stimulates B cell growth. J Exp Med. 1999;189:1747–56.
46. Mackay F, Browning JL. BAFF: a fundamental survival factor for B cells. Nat Rev Immunol. 2002;2:465–75.
47. Nossent JC, Lester S, Zahra D, et al. Polymorphism in the 5′ regulatory region of the B-lymphocyte activating factor gene is associated with the Ro/La autoantibody response and serum BAFF levels in primary Sjögren's syndrome. Rheumatology (Oxford). 2008;47:1311–6.
48. Rusakiewicz S, Nocturne G, Lazure T, et al. NCR3/NKp30 contributes to pathogenesis in primary Sjogren's syndrome. Sci Transl Med. 2013;5(195):195ra96.

49. Burbelo PD, Ambatipudi K, Alevizos I. Genome-wide association studies in Sjögren's syndrome: what do the genes tell us about disease pathogenesis? Autoimmun Rev. 2014;13(7):756–61. doi:10.1016/j.autrev.2014.02.002.
50. Adrianto I, Wang S, Wiley GB, et al. Association of two independent functional risk haplotypes in TNIP1 with systemic lupus erythematosus. Arthritis Rheum. 2012;64(11):3695–705.
51. Watford WT, Hissong BD, Bream JH, et al. Signaling by IL-12 and IL-23 and the immunoregulatory roles of STAT4. Immunol Rev. 2004;202:139–56.
52. Xu M, Mizoguchi I, Morishima N, et al. Regulation of antitumor immune responses by the IL-12 family cytokines, IL-12, IL-23, and IL-27. Clin Dev Immunol. 2010;2010.
53. Hunt KA, Zhernakova A, Turner G, et al. Newly identified genetic risk variants for celiac disease related to the immune response. Nat Genet. 2008;40(4):395–402.
54. Lessard CJ, Adrianto I, Ice JA, et al. Identification of IRF8, TMEM39A, and IKZF3-ZPBP2 as susceptibility loci for systemic lupus erythematosus in a large-scale multiracial replication study. Am J Hum Genet. 2012;90(4):648–60.
55. Gottenberg JE, Cagnard N, Lucchesi C, et al. Activation of IFN pathways and plasmacytoid dendritic cell recruitment in target organs of primary Sjögren's syndrome. Proc Natl Acad Sci U S A. 2006;103:2770–5.
56. Emamian ES, Leon JM, Lessard CJ, et al. Peripheral blood gene expression profiling in Sjögren's syndrome. Genes Immun. 2009;10:285–96.
57. Hom G, Graham RR, Modrek B, et al. Association of systemic lupus erythematosus with C8orf13-BLK and ITGAM-ITGAX. N Engl J Med. 2008;358(9):900–9.
58. Simpfendorfer KR, Olsson LM, Manjarrez Orduño N, et al. The autoimmunity-associated BLK haplotype exhibits cis-regulatory effects on mRNA and protein expression that are prominently observed in B cells early in development. Hum Mol Genet. 2012;21(17):3918–25.
59. Mells GF, Floyd JA, Morley KI, et al. Genome-wide association study identifies 12 new susceptibility loci for primary biliary cirrhosis. Nat Genet. 2011;43(4):329–32.
60. Sawcer S, Hellenthal G, Pirinen M, et al. Genetic risk and a primary role for cell-mediated immune mechanisms in multiple sclerosis. Nature. 2011;476(7359):214–9.
61. McKeigue PM, Carpenter JR, Parra EJ, et al. Estimation of admixture and detection of linkage in admixed populations by a Bayesian approach: application to African-American populations. Ann Hum Genet. 2000;64(Pt 2):171–86.
62. Hjelmervik TO, Petersen K, Jonassen I, et al. Gene expression profiling of minor salivary glands clearly distinguishes primary Sjögren's syndrome patients from healthy control subjects. Arthritis Rheum. 2005;52:1534–44.
63. Perez P, Anaya JM, Aguilera S, et al. Gene expression and chromosomal location for susceptibility to Sjögren's syndrome. J Autoimmun. 2009;33:99–108.
64. Baechler EC, Gregersen PK, Behrens TW. The emerging role of interferon in human systemic lupus erythematosus. Curr Opin Immunol. 2004;16:801–7.
65. Sozzani S, Bosisio D, Scarsi M, et al. Type I interferons in systemic autoimmunity. Autoimmunity. 2010;43:196–203.
66. Fleissig Y, Deutsch O, Reichenberg E, et al. Different proteomic protein patterns in saliva of Sjögren's syndrome patients. Oral Dis. 2009;15:61–8.
67. Hu S, Wang J, Meijer J, et al. Salivary proteomic and genomic biomarkers for primary Sjögren's syndrome. Arthritis Rheum. 2007;56:3588–600.
68. Giusti L, Baldini C, Bazzichi L, et al. Proteome analysis of whole saliva: a new tool for rheumatic diseases–the example of Sjögren's syndrome. Proteomics. 2007;7:1634–43.
69. Ryu OH, Atkinson JC, Hoehn GT, et al. Identification of parotid salivary biomarkers in Sjögren's syndrome by surface-enhanced laser desorption/ionization time-of-flight mass spectrometry and two-dimensional difference gel electrophoresis. Rheumatology (Oxford). 2006;45:1077–86.
70. Hjelmervik TO, Jonsson R, Bolstad AI. The minor salivary gland proteome in Sjögren's syndrome. Oral Dis. 2009;15:342–53.
71. Online Mendelian Inheritance in Man, OMIM®. Johns Hopkins University, Baltimore, MD. MIM Number: MIM 608710: 07/10/14. World Wide Web URL: http://omim.org/608710.

72. Barrett AW. Wegener's granulomatosis of the major salivary glands. J Oral Pathol Med. 2012;41:721–7.
73. Jagiello P, Gencik M, Arning L, et al. New genomic region for Wegener's granulomatosis as revealed by an extended association screen with 202 apoptosis-related genes. Hum Genet. 2004;114:468–77.
74. Xie G, Roshandel D, Sherva R, et al. Association of granulomatosis with polyangiitis (Wegener's) with HLA-DPB1*04 and SEMA6A gene variants: evidence from genome-wide analysis. Arthritis Rheum. 2013;65(9):2457–68.
75. Lyons PA, Rayner TF, Trivedi S, et al. Genetically distinct subsets within ANCA-associated vasculitis. New Eng. J Med. 2012;367:214–23.
76. Chung SA, Xie G, Roshandel D, et al. Meta-analysis of genetic polymorphisms in granulomatosis with polyangiitis (Wegener's) reveals shared susceptibility loci with rheumatoid arthritis. Arthritis Rheum. 2012;64(10):3463–71.
77. Alberici F, Martorana D, Bonatti F, et al. Genetics of ANCA- associated vasculitides: HLA and beyond. Clin Exp Rheumatol. 2014;32(2 Suppl 82):S90–7.
78. Martorana D, Maritati F, Malerba G, et al. PTPN22 R620W polymorphism in the ANCA-associated vasculitides. Rheumatology (Oxford). 2012;51(5):805–12.
79. Mueller A, Holl-Ulrich K, Gross WL. Granuloma in ANCA-associated vasculitides: another reason to distinguish between syndromes? Curr Rheumatol Rep. 2013;15(11):376. doi:10.1007/s11926-013-0376-5. Review.
80. Kelley JM, Monach PA, Ji C, et al. IgA and IgG antineutrophil cytoplasmic antibody engagement of Fc receptor genetic variants influences granulomatosis with polyangiitis. Proc Natl Acad Sci U S A. 2011;108(51):20736–41.
81. Kamisawa T, Funata N, Hayashi Y, et al. A new clinicopathological entity of IgG4-related autoimmune disease. J Gastroenterol. 2003;38(10):982–4.
82. Stone JH, Zen Y, Deshpande V. IgG4-related disease. N Engl J Med. 2012;366(6):539–51.
83. Stone JH, Khosroshahi A, Deshpande V, et al. Recommendations for the nomenclature of IgG4-related disease and its individual organ system manifestations. Arthritis Rheum. 2012;64(10):3061–7.
84. Zhang L, Smyrk TC. Autoimmune pancreatitis and IgG4-related systemic diseases. Int J Clin Exp Pathol. 2010;3:491–504.
85. Deshpande V, Zen Y, Chan JK, et al. Consensus statement on the pathology of IgG4-related disease. Mod Pathol. 2012;25(9):1181–92.
86. Cheuk W, Chan JK. IgG4-related sclerosing disease: a critical appraisal of an evolving clinicopathologic entity. Adv Anat Pathol. 2010;17(5):303–32.
87. Cornec D, Saraux A, Jousse-Joulin S, et al. The differential diagnosis of dry eyes, dry mouth, and parotidomegaly: a comprehensive review. Clin Rev Allergy Immunol. 2014 Jun 21.
88. Yao Q, Wu G, Hoschar A. IgG4-related Mikulicz's disease is a multiorgan lymphoproliferative disease distinct from Sjögren's syndrome: a Caucasian patient and literature review. Clin Exp Rheumatol. 2013;31(2):289–94.
89. Mavragani CP, Fragoulis GE, Rontogianni D, et al. Elevated IgG4 Serum Levels Among Primary Sjögren's Syndrome Patients: do they unmask underlying IgG4-related disease? Arthritis Care Res (Hoboken). 2014;66(5):773–7.
90. Soliotis F, Mavragani CP, Plastiras SC, et al. IgG4-related disease: a rheumatologist's perspective. Clin Exp Rheumatol. 2014;32:724–7. [Epub ahead of print].
91. Yamamoto M, Harada S, Ohara M, et al. Clinical and pathological differences between Mikulicz's disease and Sjögren's Syndrome. Rheumatology (Oxford). 2005;44(2):227–34.
92. Mahajan VS, Mattoo H, Deshpande V, et al. IgG4-related disease. Annu Rev Pathol. 2014;9:315–47. doi:10.1146/annurev-pathol-012513-104708.
93. Perez Alamino R, Espinoza LR, Zea AH. The great mimicker: IgG4-related disease. Clin Rheumatol. 2013;32(9):1267–73.
94. Zen Y, Nakanuma Y. Pathogenesis of IgG4-related disease. Curr Opin Rheumatol. 2011;23(1):114–8.

95. Umemura T, Ota M, Hamano H, et al. Association of autoimmune pancreatitis with cytotoxic T-lymphocyte antigen 4 gene polymorphisms in Japanese patients. Am J Gastroenterol. 2008;103(3):588–94.
96. Umemura T, Ota M, Hamano H, et al. Genetic association of Fc receptor-like 3 polymorphisms with autoimmune pancreatitis in Japanese patients. Gut. 2006;55(9):1367–8.
97. Ota M, Ito T, Umemura T, et al. Polymorphism in the KCNA3 gene is associated with susceptibility to autoimmune pancreatitis in the Japanese population. Dis Markers. 2011;31(4):223–9.
98. Ota M, Katsuyama Y, Hamano H, et al. Two critical genes (HLA-DRB1 and ABCF1) in the HLA region are associated with the susceptibility to autoimmune pancreatitis. Immunogenetics. 2007;59(1):45–52.
99. Chang MC, Chang YT, Tien YW, et al. T-cell regulatory gene CTLA-4 polymorphism/haplotype association with autoimmune pancreatitis. Clin Chem. 2007;53(9):1700–5.
100. Pandey JP. Genetic markers of Immunoglobulin G as potential risk factors for IgG4-related disease. J Rheumatol. 2012;39(10):2048.
101. Spagnolo P, Schwartz DA. Genetic predisposition to sarcoidosis: another brick in the wall. Eur Respir J. 2013;41(4):778–80.
102. Valeyre D, Prasse A, Nunes H, et al. Sarcoidosis. Lancet. 2014;383(9923):1155–67.
103. James DG, Sharma OP. Parotid gland sarcoidosis. Sarcoidosis Vasc Diffuse Lung Dis. 2000;17(1):27–32.
104. Online Mendelian inheritance in man, OMIM®. Johns Hopkins University, Baltimore, MD. MIM Number: MIM 181000: 03/01/2011. World Wide Web URL: http://omim.org/181000. Accessed 7 Sept 2014.
105. Rybicki BA, Iannuzzi MC, Frederick MM, et al. Familial aggregation of sarcoidosis. A case-control etiologic study of sarcoidosis (ACCESS). Am J Respir Crit Care Med. 2001;164(11):2085–91.
106. Rybicki BA, Kirkey KL, Major M, et al. Familial risk ratio of sarcoidosis in African-American sibs and parents. Am J Epidemiol. 2001;153(2):188–93.
107. Fischer A, Grunewald J, Spagnolo P, et al. Genetics of sarcoidosis. Semin Respir Crit Care Med. 2014;35(3):296–306.
108. Sverrild A, Backer V, Kyvik KO, et al. Heredity in sarcoidosis: a registry-based twin study. Thorax. 2008;63(10):894–6.
109. Spagnolo P, Grunewald J. Recent advances in the genetics of sarcoidosis. J Med Genet. 2013;50(5):290–7. doi:10.1136/jmedgenet-2013-101532.
110. Darlington P, Tallstedt L, Padyukov L, et al. HLA-DRB1* alleles and symptoms associated with Heerfordt's syndrome in sarcoidosis. Eur Respir J. 2011;38(5):1151–7.
111. Valentonyte R, Hampe J, Huse K, et al. Sarcoidosis is associated with a truncating splice site mutation in BTNL2. Nat Genet. 2005;37(4):357–64.
112. Adrianto I, Lin CP, Hale JJ, et al. Genome-Wide Association Study of African and European Americans Implicates Multiple Shared and Ethnic Specific Loci in Sarcoidosis Susceptibility. PLoS ONE. 2013;8(9):e43907
113. Cozier Y, Ruiz-Narvaez E, McKinnon C, et al. Replication of genetic loci for sarcoidosis in US black women: data from the Black Women's Health Study. Hum Genet. 2013;132(7):803–10.
114. Salem S, Gros P. Genetic determinants of susceptibility to Mycobacterial infections: IRF8, a new kid on the block. Adv Exp Med Biol. 2013;783:45–80.
115. Alcaïs A, Fieschi C, Abel L, Casanova JL. Tuberculosis in children and adults: two distinct genetic diseases. J Exp Med. 2005;202(12):1617–21.
116. Online Mendelian Inheritance in Man, OMIM®. Johns Hopkins University, Baltimore, MD. MIM Number: MIM 607948: 07/01/2013. World Wide Web URL: http://omim.org/607948.
117. Casanova JL, Abel L. Genetic dissection of immunity to mycobacteria: the human model. Annu Rev Immunol. 2002;20:581–620.

118. Cobat A, Orlova M, Barrera LF, et al. Host genomics and control of tuberculosis infection. Public Health Genomics. 2013;16(1–2):44–9.
119. Ottenhoff TH. New pathways of protective and pathological host defense to mycobacteria. Trends Microbiol. 2012;20(9):419–28.
120. Lei X, Zhu H, Zha L, Wang Y. SP110 gene polymorphisms and tuberculosis susceptibility: a systematic review and meta-analysis based on 10 624 subjects. Infect Genet Evol. 2012;12(7):1473–80.
121. de Albuquerque AC, Rocha LQ, de Morais Batista AH, et al. Association of polymorphism + 874 A/T of interferon-γ and susceptibility to the development of tuberculosis: meta-analysis. Eur J Clin Microbiol Infect Dis. 2012;31(11):2887–95.
122. Zhang Y, Jiang T, Yang X, et al. Toll-like receptor −1, −2, and −6 polymorphisms and pulmonary tuberculosis susceptibility: a systematic review and meta-analysis. PLoS One. 2013;8(5):e63357.
123. Gorlin RJ, Cohen MM, Levin LS. Branchial arch and oro-acral disorders in syndromes of the head and neck. 3rd ed. New York: Oxford University Press, 1990, pp. 641–9.
124. Online Mendelian Inheritance in Man, OMIM®. Johns Hopkins University, Baltimore, MD. MIM Number: MIM 164210: 09/23/2013. World Wide Web URL: http://omim.org/164210.
125. Gorlin RJ, et al. Oculoauriculovertebral dysplasia, in Syndromes of the Head and Neck. 2nd ed. New York: McGraw-Hill; 1976. pp. 546–52.
126. Kaye CI, Martin AO, Rollnick BR, et al. Oculoauriculovertebral anomaly: segregation analysis. Am J Med Genet. 1992;43:913–7.
127. Tasse C, Majewsk F, Bohringer S, et al. A family with autosomal dominant oculo-auriculo-vertebral spectrum. Clin. Dysmorph. 2007;16:1–7.
128. Kelberman D, Tyson J, Chandler DC, et al. Hemifacial microsomia: progress in understanding the genetic basis of a complex malformation syndrome. Hum Genet. 2001;109:638–45.
129. Online Mendelian Inheritance in Man, OMIM®. Johns Hopkins University, Baltimore, MD. MIM Number: MIM 180920: 09/23/2013. World Wide Web URL: http://omim.org/180920.
130. Gorlin RJ, Cohen MM, Levin LS. Other miscellaneous syndromes in Syndromes of the Head and Neck. 3rd ed. New York: Oxford University Press; 1990. p. 897.
131. Entesarian M, Dahlqvist J, Shashi V, et al. FGF10 missense mutations in aplasia of lacrimal and salivary glands (ALSG). Europ J Hum Genet. 2007;15:379–82.
132. Entesarian M, Matsson H, Klar J, et al. Mutations in the gene encoding fibroblast growth factor 10 are associated with aplasia of lacrimal and salivary glands. Nat Genet. 2005;37:125–8.
133. Milunsky JM, Zhao G, Maher TA, et al. LADD syndrome is caused by FGF10 mutations. Clin Genet 2006;69:349–54.
134. Online Mendelian Inheritance in Man, OMIM®. Johns Hopkins University, Baltimore, MD. MIM Number: MIM 149730: 09/10/2008. World Wide Web URL: http://omim.org/149730.
135. Hollister DW, Klein SH, Dejager HJ, et al. The lacrimo-auriculo-dento-digital syndrome. J Pediat. 1973;83:438–44.
136. Levy WJ. Mesoectodermal dysplasia: a new combination of anomalies. Am J Ophthal. 1967;63:978–82.
137. Gorlin, RJ, Cohen MM, Levin, LS. Syndromes with unusual dental findings in Syndromes of the Head and Neck. 3rd ed. New York: Oxford University Press; 1990. p. 868.
138. Francannet C, Vanlieferinghen P, Dechelotte P, et al. LADD syndrome in five members of a three-generation family and prenatal diagnosis. Genet Counsel. 1994;5:85–91.
139. Rohmann E, Brunner HG, Kayserili H, et al. Mutations in different components of FGF signaling in LADD syndrome. Nat Genet. 2006;38:414–7.
140. Milunsky JM, Zhao G, Maher TA, et al. LADD syndrome is caused by FGF10 mutations. Clin Genet. 2006;69:349–54.

141. Online Mendelian inheritance in man, OMIM®. Johns Hopkins University, Baltimore, MD. MIM Number: MIM 600343: 07/13/2012. World Wide Web URL: http://omim.org/600343. Accessed 7 Sept 2014.
142. Seifert G, Thomsen S, Donath K. Bilateral dysgenetic polycystic parotid glands: morphological analysis and differential diagnosis of a rare disease of the salivary glands. Virchows Arch A Path Anat Histol. 1981;390:273–88.
143. Batsakis JG, Bruner JM, Luna MA. Polycystic (dysgenetic) disease of the parotid glands. Arch Otolaryng Head Neck Surg. 1988;114:1146–8.
144. Ficarra G, Sapp JP, Christensen RE et al. Dysgenetic polycystic disease of the parotid gland: report of case. J Oral Maxillofac Surg 1996;54:1246–9.
145. Online Mendelian Inheritance in Man, OMIM®. Johns Hopkins University, Baltimore, MD. MIM Number: MIM 605041: 07/26/2011. World Wide Web URL: http://omim.org/605041. Accessed 7 Sept 2014.
146. Brooke HG. Epithelioma adenoides cysticum. Brit J Derm. 1892;4:269–87.
147. Spiegler E. Ueber Endotheliome der Haut. Arch Derm Syph. 1899;50:163–76.
148. Online Mendelian Inheritance in Man, OMIM®. Johns Hopkins University, Baltimore, MD. MIM Number: MIM132700: 07/26/2011. World Wide Web URL: http://omim.org/132700. Accessed 7 Sept 2014.
149. Online Mendelian inheritance in man, OMIM®. Johns Hopkins University, Baltimore, MD. MIM Number: MIM 601606: 07/26/2011. World Wide Web URL: http://omim.org/601606. Accessed 7 Sept 2014.
150. Merrick Y, Albeck H, Nielsen NH, et al. Familial clustering of salivary gland carcinoma in Greenland. Cancer 1986;57:2097–102.
151. Scheinfeld N, Hu G, Gill M, et al. Identification of a recurrent mutation in the CYLD gene in Brooke-Spiegler syndrome. Exp Derm. 2003;28:539–41.
152. Gutierrez PP, Eggermann T, Holler D, et al. Phenotype diversity in familial cylindromatosis: a frameshift mutation in the tumor suppressor gene CYLD underlies different tumors of skin appendages. J Invest Derm. 2002;119:527–31.
153. Hu G, Onder M, Gill M, et al. A novel missense mutation in CYLD in a family with Brooke-Spiegler syndrome. J Invest Derm. 2003;121:732–4.
154. Saggar S, Chernoff KA, Lodha S, et al. CYLD mutations in familial skin appendage tumours. (Letter) J Med Genet. 2008;45:298–302.
155. Online Mendelian inheritance in man, OMIM®. Johns Hopkins University, Baltimore, MD. MIM Number: MIM 305100: 07/11/2014. World Wide Web URL: http://omim.org/305100. Accessed 7 Sept 2014.
156. Pinheiro M, Freire-Maia N. Christ-Siemens-Touraine syndrome–a clinical and genetic analysis of a large Brazilian kindred. I. Affected females. II. Affected males. III. Carrier detection. Am J Med Genet. 1979;4:113–34.
157. Gorlin RJ, Cohen MM, Levin, LS. Syndromes affecting the skin and mucosa. In: Syndromes of the Head and Neck. 3rd ed. New York: Oxford University Press; 1990. pp. 451–6.
158. Wright JT, Grange DK, Richter MK. Hypohidrotic Ectodermal Dysplasia. 2003 Apr 28 [Updated 2014 May 15]. In: Pagon RA, Adam MP, Ardinger HH, et al., editors. GeneReviews® [Internet]. Seattle (WA): University of Washington, Seattle; 1993-2014. http://www.ncbi.nlm.nih.gov/books/NBK1112/.
159. Zonana J, Gault J, Davies, KJP, et al. Detection of a molecular deletion at the DXS732 locus in a patient with X-linked hypohidrotic ectodermal dysplasia (EDA), with the identification of a unique junctional fragment. Am J Hum Genet. 1993;52:78–84.
160. Pinheiro M, Ideriha MT, Chautard-Freire-Maia EA et al. Christ-Siemens-Touraine syndrome: investigations on two large Brazilian kindreds with a new estimate of the manifestation rate among carriers. Hum Genet. 1981;57:428–31.
161. Monreal AW, Zonana J, Ferguson B. Identification of a new splice form of the EDA1 gene permits detection of nearly all X-linked hypohidrotic ectodermal dysplasia mutations. Am J Hum Genet. 1998;63:380–9.

162. Cambiaghi S, Restano L, Paakkonen K, Caputo R, Kere J. Clinical findings in mosaic carriers of hypohidrotic ectodermal dysplasia. Arch Dermatol 2000;136:217–24.
163. Monreal AW, Ferguson BM, Headon DJ, et al. Mutations in the human homologue of mouse dl cause autosomal recessive and dominant hypohidrotic ectodermal dysplasia. Nat Genet. 1999;22:366–9.
164. Van der Hout AH, Oudesluijs GG, Venema A, et al. Mutation screening of the ectodysplasin-A receptor gene EDAR in hypohidrotic ectodermal dysplasia. Europ J Hum Genet. 2008;16:673–9.
165. Bal E, Baala L, Cluzeau C, et al. Autosomal dominant anhidrotic ectodermal dysplasias at the EDARADD locus. Hum Mutat. 2007;8:703–9.
166. Headon DJ, Emmal SA, Ferguson BM, et al. Gene defect in ectodermal dysplasia implicates a death domain adapter in development. Nature. 2001;414:913–6.
167. Cluzeau C, Hadj-Rabia S, Jambou M, et al. Only four genes (*EDA1, EDAR, EDARADD,* and *WNT10A*) account for 90% of hypohidrotic/anhidrotic ectodermal dysplasia cases. Hum Mutat. 2011;32:70–7.

Premalignant Lesions

Chia-Cheng Li, Zhe Li, Reshma S Menon and Sook-Bin Woo

Head and neck cancer is one of the leading causes of cancer-related morbidity and mortality [1]. In 2008, there were estimated 263,900 newly diagnosed cancer cases from the oral cavity and lips worldwide, with a 48.5% mortality rate [2]. In the United States, head and neck malignancy is the 8th leading cancer in men, with estimated 42,000 newly diagnosed cases and 8,500 deaths in 2014 [3]. Among all the cancer cases from the head and neck region, squamous cell carcinoma (SCC) is the most common malignancy (approximately 90%) [1]. In the United States, the overall 5-year survival rate of head and neck squamous cell carcinoma (HNSCC) is approximately 60% [3]. Specifically, stage I HNSCC has a survival rate of 83%, which decreases drastically to 36% for a stage IV tumor [3]. Moreover, it only takes 3 months in average for HNSCC to double in size [4], implying that a T1 tumor may progress to T3 within 2 years. Because most oral SCC arise from precancerous or dysplastic lesions that are generally visible and readily accessible to treatment, early detection and treatment of such precancerous and dysplastic lesions are likely to reduce progression to invasive SCC and attendant morbidity and mortality. In this chapter, we provide an overview of precancerous lesions from epidemiologic, clinical and histopathologic aspects, and also discuss pathobiological insights from the genetic and epigenetic standpoints. We also highlight novel technologies devoted to early detection of oral epithelial premalignancy.

C. C. Li (✉) · R. S. Menon · S. B. Woo
Department of Oral Medicine, Infection and Immunity, Harvard School of Dental Medicine, 188 Longwood Avenue, Boston, MA 02115, USA
e-mail: Chia-Cheng_Li@hsdm.harvard.edu

C. C. Li · Z. Li
Division of Genetics, Department of Medicine, Brigham and Women's Hospital and Harvard Medical School, Boston, MA, USA

S. B. Woo
Division of Oral Medicine and Dentistry, Brigham and Women's Hospital, Boston, MA, USA

S. T. Sonis (ed.), *Genomics, Personalized Medicine and Oral Disease*,
DOI 10.1007/978-3-319-17942-1_12

Introduction to Oral Epithelial Premalignancy

The World Health Organization (WHO) defined oral epithelial premalignancy as a "morphologically altered tissue in which oral cancer is more likely to occur than in its apparently normal counterpart" [5]. Oral epithelial premalignancy includes the well-recognized clinical conditions discussed below and histopathologically oral epithelial dysplasia (OED). Several clinical conditions in the oral cavity have long been known to be strongly associated with the development of SCC and those with the highest risk include leukoplakia and its variants, proliferative verrucous leukoplakia (PVL), erythroplakia and submucous fibrosis [5]. Of these, leukoplakia is the most common and it is defined as "plaque of questionable risk having excluded (other) known diseases or disorders that carry no increased risk for cancer (Fig. 1a) [5]. According to clinical characteristics, leukoplakia can be further divided into homogenous and non-homogenous groups [6]. The estimated global prevalence of leukoplakia is 2.6 % [7], with 1–5 % of them undergoing malignant transformation annually [8–10]. Depending on individual studies, 40–50 % of leukoplakia exhibits epithelial dysplasia or invasive SCC when biopsied [11–13]. Compared to homogenous leukoplakia, the potential for malignant transformation is significantly higher in non-homogenous leukoplakia, namely erythroleukoplakia (Fig. 1b) and leukoplakia with a verrucous or nodular component [6, 8, 14]. In addition, leukoplakia that occurs at the ventral tongue or floor of mouth is more prone to progress to SCC [8, 15–18]. PVL, a distinctive subset of leukoplakia, exhibits multifocality with a verrucous appearance as well as persistent and progressive clinical behavior [19]. PVL is a high-risk oral premalignancy, and 70–80 % of patients with PVL progress and develop oral SCC within a decade [19–22]. Erythroplakia is an uncommon intraoral condition that presents as a bright red, velvety, granular plaque [6, 23–25]. More than 90 % of erythroplakia cases are already dysplastic or cancerous when biopsied, and there is a 70 % malignant transformation rate [26–28].

Histopathologically, OED is evaluated based on architectural, organizational and cytologic features. OED initiates at the basal and parabasilar cell layer; the atypical

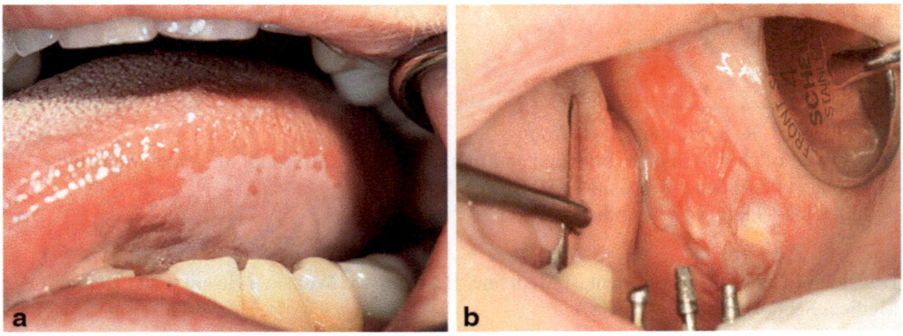

Fig. 1 Clinical appearance of oral epithelial premalignancy. **a** Homogenous leukoplakia on the left ventral tongue. **b** Erythroleukoplakia on the left buccal mucosa

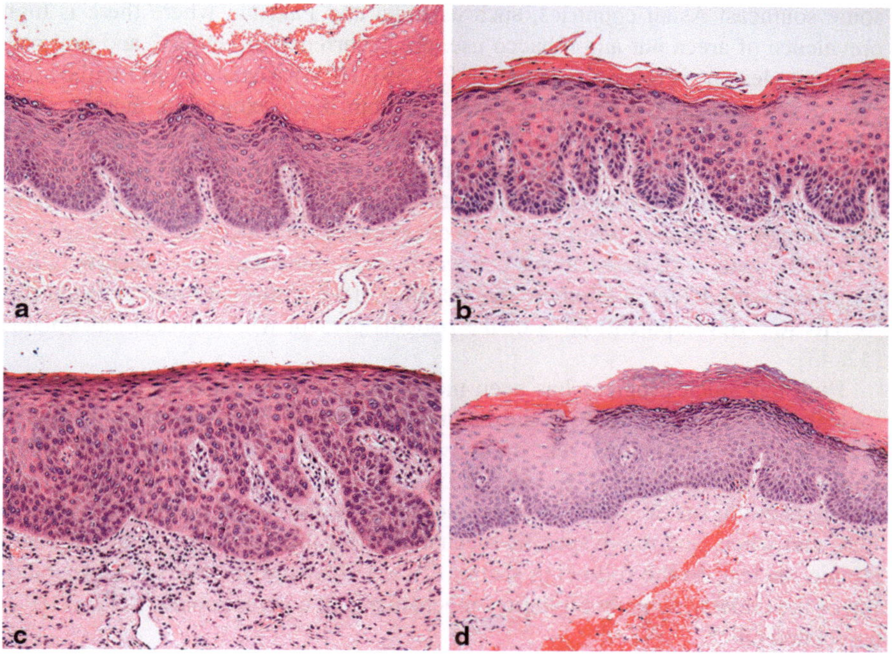

Fig. 2 Histopathologic grade of oral epithelial dysplasia. **a** Mild epithelial dysplasia (H&E, ×200). **b** Moderate epithelial dysplasia (H&E, ×200). **c** Severe epithelial dysplasia (H&E, ×200). **d** Keratosis of unknown significance, not frictional (H&E, ×200)

cytologic features include increased mitotic activity and abnormal mitotic figures, dyskeratosis, enlarged nuclei, increased nuclear-to-cytoplasmic ratio, nuclear hyperchromatism and pleomorphism, and prominent nucleoli [5, 29]. In addition, architectural disorganizations, such as verrucous architecture, bulbous or drop-shaped rete ridges, maturation disarray and discohesion, are taken into consideration [30]. Based on the severity of these atypical changes within the epithelium, the WHO grading system divides OED into three categories, mild, moderate and severe dysplasia, with carcinoma in situ as the malignant extreme (Fig. 2a–c) [31]. A binary system of low-grade and high-grade dysplasia has also been proposed [32, 33]. However, the prognostic value of such grading systems is still controversial. Studies have shown that OED is a dynamic process, and the severity of OED does not consistently correspond to the incidence of malignant transformation [18, 34]. In general, cancerous transformation occurs in 7–36 % of OED, with an annual malignant transformation rate of 1–3 % [8, 11, 18, 35].

Risk factors that initiate oral epithelial premalignancy are similar to those associated with the development of HNSCC, and can be categorized into exogenous factors or endogenous conditions [36]. Well-recognized exogenous factors include cigarette smoking, excessive alcohol consumption and areca nut chewing, as well as infectious pathogens (*e.g.*, high-risk human papillomavirus [HPV]) [37, 38]. In some southeast Asian countries, such as India and Pakistan where there is high prevalence of areca nut and tobacco use, the occurrence of HNSCC and oral premalignant lesions is relatively common [37, 39]. HPV-associated OED will be discussed in detail later in this chapter. Aside from these exogenous factors, some endogenous conditions predispose to the initiation and progression of oral epithelial premalignancy and HNSCC (*e.g.*, Li-Fraumeni syndrome and dyskeratosis congenita). Germline *TP53* mutation is identified in 70 % of families with a history of Li-Fraumeni syndrome, often leading to development of cancer at a young age [40, 41]. Individuals with dyskeratosis congenita, a rare inherited bone marrow failure syndrome with defect in telomere maintenance, have a thousand-fold elevated risk for HNSCC development; 80 % of the patients present with mucosal leukoplakia [42, 43].

Thus far, much progress has been made in the investigation of the pathobiological mechanisms of HNSCC, the diagnosis and characterization of oral epithelial premalignancy still mainly rely on conventional histology in spite of low interobserver reliability [44, 45]. Even if OED is correctly recognized with concurrence among pathologists, it is difficult to predict which lesion would progress to invasive SCC and when, and many specific mechanisms for the development of OED still remain unclear. In the following sections, we will review the current understanding of the underlying pathobiological mechanisms of conventional OED and HPV-associated OED.

Pathobiological Mechanisms of OED

Development of OED and its subsequent malignant transformation are driven by progressive accumulation of multiple genetic and/or epigenetic alterations. In this multi-step process, cells may be genetically or epigenetically altered in the beginning, but still maintain a normal cytologic phenotype; the dysplastic phenotype only becomes apparent with further alterations [46, 47]. OED and HNSCC tend to occur in a synchronous or metachronous fashion, and the concept of "field cancerization" was proposed to explain this phenomenon [48, 49]. In the completely excised specimens of both precancerous and cancerous lesions in the head and neck region, the epithelium adjacent to OED and SCC often shows abnormal hyperkeratosis and acanthosis with minimal cytologic atypia, suggesting that these lateral fields have already become "cancerized" [48–50]. In addition, one case series proposed the concept of "keratosis of unknown significance" based on "field cancerization." It was postulated that although such lesions are not yet accompanied by morphologic alterations, they may represent very early OED (Fig. 2d) [13, 50]. Several mechanisms, such as chromosomal instability (loss of heterozygosity, DNA aneuploidy and telomerase dysfunction), constitutively activated oncogenes, attenuated tumor suppressor genes and altered microenvironment, may contribute to the genetic and epigenetic dysregulations in early OED, leading to alterations of the normal cell cycle and enhancement of proliferative activity.

Chromosomal Instability Predisposes an Individual to Malignant Transformation

Solid tumors, such as HNSCC, are genetically unstable and harbor instability either at the nucleotide or the chromosome level [51]. Chromosomal instability, such as loss of heterozygosity, DNA aneuploidy and telomerase dysfunction, is associated with initiation of premalignancy and its malignant transformation [52, 53]. Current evidence indicates that chromosomal instability is widespread in the genomes of cancer cells [51–53]. Since oncogenesis is a dynamic and continuous process, it is reasonable to hypothesize that OED carries similar chromosomal instability.

Loss of Heterozygosity

Loss of heterozygosity (LOH) has been recognized as a key contributor to oncogenesis [34, 54, 55]. In a population of cells with heterozygous oncogenic mutations, LOH can be caused by various somatic mutational processes, such as point mutation, deletion, mitotic recombination or localized gene conversion, leading to loss of the corresponding wild-type allele within the daughter cells [51, 56–59]. Compared to normal oral mucosa, LOH occurs frequently at chromosomes 3p, 9p, 11q and 17p in OED and HNSCC [46, 47, 55, 60]. These loci contain critical oncogenes, tumor suppressor genes or cell-cycle regulators; for instance, *FHIT*, *CDKN2A*, *CCND1* and *TP53* [61–64]. LOH profiles have been used to predict the risk for progression of OED. LOH at chromosomes 3p and/or 9p is associated with high-risk lesions, and may be useful for surveillance of patients with an elevated risk [65]. In addition, a progression model applying the sequential occurrence of LOH has been proposed to explain the transformation of OED toward HNSCC [49, 66]. Some research has suggested that LOH at chromosomes 3p, 9p and 17p is an early oncogenic event, followed by later alterations at chromosomes 8p, 11q and 18q [46, 49, 50]. However, this progression model is not yet universally accepted, due to the lack of proper controls and longitudinal follow-up [36].

DNA Aneuploidy

Aneuploidy, defined as numeric aberrations of chromosomes, is a dynamic process, and is one of the driving forces for developing malignancies [52, 67]. In a non-dividing human somatic cell, its DNA is diploid, containing 23 pairs of chromosomes. Prior to cell division, the DNA content replicates and becomes tetraploid. During cell division, centrosomes play a key role in organizing the microtubules and balancing segregation of chromosomes into daughter cells; it is such missegregation of chromosomes that leads to DNA aneuploidy [51]. Abnormal centrosomes and altered mitotic spindles have been reported in many human cancers [68, 69]; and DNA aneuploidy is noted in both OED and HNSCC [70]. Approximately 20–45 % of OED showed DNA aneuploidy, and more than half of PVL and HNSCC were aneuploid [71–75]. Interestingly, DNA aneuploidy was observed in a significantly higher frequency in more severe OED as well as OED that eventually underwent malignant transformation [74, 76]. DNA aneuploidy was more commonly seen in OED located on the floor of mouth or ventral tongue [77]. Furthermore, HNSCC with DNA aneuploidy was associated with more aggressive clinical behavior, namely local recurrence and lymph node metastases [78–80]. As such, there is a potential application of DNA aneuploidy as a biomarker to identify the oral epithelial premalignancy at a higher risk of progressing toward HNSCC [81, 82].

Telomerase Dysfunction

Telomeres are G-rich nucleotide sequences at the terminal ends of chromosomes, and are crucial in the maintenance of genomic stability [83]. However, due to the linear structure of human chromosomes and the unidirectional function of DNA polymerase, telomeres may shorten by up to 50–100 base pairs each time cells replicate; the gradual erosion of telomeres leads to aging and senescence of somatic cells [83, 84]. Telomerase, a telomere-specific polymerase, consists of two essential subunits, human telomerase RNA (hTER) and human telomerase reverse transcriptase (hTERT) [85]. Telomerase can synthesize telomeric DNA sequences *de novo* using its RNA (hTER) as the template, thereby maintaining telomere length [86]. There is a fine balance between attrition of telomeres and activation of telomerase; hTERT activity is tightly controlled in human cells [86]. Approximately 90% of human tumors are positive for telomerase expression while most normal tissues lack telomerase activity, suggesting that cancer cells may adopt this mechanism to develop immortality by upregulating the expression level of hTERT [87–89]. In comparison with normal oral mucosa (<30% positivity), both expression and activity of telomerase are enhanced in OED and HNSCC (60–100% positivity) [90–92]. In addition, increased telomerase activity has be demonstrated even in leukoplakia without prominent OED [93]. Multiple studies have shown that telomerase activity is upregulated with increasing grade of OED, indicating a critical role of telomerase hyperactivation in oncogenesis [94–96]. In this regard, a large-scale cohort study with proper control is needed to definitively determine the relationship between telomerase activation and progression and malignant transformation of OED.

Altered Signaling Pathways Enhance Oncogenesis or Deteriorate Tumor Suppression

To initiate a premalignant lesion in the oral cavity, multiple genetic and epigenetic mutations accumulate that allow a cell to acquire dysplastic features, and ultimately become a malignant and invasive entity [47]. In OED and HNSCC, genetic and epigenetic alterations that cause upregulation of oncogenes or down-regulation of tumor suppressor genes are often seen. Epigenetic mechanisms, such as DNA hypermethylation and post-translational histone modifications, result in heritable regulations of gene expression by adjusting chromatin structures without affecting the coding sequences [97]. In normal oral mucosa, unmethylated CpG islands can be found around the promoter regions of tumor suppressor genes, representing an active transcriptional status [97]. On the other hand, hypermethylation of these loci stabilizes and silences tumor suppressor activity. DNA hypermethylation at specific loci (*e.g.*, *CDKN2A*, *MGMT* and *CDH1*) has been frequently detected in OED and HNSCC [98–100]. In addition, "normal-appearing" tissues adjacent to HNSCC also demonstrate similar hypermethylation patterns, suggesting that DNA methylation occurs early in carcinogenesis (see "field cancerization" above) [98, 101]. Beyond these genetic and epigenetic mutations, microRNAs (miRNAs) also play important roles in regulating gene expression post-transcriptionally [102]. Altered expression of miRNAs, such as miR-21, miR-345 and miR-18b, has been linked to the initiation and progression of oral precancerous and cancerous epithelium [103, 104]. However, the detailed regulatory mechanisms have not been elucidated yet. In the next section, we discuss the current understanding of the major signaling pathways that are closely related to the development of oral epithelial premalignancy.

TP53 Signaling Pathway

Tumor suppressor p53 protein, encoded by *TP53* gene on chromosome 17p, is considered a "guardian of the genome." In addition to cell cycle regulation, p53 is involved in various physiologic functions within a cell, such as cell differentiation, DNA repair and apoptosis [105]. Here we focus the discussion on the role of p53 in oncogenesis. Protein p53 can be activated by various forms of exogenous or endogenous cellular stress, such as viral replication, hypoxia, DNA damage, and oncogene activation [106]. *TP53* mutation or loss of functional p53, commonly seen at a hot spot region at codon 245 and codon 248, results in dysregulation of cell cycle and altered reactions toward cellular stress, subsequently leading to genomic instability [34, 107]. In normal oral mucosa, wild-type p53 is expressed in the basal cell layer [108]. *TP53* mutations were found in 50–80 % of HNSCC cases, and were significantly correlated with decreased overall survival rate [109–112]. Approximately 45 % of oral leukoplakia with no cytologic atypia also showed upregulation of mutated p53 [113]. In OED, suprabasal p53 expression has been noted [114], and both the percentage and intensity of mutated p53 expression were positively correlated with increasing histopathologic grade of OED [115, 116]. However, it is still controversial as to whether p53 can be used as a predictive marker for malignant progression [117–120].

Other members of the p53 family, such as p63, may also play important pathobiological roles in OED and HNSCC. Protein p63, encoded by *TP63* gene, which is located on chromosome 3q, is the homolog of p53 and shares structural and functional similarities [121]. Protein p63 coordinates with p53 to regulate cell cycle and induce apoptosis [121]. In the normal oral mucosa, p63 is expressed in the basal cell layer; this extends to the spinous layer in the dysplastic epithelium, similar to the expression pattern of p53 [116, 122]. On average, p63 overexpression was demonstrated in 10 % of homogenous leukoplakia, 5 % of nodular or speckled leukoplakia, nearly 20 % of erythroleukoplakia, and the majority of HNSCC (64.4 %) [112, 113].

However, the mechanism by which these proteins function is unclear and their prognostic value are still poorly understood.

Protein p21, a cyclin-dependent kinase (CDK) inhibitor, is the product of gene *WAF1*, *CIP1* or *SDI1* [124]. In response to DNA damages, wild-type p53 binds to the promoter region of *WAF1/CIP1* gene, inducing p21 expression which inactivates CDKs and arrests cell cycle progression [124]. Moreover, p21 is involved in cell differentiation and cell senescence, and can be seen as a tumor suppressor [125]. In HNSCC, there is no significant correlation between p21 expression and p53 accumulation [126, 127]. Although altered p21 expression in OED and HNSCC has been reported, the correlation between the degree of p21 overexpression and disease progression is still uncertain [119, 126, 128–130]. To elucidate the detailed mechanism of these cell cycle regulators in OED, further studies to dissect the corresponding signaling pathways are necessary.

Cyclin D1 Signaling Pathway

The transitions in normal cell cycle are regulated by cyclin-CDK complexes, and alteration of these protein complexes may result in failure to control cell proliferation, leading to tumor formation [131]. In many human cancers, including HNSCC, cyclin D1 is a well-known proto-oncogene, and overexpression of cyclin D1 is associated with oncogenesis. Cyclin D1 protein, encoded by *CCND1* gene on the chromosome 11q, is one of the critical cell cycle regulators [131, 132]. When complexed with CDKs (*e.g.,* CDK4 and CDK6), cyclin D1 is able to phosphorylate retinoblastoma protein (pRb) to release transcription factor E2F and activate downstream genes (*e.g., MYC*) for cell cycle progression, specifically progression from G1 to S phases [131, 133]. Overexpression of cyclin D1 has been observed in 16–65% of HNSCC, and upregulation of cyclin D1 is associated with poor clinical outcomes [134–136]. HNSCC patients who exhibited overexpression of cyclin D1 demonstrated a 5-year disease-free survival rate of 39%; low expression and no expression of cyclin D1 had survival rate of 47% and 80% respectively [134, 136]. Furthermore, increase in cyclin D1 expression correlates with the severity of the histopathologic grade of OED [123, 137]. During disease progression, overexpression of cyclin D1 was more prominent in oral leukoplakia with OED than those without OED [138]. By applying a pathway-based approach to investigate single-nucleotide polymorphisms (SNPs) of cyclin D1, studies revealed that some SNPs, such as G870A, were associated with the development of OED [139, 140]. All these findings point to cyclin D1 mutation as an early event in the conversion of normal epithelium into dysplastic epithelium; furthermore, cyclin D1 may serve as a biomarker of oral carcinogenesis [141].

Retinoblastoma Signaling Pathway

In a normal cell, cell cycle is precisely regulated, and cell cycle dysregulation is a fundamental feature of carcinogenesis [142]. Cyclin-CDK complex phosphorylates pRb to release E2F, allowing cell cycle progression from G1 phase to S phase [133]. On the other hand, CDK inhibitors, such as p16 (on chromosome 9p), can bind to CDKs and inhibit phosphorylation of pRb (on chromosome 13q), leading to G1 phase arrest [131, 133]. Both pRb and p16 are tumor suppressors, and LOH or altered expression of these two proteins has been reported in a wide range of human cancers [143].

By applying immunohistochemical techniques to study the expression patterns of pRb and p16, loss of pRb was noted at the transition from hyperplastic to dysplastic epithelium, suggesting a potential role of pRb/p16 signaling pathway in early carcinogenesis [144]. 93–100% of normal mucosa exhibits pRb and p16 expression; loss of pRb was seen in 56.4–66% of HNSCC, and 64% of OED, and lack of p16 expression in 63–67.9% of HNSCC and 59% in OED [145, 146]. However, data regarding the detection of p16 in OED and HNSCC are contradictory. Some studies showed overexpression of p16 was correlated with increasing degree of OED [147, 148]. On the other hand, a strong correlation between decreased p16 expression and severity of OED has also been reported [128, 149, 150]. It should be noted that in these earlier studies, conventional oral premalignancy/malignancy and high-risk HPV-associated lesions were not investigated separately, perhaps leading to these conflicting results.

Receptor Tyrosine Kinase Signaling Pathway

Multiple receptor tyrosine kinase transduction pathways, such as epidermal growth factor (EGF), platelet-derived growth factor (PDGF) and fibroblast growth factor (FGF), interact with each other in a regulatory network, and control cell morphology, survival, and proliferation [34]. The EGF receptor (EGFR) pathway in particular, a growth factor ligand-dependent transduction cascade, is critical for the homoeostasis of structure and function of epithelium [151]. EGFR, a transmembrane tyrosine kinase receptor, can activate many downstream signaling pathways (*e.g.*, mitogen-activated protein kinase [MAPK] and phosphoinositide 3-kinase [PI3K]/Akt) when bound with the EGF ligands [151, 152]. In addition to the growth factor ligand-dependent pathways, direct activation of downstream signaling pathways can also significantly influence cell survival and cell proliferation [152]. Aberrant activation of EGFR pathway is commonly seen in HNSCC, and is significantly associated with resistance to chemotherapeutic medications [153–155]. In OED, EGFR expression has been reported to be upregulated at both mRNA and protein levels, and was positively correlated with the histopathologic grade of premalignant lesions [153, 156–159]. Furthermore, amplification of EGFR gene in oral epithelial premalignancy was associated with a higher risk of malignant transformation within 10 years [160].

The downstream MAPK and PI3K pathways are critical for regulation of cell differentiation, proliferation, and survival [161]. Once activated by phosphorylating the kinase domains, MAPK translocates into the nucleus to activate downstream genes [162]. PI3K, a lipid kinase and another critical player in carcinogenesis, phosphorylates phosphatidylinositol (PI) within the cell membrane, further activating protein kinase B (Akt) to transmit cell survival signals [163]. Gene amplifications of the MAPK and PI3K signaling pathways were found in 30 % of high-grade OED and 43.5 % of HNSCC, while no gene amplifications were detected in the low-grade OED that never progressed [164]. PI3K/Akt signaling pathway is activated in OED and HNSCC, with 2.5 to 11-fold upregulation of PI3K at the mRNA level compared to normal oral mucosa; phosphorylated Akt protein was present only in dysplastic or cancerous oral epithelium, and correlated with poor prognosis [163, 165]. In addition, increased expression of PI synthase paralleled the histopathologic grade of OED [166]. The evidence supports a role of these receptor tyrosine kinase pathways in tumorogenesis of the head and neck region.

"Super Competition" as the Potential Mechanism to Pioneer and Dominate the Precancerized Fields

At the very earliest stage of dysplasia, the initial mutations at either the genetic or epigenetic level are difficult to track because tissues have yet to reveal phenotypic evidence of dysplasia [50, 167]. As a result, the detailed molecular mechanisms of OED remain largely elusive, especially regarding how the precancerous microenvironment is established as well as how a mutated clone pioneers and dominates the pre-cancerized zone. Currently, the most appealing hypothesis is "super competition", which was initially identified in Drosophila [168, 169]. It is generally accepted that mutations occur in stem cells; the further differentiated clones carry the same genetic or epigenetic alterations and form patches. Distinct clones of dysplastic cells compete for limited resources based on their differential fitness, namely survival capacity and replication potential. With selection, the "winning clones" with certain mutations that confer the highest fitness dominate the cancerized field, while the "losing clones" become minority by continuous apoptosis, but yet do not completely disappear, therefore maintaining tumor heterogeneity [168, 170]. As cancer advances, the selection for dominant clones may be a continuous and dynamic process. However, at present, it is not clear which genetic and/or epigenetic alterations contribute to this competitive activity.

Current Understanding of HPV-associated OED

High-risk human papillomavirus (HPV), such as HPV-16, -31, and -33, have high oncogenic potential through activation of viral E6 and E7 oncoproteins which disrupt the functions of tumor suppressor genes [171]. E6 binds and degrades p53, preventing initiation of programmed cell death; E6 also activates telomerase to promote cell cycle progression [171]. E7 binds and inactivates pRb, releasing E2F which also promotes cell cycle progression, as well as the overexpression of p16, a reliable surrogate marker for HPV-related oncogenic events [171]. In dysplastic lesions, HPV is present as a persistent infection in the form of DNA episomes while integration, resulting in development of HNSCC [171].

In general, high-risk HPV has been detected in 61 % of OED, which is significantly higher than that in normal oral mucosa (36 %) [172]. In addition, compared to the normal oral mucosa (2.8 %), the prevalence of high-risk HPV in leukoplakia and oral SCC reach 31.8 and 33.8 %, respectively [173]. However, most of the studies evaluated only the presence of high-risk HPV without the presence of p16 which is of paramount importance since all HPV infections including high-risk types, resolve spontaneously within 1 year [174]. HPV-associated OED had been previously reported as koilocytic dysplasia or bowenoid lesions [175, 176]. Similar to other dysplastic leukoplakias, this condition presents as a sharply-demarcated white plaque, most commonly on the lateral or ventral tongue and floor of mouth [38]. It has a four-fold predilection for males, and is most common in the sixth decade of life [38, 175]. These lesions have distinct histopathology. They are characterized by brightly eosinophilic parakeratin (although sometimes orthokeratin is present), and importantly, prominent karyorrhectic and apoptotic cells within the full thickness of the epithelium that also exhibits severe or high grade epithelial dysplasia (Fig. 3a, b) [38, 175]. Koilocytes are only present in small numbers. In all cases, p16 is positive

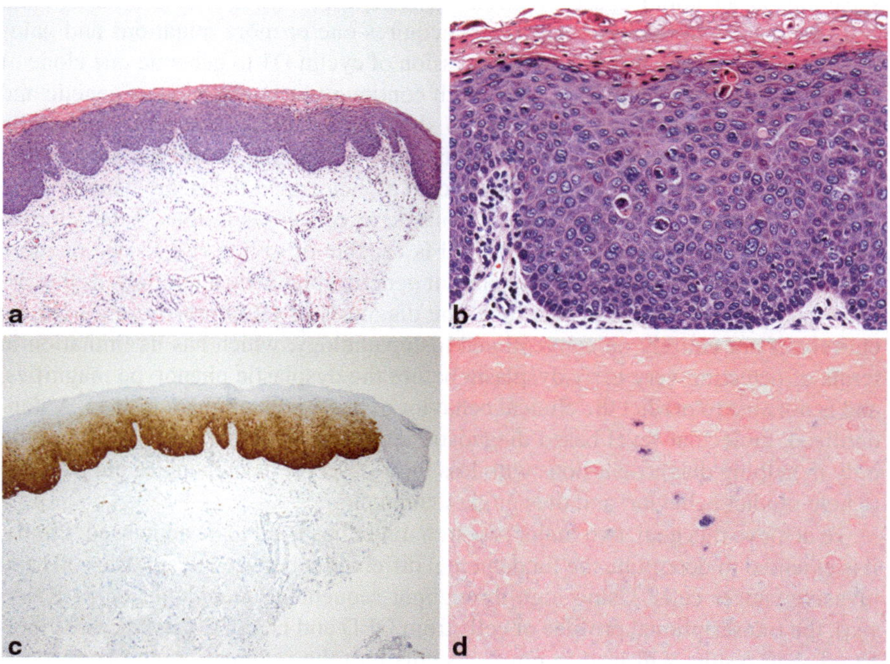

Fig. 3 HPV-associated oral epithelial dysplasia. **a** Histopathology of HPV-associated oral epithelial dysplasia (H&E, × 100). **b** Histopathology of HPV-associated oral epithelial dysplasia characterized by karyorrhexis and apoptosis (H&E, × 400). **c** Protein p16 positivity in a continuous band with an abrupt transition to nondysplastic mucosa (IHC, × 100). **d** Presence of high-risk HPV in dot-like fashion within nuclei (ISH, × 400)

in a continuous linear band involving the full thickness of the epithelium (Fig. 3c) and all cases are positive by *in-situ* hybridization (ISH) for high-risk HPV subtypes (Fig. 3d) [38]. Studies have shown variable results in the expression pattern of p16 and detection of high-risk HPV in OED, mainly because all cases of OED were examined, rather than those with specific histopathology as noted above. Furthermore, different criteria were employed, and p16 was reported as positive (0–28 %) when staining was patchy rather than a continuous band [177–179]. The prognosis of HPV-associated OED is unclear, and more well-controlled clinical studies are warranted to clarify the clinical behavior and prognosis of this entity. In one study of 20 cases, 3 of 11 cases showed progression of residual disease, and 2 of 20 (10 %) developed invasive SCC [38]. Whether these tumors will have a better prognosis as in HPV-associated oropharyngeal cancers is still unclear [180–182].

Future Perspectives and Conclusion

This chapter aims at providing a review on the current understanding of oral epithelial premalignancy from both clinical and basic research perspectives. Generally speaking, in an initial phase of carcinogenesis, a cell (most likely a tissue stem cell that resides in the basal cell layer) acquires one or more mutations and gains proliferative capacity through overexpression of cyclin D1 to generate one clone of altered differentiated cells [50, 183]. With continuous accumulation of genetic and epigenetic mutations, the clones start to show LOH, undergo rampant expansion and replace the surrounding normal epithelium without an actual invasion [184]. These clonally involved regions, also known as cancerized fields, are usually more extensive than the clinically visible dysplastic or cancerous lesions. After surgical resection, the remaining cancerized field is capable of giving rise to one or more secondary field tumors which share similar genetic/epigenetic alterations as the primary one [48]. However, at this time, the diagnosis of OED mostly relies almost entirely, if not entirely on conventional histopathology, which has its limitation in terms of detecting very early dysplasia before the dysplastic phenotype magnifies, and is not able to predict the clinical behavior of these premalignant lesions. A standardized, molecular-level based diagnostic approach would be very welcome, not only in helping diagnose lesions with less inter-observer discrepancies but possibly to help predict behavior, and thus guide treatment.

To achieve this goal, two major research directions need to be addressed. Firstly, it is essential to determine the fundamental differences between dysplastic cells and invasive cancer cells. Using high-throughput sequencing techniques (*e.g.,* RNA-seq), the transcriptional profiles of cells from OED and HNSCC can be easily compared. The results will lead to novel insights into the progression and conversion of the premalignancy to frank malignancy. In 2011, two research teams compared genomic profiles of HNSCC and corresponding normal mucosa using massively parallel sequencing (a.k.a. next-generation sequencing), and identified a novel mutation in the *NOTCH* gene, in addition to the well-known ones (*e.g.,* those in *TP53* and *CDKN2A*) [111, 185]. A similar strategy can also be applied in a carefully-designed cohort study, to systemically analyze the gene expression profiles of OED with different histopathologic grades, as well as of tumors that arise from the corresponding cancerized field of the same patient.

The second clinically significant direction would be the identification of biomarkers that correlate with disease stages, and that can be further used to predict the clinical behavior of OED. This research direction departs from genome-wide screenings, and focuses instead on alterations at the protein level. This will be especially useful in terms of patient surveillance by using noninvasive approaches to obtain samples, such as saliva [186]. Preventive therapies can be further developed to target such proteins and their pathways. Targeting cell competition regulators (*e.g., MYC*), for instance, can be just such a novel therapeutic approach [187]. In this regard, a systemic screening for biomarkers incorporating proteomic, transcriptomic, or even methylomics needs to be accomplished. The results will bring the clinical diagnostics to the next level. The integration of insights from both clinical research and basic scientific research is expected to offer a thorough understanding and effective management of oral epithelial premalignancy.

References

1. Warnakulasuriya S. Global epidemiology of oral and oropharyngeal cancer. Oral Oncol. 2009;45:309–16.
2. Jemal A, et al. Global cancer statistics. CA Cancer J Clin. 2011;61:69–90.
3. Siegel R, Ma J, Zou Z, Jemal A. Cancer Statistics, 2014. CA Cancer J Clin. 2014;64:9–29.
4. Goy J, Hall SF, Feldman-Stewart D, Groome PA. Diagnostic delay and disease stage in head and neck cancer: a systematic review. Laryngoscope. 2009;119:889–98.
5. Warnakulasuriya S, Johnson NW, van der Waal I. Nomenclature and classification of potentially malignant disorders of the oral mucosa. J Oral Pathol Med. 2007;36:575–80.
6. Sciubba JJ. Oral cancer. The importance of early diagnosis and treatment. Am J Clin Dermatol. 2001;2:239–51.
7. Petti S. Pooled estimate of world leukoplakia prevalence: a systematic review. Oral Oncol. 2003;39:770–80.
8. Silverman S, Jr, Gorsky M, Lozada F. Oral leukoplakia and malignant transformation. A follow-up study of 257 patients. Cancer. 1984;53:563–8.
9. Hsue SS, et al. Malignant transformation in 1458 patients with potentially malignant oral mucosal disorders: a follow-up study based in a Taiwanese hospital. J Oral Pathol Med. 2007;36:25–9.
10. Brouns E, et al. Malignant transformation of oral leukoplakia in a well-defined cohort of 144 patients. Oral Dis. 2014;20:19–24.
11. Schepman KP, van der Meij EH, Smeele LE, van der Waal I. Malignant transformation of oral leukoplakia: a follow-up study of a hospital-based population of 166 patients with oral leukoplakia from The Netherlands. Oral Oncol. 1998;34:270–5.
12. Lee JJ, et al. Carcinoma and dysplasia in oral leukoplakias in Taiwan: prevalence and risk factors. Oral Surg Oral Med Oral Pathol Oral Radiol Endod. 2006;101:472–80.
13. Woo SB, Grammer RL, Lerman MA. Keratosis of unknown significance and leukoplakia: a preliminary study. Oral Surg Oral Med Oral Pathol Oral Radiol. 2014;118:713–24.
14. Kramer IR, El-Labban N, Lee KW. The clinical features and risk of malignant transformation in sublingual keratosis. Br Dent J. 1978;144:171–80.

15. Waldron CA, Shafer WG. Leukoplakia revisited. A clinicopathologic study. 3256 oral leukoplakias. Cancer. 1975;36:1386–92.
16. Lumerman H, Freedman P, Kerpel S. Oral epithelial dysplasia and the development of invasive squamous cell carcinoma. Oral Surg Oral Med Oral Pathol Oral Radiol Endod. 1995;79:321–9.
17. Liu W, et al. Oral cancer development in patients with leukoplakia–clinicopathological factors affecting outcome. PLoS One. 2012;7:e34773.
18. Dost F, Le Cao K, Ford PJ, Ades C, Farah CS. Malignant transformation of oral epithelial dysplasia: a real-world evaluation of histopathologic grading. Oral Surg Oral Med Oral Pathol Oral Radiol. 2014;117:343–52.
19. Hansen LS, Olson JA, Silverman S, Jr. Proliferative verrucous leukoplakia. A long-term study of thirty patients. Oral Surg Oral Med Oral Pathol. 1985;60:285–98.
20. Silverman S, Jr, Gorsky M. Proliferative verrucous leukoplakia: a follow-up study of 54 cases. Oral Surg Oral Med Oral Pathol Oral Radiol Endod. 1997;84:154–7.
21. Cabay RJ, Morton TH, Jr, Epstein JB. Proliferative verrucous leukoplakia and its progression to oral carcinoma: a review of the literature. J Oral Pathol Med. 2007;36:255–61.
22. Bagan J, Scully C, Jimenez Y, Martorell M. Proliferative verrucous leukoplakia: a concise update. Oral Dis. 2010;16:328–32.
23. Kramer IR, Lucas RB, Pindborg JJ, Sobin LH. Definition of leukoplakia and related lesions: an aid to studies on oral precancer. Oral Surg Oral Med Oral Pathol. 1978;46:518–39.
24. Lay KM, Sein K, Myint A, Ko SK, Pindborg JJ. Epidemiologic study of 600 villagers of oral precancerous lesions in Bilugyun: preliminary report. Community Dent Oral Epidemiol. 1982;10:152–5.
25. Zain RB, et al. A national epidemiological survey of oral mucosal lesions in Malaysia. Community Dent Oral Epidemiol. 1997;25:377–83.
26. Shafer WG, Waldron CA. Erythroplakia of the oral cavity. Cancer. 1975;36:1021–8.
27. Amagasa T, et al. A study of the clinical characteristics and treatment of oral carcinoma in situ. Oral Surg Oral Med Oral Pathol. 1985;60:50–5.
28. Reichart PA, Philipsen HP. Oral erythroplakia–a review. Oral Oncol. 2005;41:551–61.
29. Woo SB. Oral pathology: a comprehensive atlas and text. 1st edn. USA: Saunders, an imprint of Elsevier Inc.; 2012.
30. Eversole LR. Dysplasia of the upper aerodigestive tract squamous epithelium. Head Neck Pathol. 2009;3:63–8.
31. Speight PM. Update on oral epithelial dysplasia and progression to cancer. Head Neck Pathol. 2007;1:61–6.
32. Kujan O, et al. Evaluation of a new binary system of grading oral epithelial dysplasia for prediction of malignant transformation. Oral Oncol. 2006;42:987–93.
33. Gale N, Zidar N, Poljak M, Cardesa A. Current views and perspectives on classification of squamous intraepithelial lesions of the head and neck. Head Neck Pathol. 2014;8:16–23.
34. Lingen MW, et al. Genetics/epigenetics of oral premalignancy: current status and future research. Oral Dis. 2011;17(Suppl. 1):7–22.
35. Pindborg JJ, Daftary DK, Mehta FS. A follow-up study of sixty-one oral dysplastic precancerous lesions in Indian villagers. Oral Surg Oral Med Oral Pathol. 1977;43:383–90.
36. Li CC, Woo SB. Understanding the pathobiology of head and neck squamous cell carcinoma. Curr Oral Health Rep. 2014;1:196–203.
37. Arduino PG, Bagan J, El-Naggar AK, Carrozzo M. Urban legends series: oral leukoplakia. Oral Dis. 2013;19:642–59.
38. Woo SB, Cashman EC, Lerman MA. Human papillomavirus-associated oral intraepithelial neoplasia. Mod Pathol. 2013;26:1288–97.
39. Warnakulasuriya S. Living with oral cancer: epidemiology with particular reference to prevalence and life-style changes that influence survival. Oral Oncol. 2010;46:407–10.
40. Friedlander PL. Genomic instability in head and neck cancer patients. Head Neck. 2001;23:683–91.

41. McBride KA, et al. Li-Fraumeni syndrome: cancer risk assessment and clinical management. Nat Rev Clin Oncol. 2014;11:260–71.
42. Alter BP, Giri N, Savage SA, Rosenberg PS. Cancer in dyskeratosis congenita. Blood. 2009;113:6549–57.
43. Scully C, Langdon J, Evans J. Marathon of eponyms: 26 Zinsser-Engman-Cole syndrome (Dyskeratosis congenita). Oral Dis. 2012;18:522–3.
44. Kujan O, et al. Why oral histopathology suffers inter-observer variability on grading oral epithelial dysplasia: an attempt to understand the sources of variation. Oral Oncol. 2007;43:224–31.
45. Abbey LM, et al. The effect of clinical information on the histopathologic diagnosis of oral epithelial dysplasia. Oral Surg Oral Med Oral Pathol Oral Radiol Endod. 1998;85:74–7.
46. Califano J, et al. Genetic progression model for head and neck cancer: implications for field cancerization. Cancer Res. 1996;56:2488–92.
47. Califano J, et al. Genetic progression and clonal relationship of recurrent premalignant head and neck lesions. Clin Cancer Res. 2000;6:347–52.
48. Slaughter DP, Southwick HW, Smejkal W. Field cancerization in oral stratified squamous epithelium; clinical implications of multicentric origin. Cancer. 1953;6:963–8.
49. Braakhuis BJ, Tabor MP, Kummer JA, Leemans CR, Brakenhoff RH. A genetic explanation of Slaughter's concept of field cancerization: evidence and clinical implications. Cancer Res. 2003;63:1727–30.
50. Tabor MP, et al. Persistence of genetically altered fields in head and neck cancer patients: biological and clinical implications. Clin Cancer Res. 2001;7:1523–32.
51. Lengauer C, Kinzler KW, Vogelstein B. Genetic instabilities in human cancers. Nature. 1998;396:643–9.
52. Sen S. Aneuploidy and cancer. Curr Opin Oncol. 2000;12:82–8.
53. Oulton R, Harrington L. Telomeres, telomerase, and cancer: life on the edge of genomic stability. Curr Opin Oncol. 2000;12:74–81.
54. El-Naggar AK, et al. Localization of chromosome 8p regions involved in early tumorigenesis of oral and laryngeal squamous carcinoma. Oncogene. 1998;16:2983–7.
55. Partridge M, et al. Allelic imbalance at chromosomal loci implicated in the pathogenesis of oral precancer, cumulative loss and its relationship with progression to cancer. Oral Oncol. 1998;34:77–83.
56. Cavenee WK, et al. Expression of recessive alleles by chromosomal mechanisms in retinoblastoma. Nature. 1983;305:779–84.
57. Hall JG. Review and hypotheses: somatic mosaicism: observations related to clinical genetics. Am J Hum Genet. 1988;43:355–63.
58. White VA, McNeil BK, Horsman DE. Acquired homozygosity. (isodisomy) of chromosome 3 in uveal melanoma. Cancer Genet Cytogenet. 1998;102:40–5.
59. Groden J, Nakamura Y, German J. Molecular evidence that homologous recombination occurs in proliferating human somatic cells. Proc Natl Acad Sci U S A. 1990;87:4315–9.
60. Rosin MP, et al. Use of allelic loss to predict malignant risk for low-grade oral epithelial dysplasia. Clin Cancer Res. 2000;6:357–62.
61. Papadimitrakopoulou V, et al. Frequent inactivation of p16INK4a in oral premalignant lesions. Oncogene. 1997;14:1799–803.
62. Tanimoto K, et al. Abnormalities of the FHIT gene in human oral carcinogenesis. Br J Cancer. 2000;82:838–43.
63. Jiang WW, Fujii H, Shirai T, Mega H, Takagi M. Accumulative increase of loss of heterozygosity from leukoplakia to foci of early cancerization in leukoplakia of the oral cavity. Cancer. 2001;92:2349–56.
64. Tsui IF, Rosin MP, Zhang L, Ng RT, Lam WL. Multiple aberrations of chromosome 3p detected in oral premalignant lesions. Cancer Prev Res (Phila). 2008;1:424–9.
65. Zhang L, et al. Loss of heterozygosity (LOH) profiles–validated risk predictors for progression to oral cancer. Cancer Prev Res (Phila). 2012;5:1081–9.

66. El-Naggar AK, et al. Sequential loss of heterozygosity at microsatellite motifs in preinvasive and invasive head and neck squamous carcinoma. Cancer Res. 1995;55:2656–9.
67. Magennis DP. Nuclear DNA in histological and cytological specimens: measurement and prognostic significance. Br J Biomed Sci. 1997;54:140–8.
68. Pihan GA, et al. Centrosome defects and genetic instability in malignant tumors. Cancer Res. 1998;58:3974–85.
69. Lingle WL, Lutz WH, Ingle JN, Maihle NJ, Salisbury JL. Centrosome hypertrophy in human breast tumors: implications for genomic stability and cell polarity. Proc Natl Acad Sci U S A. 1998;95:2950–5.
70. Klanrit P, et al. DNA ploidy in proliferative verrucous leukoplakia. Oral Oncol. 2007;43:310–6.
71. Diwakar N, Sperandio M, Sherriff M, Brown A, Odell EW. Heterogeneity, histological features and DNA ploidy in oral carcinoma by image-based analysis. Oral Oncol. 2005;41:416–22.
72. Abou-Elhamd KE, Habib TN. The flow cytometric analysis of premalignant and malignant lesions in head and neck squamous cell carcinoma. Oral Oncol. 2007;43:366–72.
73. Pentenero M, et al. DNA aneuploidy and dysplasia in oral potentially malignant disorders: association with cigarette smoking and site. Oral Oncol. 2009;45:887–90.
74. Torres-Rendon A, Stewart R, Craig GT, Wells M, Speight PM. DNA ploidy analysis by image cytometry helps to identify oral epithelial dysplasias with a high risk of malignant progression. Oral Oncol. 2009;45:468–73.
75. Gouvea AF, et al. High incidence of DNA ploidy abnormalities and increased Mcm2 expression may predict malignant change in oral proliferative verrucous leukoplakia. Histopathology. 2013;62:551–62.
76. Donadini A, et al. Oral cancer genesis and progression: DNA near-diploid aneuploidization and endoreduplication by high resolution flow cytometry. Cell Oncol. 2010;32:373–83.
77. Islam MN, Kornberg L, Veenker E, Cohen DM, Bhattacharyya I. Anatomic site based ploidy analysis of oral premalignant lesions. Head Neck Pathol. 2010;4:10–4.
78. Baretton G, et al. Prognostic significance of DNA ploidy in oral squamous cell carcinomas. A retrospective flow and image cytometric study with comparison of DNA ploidy in excisional biopsy specimens and resection specimens, primary, tumors, and lymph node metastases. Oral Surg Oral Med Oral Pathol Oral Radiol Endod. 1995;79:68–76.
79. Rubio Bueno P, Naval Gias L, Garcia Delgado R, Domingo Cebollada J, Diaz Gonzalez FJ. Tumor DNA content as a prognostic indicator in squamous cell carcinoma of the oral cavity and tongue base. Head Neck. 1998;20:232–9.
80. Hemmer J, Nagel E, Kraft K. DNA aneuploidy by flow cytometry is an independent prognostic factor in squamous cell carcinoma of the oral cavity. Anticancer Res. 1999;19:1419–22.
81. Sperandio M, et al. Predictive value of dysplasia grading and DNA ploidy in malignant transformation of oral potentially malignant disorders. Cancer Prev Res (Phila). 2013;6:822–31.
82. Siebers TJ, et al. Chromosome instability predicts the progression of premalignant oral lesions. Oral Oncol. 2013;49:1121–8.
83. Greider CW. Telomere length regulation. Annu Rev Biochem. 1996;65:337–65.
84. Watson JD. Origin of concatemeric T7 DNA. Nat New Biol. 1972;239:197–201.
85. Feng J, et al. The RNA component of human telomerase. Science. 1995;269:1236–41.
86. Greider CW, Blackburn EH. Identification of a specific telomere terminal transferase activity in Tetrahymena extracts. Cell. 1985;43:405–13.
87. De Lange T. Telomere dynamics and genome instability in human cancer. Cold Spring Harb Monogr Ser. 1995;29:265–94.
88. Nakayama J, et al. Telomerase activation by hTRT in human normal fibroblasts and hepatocellular carcinomas. Nat Genet. 1998;18:65–8.
89. Shay JW, Wright WE. Telomeres and telomerase in normal and cancer stem cells. FEBS Lett. 2010;584:3819–25.
90. Sumida T, et al. Detection of telomerase activity in oral lesions. J Oral Pathol Med. 1998;27:111–5.
91. Chang LY, et al. Telomerase activity and in situ telomerase RNA expression in oral carcinogenesis. J Oral Pathol Med. 1999;28:389–96.

92. Chen HH, et al. Expression of human telomerase reverse transcriptase (hTERT) protein is significantly associated with the progression, recurrence and prognosis of oral squamous cell carcinoma in Taiwan. Oral Oncol. 2007;43:122–9.
93. Mutirangura A, et al. Telomerase activity in oral leukoplakia and head and neck squamous cell carcinoma. Cancer Res. 1996;56:3530–3.
94. Miyoshi Y, et al. Telomerase activity in oral cancer. Oral Oncol. 1999;35:283–9.
95. Zhang L, Zhang W. Telomerase hTR and hTRT gene expression in oral precancerous lesions and squamous cell carcinomas. Chin J Dent Res. 1999;2:43–8.
96. Liao J, Mitsuyasu T, Yamane K, Ohishi M. Telomerase activity in oral and maxillofacial tumors. Oral Oncol. 2000;36:347–52.
97. Egger G, Liang G, Aparicio A, Jones PA. Epigenetics in human disease and prospects for epigenetic therapy. Nature. 2004;429:457–63.
98. Kulkarni V, Saranath D. Concurrent hypermethylation of multiple regulatory genes in chewing tobacco associated oral squamous cell carcinomas and adjacent normal tissues. Oral Oncol. 2004;40:145–53.
99. Kato K, et al. Aberrant promoter hypermethylation of p16 and MGMT genes in oral squamous cell carcinomas and the surrounding normal mucosa. J Cancer Res Clin Oncol. 2006;132:735–43.
100. Yeh KT, et al. The correlation between CpG methylation on promoter and protein expression of E-cadherin in oral squamous cell carcinoma. Anticancer Res. 2002;22:3971–5.
101. Maruya S, et al. Differential methylation status of tumor-associated genes in head and neck squamous carcinoma: incidence and potential implications. Clin Cancer Res. 2004;10:3825–30.
102. Lagos-Quintana M, Rauhut R, Lendeckel W, Tuschl T. Identification of novel genes coding for small expressed RNAs. Science. 2001;294:853–8.
103. Gorenchtein M, Poh CF, Saini R, Garnis C. MicroRNAs in an oral cancer context—from basic biology to clinical utility. J Dent Res. 2012;91:440–6.
104. Brito JA, Gomes CC, Guimaraes AL, Campos K, Gomez RS. Relationship between microRNA expression levels and histopathological features of dysplasia in oral leukoplakia. J Oral Pathol Med. 2014;43:211–6.
105. Vousden KH, Lane DP. p53 in health and disease. Nat Rev Mol Cell Biol. 2007;8:275–83.
106. Gasco M, Crook T. The p53 network in head and neck cancer. Oral Oncol. 2003;39:222–31.
107. Somers KD, et al. Frequent p53 mutations in head and neck cancer. Cancer Res. 1992;52:5997–6000.
108. Swaminathan U, Joshua E, Rao UK, Ranganathan K. Expression of p53 and Cyclin D1 in oral squamous cell carcinoma and normal mucosa: an Immunohistochemical study. J Oral Maxillofac Pathol. 2012;16:172–7.
109. van Houten VM, et al. Mutated p53 as a molecular marker for the diagnosis of head and neck cancer. J Pathol. 2002;198:476–86.
110. Poeta ML, et al. TP53 mutations and survival in squamous-cell carcinoma of the head and neck. N Engl J Med. 2007;357:2552–61.
111. Stransky N, et al. The mutational landscape of head and neck squamous cell carcinoma. Science. 2011;333:1157–60.
112. Oliveira LR, et al. Prognostic factors and survival analysis in a sample of oral squamous cell carcinoma patients. Oral Surg Oral Med Oral Pathol Oral Radiol Endod. 2008;106:685–95.
113. Nasser W, Flechtenmacher C, Holzinger D, Hofele C, Bosch FX. Aberrant expression of p53, p16INK4a and Ki-67 as basic biomarker for malignant progression of oral leukoplakias. J Oral Pathol Med. 2011;40:629–35.
114. Schoelch ML, et al. Apoptosis-associated proteins and the development of oral squamous cell carcinoma. Oral Oncol. 1999;35:77–85.
115. Brennan PA, Conroy B, Spedding AV. Expression of inducible nitric oxide synthase and p53 in oral epithelial dysplasia. Oral Surg Oral Med Oral Pathol Oral Radiol Endod. 2000;90:624–9.

116. Vered M, Allon I, Dayan D. Maspin, p53, p63, and Ki-67 in epithelial lesions of the tongue: from hyperplasia through dysplasia to carcinoma. J Oral Pathol Med. 2009;38:314–20.
117. Cruz IB, et al. p53 expression above the basal cell layer in oral mucosa is an early event of malignant transformation and has predictive value for developing oral squamous cell carcinoma. J Pathol. 1998;184:360–8.
118. Murti PR, et al. p53 expression in oral precancer as a marker for malignant potential. J Oral Pathol Med. 1998;27:191–6.
119. Kodani I, et al. Expression of minichromosome maintenance 2 (MCM2), Ki-67, and cell-cycle-related molecules, and apoptosis in the normal-dysplasia-carcinoma sequence of the oral mucosa. Pathobiology. 2001;69:150–8.
120. Shah NG, et al. Molecular alterations in oral carcinogenesis: significant risk predictors in malignant transformation and tumor progression. Int J Biol Markers. 2007;22:132–43.
121. Levrero M, et al. The p53/p63/p73 family of transcription factors: overlapping and distinct functions. J Cell Sci. 2000;113(Pt 10):1661–70.
122. Chen YK, Hsue SS, Lin LM. Expression of p63 protein and mRNA in oral epithelial dysplasia. J Oral Pathol Med. 2005;34:232–9.
123. Kovesi G, Szende B. Prognostic value of cyclin D1, p27, and p63 in oral leukoplakia. J Oral Pathol Med. 2006;35:274–7.
124. el-Deiry WS, et al. WAF1, a potential mediator of p53 tumor suppression. Cell. 1993;75:817–25.
125. Jiang H, et al. Induction of differentiation in human promyelocytic HL-60 leukemia cells activates p21, WAF1/CIP1, expression in the absence of p53. Oncogene. 1994;9:3397–406.
126. Schoelch ML, et al. Cell cycle proteins and the development of oral squamous cell carcinoma. Oral Oncol. 1999;35:333–42.
127. Xie X, Clausen OP, Boysen M. Prognostic significance of p21WAF1/CIP1 expression in tongue squamous cell carcinomas. Arch Otolaryngol Head Neck Surg. 2002;128:897–902.
128. Shintani S, et al. Expression of cell cycle control proteins in normal epithelium, premalignant and malignant lesions of oral cavity. Oral Oncol. 2002;38:235–43.
129. Choi HR, et al. Differential expressions of cyclin-dependent kinase inhibitors (p27 and p21) and their relation to p53 and Ki-67 in oral squamous tumorigenesis. Int J Oncol. 2003;22:409–14.
130. Nemes JA, Nemes Z, Marton IJ. p21WAF1/CIP1 expression is a marker of poor prognosis in oral squamous cell carcinoma. J Oral Pathol Med. 2005;34:274–9.
131. Hunter T, Pines J. Cyclins and cancer. II: Cyclin D and CDK inhibitors come of age. Cell. 1994;79:573–82.
132. Inaba T, et al. Genomic organization, chromosomal localization, and independent expression of human cyclin D genes. Genomics. 1992;13:565–74.
133. Goodger NM, Gannon J, Hunt T, Morgan PR. Cell cycle regulatory proteins–an overview with relevance to oral cancer. Oral Oncol. 1997;33:61–73.
134. Michalides R, et al. Overexpression of cyclin D1 correlates with recurrence in a group of forty-seven operable squamous cell carcinomas of the head and neck. Cancer Res. 1995;55:975–8.
135. Mineta H, Borg A, Dictor M, Wahlberg P, Wennerberg J. Correlation between p53 mutation and cyclin D1 amplification in had and neck squamous cell carcinoma. Oral Oncol. 1997;33:42–6.
136. Mineta H, et al. Cyclin D1 overexpression correlates with poor prognosis in patients with tongue squamous cell carcinoma. Oral Oncol. 2000;36:194–8.
137. Turatti E, da Costa Neves A, de Magalhaes MH, de Sousa SO. Assessment of c-Jun, c-Fos and cyclin D1 in premalignant and malignant oral lesions. J Oral Sci. 2005;47:71–6.
138. Ishida K, et al. Nuclear localization of beta-catenin involved in precancerous change in oral leukoplakia. Mol Cancer. 2007;6:62.
139. Huang M, et al. Cyclin D1 gene polymorphism as a risk factor for oral premalignant lesions. Carcinogenesis. 2006;27:2034–7.
140. Ye Y, et al. Genetic variations in cell-cycle pathway and the risk of oral premalignant lesions. Cancer. 2008;113:2488–95.

141. Ramakrishna A, et al. Cyclin D1 an early biomarker in oral carcinogenesis. J Oral Maxillofac Pathol. 2013;17:351–7.
142. McDuff FK, Turner SD. Jailbreak: oncogene-induced senescence and its evasion. Cell Signal. 2011;23:6–13.
143. Horowitz JM, et al. Frequent inactivation of the retinoblastoma anti-oncogene is restricted to a subset of human tumor cells. Proc Natl Acad Sci U S A. 1990;87:2775–9.
144. Soni S, et al. Alterations of rb pathway components are frequent events in patients with oral epithelial dysplasia and predict clinical outcome in patients with squamous cell carcinoma. Oncology. 2005;68:314–25.
145. Pande P, Mathur M, Shukla NK, Ralhan R. pRb and p16 protein alterations in human oral tumorigenesis. Oral Oncol. 1998;34:396–403.
146. Nakahara Y, et al. Alterations of Rb, p16(INK4A) and cyclin D1 in the tumorigenesis of oral squamous cell carcinomas. Cancer Lett. 2000;160:3–8.
147. Chen Q, Luo G, Li B, Samaranayake LP. Expression of p16 and CDK4 in oral premalignant lesions and oral squamous cell carcinomas: a semi-quantitative immunohistochemical study. J Oral Pathol Med. 1999;28:158–64.
148. Gologan O, Barnes EL, Hunt JL. Potential diagnostic use of p16INK4A, a new marker that correlates with dysplasia in oral squamoproliferative lesions. Am J Surg Pathol. 2005;29:792–6.
149. Soria JC, et al. Telomerase activation cooperates with inactivation of p16 in early head and neck tumorigenesis. Br J Cancer. 2001;84:504–11.
150. Bradley KT, Budnick SD, Logani S. Immunohistochemical detection of p16INK4a in dysplastic lesions of the oral cavity. Mod Pathol. 2006;19:1310–6.
151. Hynes NE, Lane HA. ERBB receptors and cancer: the complexity of targeted inhibitors. Nat Rev Cancer. 2005;5:341–54.
152. Choi S, Myers JN. Molecular pathogenesis of oral squamous cell carcinoma: implications for therapy. J Dent Res. 2008;87:14–32.
153. Grandis JR, Tweardy DJ. Elevated levels of transforming growth factor alpha and epidermal growth factor receptor messenger RNA are early markers of carcinogenesis in head and neck cancer. Cancer Res. 1993;53:3579–84.
154. Ongkeko WM, Altuna X, Weisman RA, Wang-Rodriguez J. Expression of protein tyrosine kinases in head and neck squamous cell carcinomas. Am J Clin Pathol. 2005;124:71–6.
155. Sok JC, et al. Mutant epidermal growth factor receptor (EGFRvIII) contributes to head and neck cancer growth and resistance to EGFR targeting. Clin Cancer Res. 2006;12:5064–73.
156. Bergler W, Bier H, Ganzer U. The expression of epidermal growth factor receptors in the oral mucosa of patients with oral cancer. Arch Otorhinolaryngol. 1989;246:121–5.
157. Shin DM, Ro JY, Hong WK, Hittelman WN. Dysregulation of epidermal growth factor receptor expression in premalignant lesions during head and neck tumorigenesis. Cancer Res. 1994;54:3153–9.
158. Nagatsuka H, Ishiwari Y, Tsujigiwa H, Nakano K, Nagai N. Quantitation of epidermal growth factor receptor gene amplification by competitive polymerase chain reaction in premalignant and malignant oral epithelial lesions. Oral Oncol. 2001;37:599–604.
159. Srinivasan M, Jewell SD. Evaluation of TGF-alpha and EGFR expression in oral leukoplakia and oral submucous fibrosis by quantitative immunohistochemistry. Oncology. 2001;61:284–92.
160. Taoudi Benchekroun M, et al. Epidermal growth factor receptor expression and gene copy number in the risk of oral cancer. Cancer Prev Res (Phila). 2010;3:800–9.
161. Mishima K, Inoue K, Hayashi Y. Overexpression of extracellular-signal regulated kinases on oral squamous cell carcinoma. Oral Oncol. 2002;38:468–74.
162. Marshall CJ. Specificity of receptor tyrosine kinase signaling: transient versus sustained extracellular signal-regulated kinase activation. Cell. 1995;80:179–85.
163. Massarelli E, et al. Akt activation correlates with adverse outcome in tongue cancer. Cancer. 2005;104:2430–6.
164. Tsui IF, et al. Multiple pathways in the FGF signaling network are frequently deregulated by gene amplification in oral dysplasias. Int J Cancer. 2009;125:2219–28.

165. Watanabe S, et al. Activation of PI3K-AKT pathway in oral epithelial dysplasia and early cancer of tongue. Bull Tokyo Dent Coll. 2009;50:125–33.
166. Kaur J, et al. Clinical significance of phosphatidyl inositol synthase overexpression in oral cancer. BMC Cancer. 2010;10:168.
167. Woo SB, Grammer RL, Lerman MA. Keratosis of unknown significance and leukoplakia: a preliminary study. Oral Surg Oral Med Oral Pathol Oral Radiol 2014;118:713–24.
168. Rhiner C, Moreno E. Super competition as a possible mechanism to pioneer precancerous fields. Carcinogenesis. 2009;30:723–8.
169. Struhl G, Basler K. Organizing activity of wingless protein in *Drosophila*. Cell. 1993;72:527–40.
170. Moreno E, Basler K. dMyc transforms cells into super-competitors. Cell. 2004;117:117–29.
171. Allen CT, Lewis JS, Jr, El-Mofty SK, Haughey BH, Nussenbaum B. Human papillomavirus and oropharynx cancer: biology, detection and clinical implications. Laryngoscope. 2010;120:1756–72.
172. Sugiyama M, et al. Detection of human papillomavirus-16 and HPV-18 DNA in normal, dysplastic, and malignant oral epithelium. Oral Surg Oral Med Oral Pathol Oral Radiol Endod. 2003;95:594–600.
173. Szarka K, et al. Progressive increase of human papillomavirus carriage rates in potentially malignant and malignant oral disorders with increasing malignant potential. Oral Microbiol Immunol. 2009;24:314–8.
174. Kreimer AR, et al. Incidence and clearance of oral human papillomavirus infection in men: the HIM cohort study. Lancet. 2013;382:877–87.
175. Fornatora M, Jones AC, Kerpel S, Freedman P. Human papillomavirus-associated oral epithelial dysplasia. (koilocytic dysplasia): an entity of unknown biologic potential. Oral Surg Oral Med Oral Pathol Oral Radiol Endod. 1996;82:47–56.
176. Daley T, Birek C, Wysocki GP. Oral bowenoid lesions: differential diagnosis and pathogenetic insights. Oral Surg Oral Med Oral Pathol Oral Radiol Endod. 2000;90:466–73.
177. Buajeeb W, Poomsawat S, Punyasingh J, Sanguansin S. Expression of p16 in oral cancer and premalignant lesions. J Oral Pathol Med. 2009;38:104–8.
178. Ishibashi M, et al. The prevalence of human papillomavirus in oral premalignant lesions and squamous cell carcinoma in comparison to cervical lesions used as a positive control. Int J Clin Oncol. 2011;16:646–53.
179. McCord C, et al. Association of high-risk human papillomavirus infection with oral epithelial dysplasia. Oral Surg Oral Med Oral Pathol Oral Radiol. 2013;115:541–9.
180. Fakhry C, et al. Improved survival of patients with human papillomavirus-positive head and neck squamous cell carcinoma in a prospective clinical trial. J Natl Cancer Inst. 2008;100:261–9.
181. Ang KK, et al. Human papillomavirus and survival of patients with oropharyngeal cancer. N Engl J Med. 2010;363:24–35.
182. Salazar CR, et al. Human papillomavirus-associated head and neck squamous cell carcinoma survival: a comparison by tumor site and initial treatment. Head Neck Pathol. 2014;8:77–87.
183. Izzo JG, et al. Dysregulated cyclin D1 expression early in head and neck tumorigenesis: in vivo evidence for an association with subsequent gene amplification. Oncogene. 1998;17:2313–22.
184. Tabor MP, et al. Comparative molecular and histological grading of epithelial dysplasia of the oral cavity and the oropharynx. J Pathol. 2003;199:354–60.
185. Agrawal N, et al. Exome sequencing of head and neck squamous cell carcinoma reveals inactivating mutations in NOTCH1. Science. 2011;333:1154–7.
186. Yakob M, Fuentes L, Wang MB, Abemayor E, Wong DT. Salivary biomarkers for detection of oral squamous cell carcinoma—current state and recent advances. Curr Oral Health Rep. 2014;1:133–41.
187. Dorsey K, Agulnik M. Promising new molecular targeted therapies in head and neck cancer. Drugs. 2013;73:315–25.

Oral Cancer

Srinivas Vinod Saladi and Leif W. Ellisen

Clinical and Epidemiological Features

By far the most prevalent human oral cancer is part of a group of diseases known collectively as head and neck squamous cell carcinoma (HNSCC). Other, much rarer oral cancers include oral melanoma, mucoepidermoid carcinoma, and adenoid cystic carcinoma; of these, only the latter has been the focus of significant genomic analyses and will be discussed here. HNSCC is a common cancer worldwide with an estimated 600,000 new cases diagnosed annually In the United States, 50,000 new cases are diagnosed and nearly 10,000 deaths are attributable to this disease each year. HNSCC arises from the squamous mucosa of several distinct structures in the upper aerodigestive tract (including the tongue, oral cavity and oropharynx) that exhibit distinct microscopic features and patterns of lymphatic/venous drainage. Consequently, cancers of these diverse subsites mandate distinct treatment approaches and are associated with variable outcomes. Further complexity of HNSCC is evidenced by molecular analyses revealing biologic heterogeneity that is independent of disease subsite, including tumors with particular mRNA expression profiles or alterations in DNA copy number patterns that correlate with prognosis [1–4].

Alcohol consumption and tobacco use are the two most important risk factors for most types of HNSCC. However, for Oropharyngeal SCC (OPSCC), infection with human papilloma virus (HPV) is now recognized as a key risk factor. Importantly, the overall survival for patients with HPV-positive OPSCC is markedly better than for those with HPV-negative cancers, and this effect is independent of the mode of treatment [5–7]. While the precise reason for this observation remains to be determined, proposed explanations include tumor-specific factors (unique genetic pathogenesis conferring increased treatment responsiveness) and host-specific factors

L. W. Ellisen (✉) · S. V. Saladi
Massachusetts General Hospital Cancer Center and Harvard Medical School,
GRJ-904, 55 Fruit Street, Boston, MA 02114, USA
e-mail: lellisen@mgh.harvard.edu

S. T. Sonis (ed.), *Genomics, Personalized Medicine and Oral Disease,*
DOI 10.1007/978-3-319-17942-1_13

(younger and healthier patients with better treatment tolerance, enhanced tumor-specific immune response). Indeed, before HPV-related cancers were recognized to have a distinct natural history, their rising prevalence and relatively favorable prognosis initially masked the unfortunate fact that little or no progress has been made in improving outcomes for patients with HPV-negative tumors during the past 30 years [8]. Going forward, it is hoped that a more detailed understanding of the molecular pathways that drive HSNCC initiation and progression will lead to more selective and effective therapies.

Germline Predisposition

HNSCC is considered to be a cancer induced largely by environmental carcinogens rather than germline (inherited) predisposition. Nonetheless, several germline genetic alterations have been associated with an increased risk of HNSCC. Among the most prevalent germline factors are a host of polymorphisms in genes involved in biotransformation and detoxification of carcinogens and pro-carcinogens [9]. Given the distinct pathogenesis of HPV-positive HNSCC, it is not surprising that a discrete set of genetic polymorphisms are now emerging as contributors selectively to these tumors [9]. These relatively common polymorphisms are associated with only modest (<2-fold) elevations in risk of HNSCC. In contrast, rare deleterious variants in a subset of established human tumor suppressor genes are associated with much higher risk, resulting in the clustering of HNSCC and other cancers in families. Germline mutation of *CDKN2A*, a gene frequently targeted for somatic inactivation in HNSCC, is associated with FAMMM (familial atypical multiple mole melanoma), a syndrome involving predisposition to melanoma, pancreatic cancer and HNSCC. The DNA repair-associated kinase gene *ATR* was recently shown to be mutated in a family presenting with multiple oropharyngeal carcinomas and other malignancies [10]. Dramatically increased risk for HNSCC is observed in Fanconi Anemia (FA), a syndrome associated with mutations in more than a dozen different genes that comprise an essential DNA repair pathway. Individuals with FA develop bone marrow failure and leukemia, and are at >500-fold increased risk for the development of HNSCC compared to the general population [11].

Somatic Genetic Events and Clonal Tumor Evolution

As with most cancers, the genetic pathogenesis of oral cancers involves the sequential acquisition of somatic genetic alterations in nascent tumor cells. Clonal evolution is central to this process and refers to the progressive selection of those genetic events that increase tumor survival, proliferation, and ultimately invasion and metastasis. The key somatic genetic changes contributing to tumor fitness are referred to as "driver" abnormalities. However, not all tumor-associated genetic changes contribute to pathogenesis. Some alterations, referred to as "passengers", are simply the product of tumor genome instability and are carried along in the dominant tumor clone. In some cases, distinguishing driver versus passenger mutations in a tumor can be quite challenging. This has become particularly evident as dramatic advances in DNA sequencing technology in the past decade have made possible the complete genomic characterization of human tumors. Distinguishing passengers from drivers now involves not only the traditional, functional experimental approaches, but also the bioinformatic analysis of large datasets comprising dozens to hundreds of sequenced tumors. Major efforts are now focused on optimized algorithms for these "in silico" studies, which will undoubtedly become more sophisticated and integral to the interpretation of complex tumor genomes [12, 13].

Recent genomic studies of HNSCC have provided an instructive example of the complexity associated with tumor genomic analysis. A series of papers published in 2011–2013 reported whole-exome sequencing, in some cases together with gene copy number data and gene expression profiling, of a substantial number (nearly 200 in total) of HNSCC primary tumors [14–17]. While many of the key genes targeted for mutation in HNSCC were identified in both studies, the estimates of overall mutations/tumor varied by approximately 5-fold among these studies. These differences are unlikely to be related to tumor selection or to purely technical issues with sequencing itself, and instead allude to differences in bioinformatic analysis of the sequencing data [18]. Even accounting for such differences, it is clear that a substantial fraction of detected mutations are passengers rather than drivers, which is a common finding in carcinogen-induced tumors such as HNSCC. Among the driver mutations a clear pattern emerged, revealing that HNSCC is a disease driven largely by tumor suppressor pathways including p53, Rb/CDKN2A, and NOTCH rather than by activated oncogenes. Notably however, in HPV-positive OSCC both somatic driver and passenger mutations are much less common, consistent with the viral etiology of this cancer. In particular, p53 and Rb pathway mutations are quite rare in these tumors, consistent with the ability of viral E6 and E7 proteins to inactive the respective tumor suppressors.

Major Genetic Abnormalities and Pathways

Multiple different approaches have been used to classify the functional and genomic landscape of oral HNSCC [2, 16, 19]. It is clear, however, that the major genes affected in these cancers reflect to a large degree the functional pathways that are central to homeostasis within the normal epithelium. These include cellular proliferation, squamous epithelial differentiation, cell survival, and adhesion/migration, with many of the genes impacting more than a single pathway (Fig. 1). The most recent genomic analyses of more than 300 HNSCC specimens through The Cancer Genome Atlas (TCGA) project have also pointed to a critical role for epigenetic regulation and the oxidative stress response in disease pathogenesis [12, 13]. Below are described the common genetic pathways altered in HNSCC and their potential

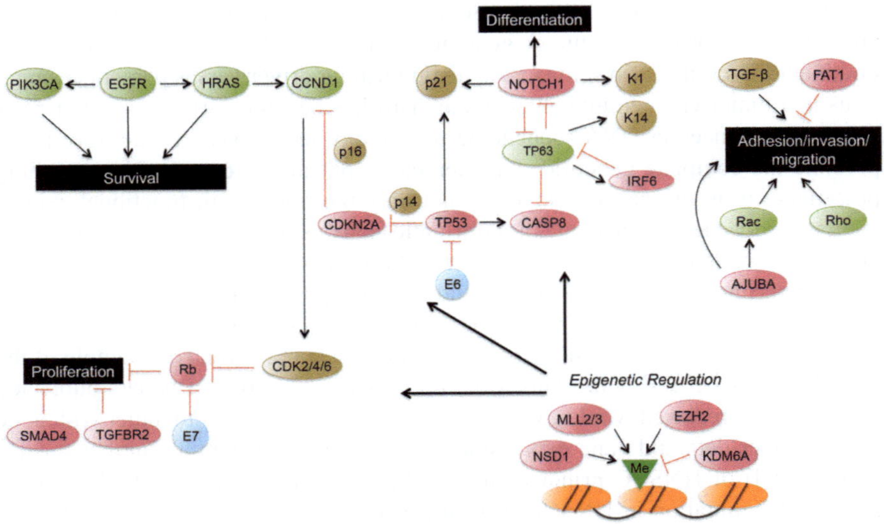

Fig. 1 Key genomic alterations and pathways in oral HNSCC. The most prevalent tumor-specific genetic alterations are shown. *Green*, indicates activating mutations; *red*, loss-of-function mutations; *brown*, other relevant factors; *blue*, viral proteins. Fundamental cellular processes/pathways controlled by these genetic events are indicated by *black boxes*. These processes and pathways are interconnected as indicated, and several factors (e.g. TGFβ signaling) may contribute to more than one process. *Bottom right*, epigenetic regulatory enzymes affecting histone methylation (Me) and nucleosome (*orange*) architecture are subject to mutations that may impact gene regulation and tumor pathogenesis in multiple ways

functional contribution in these tumors. Effective therapy for these tumors will require not only a detailed understanding of the contribution of these individual genes, but also a working knowledge of the interactions and cross-talk among these diverse functional units.

p53/Rb/CDKN2A

The *p53 (TP53)* and *Rb (RB1)* genes play an important role in cell cycle regulation and proliferation. Mutation of the *p53* tumor suppressor is the most common genetic alteration in HNSCC, occurring in approximately two-thirds of cases [2, 20]. Similar to other cancers, 75 % of these are missense mutations which cluster in the p53 DNA binding domain. Consequently, mutant p53 protein is overexpressed in these tumors and confers a dominant-negative effect as well as a gain-of-function property that is poorly understood but may contribute to invasion and metastatsis [21–23]. Even in the remaining HNSCC cases expressing wild-type p53, its function may be abrogated by various mechanisms. These include the expression of HPV E6; amplification or overexpression of MDM2, which targets p53 for proteasomal degradation; and deletion of *CDKN2A*, resulting in loss of p14/ARF protein, a negative regulator of MDM2 [24–27]. In total, the p53 pathway is inactivated in vast majority of oral HNSCC [2].

Unlike in some cancers in which it is a late event, *p53* mutation is likely the earliest genetic initiating event in oral HNSCC, as it is detectable both in normal-appearing mucosa and in precursor lesions including leukoplakia (see chapter on premalignant lesions). The role of the p53 pathway in these tumors is multi-faceted, involving effects on tumor cell proliferation, survival, and metabolism [28, 29]. As a key gatekeeper in the response to DNA damage, it has been hypothesized that p53's major clinical impact would be as a determinant of response to genotoxic therapy. However, it is now recognized that certain classes of *p53* mutation confer significant prognostic information in cases treated with surgery either with or without post-operative radiotherapy [28, 29]. While substantial efforts have focused on therapeutic targeting of the p53 pathway in a variety of cancers, among the most far advanced is the use of MDM2 inhibitors to activate wild-type p53 in the small fraction of tumors lacking *p53* mutations [30].

Inactivation of the *CDKN2A* gene, encoding the Rb pathway cell cycle regulators p16/INK4A and p14/ARF/INK4B, is also an early and common event in HNSCC pathogenesis. Mutations are found in nearly 20 % of cases, copy number loss in an additional 20–30 %, and epigenetic inactivation in many more, yielding approximately 75 % of HNSCC tumors with inactivation of this locus [14, 17, 31–33]. Distinct contributions of p16/INK4A and p14/ARF/INK4B in this disease are evidenced by the observation that loss of the former correlates with poor prognosis, whereas selective loss of the latter (e.g. by methylation with retained expression of p16) has in some studies been associated with more a favorable outcome [34–36]. Like the p53 pathway, Rb pathway function is inactivated in HPV-positive OPSCC, in this case through expression of the E7 protein. Loss of *Rb* abrogates the requirement for p16 silencing in these tumors, and as a result high-level p16 protein expression has been used as a clinical marker of HPV-associated OPSCC [37].

A discrete chromosomal region on chromosome 11q13 encompassing the *CCND1* gene (encoding Cyclin D1) is amplified in approximately 30 % of HNSCC, while overexpression is observed in nearly 80 % of cases [2, 4, 38]. It is therefore clear that *CDKN2A* inactivation and *CCND1* overexpression are not mutually exclusive events in these tumors, even though they would appear to be redundant mechanisms to activate G1 phase cell cycle progression. Further underscoring this point are the independent (and additive) contributions to poor prognosis of p16 loss and Cyclin D1 overexpression in tongue cancer [39]. Potential explanations for these observations include the distinct contributions of Cyclin D1 as a cofactor for gene transcription and for DNA repair [40–42]. Targeting cyclin-dependent kinase (CDK) function in Cyclin D1-overexpressing tumors is an attractive possibility that will soon be tested using new, selective CDK4/6 inhibitors that have shown activity in other solid tumors [43].

Notch/p63

Squamous cell carcinomas including HNSCC are characterized histologically by regions demonstrating terminal differentiation and associated cell cycle exit. Thus, it may come as no surprise that mutations affecting the normal differentiation process are central drivers of these tumors. Signaling through the Notch pathway is highly pleiotropic and has been linked in different contexts to cell fate determination, self-renewal capacity, differentiation, and cellular proliferation and survival. In the stratified epithelium Notch has a prominent role in promoting growth arrest and terminal differentiation, in part through direct regulation of suprabasal keratins and indirectly through the Wnt and hedgehog signaling pathways. A major finding from the initial exome sequencing analyses of HNSCC was the identification of *NOTCH1* mutations in up to 17 % of cases, with mutations in additional *NOTCH* family members in another 3–5 % [14, 17]. Previous studies in T-cell acute lymphoblastic leukemia (T-ALL), chronic lymphocytic leukemia (CLL) and diffuse large B-cell lymphoma (DLBCL) have identified recurrent activating mutations in *NOTCH*, implying a tumorigenic role in these cancers [44–48]. However, a distinct set of *NOTCH* mutations are observed in HNSCC including numerous frameshift and nonsense mutations, indicating a tumor suppressor role for these genes [14, 17]. The tumor suppressor role of *NOTCH* is likely to extend to HPV-positive OPSCC as well, as NOTCH signaling has also been shown to restrict the expression of HPV E6 and E7 proteins, thereby providing further selective pressure for NOTCH loss-of-function [49, 50]. These distinct, context-dependent roles for NOTCH in human cancer are supported by animal models, as activated NOTCH expression in hematopoietic cells mimics human T-ALL, whereas loss of NOTCH function in the epidermis is associated with squamous carcinogenesis [51–54].

Control of Notch function in squamous epithelia is mediated in part by the p53 family transcription factor p63 (TP63), a master regulator of epithelial development. Expression of p63 is required for embryonic development of the stratified epithelium through control of stem cell fate, proliferative potential of committed progenitor cells, lineage specification and differentiation [55]. Mechanisms for control of NOTCH signaling by p63 in this context include p63-dependent inhibition of NOTCH expression itself, antagonism of NOTCH effector genes including *Hes-1*, *Hey1/2*, and regulation of NOTCH ligands including Jagged1/2 [56–58]. The reported frequency of genomic amplification of the *p63* locus in HNSCC has varied but appears to be greater than 25 % [16], with mutations observed in up to 8 % of cases in some studies. While p63 is expressed as multiple protein isoforms, the major form associated with squamous carcinogenesis is ΔNp63α. Overexpression of this protein has been show to cooperate with HRAS expression for malignant conversion of keratinocytes [59, 60]. Additionally, genetic loss of *p63* in an animal model of HNSCC induces rapid apoptosis and tumor regression, through abrogation of a paracrine tumor/stromal signaling circuit involving ΔNp63α regulation of FGFR2 [61].

P63 regulates additional suspected genomic drivers of HNSCC, including the zinc-finger transcription factor ZNF750 and the interferon-response gene *IRF6* [62, 63]. *ZNF750*, which is subject to loss-of-function mutation in 4% of HNSCC, is required for the terminal differentiation program in keratinocytes through its regulation of KLF4. Reflecting the developmental contribution of this same genetic program, ZNF750 and KLF4 were earlier shown to be important targets of p63 in a human organotypic model of ankyloblepharon-ectodermal dysplasia-cleft palate (AEC) syndrome, which is caused by germline *p63* mutation [64]. Another factor linked to both cleft palate and tumorigenesis through p63/NOTCH signaling is IRF6, which is activated by p63 and appears to function as a negative feedback regulator of p63 protein levels. In turn, IRF6 serves as a mediator of NOTCH-dependent differentiation [64–66]. Genetic variation in the enhancer element for p63-dependent regulation of IRF6 is associated with increased risk of cleft palate, and in HNSCC IRF6 was shown to be mutated in a small (2%) but significant fraction of tumors [12].

EGFR/RAS/PIK3CA/PTEN/CASPASE8

Second in frequency only to p53/Rb pathway abnormalities, genomic events leading to deregulation of the phosphotidylinositol-3 kinase (PI3K) pathway are common in HNSCC, occurring in up to 30% of cases [15]. Most common are "hotspot" activating mutations within the catalytic or regulatory domains of the PI3K catalytic subunit gene *PIK3CA*, which occur in nearly 20% of cases [67, 68]. Additional cases harbor amplification of the *PIK3CA* gene, which residues at a locus on 3p26 that includes other amplified genes in HNSCC including *p63* [69]. The PI3K pathway may be of particular relevance to HPV-positive OPSCC, as mutations of *PIK3CA* or the PI3K regulatory subunit gene *PIK3R1* are more common in HPV-positive cases and are reported to be the only mutated cancer genes in a subset of these tumors [15, 70]. The mechanism responsible for this association remains to be elucidated, but prior data proposed specific cooperation of the PI3K pathway with HPV-dependent carcinogenesis in a cervical cancer mode [71]. The negative PI3K regulatory phosphatase gene *PTEN* is subject to loss of function through mutation, deletion or silencing in a substantial fraction of HNSCC cases. *PTEN* mutation is observed in 3–5% of cases, with allelic loss occurring in at least another 8% [15]. Notably, bi-allelic loss of *PTEN* may underestimate the frequency of its functional inactivation, given emerging evidence for a haploinsufficient phenotype for this gene [72–74]. The frequency of PI3K activity has generated much excitement in the clinical research community, as many new and potentially promising inhibitors of this pathway are now being tested in clinical trials. These include combination PI3K/mTOR kinase inhibitors, pan-PI3K inhibitors, and isoform selective PI3K inhibitors. Questions now under active investigation include which inhibitor will be most useful, in what tumor context and in combination with which other therapies.

Activating RAS mutations are observed in many cancers, with mutations in *KRAS* being by far the most prevalent overall. In HNSCC, however, *HRAS* rather than *KRAS* is targeted for mutations that affect codons 12, 13, and 61 and are observed in approximately 4–6% of cases, while *KRAS* mutations are very rare [75, 76]. The selective propensity for *HRAS* mutation in the stratified squamous epithelium is not fully explained but is consistently observed in tobacco-associated HNSCC, HPV-associated OPSCC, and carcinogen-induced animal models of HNSCC and epidermal SCC [77]. RAS signaling can activate both PI3K and mitogen-activated protein kinase (MAPK) pathways, and both pathways likely contribute to the RAS-mediated effects on tumorigenesis. Consistent with this idea, additional downstream MAPK pathway genes are subject to recurrent mutation in HNSCC including *ERK1/2, MEK1/2* and *RAF* family members [15]. RAS proteins have been proven difficult to target directly for therapeutic purposes, but multiple indirect approaches are now being tested to address this unmet need. Some of these include targeting the downstream components of the pathway (e.g. combining PI3K and MAPK inhibitors) or taking advantage of certain synthetic lethal dependencies that result from RAS mutational activation, as demonstrated in preclinical models [78–80].

In normal keratinocytes, MAPK and PI3K pathways are induced by ligand-dependent activation of upstream receptor tyrosine kinases such as EGFR. Thus, the frequent genetic activation of these downstream pathways in HNSCC may in part explain why genetic activation of *EGFR* itself is relatively uncommon, and why therapeutic approaches targeting this receptor have had only a modest impact. Mutation of *EGFR* itself is rare in these cancers, and genetic amplification is somewhat more frequent but of variable magnitude [81, 82]. Notably, although copy number gains of *EGFR* have been shown to confer a poor prognosis in HNSCC [82, 83], no correlation has been observed between this event and response to EGFR-directed therapy using the small molecule inhibitors erlotinib and gefinitinb or the therapeutic monoclonal antibody cetuximab [84–88]. Future approaches that target both EGFR and an activated downstream PI3K/MAPK pathway may be one strategy to improve upon the limited success observed with these drugs when used as single agents. Alternatively, targeting additional, parallel RTK pathways such as FGFR1/2/3 that are activated in a subset of these tumors may prove fruitful and is currently under investigation [61, 89, 90].

The RTK/RAS/PI3K pathways described above have a central role in tumor cell survival. Other genetic events thought to promote survival in HNSCC include loss-of-function mutations in *CASP8*, seen in 9% of cases [17], which encodes a pro-apoptotic caspase that initiates a cascade of proteolytic events leading to apoptosis. Missense mutations in *FBXW7*, found in HNSCC and several other tumors, lead to stabilization of MCL-1, a key survival factor of the Bcl-2 family [12]. BCL2 itself is highly overexpressed in some HNSCC tumors and correlates with poor prognosis [91]. Notably, HDAC inhibitors have been shown to down-regulate Mcl-1 expression in HNSCC, although neither HDAC inhibitors nor Bcl-2 inhibitory BH3 domain mimetics have shown clinical activity as single agents in HNSCC [92]. Nevertheless, the combination of these two drugs was shown to have a synergistic effect in preclinical work, particularly in *FBXW7*-mutant squamous carcinomas [93].

TGFβ/SMAD/FAT1

The TGFβ/SMAD pathway is well established to function as a suppressor of tumor initiation, and multiple components of this pathway are mutated in human cancer. The TGFβ receptor gene *TGFBR2* is the most commonly inactivated gene in the pathway in HNSCC, occurring through genomic loss or less commonly through mutations that are observed in 3% of tumors [12, 13, 94]. Similarly, the TGF-β effector genes *SMAD2* and *SMAD4* are mutated in a small subset of HNSCC cell lines; *SMAD4* mutations are seen in 2% of primary tumors, but genomic loss of the *SMAD4* locus on chromosome 18q is a much more common event [95, 96]. In addition to its role in tumor suppression, TGFβ pathway is a pleiotropic regulator in cancer, as pathway activation in the later stages of carcinogenesis promotes metastasis. Accordingly, in mouse models conditional genetic deletion of *SMAD4* in the stratified epithelium led to HNSCC initiation in association with genomic instability, but also to increased inflammation and tumor progression associated with activation of TGF-β1 and other SMADs [97]. Epithelial-specific deletion of *TGFBR2* alone in a mouse model was not sufficient to induce HNSCC, but this genetic event did cooperate with activated RAS, giving rise to squamous carcinomas that metastasized to local lymph nodes [98].

A notable new genetic instigator of oral cancer identified through exome sequencing studies is *FAT1*, which encodes a member of the cadherin family of cell membrane proteins. Multiple studies have demonstrated that loss-of-function mutations in this gene are common in HNSCC, occurring in 20% of tumors in one large study [12, 99]. Although its precise contribution to tumorigenesis remains to be determined, as a cadherin-like protein FAT1 may play a role in maintaining cell polarity and mediating cell-cell contact. In this way, loss of FAT1 might affect cellular adhesion, growth and migration. One recent study has implicated FAT1 as an anatagonist of Wnt signaling, a well-established tumor driver pathway [100]. Two other genes with roles in cytoskeletal organization, adhesion and migration that are significantly mutated are *RAC1* (3%) and *RHOA* (2%) [12, 13]. RAC1 and RHOA belong to the family of Ras-related GTP binding proteins that are known to regulate the actin cytoskeleton. In addition to its mutation, RAC1 was found to be highly upregulated in HNSCC cell lines [101, 102]. Another possible contributor to deregulated RAC1 and adhesion in HNSCC is the LIM domain-containing protein AJUBA. Expression of AJUBA normally regulates RAC1 activation and is required for E-cadherin-dependent adhesion [103], and predominantly loss-of-function mutations in *AJUBA* are observed in 5% of HNSCC (but rarely in other tumor types) [12]. Additional, even less common mutations in other genes affecting cell polarity have been described in HNSCC, further highlighting the importance of membrane and cytoskeletal organization and signaling to this disease [19].

Alterations in the Epigenome and Epigenetic Regulators

The epigenetic deregulation of gene expression has long been recognized as a contributor to oral cancer pathogenesis. Focal DNA methylation of CpG residues within the regulatory regions of key genes including *CDKN2A*, *CCNA1* (encoding cyclin A1), *MGMT* (encoding a repair protein) and the death-associated kinase gene *DAPK1* contributes to their silencing in HNSCC [16, 35]. The diverse roles of these genes in cellular homeostasis speaks to the broad contribution of this epigenetic mechanism in tumorigenesis.

More recently, it has been recognized that epigenetic regulatory factors themselves are frequently mutated in HNSCC. Thus, global epigenetic deregulation is hallmark of this and several other cancers. In particular, mutations are observed in multiple enzymes that function as methyltransferases targeting particular lysines on histone proteins. This pattern of methylated lysines controls transcription through its effects on nucleosome organization and co-factor recruitment. Five such lysine methyltransferases mutated in HNSCC include *MLL2* (11–18 %), *MLL3* (7.3 %), *PRDM9* (6.5–11 %), *NSD1* (10 %), and *EZH2* (0.3–6 %) [12–14, 17]. The majority of these mutations are loss-of-function, indicating a potential tumor suppressor role for these genes. However, inactivating mutations are also observed in the lysine demethylase gene *KDM6A* in HNSCC (2 %), alluding to the complex functional interplay between opposing chromatin modifying enzyme pathways. Two genes enocoding chromatin remodeling factors, *ARID2* (3 %) and *PBRM1* (2 %), which are involved in related mechanisms for gene regulation are also subject to rare but likely significant mutations in HNSCC [12].

Recurrent mutations in lysine methyltransferase and other epigenetic regulatory genes are now known to occur in a variety of solid and liquid tumors. Establishing their precise contribution in tumorigenesis nonetheless remains challenging because of the global nature of epigenetic deregulation resulting from their mutation. For the same reason, developing successful therapeutic approaches that selectively target tumors with these mutations is not straightforward. "First-generation", non-selective inhibitors of DNA methylation (e.g. 5-aza-cytindine) or histone acetylation (e.g. vorinostat) have not been particularly successful in HNSCC [92]. Recently, more selective inhibitors of enzymes including EZH2 and HDAC1/2 have been developed and are now entering clinical trials [93, 104]. However, as noted above many of the HNSCC-associated mutations in epigenetic regulatory enzymes confer loss of function, which complicates the direct application of an inhibitor of the mutant enzyme. On the other hand, because this regulation involves reciprocal enzymatic programs (e.g. a demethylase opposing a methyltransferase) it may indeed be possible to infer an appropriate enzyme-targeted therapeutic strategy from an inactivating mutation. How such therapies may be effectively combined with existing approaches is another promising area of investigation.

Additional Genes/Pathways

An emerging pathway relevant to HNSCC and related cancers is the anti-oxidant response through the NRF2 transcription factor. In normal cells NRF2 controls the expression of genes involved in protection from oxidative damage, endogenous and exogenous toxins. In this way NRF2 activation protects against conditions including neurodegeneration, cardiovascular disease and photo-toxicity [105]. However, constitutive activation of NRF2 by a variety of mechanisms likely contributes to the pathogenesis of several squamous carcinomas including HNSCC, lung and esophageal SCC. NRF2 is negatively regulated by KEAP1, a substrate adaptor for the E3 ubiquitin ligase that promotes proteasome-dependent degradation of NRF2 [106]. Missense mutations in *NFE2L2,* the gene encoding NRF2, are observed in 5 % of HNSCC (and 15 % of lung SCC) and cluster in two domains required for KEAP1 substrate recognition [12]. Consequently, mutant NRF2 is stabilized and translocates to the nucleus to regulate gene expression. Interestingly, corresponding substrate recognition-disrupting mutations in *KEAP1* are observed in > 10 % of lung squamous and adenocarcinomas but not in HNSCC [12, 106]. The therapeutic implications of these findings are somewhat complex given that transient activation of NRF2 is not carcinogenic but instead protective against chemical carcinogenesis in some settings [105]. Thus, it remains to be determined whether inhibition of NRF2 will prove to be a viable therapeutic strategy in HNSCC or other cancers.

Adenoid Cystic Carcinoma

Adenoid cystic carcinoma (ACC) is a rare malignancy of the salivary gland that is characterized by frequent perineural invasion, local recurrence and ultimately metastatic dissemination [107]. While the paucity of cases had hampered efforts to discover the molecular underpinnings of this disease, several reports of cytogenetic, copy number and genomic sequencing analyses have now revealed key driver pathways. Multiple studies have now identified a common translocation resulting in the fusion of the *MYB* transcription factor gene on chromosome 6q24 to the *NFIB* gene on chromosome 9, resulting in production of a fusion protein [108]. *MYB* plays a role in multiple developmental processes, and although its precise contribution ACC is unknown, it is proposed that the fusion is highly overexpressed due to deletion of microRNA (miR) binding sites within the MYB mRNA [108]. More recent, whole-exome and whole-genome sequencing studies of ACC have also identified mutations in established cancer genes including *PIK3CA*, *FGFR2/4*, *ATM* and others [109, 110]. Notably, mutations in both *NOTCH1* and *NOTCH2* have been identified in ACC. However, these include both activating and truncating mutations, and therefore the contribution of NOTCH as an oncogene versus tumor suppressor in this disease may be context-specific. A recurrent theme in the mutational landscape of ACC is the involvement of chromatin remodeling genes, including *SMARCA2*, *ARID1A*, *KDM6A* and others [109, 110]. Substantial epigenetic deregulation of gene expression has also been reported in these tumors [111]. Potential near-term therapeutic implications of these findings include the application of inhibitors of the FGFR/PI3K pathway, epigenetic therapy as discussed above, and possibly Notch inhibitors in a subset of tumors.

Conclusions

Although the genomic landscape of oral cancers including HNSCC is complex, systematic studies within the past few years have revealed a limited number of functional nodes that are deregulated as a result of tumor-associated mutations. HPV-associated tumors exhibit virally-mediated inactivation of p53 and Rb/CDKN2A pathways and few, but notable, other mutations including in the *PIK3CA* gene, a potential therapeutic target. In contrast, poorer-prognosis, alcohol and tobacco-associated HNSCC harbor frequent mutations in these and other tumor suppressors and less commonly, in driver oncogenes. Thus, a central challenge in molecular-directed treatment of this disease is identifying specific tumor vulnerabilities resulting from genetic loss-of-function. Understanding how such targeted approaches may be combined or sequenced with traditional therapeutic approaches is another area in which progress is anticipated in the coming years. The limited heterogeneity and relative genomic stability of HPV-positive tumors are likely to make them an increasingly curable subset. Conversely, for the genetically more complex carcinogen-induced tumors, de novo and acquired treatment resistance are likely to emerge as key challenges to the successful implementation of genotype-directed therapy.

References

1. Hermsen M, Guervos MA, Meijer G, et al. New chromosomal regions with high-level amplifications in squamous cell carcinomas of the larynx and pharynx, identified by comparative genomic hybridization. J Pathol. 2001;194(2):177–82.
2. Leemans CR, Braakhuis BJ, Brakenhoff RH. The molecular biology of head and neck cancer. Nature Rev Cancer. 2011;11(1):9–22.
3. Slebos RJ, Yi Y, Ely K, et al. Gene expression differences associated with human papillomavirus status in head and neck squamous cell carcinoma. Clinical Cancer Res: Official J Am Assoc Cancer Res. 2006;12(3 Pt 1):701–9.
4. Smeets SJ, Braakhuis BJ, Abbas S, et al. Genome-wide DNA copy number alterations in head and neck squamous cell carcinomas with or without oncogene-expressing human papillomavirus. Oncogene. 2006;25(17):2558–64.
5. Fakhry C, Westra WH, Li S, et al. Improved survival of patients with human papillomavirus-positive head and neck squamous cell carcinoma in a prospective clinical trial. J Natl Cancer Inst. 2008;100(4):261–9.
6. Licitra L, Perrone F, Bossi P, et al. High-risk human papillomavirus affects prognosis in patients with surgically treated oropharyngeal squamous cell carcinoma. J Clin Oncol. 2006;24(36):5630–6.
7. Posner MR, Hershock DM, Blajman CR, et al. Cisplatin and fluorouracil alone or with docetaxel in head and neck cancer. N Engl J Med. 2007;357(17):1705–15.
8. Chaturvedi AK, Engels EA, Pfeiffer RM, et al. Human papillomavirus and rising oropharyngeal cancer incidence in the United States. J Clin Oncol. 2011;29(32):4294–301.
9. Lacko M, Braakhuis BJ, Sturgis EM, et al. Genetic susceptibility to head and neck squamous cell carcinoma. Int J Radiat Oncol Biol Phys. 2014;89(1):38–48.
10. Tanaka A, Weinel S, Nagy N, et al. Germline mutation in ATR in autosomal-dominant oropharyngeal cancer syndrome. Am J Hum Genet. 2012;90(3):511–7.

11. Kutler DI, Auerbach AD, Satagopan J, et al. High incidence of head and neck squamous cell carcinoma in patients with Fanconi anemia. Arch Otolaryngol. Head Neck Surg. 2003;129(1):106–12.
12. Lawrence MS, Stojanov P, Mermel CH, et al. Discovery and saturation analysis of cancer genes across 21 tumour types. Nature. 2014;505(7484):495–501.
13. Zack TI, Schumacher SE, Carter SL, et al. Pan-cancer patterns of somatic copy number alteration. Nat Genet. 2013;45(10):1134–40.
14. Agrawal N, Frederick MJ, Pickering CR, et al. Exome sequencing of head and neck squamous cell carcinoma reveals inactivating mutations in NOTCH1. Science. 2011;333(6046):1154–7.
15. Lui VW, Hedberg ML, Li H, et al. Frequent mutation of the PI3K pathway in head and neck cancer defines predictive biomarkers. Cancer Disc. 2013;3(7):761–9.
16. Pickering CR, Zhang J, Yoo SY, et al. Integrative genomic characterization of oral squamous cell carcinoma identifies frequent somatic drivers. Cancer Disc. 2013;3(7):770–81.
17. Stransky N, Egloff AM, Tward AD, et al. The mutational landscape of head and neck squamous cell carcinoma. Science. 2011;333(6046):1157–60.
18. Rothenberg SM, Ellisen LW. The molecular pathogenesis of head and neck squamous cell carcinoma. J Clin Invest. 2012;122(6):1951–7.
19. Rothenberg SM, Mohapatra G, Rivera MN, et al. A genome-wide screen for microdeletions reveals disruption of polarity complex genes in diverse human cancers. Cancer Res. 2010;70(6):2158–64.
20. Olivier M, Hollstein M, Hainaut P. TP53 mutations in human cancers: origins, consequences, and clinical use. Cold Spring Harb Perspect Biol. 2010;2(1):a001008.
21. Petitjean A, Mathe E, Kato S, et al. Impact of mutant p53 functional properties on TP53 mutation patterns and tumor phenotype: lessons from recent developments in the IARC TP53 database. Human Mutat. 2007;28(6):622–9.
22. Lang GA, Iwakuma T, Suh YA, et al. Gain of function of a p53 hot spot mutation in a mouse model of Li-Fraumeni syndrome. Cell. 2004;119(6):861–72.
23. Olive KP, Tuveson DA, Ruhe ZC, et al. Mutant p53 gain of function in two mouse models of Li-Fraumeni syndrome. Cell. 2004;119(6):847–60.
24. Brown CJ, Lain S, Verma CS, et al. Awakening guardian angels: drugging the p53 pathway. Nature Rev Cancer. 2009;9(12):862–73.
25. Millon R, Muller D, Schultz I, et al. Loss of MDM2 expression in human head and neck squamous cell carcinomas and clinical significance. Oral Oncol. 2001;37(8):620–31.
26. Scheffner M, Werness BA, Huibregtse JM, et al. The E6 oncoprotein encoded by human papillomavirus types 16 and 18 promotes the degradation of p53. Cell. 1990;63(6):1129–36.
27. Werness BA, Levine AJ, Howley PM. Association of human papillomavirus types 16 and 18 E6 proteins with p53. Science. 1990;248(4951):76–9.
28. Lindenbergh-van der Plas M, Brakenhoff RH, Kuik DJ, et al. Prognostic significance of truncating TP53 mutations in head and neck squamous cell carcinoma. Clin Cancer Res. 2011;17(11):3733–41.
29. Poeta ML, Manola J, Goldwasser MA, et al. TP53 mutations and survival in squamous-cell carcinoma of the head and neck. N Engl J Med. 2007;357(25):2552–61.
30. Essmann F, Schulze-Osthoff K. Translational approaches targeting the p53 pathway for anti-cancer therapy. Br J Pharmacol. 2012;165(2):328–44.
31. Perez-Sayans M, Suarez-Penaranda JM, Gayoso-Diz P, et al. p16(INK4a)/CDKN2 expression and its relationship with oral squamous cell carcinoma is our current knowledge enough? Cancer Lett. 2011;306(2):134–41.
32. Reed AL, Califano J, Cairns P, et al. High frequency of p16 (CDKN2/MTS-1/INK4A) inactivation in head and neck squamous cell carcinoma. Cancer Res. 1996;56(16):3630–3.
33. Smeets SJ, Brakenhoff RH, Ylstra B, et al. Genetic classification of oral and oropharyngeal carcinomas identifies subgroups with a different prognosis. Cell Oncol. 2009;31(4):291–300.
34. Dominguez G, Silva J, Garcia JM, et al. Prevalence of aberrant methylation of p14ARF over p16INK4a in some human primary tumors. Mutat Res. 2003;530(1–2):9–17.

35. Ogi K, Toyota M, Ohe-Toyota M, et al. Aberrant methylation of multiple genes and clinicopathological features in oral squamous cell carcinoma. Clinical Can Res. 2002;8(10):3164–71.
36. Sailasree R, Abhilash A, Sathyan KM, et al. Differential roles of p16INK4A and p14ARF genes in prognosis of oral carcinoma. Cancer Epidemiol Biomarkers Prev. 2008;17(2):414–20.
37. Schache AG, Liloglou T, Risk JM, et al. Evaluation of human papilloma virus diagnostic testing in oropharyngeal squamous cell carcinoma: sensitivity, specificity, and prognostic discrimination. Clin Cancer Res. 2011;17(19):6262–271.
38. Sheu JJ, Hua CH, Wan L, et al. Functional genomic analysis identified epidermal growth factor receptor activation as the most common genetic event in oral squamous cell carcinoma. Cancer Res. 2009;69(6):2568–76.
39. Bova RJ, Quinn DI, Nankervis JS, et al. Cyclin D1 and p16INK4A expression predict reduced survival in carcinoma of the anterior tongue. Clin Cancer Res. 1999;5(10):2810–9.
40. Fu M, Rao M, Bouras T, et al. Cyclin D1 inhibits peroxisome proliferator-activated receptor gamma-mediated adipogenesis through histone deacetylase recruitment. J Biol Chem. 2005;280(17):16934–41.
41. Jirawatnotai S, Hu Y, Michowski W, et al. A function for cyclin D1 in DNA repair uncovered by protein interactome analyses in human cancers. Nature. 2011;474(7350):230–4.
42. Musgrove EA, Caldon CE, Barraclough J, et al. Cyclin D as a therapeutic target in cancer. Nat Rev Cancer. 2011;11(8):558–72.
43. Dickson MA. Molecular pathways: CDK4 inhibitors for cancer therapy. Clin Cancer Res. 2014;20(13):3379–83.
44. Ellisen LW, Bird J, West DC, et al. TAN-1, the human homolog of the Drosophila notch gene, is broken by chromosomal translocations in T lymphoblastic neoplasms. Cell. 1991;66(4):649–61.
45. Lee SY, Kumano K, Nakazaki K, et al. Gain-of-function mutations and copy number increases of Notch2 in diffuse large B-cell lymphoma. Cancer Sci. 2009;100(5):920–6.
46. Malecki MJ, Sanchez-Irizarry C, Mitchell JL, et al. Leukemia-associated mutations within the NOTCH1 heterodimerization domain fall into at least two distinct mechanistic classes. Mol Cell Biol. 2006;26(12):4642–51.
47. Puente XS, Pinyol M, Quesada V, et al. Whole-genome sequencing identifies recurrent mutations in chronic lymphocytic leukaemia. Nature. 2011;475(7354):101–5.
48. Weng AP, Ferrando AA, Lee W, et al. Activating mutations of NOTCH1 in human T cell acute lymphoblastic leukemia. Science. 2004;306(5694):269–71.
49. Talora C, Cialfi S, Segatto O, et al. Constitutively active Notch1 induces growth arrest of HPV-positive cervical cancer cells via separate signaling pathways. Exp Cell Res. 2005;305(2):343–54.
50. Talora C, Sgroi DC, Crum CP, et al. Specific down-modulation of Notch1 signaling in cervical cancer cells is required for sustained HPV-E6/E7 expression and late steps of malignant transformation. Genes Dev. 2002;16(17):2252–63.
51. Chen J, Jette C, Kanki JP, et al. NOTCH1-induced T-cell leukemia in transgenic zebrafish. Leukemia. 2007;21(3):462–71.
52. Dotto GP. Notch tumor suppressor function. Oncogene. 2008;27(38):5115–23.
53. Nicolas M, Wolfer A, Raj K, et al. Notch1 functions as a tumor suppressor in mouse skin. Nat Genet. 2003;33(3):416–21.
54. Pear WS, Aster JC, Scott ML, et al. Exclusive development of T cell neoplasms in mice transplanted with bone marrow expressing activated Notch alleles. J Exp Med. 1996;183(5):2283–91.
55. Koster MI. p63 in skin development and ectodermal dysplasias. J Invest Dermatol. 2010;130(10):2352–8.
56. Nguyen BC, Lefort K, Mandinova A, et al. Cross-regulation between Notch and p63 in keratinocyte commitment to differentiation. Genes Dev. 2006;20(8):1028–42.
57. Sasaki Y, Ishida S, Morimoto I, et al. The p53 family member genes are involved in the Notch signal pathway. J Biol Chem. 2002;277(1):719–24.

58. Wu G, Nomoto S, Hoque MO, et al. DeltaNp63alpha and TAp63alpha regulate transcription of genes with distinct biological functions in cancer and development. Cancer Res. 2003;63(10):2351–7.
59. Ha L, Ponnamperuma RM, Jay S, et al. Dysregulated DeltaNp63alpha inhibits expression of Ink4a/arf, blocks senescence, and promotes malignant conversion of keratinocytes. PLoS One. 2011;6(7):e21877.
60. Keyes WM, Wu Y, Vogel H, et al. p63 deficiency activates a program of cellular senescence and leads to accelerated aging. Genes Dev. 2005;19(17):1986–99.
61. Ramsey MR, Wilson C, Ory B, et al. FGFR2 signaling underlies p63 oncogenic function in squamous cell carcinoma. J Clin Invest. 2013;123(8):3525–38.
62. Moretti F, Marinari B, Lo Iacono N, et al. A regulatory feedback loop involving p63 and IRF6 links the pathogenesis of 2 genetically different human ectodermal dysplasias. J Clin Invest. 2010;120(5):1570–7.
63. Sen GL, Boxer LD, Webster DE, et al. ZNF750 is a p63 target gene that induces KLF4 to drive terminal epidermal differentiation. Dev Cell. 2012;22(3):669–77.
64. Zarnegar BJ, Webster DE, Lopez-Pajares V, et al. Genomic profiling of a human organotypic model of AEC syndrome reveals ZNF750 as an essential downstream target of mutant TP63. Am J Human Genet. 2012;91(3):435–43.
65. Restivo G, Nguyen BC, Dziunycz P, et al. IRF6 is a mediator of Notch pro-differentiation and tumour suppressive function in keratinocytes. EMBO J. 2011;30(22):4571–85.
66. Thomason HA, Zhou H, Kouwenhoven EN, et al. Cooperation between the transcription factors p63 and IRF6 is essential to prevent cleft palate in mice. J Clin Invest. 2010;120(5):1561–9.
67. Qiu W, Schonleben F, Li X, et al. PIK3CA mutations in head and neck squamous cell carcinoma. Clin Cancer Res. 2006;12(5):1441–6.
68. Qiu W, Tong GX, Manolidis S, et al. Novel mutant-enriched sequencing identified high frequency of PIK3CA mutations in pharyngeal cancer. Int J Cancer (Journal international du cancer). 2008;122(5):1189–94.(
69. Hibi K, Trink B, Patturajan M, et al. AIS is an oncogene amplified in squamous cell carcinoma. Proc Natl Acad Sci U S A. 2000;97(10):5462–7.
70. Nichols AC, Palma DA, Chow W, et al. High frequency of activating PIK3CA mutations in human papillomavirus-positive oropharyngeal cancer. JAMA Otolaryngol—Head Neck Surg. 2013;139(6):617–22.
71. Henken FE, Banerjee NS, Snijders PJ, et al. PIK3CA-mediated PI3-kinase signalling is essential for HPV-induced transformation in vitro. Mol Cancer. 2011;10:71.
72. Berger AH, Knudson AG, Pandolfi PP. A continuum model for tumour suppression. Nature. 2011;476(7359):163–9.
73. Okami K, Wu L, Riggins G, et al. Analysis of PTEN/MMAC1 alterations in aerodigestive tract tumors. Cancer Res. 1998;58(3):509–11.
74. Shao X, Tandon R, Samara G, et al. Mutational analysis of the PTEN gene in head and neck squamous cell carcinoma. Int J Cancer (Journal international du cancer). 1998;77(5):684–8.
75. Anderson JA, Irish JC, Ngan BY. Prevalence of RAS oncogene mutation in head and neck carcinomas. J Otolaryngol. 1992;21(5):321–6.
76. Saranath D, Chang SE, Bhoite LT, et al. High frequency mutation in codons 12 and 61 of H-ras oncogene in chewing tobacco-related human oral carcinoma in India. Br J Cancer. 1991;63(4):573–8.
77. Anderson JA, Irish JC, McLachlin CM, et al. H-ras oncogene mutation and human papillomavirus infection in oral carcinomas. Arch Otolaryngol—Head Neck Surg. 1994;120(7):755–60.
78. Barbie DA, Tamayo P, Boehm JS, et al. Systematic RNA interference reveals that oncogenic KRAS-driven cancers require TBK1. Nature. 2009;462(7269):108–12.
79. Scholl C, Frohling S, Dunn IF, et al. Synthetic lethal interaction between oncogenic KRAS dependency and STK33 suppression in human cancer cells. Cell. 2009;137(5):821–34.
80. Singh A, Greninger P, Rhodes D, et al. A gene expression signature associated with "K-Ras addiction" reveals regulators of EMT and tumor cell survival. Cancer Cell. 2009;15(6):489–500.

81. Taoudi Benchekroun M, Saintigny P, Thomas SM, et al. Epidermal growth factor receptor expression and gene copy number in the risk of oral cancer. Cancer Prev Res. 2010;3(7):800–9.
82. Temam S, Kawaguchi H, El-Naggar AK, et al. Epidermal growth factor receptor copy number alterations correlate with poor clinical outcome in patients with head and neck squamous cancer. J Clin Oncol. 2007;25(16):2164–70.
83. Chung CH, Ely K, McGavran L, et al. Increased epidermal growth factor receptor gene copy number is associated with poor prognosis in head and neck squamous cell carcinomas. J Clin Oncol. 2006;24(25):4170–6.
84. Cohen EE, Lingen MW, Martin LE, et al. Response of some head and neck cancers to epidermal growth factor receptor tyrosine kinase inhibitors may be linked to mutation of ERBB2 rather than EGFR. Clin Cancer Res. 2005;11(22):8105–8.
85. Cohen EE, Rosen F, Stadler WM, et al. Phase II trial of ZD1839 in recurrent or metastatic squamous cell carcinoma of the head and neck. J Clin Oncol. 2003;21(10):1980–7.
86. Kirby AM, A'Hern RP, D'Ambrosio C, et al. Gefitinib (ZD1839, Iressa) as palliative treatment in recurrent or metastatic head and neck cancer. Br J Cancer. 2006;94(5):631–6.
87. Soulieres D, Senzer NN, Vokes EE, et al. Multicenter phase II study of erlotinib, an oral epidermal growth factor receptor tyrosine kinase inhibitor, in patients with recurrent or metastatic squamous cell cancer of the head and neck. J Clin Oncol. 2004;22(1):77–85.
88. Wirth LJ, Haddad RI, Lindeman NI, et al. Phase I study of gefitinib plus celecoxib in recurrent or metastatic squamous cell carcinoma of the head and neck. J Clin Oncol. 2005;23(28):6976–81.
89. Gaykalova DA, Mambo E, Choudhary A, et al. Novel insight into mutational landscape of head and neck squamous cell carcinoma. PLoS One. 2014;9(3):e93102.
90. Li H, Wawrose JS, Gooding WE, et al. Genomic analysis of head and neck squamous cell carcinoma cell lines and human tumors: a rational approach to preclinical model selection. Mol Cancer Res: MCR. 2014;12(4):571–82.
91. Michaud WA, Nichols AC, Mroz EA, et al. Bcl-2 blocks cisplatin-induced apoptosis and predicts poor outcome following chemoradiation treatment in advanced oropharyngeal squamous cell carcinoma. Clin Cancer Res. 2009;15(5):1645–54.
92. Blumenschein GR, Jr., Kies MS, Papadimitrakopoulou VA, et al. Phase II trial of the histone deacetylase inhibitor vorinostat (Zolinza, suberoylanilide hydroxamic acid, SAHA) in patients with recurrent and/or metastatic head and neck cancer. Invest New Drugs. 2008;26(1):81–7.
93. He L, Torres-Lockhart K, Forster N, et al. Mcl-1 and FBW7 control a dominant survival pathway underlying HDAC and Bcl-2 inhibitor synergy in squamous cell carcinoma. Cancer Disc. 2013;3(3):324–37.
94. Wang D, Song H, Evans JA, et al. Mutation and downregulation of the transforming growth factor beta type II receptor gene in primary squamous cell carcinomas of the head and neck. Carcinogenesis. 1997;18(11):2285–90.
95. Levy L, Hill CS. Alterations in components of the TGF-beta superfamily signaling pathways in human cancer. Cytokine Growth Factor Rev. 2006;17(1–2):41–58.
96. Qiu W, Schonleben F, Li X, et al. Disruption of transforming growth factor beta-Smad signaling pathway in head and neck squamous cell carcinoma as evidenced by mutations of SMAD2 and SMAD4. Cancer Lett. 2007;245(1–2):163–70.
97. Bornstein S, White R, Malkoski S, et al. Smad4 loss in mice causes spontaneous head and neck cancer with increased genomic instability and inflammation. J Clin Invest. 2009;119(11):3408–19.
98. Han G, Lu SL, Li AG, et al. Distinct mechanisms of TGF-beta1-mediated epithelial-to-mesenchymal transition and metastasis during skin carcinogenesis. J Clin Invest. 2005;115(7):1714–23.
99. Nakaya K, Yamagata HD, Arita N, et al. Identification of homozygous deletions of tumor suppressor gene FAT in oral cancer using CGH-array. Oncogene. 2007;26(36):5300–8.
100. Morris LG, Kaufman AM, Gong Y, et al. Recurrent somatic mutation of FAT1 in multiple human cancers leads to aberrant Wnt activation. Nat Genet. 2013;45(3):253–61.

101. Patel V, Rosenfeldt HM, Lyons R, et al. Persistent activation of Rac1 in squamous carcinomas of the head and neck: evidence for an EGFR/Vav2 signaling axis involved in cell invasion. Carcinogenesis. 2007;28(6):1145–52.
102. Yap LF, Jenei V, Robinson CM, et al. Upregulation of Eps8 in oral squamous cell carcinoma promotes cell migration and invasion through integrin-dependent Rac1 activation. Oncogene. 2009;28(27):2524–34.
103. Nola S, Daigaku R, Smolarczyk K, et al. Ajuba is required for Rac activation and maintenance of E-cadherin adhesion. J Cell Biol. 2011;195(5):855–71.
104. McCabe MT, Ott HM, Ganji G, et al. EZH2 inhibition as a therapeutic strategy for lymphoma with EZH2-activating mutations. Nature. 2012;492(7427):108–12.
105. Jaramillo MC, Zhang DD. The emerging role of the Nrf2-Keap1 signaling pathway in cancer. Genes Dev. 2013;27(20):2179–91.
106. Hast BE, Cloer EW, Goldfarb D, et al. Cancer-derived mutations in KEAP1 impair NRF2 degradation but not ubiquitination. Cancer Res. 2014;74(3):808–17.
107. Adelstein DJ, Koyfman SA, El-Naggar AK, et al. Biology and management of salivary gland cancers. Semin Radiat Oncol. 2012;22(3):245–53.
108. Persson M, Andren Y, Mark J, et al. Recurrent fusion of MYB and NFIB transcription factor genes in carcinomas of the breast and head and neck. Proc Natl Acad Sci U S A. 2009;106(44):18740–4.
109. Ho AS, Kannan K, Roy DM, et al. The mutational landscape of adenoid cystic carcinoma. Nat Genet. 2013;45(7):791–8.
110. Stephens PJ, Davies HR, Mitani Y, et al. Whole exome sequencing of adenoid cystic carcinoma. J Clin Invest. 2013;123(7):2965–8.
111. Shao C, Sun W, Tan M, et al. Integrated, genome-wide screening for hypomethylated oncogenes in salivary gland adenoid cystic carcinoma. Clin Cancer Res. 2011;17(13):4320–30.

Bisphosphonate Related Osteonecrosis of the Jaw

Athanasios Zavras

Forty five (45) years. According to the World Health Organization Global Health Observatory, this is the number of years that a boy born in 2012 in Sierra Leone is expected to live. In contrast, life expectancy in Switzerland is 81 years. This 35+ years differential in years lived demonstrates beyond statistics what economic development, scientific advances, modern medicine and public health have achieved in just few decades [1]. In contrast, the life expectancy data in Africa clearly demonstrates that parts of the World are in dire need for such basic development, medical services and basic public health infrastructure.

Personalized medicine holds the promise of major medical breakthroughs and can result in improved patient outcomes. While personalized medicine may seem a luxury for millions of individuals today, experience shows that ever improving scientific discovery becomes the major force to "democratize expertise" that will ultimately allow everyone to enjoy the benefits of personalized medicine.

In many occasions currently, new personalized treatments require molecular diagnostics via specialized laboratory analytics and biologic treatments with costs exceeding two hundred thousand dollars per year. However, costs related to genomics are drastically reduced and this scale of reduction is unprecedented. When the Human Genome Project started the cost of sequencing the human genome of a single individual was estimated at $ 1 billion. [2] Few years into the process, today sequencing costs about $ 1000 per genome, with projections for further 5-fold reductions in a decade.

A. Zavras (✉)
Department of Pediatric Dentistry, Boston University, Boston, USA
e-mail: zavras@bu.edu

Henry M. Goldman School of Dental Medicine, 100 East Newton Street, Suite 706, Boston, MA 02118, USA

S. T. Sonis (ed.), *Genomics, Personalized Medicine and Oral Disease*,
DOI 10.1007/978-3-319-17942-1_14

While early utilization of personalized medicine is currently limited to the Developed World, the promise of evidence-based, inexpensive, public health interventions that harness the principles of personalized medicine across the Globe is realistic.

As an idea, personalized medicine and the notion that genetic inheritance causes disease is around since Sir Archibald Garrod's observations of patients with alkaptonuria in 1902 [3]. However, it is only in the post Human Genome Project era that advances in biology, high throughput genomics and computer science allowed for accelerated development of the first few case studies, success stories that have motivated the medical community to fully embrace the concept. While in its infancy, personalized medicine was not embraced by all key stakeholders, today personalized medicine is widely considered a big wave of medical breakthrough, a good example of human ingenuity and perseverance that allows for a better quality of life and the control of diseases that were notoriously difficult to treat.

Cancer in particular is a major challenge. Before personalized medicine, the traditional histology-based phenotyping was partly responsible for failed treatments. Patients were classified in broad categories based on similar raw characteristics in anatomy or histology of the disease. Today, technology allows for full genome scans of the tumor and for selection of treatments that address certain etiologic pathways. While significant progress has been made, major challenges still exist due to a mutational landscape that does not remain constant over time, mutations in the cancer's genome that render treatments obsolete within few months.

Beyond treatment, technology today also allows for genotyping and risk assessment for individualized prevention. Where in the past epidemiologists relied on broad risk categories to classify risk (for example, smoking status), today complicated algorithms are constructed to incorporate the synergies between life style risk factors and inherited or acquired genetic risk factors. The risk assessment algorithms sub-classify patients into more defined strata of etiology based on biologic pathways, and thus predict with reasonable accuracy who is susceptible to develop disease. Some of these risk assessment tools have been placed in clinical practice. Various others are currently in clinical development and are expected to reduce both the incidence and the overall burden of disease in the future.

Some of the first applications of personalized medicine involved cancers of the breast and lung, targeting specific tumor mutations. Such targeted therapies seem to have fewer side effects and are better tolerated as they target a small population of affected cells. Yet, the majority of chemotherapies today still involve a broad-based target. As such, cancer chemotherapy is known to have multiple adverse effects. Applying the concept of personalized medicine to identify who will develop toxicity and adverse effects has the potential to reduce morbidity and mortality.

One such adverse effect of chemotherapy is osteonecrosis of the jaw, widely known by its initials as ONJ. ONJ has been linked to anti-resorptive bone chemotherapies such as intravenous bisphosphonates and denosumab.

Bisphosphonates (BPs) are non-metabolized analogues of inorganic pyrophosphate [4] that are used for the treatment of various types of osteolysis. They are administered orally or intravenously. The most commonly orally administered BPs are alendronate, risedronate and ibandronate; they are used in the treatment of osteoporosis, osteopenia, Paget's disease and rheumatoid arthritis. The most commonly intravenously administered BPs are zoledronic acid and pamidronate; they are used in the clinical management of hypercalcaemia of malignancy, in osteolytic lesions of multiple myeloma and in metastatic bone cancer. Osteonecrosis of the jaw (ONJ) is a rare condition first described among users of intravenous bisphosphonates [5]. Today the list of chemotherapies that can cause ONJ has been expanded to include denosumab and other anti-resorptive medications. As more medications are added to the list, the definition and the name of the condition itself has been changed from osteomyelitis to bisphosphonare related osteonecrosis of the jaw (BRONJ) to anti-resporptive osteonecrosis of the jaw (ARONJ) to medication-induced osteonecrosis of the jaw (MIONJ). However, the great majority of cases are still attributed to intravenous bisphosphonates.

Clinically, ONJ is characterized by exposed necrotic bone lesions that persist for more than 8 weeks [6]. The lesions range from small asymptomatic areas of exposed necrosis to large painful symptomatic lesions that extent to both jaws and remain refractory to treatment. Clinical management, and its resulting morbidity, depends on the severity of the symptoms and the extent of the bony destruction. Small asymptomatic lesions may be left alone to heal or are managed with long term antibiotic therapy, whereas symptomatic and infected large sequestrations extending extra-orally onto the face may lead to radical resections of both jaws [7]. Osteonecrosis may be particularly disruptive in already vulnerable cancer patients, compromising their quality of life and adding levels of complexity to their already complex treatment plans. [8]

Epidemiology and Time-to-Event Statistics

While ONJ has been reported among users of oral bisphosphonates, the majority of the reported ONJ patients have received zoledronic acid. In a review of 71 published case series we found that the average time-to-event after zoledronic acid initiation was 1.8 years, with a minimum time-to-event of 10 months. For pamidronate, the mean time was 2.8 years, and the minimum was 18 months. Oral bisphosphonates had an average time-to-event of 4.6 years, with a minimum of 3 years [9]. Factors that seem to affect ONJ and time to event were invasive dental procedures, and other comorbid factors such as increased age, rheumatoid arthritis, diabetes, use of corticosteroids, Vitamin D deficiency, and more.

Our results were consistent with the general notion that zoledronic acid constitutes the most potent bisphosphonate [10, 12], perhaps related to its high effectiveness in inhibiting bone turnover [13]. According to Ortega et al. [14] each additional administration of zoledronic acid is related to a 10 % increase in the risk of developing ONJ, while according to Cafro et al. [15] 12 consecutive doses of zoledronic acid double the risk of developing ONJ.

Not only was zoledronic acid the most prevalent BP among ONJ cases, but it was also associated with the shortest time-to-event [10–12, 16–17]. Because ONJ was developed rapidly in some cases, we searched for reasons that would lead to the outliers. Invasive dental procedures seem to be a major predisposing factor for the onset of the symptoms in these cases. Mavrokkokki et al. [10] report that 73 % of all cases of jaw osteonecrosis appeared after tooth extraction, but this report did not provide more specific details on the outliers. Several reports [12, 16–18] refer to tooth removal before the onset of the disease. While surgical interventions to the jaw bones might expedite the development or appearance of ONJ, there are reports of ONJ appearing spontaneously after few months of BP use. For example, Saussez et al. [19] reported a case of a multiple myeloma patient who presented with spontaneous jaw osteonecrosis after receiving zoledronic acid for 4 months.

A frequently reported confounding factor is the use of corticosteroids [20, 21] that demonstrate anti-angiogenic and immunosuppressive properties [22]. Corticosteroids could have an impact on wound healing after any surgical intervention to the bone ultimately leading to jaw osteonecrosis. Chemotherapeutic agents are also reported as confounding factors. In particular, dexamethasone has been described as commonly used among the cases studied by Marx et al. [23]. It is also documented experimentally that the co-administration of zoledronic acid with dexamethasone in rats results in the development of bone changes that are similar to those presented in jaw osteonecrosis in humans [24]. Additional factors that can have an additive impact on ONJ onset are rheumatoid arthritis, renal insufficiency [25, 26], alcohol [19, 23, 27], smoking [19, 26, 28], obesity [28], age [23, 29], glucocorticoids [13, 17, 23, 26], anaemia [17, 26, 30], diabetes [14, 27, 31], surgical periodontal therapy, trauma of mandibular/maxillary tori, dental caries, root canal treatments, ill-fitting dentures and dental infections [32, 33]. Diabetes might also be a contributing risk factor since it has been shown in mice that zoledronic acid administration inhibits endothelial cell function and angiogenesis, processes already impaired in diabetic patients [31]. Increased age seems to be associated with ONJ. It has been claimed that there is a 10 % increase in the risk of developing ONJ per decade of age [29]. Interestingly, no case of BONJ has been reported to date in pediatric patients exposed to intravenous BP therapy due to bone disorders such as osteogenesis imperfect [34, 35]. Further studies focusing on these factors are needed to better understand the complex etiology of osteonecrosis of the jaw.

Independent from bisphosphonate treatment, Greuter et al. [36] reported a case in which clinical symptoms were compatible with those of ONJ and appeared 2 months after treatment with bevacizumab and extraction of two teeth because of a dental infection. Bevacizumab is a humanized monoclonal antibody approved for breast and lung cancer treatment that binds to endothelial growth factor and inhibits angiogenesis. The patient was a 63-year-old woman with breast cancer. The patient had no history of bisphosphonate use, received corticosteroids, had diabetes and was a smoker. Estilo et al. [37] reported two cases of jaw osteonecrosis developed after 2 and 4 months of bevacizumab treatment respectively. Both patients had intact dentitions and no evidence of infection.

Jaw osteonecrosis may also be triggered by Herpes zoster infection [38–40]. Sporadic cases of Herpes-infected patients developing jaw osteonecrosis without concurrent bipshonate treatment have previously been reported especially in patients with concurrent HIV or cytomegalovirus (CMV) infection.

There are only few well-controlled epidemiologic studies of ONJ, and its pathogenesis remains obscure. ONJ today is considered to be multifactorial. Bisphosphonates are absorbed by the skeleton, especially by active remodeling sites and bind to bone minerals [41]. Because of the high bone turnover in the jaws, bisphosphonates tend to concentrated there [23]. Zoledronic acid, pamidronate, alendronate and risedronate are nitrogen-containing bisphosphonates involved in the inhibition of the mevalonate pathway in osteoclasts [42] through their action on farnesyl pyrophosphate synthase (FPPS), an enzyme involved in the biosynthesis of cholesterol and other sterols [41]. The main function of bisphosphonates is to decrease the rate of bone resorption, ultimately reducing the rate of bone remodeling. They suppress osteoclast activity by preventing their development, diminishing their lifespan and increasing their apoptosis [43]. In addition, there are studies conducted both *in vitro* and *in vivo*, which support the hypothesis that bisphosphonates interfere with angiogenesis since it is thought that they possess anti-angiogenic properties [17, 23]. Another theory is that the oral microbial flora stimulated by oral surgery could possibly trigger ONJ related phenomena [44]. According to Allen [45], in a series of studies in healthy dogs on zoledronic acid treatment, non-viable regions of osteocytes were observed in the alveolar bone 3 months after drug administration. However the authors were not able to experimentally "produce" clinical lesions in their animal model. To our knowledge the only two animal models of osteonecrosis of the jaw are those of Sonis et al. [24] and of Nishimura et al. [46]. In the first case, zoledronic acid together with dexamethasone caused non-healing lesions resembling ONJ in the sites of extracted teeth of rats, while Nishimura et al. were able to create clinical ONJ lesions in a model of Vitamin D deficient rats treated with zoledronic acid.

Interestingly the Vitamin D pathway has been proposed as one possible biologic pathway leading to BONJ in humans. Ardine et al. [47] after studying patients with bone metastatic breast cancer on long term zoledronic acid treatment observed that low serum calcium and elevated levels of parathyroid hormone may be significantly correlated to ONJ development. These results are consistent with data from Berruti et al. [48, 49], who suggest that hypocalcaemia may predispose prostate cancer patients to ONJ development, while secondary parathyroidism may play a significant role as well. The coexistence of hypόvitaminosis D may also contribute to both the obstruction of bone repair and soft tissue destruction through an immune response that involves gamma delta T cells. Genetic predisposition may also play an important role in the development of BONJ. Genetic polymorphisms that might lead to hypocalcaemia or hyperparathyroidism, for example in the Vitamin D Receptor gene (VDR) or in the parathyroid (PTH1) gene, may explain in the future how the Vitamin D pathway is involved in ONJ. Interestingly, certain VDR genotypes have been implicated in the metabolic syndrome or in obesity via interactions with the Insulin Like Growth factor system (IGF), which may also explain epidemiologic findings associating obesity with ONJ in humans [28]. Therefore, tailoring vitamin D and calcium supplementation might prove helpful in reducing the frequency of ONJ.

While a lot of attention has been given to the hard tissue effects of bisphosphonates, osteonecrosis of the jaw is defined as exposed bone that persists for several weeks. Morphologically, exposed bone requires an overlying non-healing mucosa for ONJ to occur. Thus it is of interest to understand how bisphosphonates affect the soft tissues. Allegra et al. [50] studied the effect of bisphosphonates on endothelial cells in multiple myeloma patients. Their data are in agreement with the hypothesis that bisphosphonate interfere with angiogenic events since they documented that these drugs could also have suppressive and anti-angiogenic effects on endothelial cells. The reduced levels of angiogenic factors could cause endothelial dysfunction and reduced endothelial cell proliferation, thus leading to jaw osteonecrosis. Another study that looked into soft tissue effects was the one by Landesberg et al., describing the effect of pamidronate on oral mucosal cells. After isolation of oral keratinocytes from mice and exposure of these cells to pamidronate, inhibition of oral mucosal cell proliferation as well as inhibition of wound healing was observed. Therefore, this might be a significant promoting factor for BONJ onset. According to a study by Deng et al. [51], alendronate was found to augment the pathogenesis of periodontal bacteria by promoting IL-1β production, thus contributing to the development of inflammatory side effects including osteonecrosis of the jaw.

Our previous review clearly demonstrated that ONJ risk increases as time on the medications (and thus, cumulative dose) increased, pointing to a toxic effect. In 2013, a population pharmacokinetic model of pamidronate accumulation in plasma and bone was calculated using the Pmetrics® package. This model showed a derived toxic bone BP threshold of 0.2 mM, with mean bone BP in ONJ cases higher than in controls (0.20 vs. 0.10 mM, $P<0.001$). ONJ was also associated with longer duration BP therapy (5.3 vs. 2.7 years, $P<0.001$), older age (76 vs. 70 years, $P<0.001$), and Asian race (49 vs. 14 %, $P<0.001$) [52]. Corroborating our hypothesis of a toxic effect, the authors concluded that bisphosphonates accumulate differently and those reaching a high, toxic level are predisposed to ONJ development. However, it is currently unclear what biologic mechanism controls such concentration in the jaws. All being equal, it still remains impossible to predict accurately who will develop ONJ based solely on dose or duration of use. Therefore, it is imperative to identify the biologic mechanism that leads to higher bone accumulation.

Genome-Wide Pharmacogenetics of Osteonecrosis of the Jaw

Pharmacogenetics may offer a plausible solution to mitigate the problem. Pharmacogenetics is the science of discovering the genetic components of drug response, for both efficacy and safety. If a pharmacogenetic test is indeed validated, candidates for anti-resorptive medications will be tested prior to therapy initiation. Positive results will lead to prescription of an alternative therapy. A negative result will allow eligible patients to experience the many benefits of anti-resorptive medications without fear of the serious adverse effect. Thus, the scientific opportunity of a pharmacogenetic test is greater than just preventing osteonecrosis in those susceptible.

In the recent past we conducted a pilot pharmacogenetic case-control genome-wide association study of ONJ [53]. Genome-wide association studies do not require a formal hypothesis. Rather they are agnostic in their approach, letting the actual data dictate the responsible biologic pathways. The study was approved by the Institutional Review Boards of Harvard Medical School and Columbia University Medical Center, and was funded by the National Institute of Dental and Craniofacial Research in Bethesda. We recruited and analyzed DNA from 30 white ONJ cases and 17 non affected bisphosphonate users (treatment-tolerant controls). Cases had intraoral necrotic lesions that persisted for more than 8 weeks; controls were treatment tolerant bisphosphonates users who had not developed osteonecrosis at the time of enrolment. In what follows, we refer to the original study group as the *ONJ Group*. Saliva was collected, as well as demographic and risk factor data. High throughput genotyping was performed using the Human Omni Express 12v1 Illumina chip. The Human Omni Express 12v1.0 Beadchip captures 731,442 markers, representing more than 91 % of human variation for major alleles with frequencies above 5 % in Caucasians.

To increase the power of the study to detect possible true associations, we subsequently augmented the controls with publically available data from population based controls and a breast cancer GWAS that is published in dbGAP (dbGAP PHS000210.v1 study). More specifically, we selected Caucasian subjects from POPRES [54], the Wellcome Trust Case Control Consortium (WTCCC) [55], the Illumina iControlDB [56] and the international Serious Adverse Events Consortium (iSAEC) [57] collections. All subjects, except the ones from iControlDB, were genotyped using Illumina 1 M or 1 M-duo chips; the subjects from the iControlDB were genotyped using Illumina 550 K chip. After applying standard quality control procedures, the controls were combined with the initial ONJ group to produce the "*population control group*". The effect of population structure was assessed through Principal Components Analysis (PCA), using the smartPCA program from the EIGENSTRAT package (version 3.0) [58]. SNPs from known regions of long-range linkage disequilibrium (LD) were removed before running the PCA [59]. As the *population control group* contained a disproportionally large set of north European subjects, we chose a fixed cases/control ratio of about 1/60, selecting the best genetically matched controls based on eigenvalues of the significant PC axes.

Additionally, in order to test the effect of possible confounding factors, which may be related to either BPs exposure or clinical diagnosis, we downloaded a set of treatment-tolerant cancer sample data from the phs000210.v1.p1 cohort from dbGAP [60]. This cohort is composed of 878 breast cancer patients genotyped by the Illumina 610 K chip. We selected 107 subjects, who had been BP users, by reviewing the patient's clinical data (form: pht000898.v1.p1.c1) included in the downloaded datasets. After selecting the Caucasian BP users by PCA, the controls were combined with the initial ONJ treatment-tolerant controls to produce the *treatment-tolerant group*.

Imputation Imputation was carried out using IMPUTE2 (version February 2009), with data from the 1000 Genomes Project (112 individuals, release number Mar 2010) and HapMap III (Jun 2010, all ethnicity) as the reference panels [61]. SNPs with poor quality were pruned before the imputation to avoid false positives. We divided the genome in 5000 bp long segments and we used the "*png-miss*" option to fill in the missing genotypes. We used an ethnic mixed panel to improve the quality of the imputation for rare variants [62]. We retained imputed genotypes with: (1) posterior probability greater than 0.9 [61]; (2) no significant difference in missingness between cases and controls (Chi-sq test P-value >0.05); and (3) no significant deviation of the HWE (P-value >0.05)

Statistical Analysis We conducted statistical tests using the PLINK software. We tested the association of single SNPs using logistic regression with the PCA eigenvalues as covariates under an additive model. We set the genome-wide significance p-value threshold to 5×10^{-8}, to correct for multiple testing (Bonferroni correction). When top results were imputed we assessed the accuracy of the imputation manually, confirming that the quality of the signal intensity was within range of acceptance for all the SNPs in the haplotype generating the imputed genotype. For the candidate gene analysis we reviewed the literature and identified genes possibly involved in the pathogenesis of ONJ, including the Insulin-like Growth Factor (IGF) gene family and genes belonging to the vitamin D metabolism. We also included ADME genes from a list compiled specifically for pharmacogenetic studies, as described by Ahmadi et al. in Nature Genetics of 2005. Appropriate Bonferroni correction was applied to the candidate gene analyses.

Copy Number Variations Analysis We inferred copy number variations (CNVs) from SNP chip data using the PennCNV software (April 2009 version) [63]. To ensure the accuracy of CNV calling we applied stringent sample and CNV filtering procedures. We included all samples that had a log2 ratios standard deviation <0.5, maximum number of total CNV calls <50, bioaccumulation factor (BAF) median >0.55 or <0.45, BAF drift >0.01 or waviness factor (WF) >0.05 or <-0.05 (recommended parameters). Additionally, to ensure high-confidence CNVs, we excluded individual CNVs with PennCNV-generated confidence score <10; those with calls based on fewer than 10 SNPs/CNV probes; and those with span within 1 Mb from centromeres or telomeres.

We performed burden and common CNVs association analysis where any CNV, which was present in at least three subjects, was considered to be common. Associations were tested with the PLINK software using a two tails permuted Fisher's exact test (10^5 permutations). Duplications and deletions were analyzed separately [64]. We also investigated singleton CNVs >500 kb to find evidence for individual predisposition to ONJ. We adopted a coverage cut off, excluding all CNVs that had coverage <20 genetic markers. Finally, we selected the top 150 genes most frequently involved in CNVs (both duplications and deletions). We used the David software [65] to perform enrichment analysis (Fisher Exact test) of the Kyoto Encyclopedia of Genes and Genomes (KEGG) Pathway Database (released December 2010) [66].

The results of this study first appeared in The Oncologist in 2012 [53], and are re-printed here with permission from AlphaMed Press (Rightslink Copyright License 3515370690803).

The first phase of the analysis involved Discovery in the original study group.

Population Structure and Selection of Genetically Matched Controls We applied Principal Components Analysis (PCA) to expose population structure of the *ONJ group* and to find additional genetically matched population controls from publicly available collections. In order to confirm the self-reported ethnicity of the members of the *ONJ group* we merged their genotype data with that of 987 HapMap III subjects, which included subjects from 11 populations [67]. We found 6 individuals not clustering with the Caucasian HapMap III samples (CEU and TSI). For the remaining 47 Caucasian subjects, we attempted to refine ethnicity resolution by clustering them with population controls from the POPRES, WTCCC and iSAEC collections, representing several European sub-populations. This analysis showed that the Caucasian ONJ study subjects clustered with individuals of northwestern, southern and eastern European descent. To further increase the number of Eastern European controls we added 2978 Caucasian samples from the iControlDB dataset. We formed the "*population control group*" by selecting the 60 closest controls for each case, based on the eigenvalues of the first six principal components. The case/control ratio was chosen to maximize the total number of controls while keeping the ratio comparable among the three major clusters. In addition to the *population control group* we identified publically available GWAS data on a set of breast cancer patients who had been treated with bisphosphonates without developing ONJ. We used PCA to select 101 Caucasian treatment-tolerant cancer subjects out of the 107 drug-exposed controls from the phs000210.v1.p1 cohort [60] and we merged them with the 17 Caucasian ONJ treatment-tolerant controls in a "*treatment-tolerant group*". PCA analysis showed that all the three major ethnicity clusters (northwestern, southern and eastern Europeans) were equally balanced in the *treatment-tolerant group*.

The *Caucasian ONJ group* contained 30 Caucasian cases and 17 Caucasian treatment exposed controls. In total, 631,507 SNPs passed quality control. In order to maximize the power of the study, we grouped each of the 30 Caucasian ONJ cases with 60 of their closest genetically-matched population controls, resulting in a study sample comprising 30 cases and 1743 controls (724 males, 1049 females). As not all population controls were genotyped with the same chip, only 287,434 SNPs were shared by all subjects. We imputed this dataset utilizing reference panels from HapMap 3 and the 1000 genomes project. In total 3,542,142 markers passed quality control procedures specific for the imputation. We tested the association of single SNPs using logistic regression with the first six eigen scores as covariates under an additive model. Rs17024608, located in an intron of gene RBMS3, was found significantly associated with ONJ $P\text{-value}=7.4\times10^{-8}$; individuals positive for rs17024608 had a 5+-fold increase in their risk of developing ONJ as compared with negative individuals. We note that this SNP is present in all genotyping platforms, but 2% of the controls had a missing call for this marker; consequently, only the missing genotypes were predicted by the imputation. Table 1 summarizes the top findings from the logistic regression on the imputed data. Figures 1 and 2 respectively present the quantile-quantile plot (QQ plot) of logistic regression and the Manhattan plot of the region (±1 M bp) surrounding rs17024608.

The *treatment-tolerant group* contained 118 treatment-tolerant controls (101 individuals from breast cancer cohort and 17 from *ONJ group*). There was no

Table 1 BRONJ Genome Wide Association Study: top associated SNPs. We performed a GWAS on 30 Caucasian cases and 1743 genetically-matched controls with more than 3 million of markers, applying a logistic regression statistic. The table shows the characteristics of the top associated SNPs

SNP	CHR	Position	AA	Closest gene	Function	OR (CI95%)	PV	MAF cases	MAF Ctls
rs17024608	3	29929694	G	RBMS3	Intron	5.8 (3.0–11.0)	7.47–08	0.28	0.08
rs5768434	22	46977516	T	FAM19A5	Intergenic	12.6 (4.9–32.2)	1.17–07	0.12	0.01
rs11064477	12	6944626	A	PHB2	Intergenic	21.7 (6.5–71.9)	5.16–07	0.09	0.01
12–7016684	12	7016684	T	C1S	Intergenic	21.1 (6.4–69.8)	5.85–07	0.09	0.01
8-58133986	8	58133986	T	IMPAD1	Intergenic	7.3 (3.1–16.9)	3.10–06	0.16	0.04
rs1886629	1	194421521	C	KCNT2	Intergenic	3.6 (2.1–6.5)	5.53–06	0.32	0.10
rs7588295	2	166115757	G	CSRNP3	Intronic	8.6 (3.3-22.17)	6.24–06	0.10	0.01
rs4431170	4	165504024	G	MARCH1	Intronic	5.1 (2.5–10.6)	7.28–06	0.20	0.05
rs7740004	6	120897902	A	C6orf170	Intergenic	5.9 (2.7–13.0)	7.87–06	0.15	0.03
rs11189381	10	99553188	C	SFRP5	Intergenic	6.8 (2.9–15.8)	8.17–06	0.17	0.02
rs12903202	15	56094085	G	ALDH1A2	Intronic	4.0 (2.1–7.4)	9.15–06	0.27	0.08
rs17751934	18	47455812	T	MEX3C	Intergenic	5.0 (2.4–10.1)	9.16–06	0.18	0.04
11-23990403	11	23990403	C	LUZP2	Intergenic	12.7 (4.0–36.8)	9.94–06	0.08	0.01

MAF minor allele frequency, *OR* Odds Ratio, OR is expressed with a confidence interval of 95%, *Ctls* controls, *CHR* chromosome, *PV* logistic regression *P*-value, *AA* ancestral allele

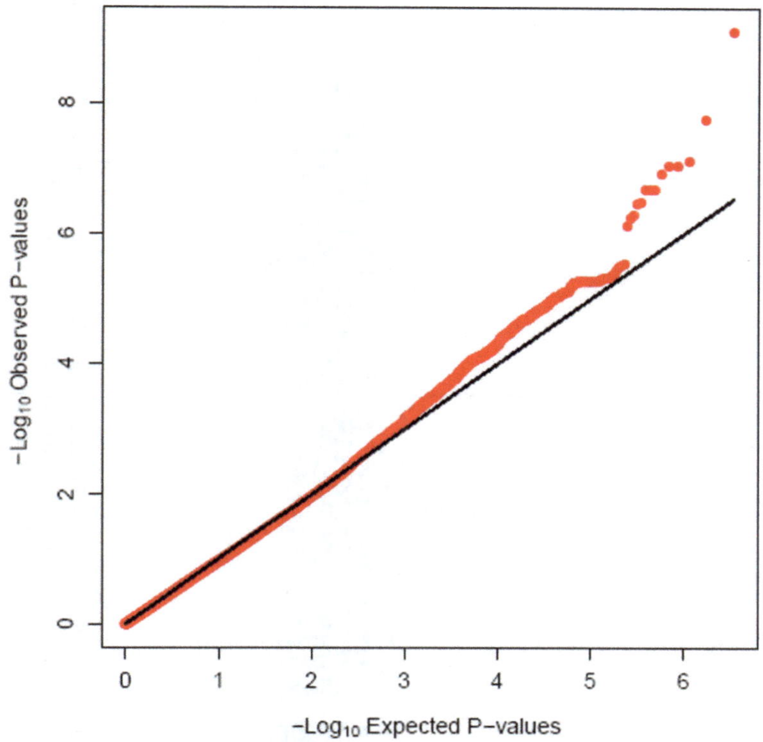

Fig. 1 QQ plot for logistic regression on the Population control group. On the x axis is −log10 of the expected *P*-values of an equally sized set of SNPs under a uniform distribution. The y axis is −log10 of the observed *P*-values. *Black solid lines* denote the uniform null distribution. The bulk of the values (*red dots*) closely follow the expectation under the null model (*black line*) showing that there is no systematic artifact of population stratification. The tail end shows significant deviation from null model illustrating that there are a few observed significant associations

difference in minor allele frequency of rs17024608 between the *treatment-tolerant* and the *population controls* (Fisher Exact test, *P*-value = 0.2). This finding is consistent with the hypothesis that this SNP is truly associated with the ONJ phenotype as its association is unlikely to be due to confounding factors related to the BPs exposure or clinical diagnosis.

SNPs in Candidate Genes Previous work suggests inherited genetic variations in the Insulin-like Growth Factor (IGF) gene family or in genes associated with drug Absorption, Distribution, Metabolism and Excretion (ADME) may play a role in the pathophysiology of ONJ. With regard to the IGF gene family, we examined 1083 SNPs located within 20 kb downstream and upstream from the longer transcripts of 40 putative causal genes (called *putative causal gene list*). We also examined a list of 4564 SNPs compiled for phamacogenetic studies related to drug- Absorption, Distribution, Metabolism and Excretion (ADME) genes (called *ADME list*).

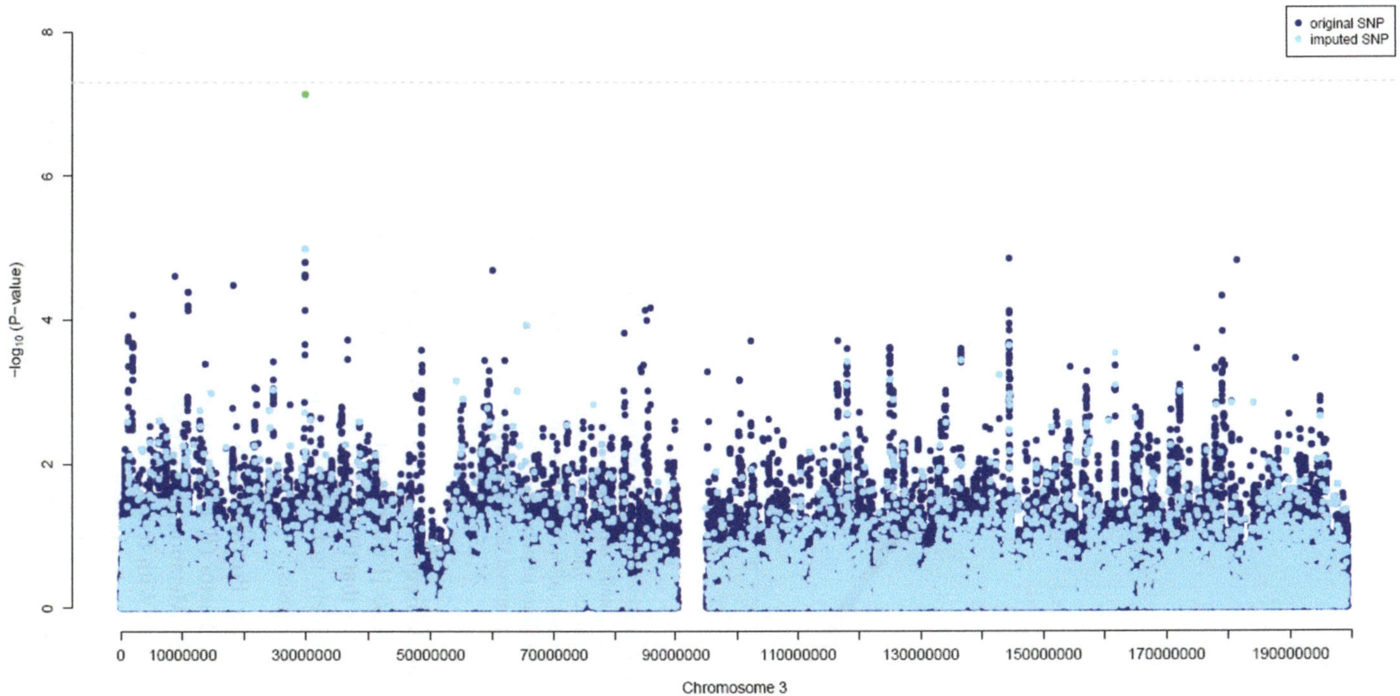

Fig. 2 Manhattan Plot of the region surrounding rs17024608 (about ± 1 Mb). Each *dot* represents a SNP. The *x* axis represents the position of the SNP on chromosome. The *y* axis represents the -$\log_{10}$ of logistic regression *P*-value of the SNP in the case-control association study. *Light blue dots* represent the imputed markers; *Dark blue dots* represent the genotyped markers. rs17024608 was marked in *green* with the *P*-value just below the genome-wide threshold (*dashed line*). For this marker we were able to impute the genotype for 2 % of controls who had missing value for this marker

The same two panels of markers were inspected in the *population control group*. No SNP reached significance after Bonferroni correction. In the *putative causal gene list*, the most significantly associated SNP was rs11934877, located in the intronic region of *IGFBP7* gene (OR = 2.9; 95 % CI: 1.7–5.2, *p*-value = 0.00022). In the ADME list, rs1678387, intronic within gene ABCC4, was the top associated with a borderline significance after multiple testing correction (OR = 5.3; 95 % CI: 2.4–11.4, *P*-value = 2.0×10^{-5}). For the SNPs (or their proxies), there was no difference in MAF between population and drug exposed control subjects (Table 2).

CNV Association Analysis All analyses were performed on the initial 52 Caucasian subjects. Fifty two individuals (33 cases and 19 controls) passed our stringent quality-control criteria for CNV calling; 431 CNVs were identified, of which 71 were duplications and 360 were deletions. Cases and controls did not differ significantly in their rate of CNVs for both deletions and duplications. After multiple-test correction, none of the common CNVs had a significant association. We found two unique oversized (greater than 700 kb) duplications in cases, and none in controls. The duplications were found on chromosomes 2 (925,407 bp; starting on rs4850234 and ending on rs16837705) and chromosome 22 (730,236 bp; starting on rs6003971 and ending on rs2845421) respectively.

Oversized singleton CNVs, as the ones predicted, might explain individual predisposition to the phenotype. In particular, the 730 kb heterozygous duplication covers the SLC7A4 gene; the gene codes for a solute carrier transporter, which may be relevant for the bioavailability of BPs. The most enriched KEGG pathways in genes within CNVs were the Notch signaling pathway involving 5 genes (NOTCH1, PSEN1, NUMB, NOTCH4 and DVL1) with an enrichment score of 6 and a fishers exact *p*-value of 0.009; the retinol pathway involved 5 genes (CYP3A7, CYP2C19, ADH6, CYP2A6, CYP2A7) with an enrichment score of 5.2 and a *p*-value of 0.014; P450 pathway involved 5 genes (CYP3A7, CYP2C19, ADH6, CYP2A6, CYP2A7) with an enrichment score of 4.6 and a *p*-value of 0.023; and the drug metabolism pathway that involved 4 genes (CES2, CYP3A7, CYP2A6, CYP2A7) with an enrichment score of 1.0 and a *p*-value of 0.039.

Discussion on Genomics and Etiology of ONJ

Osteonecrosis of the jaw is a serious adverse effect of bisphosphonates, especially among cancer patients. In this vulnerable group, osteonecrosis of the jaw (ONJ) negatively affects the patients' quality of life [68]. Differences in ONJ incidence among ethnic groups and previously published results from pharmacogenetic studies indicate that genetic factors might be central in the ONJ predisposition, besides the known risk factors. The goal of our genome wide association study was to identify high penetrance genetic biomarkers associated with ONJ. By using genetically matched population controls we were able to identify rs17024608 in RBMS3 as significantly associated with the risk of osteonecrosis, controlling for

Table 2 Top associated SNPs from the candidate genes analysis. We inspected lists of SNPs belonging to genes putatively involved in the etiology of BRONJ. The table shows Odds Ratio (OR) and *P*-value (PV) of the top associated SNPs and compares Minor Allele Frequency (MAF) between population ($n=1743$) and treatment-tolerant controls ($n=118$)

SNP	CHR	Position	Closest Gene	MAF Cases	OR (CI 95 %)	PV	MAF population controls	MAF treatment-tolerant controls	SNP proxy
rs11934877	4	57635783	IGFBP7	0.3	2.9(1.6-5)	0.0002	0.1	0.1	NA
rs1678387	13	94515907	ABCC4	0.1	5.3 (2.4–11.4)	2.0^{-5}	0.037	0.04	rs1189437

MAF minor allele frequency, *OR* Odds Ratio, OR is expressed with a confidence interval of 95 %, *PV P*-value from logistic regression

multiple comparisons. Furthermore, since there is no statistical difference in MAF between the treatment-tolerant and general population controls, the association of rs17024608 with the ONJ is unlikely to be due to potential confounding factors related either to the BPs exposure or clinical diagnosis.

The MAF of rs17024608 in our control population matches with that previously reported for the Caucasian population (MAF = 0.09) in dbSNP and it is comparable among European countries. However, the risk allele is less frequent in the African population. This may partly explain why ONJ seems to be more frequent in Caucasians than Africans.

Independent biological evidence suggests that the RBMS3 might have a pivotal role in ONJ etiology. RBMS3 is a binding protein for Prx1, a homeobox transcriptional factor that up-regulates collagen type I in fibroblasts [69]. Type I collagen, coded by the COL1A gene family, is the main part of the bone matrix. Mutations in those genes produce genetic bone disorders characterized by fragile bones as *Osteogenesis imperfecta* [70]. Variations on RBMS3 (rs10510628) and on COLA1 (rs1800012) have previously been associated with decrease in bone mass and osteoporotic fractures, linking both genes with bone turnover [71, 72]. One of the possible ONJ etiopathogenic mechanisms assumes that it can be caused by bisphosphonate-associated suppressed bone turnover, which leads to decreased blood flow, bone cell necrosis and apoptosis [73]. Recently it has also been shown that BP down regulated collagen type I synthesis in human gingival fibroblasts and osteoblasts [74]. Hence, our finding suggests that ONJ genetic susceptibility could affect the bone turnover, enhancing the bisphosphonate-associated suppressed effect on bone apposition.

Our study did not identify relevant signal on the MHC region. HLA haplotype variation is often associated with adverse drug reactions that have an immune-related pathogenesis [75–77]. HLA variants are mainly related to a drug-specific predisposition and could be detected also by GWAS with a small number of affected cases [75–77]. Given the absence of such a signal, we could speculate that this ADR is more likely to be a toxic ADR, also corroborated by the fact that patients exposed to higher cumulative doses of bisphosphonates are at a greater risk to develop ONJ. Co-occurring mutations on ADME genes might further augment BPs intrinsic toxic effect, enhancing the drug bioavailability. Indeed, the candidate SNP analyses led to interesting signals related to *ABCC4*, for which there is no significant difference in MAF between exposed and non-exposed control populations. The signal on ABCC4 is particularly intriguing. *ABCC4* codes for Multidrug Resistance transporter, (MRP4/ABCC4); these transporters efflux endogenous and xenobiotic substrates out of cells, having a protective role especially in the bone marrow, spleen, thymus, and gastrointestinal tract [78]. Inherited variation in these genes has been associated with the occurrence of toxic serious adverse events (e.g. cyclophosphamide-induced leucopenia/neutropenia) [79]. Currently there is no published information on MRP4 in BPs. Moreover, CNVs may also predispose to the phenotype by disrupting genes belonging to drug metabolism pathways (CYP3A7, CYP2A7, CYP3A6, SLC7A4).

Finally IGFBP7, from the putative causal gene list analysis, showed a suggestive association with the phenotype. IGFs, especially *IGF1* with its tyrosine kinase domain, are growth factors with potent signal transduction capabilities. Insulin-like growth factors are molecules with an important role in normal growth and development. *IGF1* deficient children fail to achieve appropriate height and pharmacologic therapies now exist to correct such deficiencies [80].

IGF1 and *IGF2* are able to influence the replication and differentiation of bone cells through activation of their receptors, especially *IGF1R* whereas IGF-binding proteins (IGFBPs), produced by bone cells, compete with the receptors in binding the ligands and thus affect the bioavailability of IGF1 and IGF2 [81, 82].

While the RBMS3 finding is compelling and was tested against different sets of controls, the reader should note that more research is needed to validate the finding in an independent set of study participants. With evolving observations about the effects of osteonecrosis in different races beyond Caucasians, validation studies are also needed in subjects of various lineages, and in particular in subjects of Chinese Han origin or descent.

Lessons Learned

These and other select markers (work in progress) provide encouraging data for a pharmacogenetic test. The data seems to suggest that osteonecrosis is a complex disease. Bisphosphonates are attracted in areas of rapid bone remodelling and they target and eventually eliminate the osteoclast. Local inflammation or trauma (such as the one sustained during a dental extraction) leads to a release of BPs in the tissues, soft and hard. In such a scenario, susceptible individuals may be affected in two ways; first, if fibroblasts, osteoblasts and osteocytes have no functional transport system, the toxic effect of the drug may also lead to their elimination, depleting the area of any living cells. The second pathway has to do with a genetically-controlled diminished production of collagen. Now, not only the bone is not absorbed because the osteoclast has been neutralized, but no bone apposition takes place either, leading to necrosis. While this model is intuitive, data-supported and compatible with many animal studies showing histologic evidence of dead osteocytes in the affected area, the model and the candidate genes we identified require independent validation.

We hope that the early lessons learned from our quest to discover a molecular risk assessment set of markers for medication-induced osteonecrosis of the jaw will assist others to navigate the era of personalized medicine, and to make valuable contributions to Dentistry. Education is of paramount importance. To our knowledge, one of the very first attempts to personalized medicine education in Dentistry goes back to 1994–1995 where this author designed and taught a graduate-level course at Tufts School of Dental Medicine entitled "Molecular Epidemiology". That first course covered some of the first applications and promises of personalized medicine in the field of Dentistry. Tremendous progress has been made since 1995. Today, the term "personalized medicine" is widely recognized as a model of care that must be fully embraced and utilized in Dentistry to maintain the oral health of the population. However, research shows that currently active healthcare professionals are not comfortable with genomics and personalized medicine, nor do they offer consultations related to personalized medicine. Our recommendation is to increase dental training to a 5 year curriculum that incorporates increasing levels of sophistication around genomics and other technologies that are destined to shape how we practice Dentistry in the near future.

While the model is promising, the road to broad clinical application of genomics in diagnosing and treating oral disease is not straight forward. System wide efforts are required on multiple fronts, from preparing the workforce of tomorrow to shifting research funding away from purely basic science to applied molecular epidemiology and genomics, to re-engineering of medical and dental reimbursement and insurance towards genomics-based risk assessment. One of the absolute requirements for the implementation of personalized oral health is a new generation of scientists that can work in cross-disciplinary teams, bringing diverse talents to the processes of genomic discovery and clinical validation of diagnostics and pathway-based therapeutics. Equally important, as mentioned previously, we need a new generation of dentists that are able to understand the language of genomics and are able to embrace and practice personalized oral health. With regards to research in an era of public finance stresses, the post Human Genome Project era requires careful re-organization of the clinical development process. With tens of millions of polymorphisms, methylation sites and proteomic variation discovered today, it is realistically impossible to sustain the current model of requiring three levels of clinical trials to validate and approve applications for clinical use. A new partnership between insurance companies (with vast data assets that can be utilized efficiently), academic researchers, and biopharmaceutical companies must emerge. Dentistry in the future must be re-tooled to facilitate the capture, analysis and chairside interpretation of DNA for the benefit of our patients. We envision multiple possibilities for the dental practice of the future to offer clinical applications that will allow our patients to prevent dental caries and periodontal disease and to identify head and neck cancer before it becomes clinically evident. The future is bright.

References

1. World Health Organization. Global Health Observatory. Life expectancy by country. http://apps.who.int/gho/data/node.main.688. Accessed 24 Nov 2014.
2. Personalized Medicine Coalition. The case for personalized medicine. Figure 1, p. 5. http://www.personalizedmedicinecoalition.org/Userfiles/PMC-Corporate/file/pmc_the_case_for_personalized_medicine.pdf. Accessed 25 Nov 2014.
3. Garrod AE. Incidence of alkaptonuria: a study in chemical individuality. Lancet. 1902;2:653–6.
4. Dimitrakopoulos I, Magopoulos C, Karakasis D. Bisphosphonate-induced avascular osteonecrosis of the jaws: a clinical report of 11 cases. Int J Oral Maxillofac Surg. 2006;35(7):588–93.
5. Migliorati CA, Schubert MM, Peterson DE, Seneda LM. Bisphosphonate-associated osteonecrosis of mandibular and maxillary bone: an emerging oral complication of supportive cancer therapy. Cancer. 2005;104(1):83–93.

6. Advisory Task Force on Bisphosphonate-Related Osteonecrosis of the Jaws. 2007. American association of oral and maxillofacial surgeons position paper on bisphosphonate-related osteonecrosis of the jaws. J Oral Maxillofac Surg. 2007;65:369–76.
7. Cartsos VM, Seamanduras A, Koo S, Zavras AI. Implications of bisphosphonate use in dentistry. Analecta Periodontol. 2009;20:1–15.
8. Zavras AI. The impact of bisphosphonates on oral health: lessons from the past and opportunities for the future. Ann N Y Acad Sci. 2011;1218:55–61.
9. Palaska PK, Cartsos V, Zavras AI. Bisphosphonates and time to osteonecrosis development. Oncologist. 2009;14:1154–66.
10. Mavrokokki T, Cheng A, Stein B, Goss A. Nature and frequency of bisphosphonate-associated osteonecrosis of the jaws in Australia. J Oral Maxillofac Surg. 2007;65(3):415–23.
11. Hoff AO, Toth BB, Altundag K, Johnson MM, Warneke CL, Hu M, Nooka A, Sayegh G, Guarneri V, Desrouleaux K, Cui J, Adamus A, Gagel RF, Hortobagyi GN. Frequency and risk factors associated with osteonecrosis of the jaw in cancer patients treated with intravenous bisphosphonates. J Bone Miner Res. 2008;23(6):826–36.
12. Migliorati CA, Siegel MA, Elting LS. Bisphosphonate-associated osteonecrosis: a long-term complication of bisphosphonate treatment. Lancet Oncol. 2006;7(6):508–14.
13. Bamias A, Kastritis E, Bamia C, Moulopoulos LA, Melakopoulos I, Bozas G, Koutsoukou V, Gika D, Anagnostopoulos A, Papadimitriou C, Terpos E, Dimopoulos MA. Osteonecrosis of the jaw in cancer after treatment with bisphosphonates: incidence and risk factors. J Clin Oncol. 2005;23(34):8580–7.
14. Ortega C, Faggiuolo R, Vormola R, Montemurro F, Nanni D, Goia F, Aglietta M. Jaw complications in breast and prostate cancer patients treated with zoledronic acid. Acta Oncol. 2006;45(2):216–7.
15. Cafro AM, Barbarano L, Nosari AM, D'Avanzo G, Nichelatti M, Bibas M, Gaglioti D, Taroni A, Riva F, Morra E, Andriani A. Osteonecrosis of the jaw in patients with multiple myeloma treated with bisphosphonates: definition and management of the risk related to zoledronic acid. Clin Lymphoma Myeloma. 2008;8(2):111–6.
16. Merigo E, Manfredi M, Meleti M, Guidotti R, Ripasarti A, Zanzucchi E, D'Aleo P, Corradi D, Corcione L, Sesenna E, Ferrari S, Poli T, Bonaninil M, Vescovi P. Bone necrosis of the jaws associated with bisphosphonate treatment: a report of twenty-nine cases. Acta Biomed. 2006;77(2):109–17.
17. Purcell PM, Boyd IW. Bisphosphonates and osteonecrosis of the jaw. Med J Aust. 2005;182(8):417–8.
18. Soileau KM. Oral post-surgical complications following the administration of bisphosphonates given for osteopenia related to malignancy. J Periodontol. 2006;77(4):738–43.
19. Saussez S, Javadian R, Hupin C, Magremanne M, Chantrain G, Loeb I, Decaestecker C. Bisphosphonate-related osteonecrosis of the jaw and its associated risk factors: a Belgian case series. Laryngoscope. 2009;119(2):323–9.
20. Boonyapakorn T, Schirmer I, Reichart PA, Sturm I, Massenkeil G. Bisphosphonate-induced osteonecrosis of the jaws: prospective study of 80 patients with multiple myeloma and other malignancies. Oral Oncol. 2008;44(9):857–69.
21. García Sáenz JA, López Tarruella S, García Paredes B, Rodríguez Lajusticia L, Villalobos L, Díaz Rubio E. Osteonecrosis of the jaw as an adverse bisphosphonate event: three cases of bone metastatic prostate cancer patients treated with zoledronic acid. Med Oral Patol Oral Cir Bucal. 2007;12:E351–6.
22. Pires FR, Miranda ÁMMA, Cardoso ES, Cardoso AS, Fregnani ER, Pereira CM, et al. Oral avascular bone necrosis associated with chemotherapy and biphosphonate therapy. Oral Dis. 2005;11:365–9.
23. Marx RE, Sawatari Y, Fortin M, Broumand V. Bisphosphonate-induced exposed bone (osteonecrosis/osteopetrosis) of the jaws: risk factors, recognition, prevention, and treatment. J Oral Maxillofac Surg. 2005;63:1567–75.

24. Sonis ST, Watkins BA, Lyng GD, Lerman MA, Anderson KC. Bony changes in the jaws of rats treated with zoledronic acid and dexamethasone before dental extractions mimic bisphosphonate-related osteonecrosis in cancer patients. Oral Oncol. 2009;45(2):164–72.
25. Carter G, Goss AN, Doecke C. Bisphosphonates and avascular necrosis of the jaw: a possible association. Med J Aust. 2005;182(8):413–5.
26. Fehm T, Beck V, Banys M, Lipp HP, Hairass M, Reinert S, Solomayer EF, Wallwiener D, Krimmel M. Bisphosphonate-induced osteonecrosis of the jaw (ONJ): incidence and risk factors in patients with breast cancer and gynecological malignancies. Gynecol Oncol. 2009;112(3):605–9.
27. Van den Wyngaert T, Huizing MT, Vermorken JB. Osteonecrosis of the jaw related to the use of bisphosphonates. Curr Opin Oncol. 2007;19(4):315–22.
28. Wessel JH, Dodson TB, Zavras AI. Zoledronate, smoking, and obesity are strong risk factors for osteonecrosis of the jaw: a case-control study. J Oral Maxillofac Surg. 2008;66(4):625–31.
29. Badros A, Weikel D, Salama A, Goloubeva O, Schneider A, Rapoport A, Fenton R, Gahres N, Sausville E, Ord R, Meiller T. Osteonecrosis of the jaw in multiple myeloma patients: clinical features and risk factors. J Clin Oncol. 2006;24(6):945–52.
30. Walter C, Al-Nawas B, Grötz KA, Thomas C, Thüroff JW, Zinser V, Gamm H, Beck J, Wagner W. Prevalence and risk factors of bisphosphonate-associated osteonecrosis of the jaw in prostate cancer patients with advanced disease treated with zoledronate. Eur Urol. 2008;54(5):1066–72.
31. Khamaisi M, Regev E, Yarom N, Avni B, Leitersdorf E, Raz I, Elad S. Possible association between diabetes and bisphosphonate-related jaw osteonecrosis. J Clin Endocrinol Metab. 2007;92(3):1172–5.
32. Levin L, Laviv A, Schwartz-Arad D. Denture-related osteonecrosis of the maxilla associated with oral bisphosphonate treatment. J Am Dent Assoc. 2007;138(9):1218–20.
33. Yarom N, Yahalom R, Shoshani Y, Hamed W, Regev E, Elad S. Osteonecrosis of the jaw induced by orally administered bisphosphonates: incidence, clinical features, predisposing factors and treatment outcome. Osteoporos Int. 2007;18(10):1363–70.
34. Brown JJ, Ramalingam L, Zacharin MR. Bisphosphonate-associated osteonecrosis of the jaw: does it occur in children? Clin Endocrinol (Oxf). 2008;68(6):863–7.
35. Chahine C, Cheung MS, Head TW, Schwartz S, Glorieux FH, Rauch F. Tooth extraction socket healing in pediatric patients treated with intravenous pamidronate. J Pediatr. 2008;153(5):719–20.
36. Greuter S, Schmid F, Ruhstaller T, Thuerlimann B. Bevacizumab-associated osteonecrosis of the jaw. Ann Oncol. 2008;19(12):2091–2.
37. Estilo CL, Fornier M, Farooki A, Carlson D, Bohle G 3rd, Huryn JM. Osteonecrosis of the jaw related to bevacizumab. J Clin Oncol. 2008;26(24):4037–8.
38. Feller L, Wood NH, Raubenheimer EJ, Meyerov R, Lemmer J. Alveolar bone necrosis and spontaneous tooth exfoliation in an HIV-seropositive subject with herpes zoster. SADJ. 2008;63(2):106–10.
39. Siwamogstham P, Kuansuwan C, Reichart PA. Herpes zoster in HIV infection with osteonecrosis of the jaw and tooth exfoliation. Oral Dis. 2006;12(5):500–5.
40. Meer S, Coleman H, Altini M, Alexander T. Mandibular osteomyelitis and tooth exfoliation following zoster-CMV co-infection. Oral Surg Oral Med Oral Pathol Oral Radiol Endod. 2006;101(1):70–5.
41. Papapoulos SE. Bisphosphonates: how do they work? Best Pract Res Clin Endocrinol Metab. 2008;22(5):831–47.
42. Cetiner S, Sucak GT, Kahraman SA, Akı SZ, Kocakahyaoglu B, Gultekin SE, Cetiner M, Haznedar R. Osteonecrosis of the jaw in patients with multiple myeloma treated with zoledronic acid. J Bone Miner Metab. 2009;27(4):435–43.
43. Landesberg R, Cozin M, Cremers S, Woo V, Kousteni S, Sinha S, Garrett-Sinha L, Raghavan S. Inhibition of oral mucosal cell wound healing by bisphosphonates. J Oral Maxillofac Surg. 2008;66(5):839–47.

44. Hellstein JW, Marek CL. BP osteochemonecrosis (bis-phossy jaw): is this phossy jaw of the 21st century? J Oral Maxillofac Surg. 2005;63(5):682–9.
45. Allen MR. Animal models of osteonecrosis of the jaw. J Musculoskelet Neuronal Interact. 2007;7(4):358–60.
46. Nishimura I, Hokugo A, Adams J.S, Garrett N, Geffen D. Vitamin D-insufficient, bisphosphonate-treatment rats developed osteonecrosis of the jaw. http://iadr.confex.com/iadr/2008Dallas/techprogram/abstract_101636.htm. Accessed 7 June 2015.
47. Ardine M, Generali D, Donadio M, Bonardi S, Scoletta M, Vandone AM, Mozzati M, Bertetto O, Bottini A, Dogliotti L, Berruti A. Could the long-term persistence of low serum calcium levels and high serum parathyroid hormone levels during bisphosphonate treatment predispose metastatic breast cancer patients to undergo osteonecrosis of the jaw? Ann Oncol. 2006;17(8):1336–7.
48. Berruti A, Ortega C, Fusco V, Piemonte Oncology Network (Rete Oncologica di Piemonte e Valle d'Aosta). Re: Christan Walter, Bilal Al-Nawas, Knut A. Grötz, et al. Prevalence and risk factors of bisphosphonate-associated osteonecrosis of the jaw in prostate cancer patients with advanced disease treated with Zoledronate. Eur Urol. 2008;54:1066–72.
49. Berruti A, Dogliotti L, Tampellini M, Lipton A, Hirsh V, Saad F, Liati P, Shirina N, Cook R, Hei YL. Effect of zoledronic acid (Z) treatment based on serum parathyroid hormone (PTH) levels in patients (pts) with malignant bone disease. J Clin Oncol. 2006 ASCO Annual Meeting Proceedings (Post-Meeting Edition);24(18S) (June 20 Supplement), 2006:8610.
50. Allegra A, Oteri G, Nastro E, Alonci A, Bellomo G, Del Fabro V, Quartarone E, Alati C, De Ponte FS, Cicciù D, Musolino C. Patients with bisphosphonates-associated osteonecrosis of the jaw have reduced circulating endothelial cells. Hematol Oncol. 2007;25(4):164–9.
51. Deng X, Tamai R, Endo Y, Kiyoura Y. Alendronate augments interleukin-1beta release from macrophages infected with periodontal pathogenic bacteria through activation of caspase-1. Toxicol Appl Pharmacol. 2009;235(1):97–104.
52. Sedghizadeh PP, Jones AC, LaVallee C, Jelliffe RW, Le AD, Lee P, Kiss A, Neely M. Population pharmacokinetic and pharmacodynamic modeling for assessing risk of bisphosphonate-related osteonecrosis of the jaw. Oral Surg Oral Med Oral Pathol Oral Radiol. 2013;115(2):224–32.
53. Nicoletti P, Cartsos VM, Palaska PK, Shen Y, Floratos A, Zavras AI. Genomewide pharmacogenetics of bisphosphonate-induced osteonecrosis of the jaw: the role of RBMS3. Oncologist. 2012. (Epub ahead of print).
54. Nelson MR, et al. The population reference sample, POPRES: a resource for population, disease, and pharmacological genetics research. Am J Hum Genet. 2008;83:347–58.
55. Wellcome Trust Case Control Consortium. Genome-wide association study of 14,000 cases of seven common diseases and 3000 shared controls. Nature. 2007;447(7145):661–78.
56. http://www.illumina.com/science/icontroldb.ilmn.
57. Shen Y, Nicolletti P, Floratos A. et al. Genome-wide association study of serious blistering skin rash caused by drugs. Pharmacogenomics J. 2011;2:96–104.
58. Price AL, et al. Principal components analysis corrects for stratification in genome-wide association studies. Nat Genet. 2006;38:904–9.
59. Novembre J, et al. Genes mirror geography within Europe. Nature. 2008;456(7218):98–101.
60. Ingle JN, et al. Genome-wide associations and functional genomic studies of musculoskeletal adverse events in women receiving aromatase inhibitors. J Clin Oncol. 2010;28(31):4674–82.
61. Marchini J, Howie B. Genotype imputation for genome-wide association studies. Nat Rev Genet. 2010;11(7):499–511.
62. Howie BN, Donnelly P, Marchini J. A flexible and accurate genotype imputation method for the next generation of genome-wide association studies. PLoS Genet. 2009;5(6):e1000529.
63. Wang K, et al. PennCNV: an integrated hidden Markov model designed for high-resolution copy number variation detection in whole-genome SNP genotyping data. Genome Res. 2007;17(11):1665–74.
64. Need AC, et al. A genome-wide investigation of SNPs and CNVs in schizophrenia. PLoS Genet. 2009;5(2):e1000373.

65. Sherman BT, et al. DAVID Knowledgebase: a gene-centered database integrating heterogeneous gene annotation resources to facilitate high-throughput gene functional analysis. BMC Bioinform. 2007;8:426.
66. Kanehisa M, et al. KEGG for representation and analysis of molecular networks involving diseases and drugs. Nucleic Acids Res. 2010;38(suppl 1):D355–60.
67. Altshuler DM, Gibbs RA, et al. Integrating common and rare genetic variation in diverse human populations. Nature. 2010. 467:52–8.
68. Miksad RA, et al. Quality of life implications of bisphosphonate-associated osteonecrosis of the jaw. Oncologist. 2011;16(1):121–32.
69. Fritz D, Stefanovic B. RNA-binding protein RBMS3 is expressed in activated hepatic stellate cells and liver fibrosis and increases expression of transcription factor Prx1. J Mol Biol. 2007;371(3):585–95.
70. Rauch F, Glorieux FH. Osteogenesis imperfecta. The Lancet. 2004;363(9418):1377–85.
71. Kiel D, et al. Genome-wide association with bone mass and geometry in the Framingham Heart Study. BMC Med Genet. 2007;8(Suppl 1):S14.
72. Qureshi AM, et al. COLIA1 Sp1 polymorphism predicts response of femoral neck bone density to cyclical etidronate therapy. Calcif Tissue Int. 2002;70(3):158–63.
73. Rizzoli R, et al. Osteonecrosis of the jaw and bisphosphonate treatment for osteoporosis. Bone. 2008;42(5):841–7.
74. Simon M, et al. Expression profile and synthesis of different collagen types I, II, III, and V of human gingival fibroblasts, osteoblasts, and SaOS-2 cells after bisphosphonate treatment. Clin Oral Investig. 2010;14(1):51–8.
75. Chung WH, et al. Medical genetics: a marker for Stevens-Johnson syndrome. Nature. 2004;428(6982):486.
76. Daly AK, et al. HLA-B*5701 genotype is a major determinant of drug-induced liver injury due to flucloxacillin. Nat Genet. 2009;7:816–9.
77. Lonjou C, et al. A European study of HLA-B in Stevens-Johnson syndrome and toxic epidermal necrolysis related to five high-risk drugs. Pharmacogenet Genomics. 2008;18(2):99–107.
78. Abla N, et al. The Human Multidrug Resistance Protein 4 (MRP4, ABCC4): functional analysis of a highly polymorphic gene. J Pharmacol Exp Ther. 2008;325(3):859–68.
79. Low SK, et al. Association study of genetic polymorphism in ABCC4 with cyclophosphamide-induced adverse drug reactions in breast cancer patients. J Hum Genet. 2009;54(10):564–71.
80. http://www.fda.gov/cder/drug/InfoSheets/patient/mecaserminPIS.htm, P.I.S.M.r.o.m.a.I.C.b. f.e.a.t.F.w.s.
81. Andreassen TT, Oxlund H. The effects of growth hormone on cortical and cancellous bone. J Musculoskelet Neuronal Interact. 2001;2(1):49–58.
82. Gordeladze JO, Reseland JE, Drevon CA. Pharmacological interference with transcriptional control of osteoblasts: a possible role for leptin and fatty acids in maintaining bone strength and body lean mass. Curr Pharm Des. 2001;7(4):275–90.

Gene Therapy for Xerostomia

Bruce J. Baum

Introduction

The path leading to the first-in-human gene therapy to correct radiation-induced salivary hypofunction was a long and circuitous one, with its origins at the National Institute on Aging (NIA) in Baltimore, MD. In 1981, while serving as a senior investigator at NIA, I published a paper in the Journal of Dental Research [1] that examined parotid gland saliva production in healthy adults across the human lifespan. Although at the time there was a general impression that salivary gland function decreased with age, i.e., salivary hypofunction was an inevitable consequence of growing old, my results were quite different. The study, which was conducted through the NIA's Baltimore Longitudinal Study of Aging, showed clearly that parotid salivary gland performance remains fairly constant across the lifespan, in both men and women, a finding later confirmed by others in cross-sectional and longitudinal studies. These results implied that the salivary dysfunction shown to occur in elders by others actually was a result of disease and its sequelae, and thus something that could, theoretically, be corrected.

In 1982, I left NIA to join the then National Institute of Dental Research (NIDR) and carried with me the notion that salivary gland function did not normally change during physiological aging. At NIDR I began to focus on salivary gland dysfunction. Together with my long-time colleague Philip Fox, in 1982–1983 we established what to our knowledge was the first Dry Mouth Clinic to study salivary dysfunction [2, 3]. The two major clinical conditions leading to salivary hypofunction are therapeutic radiation for head and neck cancers and Sjögren's syndrome, and such individuals formed the majority of the patients seen by us. Many patients had little to no unstimulated parotid saliva secretion, but could be stimulated to secrete some parotid saliva, and recognize the increased moisture in their mouth,

B. J. Baum (✉)
Gene Transfer Section, National Institute of Dental and Craniofacial Research,
National Institutes of Health, Bethesda, MD 20892, USA
e-mail: bjbaum@mac.com

S. T. Sonis (ed.), *Genomics, Personalized Medicine and Oral Disease*,
DOI 10.1007/978-3-319-17942-1_15

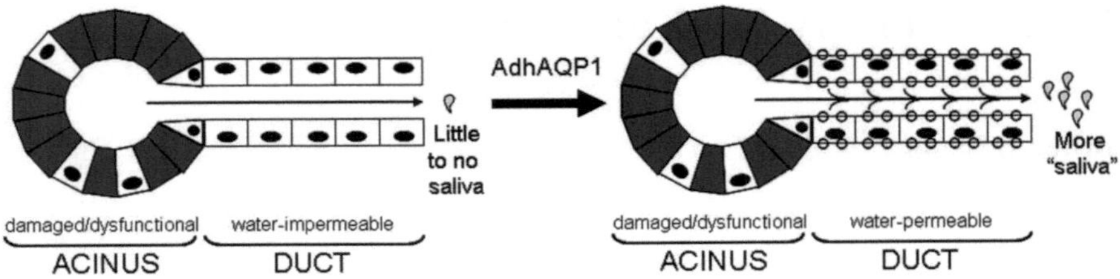

Fig. 1 Schematic depiction of the gene transfer strategy being employed in the AdhAQP1 clinical trial. See text for details. (This figure was originally published in Oral Oncology [7] and is reprinted with permission) This experimental strategy was initially tested in rats [8]

by a simple clinical stimulation tool, swabbing their tongue with 2% citric acid. The fact that such glands could respond to this gustatory stimulation meant that they must have had remaining functional secretory epithelium (Fig. 1; left panel; acinar cells) present, along with intact neurotransmitter signaling. Based on this observation, we began an effort to develop a convenient treatment for these dry mouth patients using pilocarpine by mouth [4, 5], an effort that eventually led [6] to FDA approval of pilocarpine as the first drug specifically for use in the treatment of radiation-induced dry mouth.

Unfortunately, however, pilocarpine and another similar drug that soon became available (cevimeline) did not benefit all patients with a radiation-induced dry mouth. Such patients typically had a poor or no response to stimulation, i.e., were categorized as grades 2–4 by the criteria of the Radiation Therapy Oncology Group (RTOG; Cox et al. [9]). Patients in RTOG grades 2–4 are not homogenous, despite their common phenotype of having too little saliva. Patients in grades 2 and 3 have some remaining salivary epithelial tissue, while for patients in grade 4 the salivary glands have essentially been replaced by fibrotic tissue.

First Steps Towards Gene Therapy

In the late 1980s, after several years of considering possible novel approaches to help patients unresponsive to pilocarpine, I began to explore the potential of using gene therapy specifically to treat patients in RTOG grades 2 and 3. The impetus for this, in large part, stemmed from my recognition that gene therapy was being considered for use in the lungs for cystic fibrosis patients, i.e., *in vivo* gene transfer in the lung could be readily accomplished [10, 11]. Since I had worked, on lung cells during my post-doctoral research fellowship (e.g., Baum et al. [12]), I knew that many features of pulmonary epithelial biology were similar to salivary epithelial biology, i.e., electrolyte and fluid transport, protein secretion. I reasoned that salivary glands also would be good targets for *in vivo* gene transfer using vectors that had been already shown to work in the lung [13]. The journey to make this a reality has been described previously in detail [14, 15] and the key steps are described briefly below (see Table 1; modified from Samuni and Baum [14]).

Table 1 Steps in taking a potential gene therapy from the bench to the clinic[a]

Identifying a clinical problem without conventional therapy
Understanding the biology of the intended target tissue
Assessing the risk/benefit ratio for using viral and non-viral vectors
Choosing the vector to be used (herein, a viral vector)
Understanding the biology of the vector to be used
Determining the availability of suitable in vitro and in vivo (small and large animal) models for testing the idea
Constructing the vector to be used
Functional testing of the vector in vitro
Efficacy studies with the vector in a small animal disease model
Efficacy studies with the vector in a large animal disease model
Conducting a toxicology and biodistribution study with the vector in small animals (GLP level[b])
Developing a clinical protocol
Undergoing required reviews and receiving approval of the clinical protocol[c]
Producing a clinical grade gene transfer vector (GMP[d])
Establishing the infrastructure required to support the study
Beginning subject enrollment

[a]Modified from Samuni and Baum [14]
[b]GLP, Good Laboratory Practice as defined by the US Food and Drug Administration
[c]For the AdhAQP1 clinical study discussed herein, five separate reviews were required: NIDCR Institutional Review Board, NIH Biosafety Committee, US Recombinant DNA Advisory Committee, US Food and Drug Administration, and the study's Data Safety and Monitoring Board
[d]GMP, Good Manufacturing Practice as defined by the US Food and Drug Administration

There were, however, a few large hurdles that I needed to pass before the journey could begin in earnest. The first was to overcome the notion that gene therapy could be used to treat a quality of life disorder, and for that matter one that was oral/dental in scope, because at the time the field was almost entirely focused on correcting single gene mutations causing disease, e.g., cystic fibrosis [10, 11], and cancers refractory to conventional treatment, e.g., malignant melanoma [16]. However, for any condition lacking an effective conventional treatment, it seemed reasonable to me to employ any potentially beneficial approach, including gene therapy, and people eventually accepted this. Another major hurdle was my complete lack of training in molecular biology and virology. The former was corrected through a short didactic course, followed by a 6-month sabbatical working in the research group of a NIH colleague. The latter was achieved through collaboration with my former post-doctoral mentor, Ronald Crystal then at the National, Heart, Lung and Blood Institute, and whose research group was quite active in developing gene therapy as a treatment for cystic fibrosis [11]. The final major hurdle was building a team of colleagues who could bring the ideas to fruition. I had the enormous good fortune to recruit two superb post-doctoral fellows to start the endeavor: Brian O'Connell (now a professor at Trinity College, Dublin) and Christine Delporte (now a professor at the Free University of Brussels). Brian spent his first year at NIH working within the Crystal group, and was able to bring back to our laboratory all of the techniques we needed to use recombinant serotype 5 adenoviral (Ad5) vectors to accomplish salivary gland gene transfer [17]. Christine took the lead on the radiation damage project and successfully showed proof of concept for the idea in an irradiated rat model (Delporte et al. [8]; see below).

Preclinical Proof of Concept

The general strategy underlying the gene therapy developed for "repairing" radiation damaged salivary glands is shown in Figs. 1 (organ level) and 2 (cellular level). Following radiation therapy, the primary salivary cells damaged are the fluid secreting acinar cells, with the non-fluid secreting duct cells being relatively spared [7]. We reasoned that that in the absence of significant numbers of acinar cells, duct cells were capable of generating an osmotic gradient, lumen > interstitum, but since duct cells lacked a facilitated water permeability pathway, i.e., a water channel, little to no water could be secreted [18]. Serendipitously, about the time we were developing our gene therapy strategy, Peter Agre discovered the archetypal water channel, aquaporin-1 (AQP1; Preston and Agre [19]). Thus, we hypothesized that transfer of the AQP1 cDNA into surviving duct cells would allow increased fluid secretion from an irradiation-damaged salivary gland.

Christine Delporte constructed an Ad5 vector encoding human (h) AQP1, AdhAQP1, and initially tested its function *in vitro* in an epithelial cell monolayer. After showing that AdhAQP1 transduction of the cells led to functional water channel expression, she tested the hypothesis *in vivo* with irradiated rats [8]. At 4 months following a single dose of 21 Gy to the ventral surface of the neck, either a control Ad5 vector or AdhAQP1 was administered to the rats' submandibular glands. Three days after administration of the control vector, we measured a ~65 % reduction in pilocarpine-stimulated whole saliva secretion compared to that seen with sham-irradiated rats (Table 2). Conversely, rats administered AdhAQP1 showed levels of saliva secretion no different from the sham-irradiated animals (Table 2). These studies demonstrated, for the first time, proof of concept that the AdhAQP1 strategy could be effective in "repairing" irradiation-damaged salivary glands. It must be emphasized, however, that these studies did not prove our hypothesized mechanism of action (above, Vitolo and Baum [18]); something we still do not understand fully. However, AdhAQP1 delivery to irradiated salivary glands clearly worked; it led to increased saliva production in rats when compared to results seen after the control vector was administered.

A critical step in translating any proposed biological therapy to the clinic is to show scalability to a large animal model. With the collaboration of a former postdoctoral fellow, Songlin Wang, who is now a professor at Capital Medical University in Beijing, we tested the AdhAQP1 strategy in miniature pigs (weighing 30–40 kg versus 300–400 g rats; Shan et al. [20]). Individual parotid glands were irradiated with a single dose of 20 Gy and 4 months later their secretion was reduced by ~80 %. Three days after administering AdhAQP1 to the irradiated parotid glands,

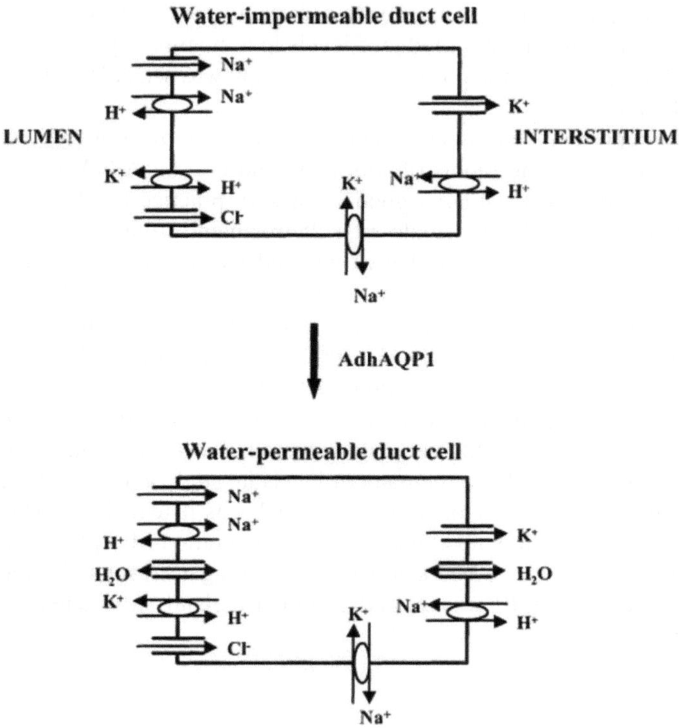

Fig. 2 Hypothesized process by which AdhAQP1 facilitates fluid secretion from irradiated salivary glands. This is based on the experiments presented in Delporte et al. [8]. Surviving duct cells are presented in a simplified form, with only ion channels and ion transporters depicted (*top*). The lumen is to the *left* of the cell shown, and the interstitium is to the *right*. The cell shown is water impermeable. After transduction of this cell with AdhAQP1 (*bottom*), the water channel aquaporin-1 is inserted into the apical and basal membranes providing a pathway by which water can flow in response to an osmotic gradient. We have hypothesized that this gradient would be generated by movement of K^+ and HCO_3) into the lumen, i.e. lumen > interstitium, though this is still unproven and may not be correct. (This figure is reprinted from, and the legend slightly modified from, Vitolo and Baum in Oral Diseases [18] and is reprinted with permission)

Table 2 Effect of AdhAQP1 transduction on secretion by rat submandibular glands irradiated with 21 Gy. (Modified from Delporte et al. [8])

Saliva secretion after viral transduction, ml/100 g body weight per 15 min		
Vector used		
	Addl312	AdhAQP1
Sham-irradiation	36.6 ± 6.8^{a} ($n=4$)	28.4 ± 8.0^{c} ($n=6$)
21 Gy irradiation	$13.2 \pm 3.7^{a,b,c}$ ($n=6$)	30.6 ± 3.5^{b} ($n=9$)

Animals were either sham-irradiated or their salivary glands were exposed to a single radiation dose of 21 Gy. Four months later, either a control viral vector (Addl312) or AdhAQP1 was administered via retrograde ductal instillation to the submandibular glands. Three days later whole saliva was collected following pilocarpine stimulation. The numbers in parentheses represent the number of animals in each treatment group. Values are mean ± SEM; values with the same superscript are significantly different, $P<0.05$, by a Student's t test

parotid saliva output was nearly that seen prior to radiation (~80 %). Conversely, little change was seen from the radiation-reduced levels in miniature pigs administered the control Ad5 vector. Thus, the proposed gene therapy was scalable and, we thought, might work in humans.

Another critically important step in developing any gene therapy strategy is to evaluate its safety (toxicology and biodistribution). We did this for AdhAQP1 in a small animal study, and used the FDA's good laboratory practice (GLP) standards. This study (100 male and 100 female rats; 25/treatment group) was conducted in collaboration with our colleagues Rick Irwin and Molly Vallant at the US National Toxicology Program of the National Institute of Environmental Health Sciences [21]. We examined three AdhAQP1 vector doses that we thought would likely bracket the doses to be used clinically (2×10^8–2×10^{11} vector particles [vp]/gland), as well as a control (saline administration). Administration of the vector to a single submandibular gland resulted in no animal mortality or morbidities, and no adverse signs of clinical toxicity. Additionally, there were no vector-associated effects on either water consumption by, or hematocrit levels in, study animals. Three days after delivery of the highest vector dose, AdhAQP1 was detected primarily in the targeted gland, and there was no evidence of the generation of replication-competent adenovirus in saliva or blood samples. The aggregate results showed that localized delivery of AdhAQP1 to mammalian salivary glands was generally safe [21].

Clinical Study

The clinical protocol, "Open-label, dose-escalation study evaluating the safety of a single administration of an adenoviral vector encoding human aquaporin-1 to one parotid salivary gland in individuals with irradiation-induced parotid salivary hypofunction", was simultaneously submitted to both the National Institute of Dental and Craniofacial Research Institutional Review Board (IRB) and the US Recombinant DNA Advisory Committee, and reviewed in late 2005. In 2006, both of those committees gave their approval (NIH clinical protocol #06-D-0206), as did the FDA (FDA investigational new drug number 13,102) and the NIH Biosafety Committee. In early 2007, an independent Data and Safety Monitoring Board also approved the study and allowed us to begin to enroll patients. The Belfer Gene Therapy Core of Cornell University—Weill Medical School constructed the clinical grade vector according to the FDA's good manufacturing practice (GMP) standards.

Fifteen subjects were approved per the protocol; three in each of five vector dose groups (4.8×10^7, 2.9×10^8, 1.3×10^9, 5.8×10^9, and 3.5×10^{10} vp; to be administered to a single parotid gland). However, only 11 individuals could be treated prior to the expiration date of the vector; 3 each in the first 3 dose groups and 2 in the fourth group. The IRB required us to obtain a vector expiration date before protocol approval, and that date was defined as 5 years from the start of vector production, May 2006, by the vector production facility, i.e., we could only use the clinical grade AdhAQP1 in subjects until May 2011. Unfortunately, I had been unrealistically optimistic in planning the time required to establish the necessary study infrastructure. Along with more typical difficulties associated with a "first in human" demanding study, including 12 in-patient or outpatient visits during a 1-year period with the associated difficulties in subject recruitment, the end result was that we did not have sufficient time to treat all 15 approved subjects [22]. This, however, proved to be fortuitous, as both study subjects treated with the highest vector dose (5.8×10^9 vp), and 1 subject administered 1.3×10^9 vp, exhibited a significant inflammatory response to the AdhAQP1 and no objective or subjective benefits. Thus, it seems reasonable to posit that administration of the higher approved vector dose would have resulted in significant inflammation in the targeted glands and no subject benefits.

Five of the eleven treated subjects benefitted from the administration of AdhAQP1 to an irradiation-damaged parotid gland. Each of them exhibited increased parotid saliva flow rates in the targeted gland, as well as a significant improvement in their xerostomic symptoms (Fig. 3; Baum et al. [22]). These 5 subjects received either 4.8×10^7 ($n=1$), 2.9×10^8 ($n=2$) or 1.3×10^9 ($n=2$) vp to their targeted gland. The remaining 6 subjects experienced no benefit from vector delivery. As indicated above, significant inflammation was observed in 3 subjects. Another subject (receiving 4.8×10^7 vp) had an undetectable latent Ad5 infection of the targeted gland, with subsequent lysis of the vector-transduced cells. This previously has been described extensively [23]. Interestingly, that subject was never viremic, with vector and wild type Ad5 only detected in saliva from the targeted gland; never in the serum. One subject (2.9×10^8 vp) had highly unusual $^{99m}TcO_4$ uptake and release patterns on scintiscan, suggesting the presence of major functional deficits in the remaining glandular epithelial tissue. As to the last subject, who received 4.8×10^7 vp, we have not yet been able to understand the reason for the absence of benefit.

While this study showed the benefit of hAQP1 cDNA transfer for a subset of subjects with irradiation-damaged parotid glands, this was a phase I/II study, primarily focused on the safety of vector delivery. Thus, it is most important to recognize that all subjects tolerated vector delivery and associated procedures well over the entire study period. There were no deaths, serious adverse events, or dose-limiting toxicities and, generally, few adverse events occurred (Table 3). Over the first 42 days post-vector delivery a total of only 65 adverse events were observed [22]. These were all judged as either mild (~91%) or moderate (~9%). Furthermore, most (>75%) were considered to be unrelated or unlikely related to the vector treatment. Of the remaining adverse events, ten were considered possibly related, four probably related, and one definitely related to vector treatment, and all of these were mild or moderate. The last five adverse events (those probably and definitely related), in fact occurred in a single patient (administered 1.3×10^9 vp; who was also the only female treated). She experienced a significant inflammatory response (flocculent swelling) to vector in the targeted gland, which resolved without treatment. Additionally, we found no consistent or systematic changes in all of the clinical chemistry and hematology parameters measured.

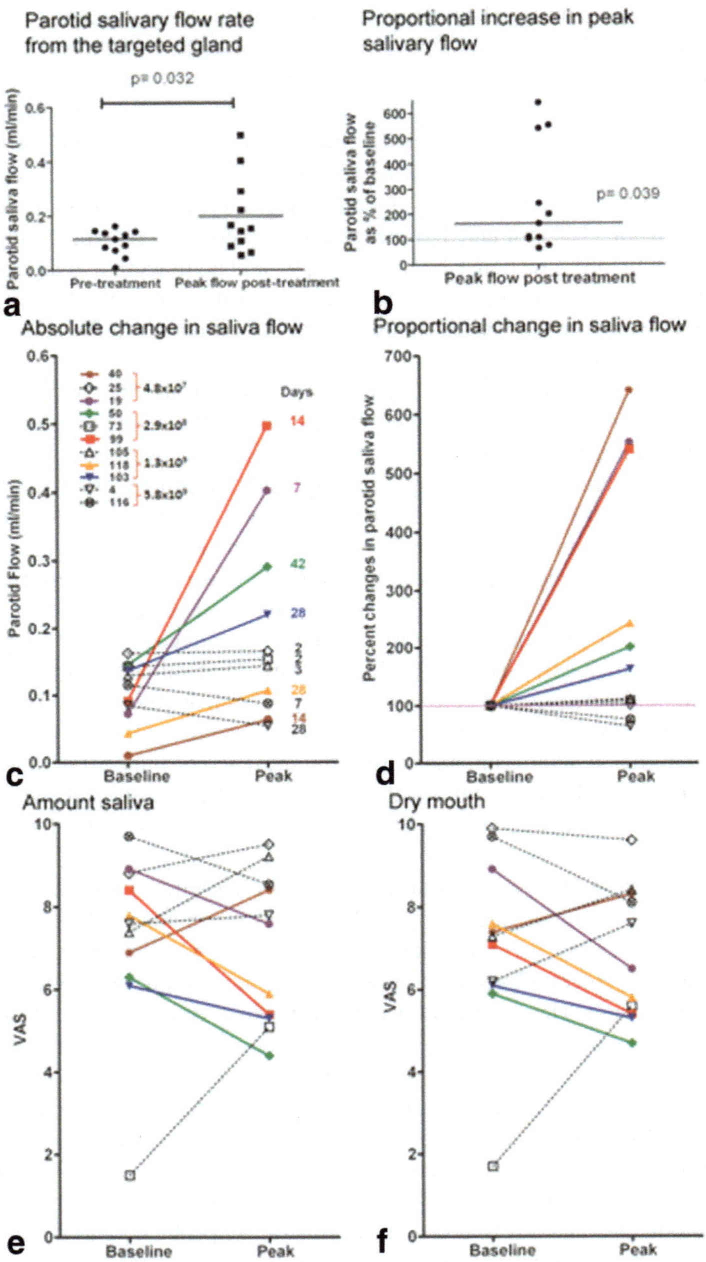

Fig. 3 Summary of AdhAQP1 trial clinical response data. Clinical responses following vector delivery as measured by **a** absolute parotid salivary flow rate from the targeted gland and **b** the proportional increase in peak parotid salivary flow shown as the percent of baseline. Significance was determined using the Wilcoxon matched pair rank test for the change in absolute values. The Wilcoxon signed rank test was used to test if the peak proportional increase in parotid salivary flow was significantly different from the baseline (100 %). Individual changes in parotid salivary

Table 3 Summary of adverse events through day 42. (Modified from Baum et al. [22])

Dose Tier (*n*)	Grade 1 (mild)	Grade 2 (moderate)	Grade 3 (severe)
1 (3)	18[a]	2	0
2 (3)	19[b]	3	0
3 (3)	19[c]	1	0
4 (2)	*3*	*0*	0
Total (%)	59 (90.8 %)	6 (9.2 %)	0 (0 %)

Data shown are the number of adverse events (Grades 1, 2 or 3) recorded in each dosing tier (see text for dosing information). The percentages shown are of the total number of adverse events, i.e., 65

[a]5 were judged possibly related to treatment

[b]3 were judged possibly related to treatment

[c]2 were judged possibly, 4 probably and 1 definitely related to treatment (all with a single subject in the third dose tier)

All other adverse events (50/65; 76.9 %) were judged as unlikely related or unrelated to treatment

Currently, my former colleagues at the National Institute of Dental and Craniofacial Research are developing a new clinical trial, for individuals with radiation damaged parotid glands, using the same hAQP1 cDNA as used in protocol 06-D-0206, but employing a much less inflammatory viral vector, based on the serotype 2 adeno-associated virus (AAV2 [24, 25]) for targeted gene delivery. Importantly for all of the subjects who were administered AdhAQP1 in the study described above, but received no benefits, they are theoretically eligible candidates for this new study. Additionally, since the AdhAQP1 study showed that viral vector-mediated gene transfer to a salivary gland can be performed safely, and with some benefit, it seems reasonable to think that other applications for salivary gland gene transfer, for both local and systemic use (e.g., [26–30]) should find their way to the clinic before too long.

Acknowledgements I have been extremely fortunate since the outset of the AdhAQP1 project to work with so many wonderful and talented colleagues and collaborators. I am sincerely appreciative to a great many people, but in particular I would like to thank Profs. Brian O'Connell, Christine Delporte and Songlin Wang for their critical contributions to the pre-clinical development of the AdhAQP1 strategy. Dr. Changyu Zheng and Ms. Corinne Goldsmith also played important roles in the pre-clinical studies, as well as were key members of the clinical study team. Drs. Ilias Alevizos,

flow are shown in **c** for absolute salivary flow rates and in **d** for proportional changes compared to baseline. Coding for individual subjects is shown as indicated in the panel (**c**) insert. All subjects shown in *black* were considered non-responders (<50 % increase in salivary flow rate). All subjects shown in colors were considered responders (at least a 50 % increase in parotid salivary flow rate following AdhAQP1 administration). The days indicated to the right of each peak data point correspond to the days on which that peak parotid flow rate was observed. Visual analogue scale (*VAS*) results from all subjects, at baseline and peak time of parotid salivary flow, are shown for both the amount of saliva perceived (**e**); rate how much saliva is in your mouth) and dryness of their mouth (**f**); rate the dryness in your mouth). Note that lower VAS results indicate an improvement in symptoms. The colors and symbols used to identify individual subjects are identical to those shown in panel **c**. (This figure is reprinted from, and the legend slightly modified from, Baum et al. [22] in the Proceedings of the National Academy of Sciences (USA))

Ana Cotrim, Gabor Illei and Nikolai Nikolov, all core members of the clinical study team, were instrumental to the success of the clinical study, as were the two extremely talented Research Nurses for this study; Mss. Linda McCullagh and Shuying Liu. Additionally, I had the remarkable good fortune to work throughout most of my career in the Intramural Research Program of the National Institute of Dental and Craniofacial Research, which supported my studies from 1982 to 2011. Finally, I am especially indebted to the 11 research subjects, who participated in this first in human, salivary gland gene therapy study.

References

1. Baum BJ. Evaluation of stimulated parotid flow rate in different age groups. J Dent Res. 1981;60:1292–6.
2. Fox PC, van der Ven PF, Sonies BC, et al. Xerostomia: evaluation of a symptom with increasing significance. J Am Dent Assoc. 1985;110:519–25.
3. Fox PC, Busch KA, Baum BJ. Subjective reports of xerostomia and objective measures of salivary gland performance. J Am Dent Assoc. 1987;115:581–4.
4. Fox PC, van der Ven PF, Baum BJ, et al. Pilocarpine for the treatment of xerostomia associated with salivary gland dysfunction. Oral Surg Oral Med Oral Pathol. 1986;61:243–8.
5. Fox PC, Atkinson JC, Macynski AA, et al. Pilocarpine treatment of salivary gland hypofunction and dry mouth (xerostomia). Arch Int Med. 1990;151:1149–52.
6. Johnson JT, Ferretti GA, Nethery WJ, et al. Oral pilocarpine for post-irradiation xerostomia in patients with head and neck cancer. N Engl J Med. 1993;329:390–5.
7. Baum BJ, Zheng C, Alevizos I, et al. Development of a gene transfer-based treatment for radiation-induced salivary hypofunction. Oral Oncol. 2010;46:4–8.
8. Delporte C, O'Connell BC, He X, et al. Increased fluid secretion after adenoviral-mediated transfer of the aquaporin-1 cDNA to irradiated rat salivary glands. Proc Natl Acad Sci U S A. 1997;94:3268–73.
9. Cox JD, Stetz J, Pajak TF. Toxicity criteria of the Radiation Therapy Oncology Group (RTOG) and the European Organization for Research and Treatment of Cancer (EORTC). Int J Radiat Oncol Biol Phys. 1995;31:1341–6.
10. Whitsett JA, Dey CR, Stripp BR, et al. Human cystic fibrosis transmembrane conductance regulator directed to respiratory epithelial cells in transgenic mice. Nat Genet. 1992;2:13–20.
11. Yoshimura K, Rosenfeld MA, Nakamura H, et al. Expression of the human cystic fibrosis transmembrane conductance regulator gene in the mouse lung after in vivo intratracheal plasmid mediated gene transfer. Nucleic Acids Res. 1992;20:3233–40.
12. Baum BJ, Moss J, Breul SD, et al. Association in normal human fibroblasts of newly elevated levels of adenosine 3',5'-monophosphate with a selective decrease in collagen production. J Biol Chem. 1978;253:3391–4.
13. Baum BJ, O'Connell BC. The impact of gene therapy on dentistry. J Am Dent Assoc. 1995;126:179–89.
14. Samuni Y, Baum BJ. Gene delivery in salivary glands: from the bench to the clinic. Biochem Biophys Acta. 2011;1812:1515–21.
15. Baum BJ. Salivary gland gene therapy: personal reflections. J Oral Biosci. 2014;56:38–42.
16. Rosenberg SA, Aebersold P, Cornetta K, et al. Gene transfer into humans—immunotherapy of patients with advanced melanoma, using tumor-infiltrating lymphocytes modified by retroviral gene transduction. N Engl J Med. 1990;323:570–8.
17. Mastrangeli A, O'Connell BC, Aladib W, et al. Direct in vivo adenovirus-mediated gene transfer to salivary glands. Am J Physiol. 1994;266:G1146–55.
18. Vitolo JM, Baum BJ. The use of gene transfer for the protection and repair of salivary glands. Oral Dis. 2002;8:183–91.

19. Preston G, Agre P. Isolation of the cDNA for erythrocyte integral membrane protein of 28 kilodaltons: member of an ancient channel family. Proc Natl Acad Sci U S A. 1991;88:11110–4.
20. Shan Z, Li J, Zheng C, et al. Increased fluid secretion after adenoviral-mediated transfer of the human aquaporin-1 cDNA to irradiated miniature pig parotid glands. Mol Ther. 2005;11:444–51.
21. Zheng C, Goldsmith CM, Mineshiba F, et al. Toxicity and biodistribution of a first generation recombinant adenoviral vector, encoding aquaporin-1, after retroductal delivery to a single rat submandibular gland. Hum Gene Ther. 2006;17:1122–33.
22. Baum BJ, Alevizos A, Zheng C, et al. Early responses to adenoviral-mediated transfer of the aquaporin-1 cDNA for radiation-induced salivary hypofunction. Proc Natl Acad Sci U S A. 2012;109:19403–7.
23. Zheng C, Nikolov NP, Alevizos I, et al. Transient detection of E1-containing adenovirus in saliva after delivery of a first generation adenoviral vector to human parotid gland. J Gene Med. 2010;12:3–10.
24. Braddon VR, Chiorini JA, Wang S, et al. Adenoassociated virus mediated transfer of a functional water channel into salivary epithelial cells in vitro and in vivo. Hum Gene Ther. 1995;9:2777–85.
25. Gao R, Yan X, Zheng C, et al. AAV2-mediated transfer of the human aquaporin-1 cDNA restores fluid secretion from irradiated miniature pig parotid glands. Gene Ther. 2011;18:38–42.
26. Zheng C, Cotrim AP, Sunshine AN, et al. Prevention of radiation-induced oral mucositis after adenoviral vector-mediated transfer of the keratinocyte growth factor cDNA to mouse submandibular glands. Clin Cancer Res. 2009;15:4641–8.
27. Timiri Shanmugam PS, Dayton RD, Palaniyandi S, et al. Recombinant AAV9-TLK1B administration ameliorates fractionated radiation-induced xerostomia. Hum Gene Ther. 2013;24:604–12.
28. Voutetakis A, Kok MR, Zheng C, et al. Reengineered salivary glands are stable endogenous bioreactors for systemic gene therapeutics. Proc Natl Acad Sci U S A. 2004;101:3053–8.
29. Passineau MJ, Fahrenholz T, Machen L, et al. α-Galactosidase A expressed in the salivary glands partially corrects organ biochemical deficits in the fabry mouse through endocrine trafficking. Hum Gene Ther. 2011;22:293–301.
30. Rowzee AM, Perez-Riveros PJ, Zheng C, et al. Expression and secretion of human proinsulin-B10 from mouse salivary glands: implications for treatment of type 1 diabetes mellitus. PLoS One. 2013;8(3):e59222.

Gene Therapy for Mucositis

Joel S. Greenberger, Michael W. Epperly, Peter Wipf, Song Li, Valerian Kagan and Xiang Gao

Introduction

Ionizing irradiation-induced mucositis of the oral cavity remains a major toxicity in the treatment of head and neck cancers. Modern surgical techniques have facilitated removal of oral cavity and oropharyngeal cancers with negative resection margins, in operable patients. Many of these patients require post-operative chemoradiotherapy. In patients felt to be marginally operable, pre-operative chemoradiotherapy is a modality used in some protocols to facilitate subsequent surgical resection. In those patients felt to be non-surgical candidates, definitive chemoradiotherapy requires multiple weeks of external beam radiotherapy supplemented by either brachytherapy implant or stereotactic radiosurgical boost. In those patients, who suffer local recurrence, further surgery and/or chemoradiotherapy are often indicated. For most head and neck cancer patients, radiotherapy remains a major part of their treatment program. Ionizing irradiation doses required to sterilize microscopic disease in the oral cavity usually induce significant mucositis. At doses exceeding 3 Gy to significant volumes, ulceration of the mucosal barrier is followed by opportunistic infection, usually yeast, subsequent exacerbation of irradiation-induced

J. S. Greenberger (✉) · M. W. Epperly
Department of Radiation Oncology, University of Pittsburgh Medical Center, 5150 Centre Avenue, Rm. 533, Pittsburgh, PA 15232, USA
e-mail: greenbergerjs@upmc.edu

P. Wipf
Department of Chemistry and Center for Chemical Methodologies and Library Development, University of Pittsburgh, Pittsburgh, PA, USA

S. Li · X. Gao
Department of Pharmaceutical Science, University of Pittsburgh, Pittsburgh, PA, USA

V. Kagan
Department of Environmental and Occupational Health, School of Public Health, University of Pittsburgh, Pittsburgh, PA, USA

S. T. Sonis (ed.), *Genomics, Personalized Medicine and Oral Disease*,
DOI 10.1007/978-3-319-17942-1_16

inflammation, and significant pain and suffering by patients. Palliative management of pain and dysphasia is a common practice in modern clinical radiotherapy. In the last two decades, multiple agents have been tested for potential capacity to decrease or eliminate irradiation-induced mucositis. Many of these agents have been tested in clinical trials including keratinocyte growth factor (KGF), granulocyte macrophage, stimulating factor, pentoxyfillene, and flagellin, and some have proved successful.

Amifostine, WR2729 (Ethanol) was developed as an antioxidant, free radical scavenger, and tested in protocols of reduction of irradiation-induced mucositis in head and neck cancer. Marginal effectiveness was demonstrated scoring reduction in mucositis; however, there was significant prevention of xerostomia due to concentration of drug in the salivary glands.

Another approach toward ameliorating irradiation-induced oral cavity and oropharyngeal mucositis has been use of targeted mitochondrial stabilization agents. The first such agent tested was MnSOD-Plasmid Liposomes (MnSOD-PL) [1–14]. We reported success with this modality in animal models [1–14], completed a phase I clinical trial for esophagus radioprotection in lung cancer patients [15], and have designed a three-arm protocol currently undergoing IRB review at the University of Pittsburgh Cancer Institute. The success of mitochondrial targeted gene therapy has led to evolution of this technology substituting a small molecule mitochondrial targeted ROS scavenger for the MnSOD Liposomes [16–22, 23]. Analysis of the experimental data and evolution of the therapeutic approach has revealed a model system that can similarly be applied to amelioration of irradiation-induced toxicity in other organs and organ systems.

Oral Cavity/Oropharyngeal Radiation Mucositis

Squamous cell carcinoma of the head and neck remains a major cancer in the United States. Over 140,000 cases per year are expected, and over a 120,000 deaths anticipated with an incidence of 180,000 cases a year. A recent discovery of a role of Human Papillomal Virus (HPV-7) in the pathogenesis of oral cavity/oropharyngeal cancer and the role of HPV oncogenes as tumor promoters in addition to other mutations discovered in head and neck cancers emphasizes a multi-factor causation of this severe malignancy. Tobacco products, prominently cigarette smoking, and other forms of tobacco ingestion continue to be a major component in carcinogenesis of head and neck cancers.

Treatment of squamous cell carcinomas of the head and neck has been greatly aided by improvements in surgery, Radiation Oncology, and Medical Oncology. Conservative surgical techniques have provided increased capacity to remove gross tumor leaving microscopic margins that can be managed with radiation therapy and chemotherapy. Radiotherapy techniques have advanced significantly with the advent of Intensity Modulated Radiotherapy (IMRT) and Stereotactic Radiosurgery. The latter category treatment delivery is facilitated by improvements in image guided radiotherapy, on-line portal imaging, and sophisticated 3-dimensional treatment planning software, which can tightly control the multi-leaf collimators on modern linear accelerators to define dose distributions that can deliver radio-curable doses of ionizing irradiation to tumor volumes while sparing high doses to critical structures.

For most head and neck cancers, sparing of the contralateral parotid gland has greatly decreased the incidence of xerostomia. Advances in Medical Oncology have included the highly successful combination of Taxol and Carboplatinum. Other potent agents including Etoposide and Navelbine added to the armamentarium of combined modality radiotherapy and chemotherapy. With these improvements in all three elements of the combined modality approach, rates of local recurrence for T2 and T3 squamous cell carcinoma of the head and neck have greatly decreased. Effective chemotherapy also limits the incidence of distant metastasis. Treatment of metastatic disease with SRS techniques has provided for increased quality of life and expected greater longevity of patients presenting with squamous cell carcinoma of the head and neck. Future additions of targeted therapies including those that specifically provide radiosensitization of tumors and/or targeting of specific mutations or growth factor receptor abnormalities in tumor cells will also improve quality of life and longevity.

Despite these technical advancements in radiotherapy, the toxicity of chemoradiotherapy remains a prominent concern in the management of patients with head and neck cancer [24]. Among the toxicities, radiation-induced mucositis of the oral cavity and oropharynx remains the prominent dose limiting toxicity [25–27]. In those patients recovering from the acute reaction of chemoradiotherapy, late radiation fibrosis with clinical onset 6 months to 2 years after completion of chemoradiotherapy remains a prominent late toxicity. Efforts to improve quality of life and survival in head and neck cancer patients have also focused on the development of radiation protectors/mitigators to focus on the optimization of the therapeutic ratio by decreasing normal tissue toxicity [28–34].

There have been several approaches to the development of radioprotectors for the oral cavity and oropharynx, focusing on the multiple components of the induction and exacerbation of radiation mucositis [30–39]. These strategies have focused on the components of the pathophysiology of radiation mucositis, namely normal tissue barrier breakdown, histologic parameters of edema tissue breakdown, and mucosal ulceration, opportunistic infection, and the sequella of weight loss dehydration and malnutrition [26, 27]

The Logic of Mitochondrial Targeted Radiation Protection

Ionizing irradiation induces free radicals from the ionization of water and oxygen in cells in culture and in vivo. These free radicals are consumed by the antioxidant stores in cells within fractions of a second. They produce DNA double strand breaks in the nucleus, which activate a cascade of events occurring within minutes, including initial phosphorylation of ataxia telangiectasia mutations kinase (ATM kinase) activation of the Fanconi Anemia (FA) DNA repair platform, and initiation of homologous recombination and non-homologous end-joining pathways for repair of these double strand breaks. Coincident with the initiation of DNA repair pathways is the activation of a communication system from the nucleus to the mitochondria. Stress activated protein kinases (SAP kinases) including JNK1, p38 are transported to the mitochondrial membrane [12, 40]. Furthermore, pro-apoptotic proteins including BAX, accumulate in the mitochondrial membrane.

Many of these communication mechanisms are dependent upon p53 and p21. Ionizing irradiation induces cell cycle arrest at both G1 and G2/M causing cell cycle arrest, lack of proliferation, and entry of cells into a quiescent state [11, 20, 41]. The p53 dependent signaling pathways have been shown to determine either cell death through apoptosis or effective nuclear DNA repair and cell survival.

Prominent in the initial DNA damage response is the activation of transcription of multiple inflammatory cytokines, which are released into the extracellular matrix locally, and into the circulatory system [7, 42]. Inflammatory cytokines exacerbate DNA damage response, many of which can cause DNA damage through extracellular signaling pathways such as TNFα, Interleukin-1 (IL-1), and others [11, 20, 41]. Finally, dying cells within tissues release nucleotides, proteins, and other cell debris, which stimulates a further inflammatory response including the arrival into tissues of macrophages and neutrophils as well as several categories of lymphocytes. These inflammatory cells establish a toxic microenvironment in which further pro-apoptotic events occur and the cycle of cell death is escalated [43]. In fractionated radiotherapy, this damage response and natural repair processing occurs after retraction, but is then aggravated by the second radiotherapy fraction. Cumulative damage to the tissue resulting in mucositis, ulceration, and tissue barrier breakdown results when the process of continuous damage counterbalances cellular efforts to combat irradiation damage.

Early in the tissue response to ionizing irradiation is regulation of the antioxidant stores in an attempt to neutralize free radicals and quiet the oxidative stress response itself. Cells maintain an antioxidant store capacity including glutathione, which is consumed rapidly after irradiation [44]. Consumption of antioxidant stores results in rapid induction of apoptosis [44]. A prominent antioxidant is the enzyme Manganese Superoxide Dismutase. MnSOD, mitochondrial targeted antioxidant enzyme (SOD2) is distinct from two other SOD forms, both cellular cytoplasm Cu/Zn superoxide dismutase (SOD1) and extracellular SOD3. The importance of mitochondrial targeting of MnSOD for radioprotection was demonstrated in an experiment modifying the mitochondrial targeting sequence of the transgene product in 32D cl 3 hematopoietic cells in culture [40]. Removing the mitochondrial targeting sequence from the MnSOD transgene resulted in no radioprotection. In contrast, slicing the mitochondrial targeting sequence to the Cu/ZnSOD transgene resulted in mitochondrial localization and radioprotection [40]. MnSOD converts superoxide into hydrogen peroxide, another potent oxidant [19, 21, 45–48]. Hydrogen peroxide is neutralized in cells by either glutathione peroxidase or catalase turning it to water [49]. The precise role of MnSOD in the cascade of generation of additional radical oxygen species including hydrogen peroxide is not fully understood.

However, MnSOD appears to be critical for proper functioning of the antioxidant pathway. MnSOD deficient mice do not survive past 1–2 days after birth [6]. The recent use of Doxycycline inducible MnSOD in MnSODtet/tet mouse cells has demonstrated the dependence of MnSOD levels on radiation resistance [50]. MnSOD heterozygote+/− mice demonstrate reduced radiation resistance, which can be restored by MnSOD-PL gene therapy [50]. Most importantly, antioxidants can defend against irradiation-induced killing of normal tissues, and depletion of antioxidant stores is reduced in cells overexpressing MnSOD [44], and is accelerated in cells in animals deficient in MnSOD [29]. This large body of background information led to the design of pre-clinical and clinical trials for the use of MnSOD-PL in clinical radioprotection.

Molecular Mechanism of MnSOD Transgene Mediated Irradiation Protection

The molecular biology of events at the mitochondrial membrane following irradiation has been elegantly elucidated in the past decade.

Radiation-induced cellular apoptosis has been shown to be an event which follows ineffective mitophagy. Damaged mitochondria through oxidative changes between the inner and outer mitochondrial membrane, principally the oxidation of cardiolipin with its transport to the outer mitochondrial membrane, elicit the program of mitophagy, with removal of such damage to the mitochondria. Ineffective mitophagy, by overload of the mechanism for this corrective action, leads to apoptosis trauma, a signal at the cell membrane mediated through phosphatidyl-serine [22, 51, 52].

Molecular events, which inhibit or reduce the transport of oxidized cardiolipin from the inner to outer mitochondrial membrane ameliorate irradiation-induced cell damage. The transport of oxidized cardiolipin also results in its ineffective binding to cytochrome C. Cytochrome C is 70 % bound to cardiolipin in the natural cellular state, and its release from cardiolipin along with the oxidized cardiolipin migration to the outer mitochondrial membrane, resulting in mitochondrial membrane permeability and leakage of cytochrome C into the cytoplasm [48]. Cytochrome C initiates activation of the caspase pathway through caspase-3 and leads to cleavage and activation of poly-ADP-ribosyl-preliminase, and DNA fragmentation from apoptosis [12, 40]. Accumulation of apoptotic cells in tissues results in the cascade of inflammatory events and tissue damage as described above. Therefore, interruption of the process of accumulating mitophagy, apoptosis, and tissue damage can be initiated by stabilization of the mitochondrial membrane. This stabilization can be achieved through targeting an antioxidant to the mitochondria. The first of these experimental approaches were carried out with MnSOD-PL gene therapy.

Animal Models of Irradiation-Induced Mucositis

Several assays have been developed for quantitation of irradiation-induced toxicity to the oral cavity and oropharynx including measurement of alteration in mucosal tissues and salivary gland output. These highly quantitative histopathologic markers have been supplemented recently with assays for irradiation response biomarkers. The irradiation-induced expression of RNA and protein for inflammatory cytokines including TGFβ, IL-1, and TNFα have been utilized as a marker for irradiation-induced toxicity. Induction of RNA transcripts for radiation response genes including the gene promoter binding proteins NFκβ, Nrf2, AP1, and SP1 have also been utilized. Biomarkers have also included quantitation of mRNA induced by irradiation and mediating the production of antioxidant proteins themselves including MnSOD RNA transcripts [53].

Initial studies using C57BL/6NHsd mice demonstrated the highly reproducible and quantitative induction of radiation mucositis and reduction in salivary gland function in vivo [35, 38, 54]. In these initial studies, intraoral administration of MnSOD-PL was shown to reduce histopathologic as well as other biomarkers of irradiation damage. Most importantly, animal survival increased and parameters of objective radiation toxicity including dehydration and weight loss were greatly reduced. For translation of these animal studies to a clinical trial, it was particularly important to correlate the in vitro parameters of irradiation and killing with those same events in vivo. Studies in the mouse model demonstrated mitochondrial targeting of an epitope-tagged MnSOD transgene [55] confirming the mechanism of radiation protection and lack of protection by plasmid liposomes containing other transgenes including beta-galactosidase, green fluorescent protein, and a non-mitochondrial targeted catalase. Further confirming the importance of mitochondrial targeting, a mitochondrial targeted catalase was shown to be effective as a radioprotector in vivo [49]. Delivery of the MnSOD transgene was shown to be effective when administered by plasmid liposomes, minicircle plasmid, or viral vectors including adenovirus, and Herpes virus [56]. MnSOD gene therapy protection of tissues in animal models was shown to be effective not only in the oral cavity/oropharynx, but also in the lung [1, 57], esophagus [10, 58, 59], bladder [13], intestine [56], and systemically when delivered to total body irradiated mice [60, 61].

In important control experiments, it was demonstrated that administration of MnSOD protein alone was not as effective as administration of the transgene. These studies led to the conclusion that the concentration of gene product at the mitochondrial membrane was more effective when delivered intracellularly via entry of transgene to the nucleus, transcription of transgene message, and then translation of transgene product in the cytoplasm with transport to the mitochondria, rather than delivery of the protein through the circulation or across the plasma membrane into the cytoplasm for anticipated concentration in the mitochondria.

MnSOD-Plasmid Liposome Gene Therapy for Radiation Protection

MnSOD-plasmid liposome radioprotective gene therapy was demonstrated to be effective in multiple models of organ specific radioprotection [62]. These included lung [1, 3, 5, 6, 8, 42, 57], esophagus [4, 7, 10, 58, 63–65], oral cavity/oropharynx [29, 35, 38, 54], bladder [13], intestine [56], and some very interesting studies of total body irradiation [60, 61]. During the evaluation of data from these studies, it became obvious that the formulation/emulsion utilized to get MnSOD-plasmid into appropriate solution for organ specific delivery was a critical part of the therapeutic paradigm.

The first clinical trial of MnSOD-plasmid liposomes for radiation protection was in the esophagus [15]. Patients with non-small cell lung cancer, felt to be surgically unresectable (stages IIIA and IIIB) received MnSOD-plasmid liposomes (30 ml in a novel liposomal mixture (JVRS100), initially produced by Valentis Corporation), and in a three tiered Phase I Clinical Trial consisting of 0.3 mg, 3.0 mg, and in a third cohort, 30 mg plasmid liposomes. There was no toxicity demonstrated in patients on this clinical trial and part of the protocol required esophagoscopies and biopsies of esophageal tissue at three levels (cervical, mid-esophagus, GE junction) to demonstrate the extent of transgene uptake in the tissue initially, and then one month after completion of the 7 ½ weeks clinical trial, demonstrating that the transgene had cleared. These patients received 7 ½ weeks conformal multifield radiotherapy to lung corrected doses around 70.0 Gy and combination chemotherapy consisting of Carboplatinum and Taxol. This clinical regimen for radiotherapy is associated with a 33 % significant esophagitis frequency. The patients in the Phase I Clinical Trial of MnSOD-PL demonstrated no significant esophagitis, and in addition, no other toxicities. One of the patients in the third cohort remains disease free with some irradiation related esophageal stricture now over two years since completion of clinical radiotherapy. This patient suffered vocal cord paresis after the final esophagoscopy, but this cleared for six months, and she remains disease free at last followup. The Phase II portion of this clinical trial is in progress, and has highlighted one of the associated complications of using plasmid liposome gene therapy, namely cost.

The expense of producing clinical grade plasmid has increased significantly since the initiation of the Phase I Clinical Trial over 4 years ago. The cost of production of plasmid liposomes and the general concern for delivering transgene to normal tissues for radioprotection remains a concern for some clinical trial coordinators and patients. Therefore, research pushed forward to develop an alternative for plasmid liposomes, namely using a small molecule radioprotector that could “mimic” the actions of the MnSOD transgene (Fig. 1).

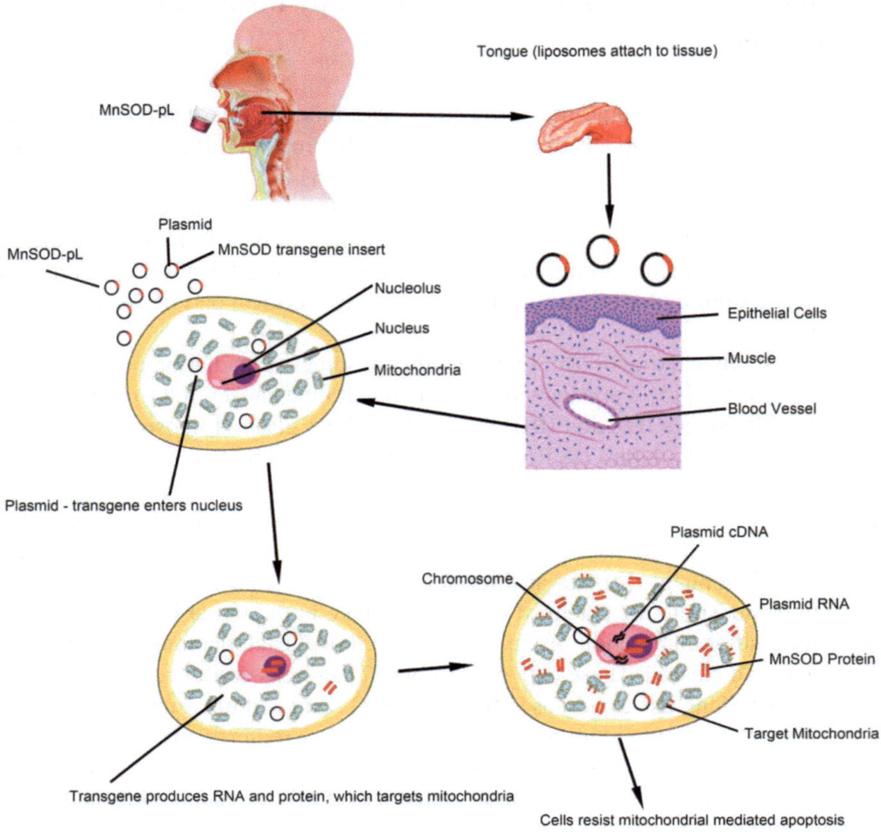

Fig. 1 MnSOD-PL gene therapy for radioprotection of the oral cavity/oropharynx

Transition to Small Molecule Radioprotectors, the Role of Formulations for Drug Delivery

Superoxide dismutase mimics have been in research focus for many years. Initial studies using water soluble SOD mimics with the C. Elegans model indicated that this category of drugs had an anti-aging effect when studied in culture systems. Since MnSOD transgene has been demonstrated to ameliorate both the acute and late effects of ionizing irradiation, studies were carried out in a zebra fish model attempting to demonstrate a decreased late radiation fibrosis in the fish tail by administration of a SOD mimic in the water [66]. Several categories of SOD mimics have been developed and had biological radioprotective effects in several in vitro systems, however, their effectiveness in vivo has been inconsistent. The search for more effective small molecule radioprotectors next switched to a class of molecules called nitroxides.

GS-Nitroxide Small Molecule Radioprotective Therapy

The attractiveness of nitroxides as radioprotectors grows largely from the redox chemistry of the molecule. Nitroxides cycle continuously between hydroxyl amine moieties back to nitroxides, and in each case the stoichiometry indicated consumption of a free radical [16, 17, 67, 68]. This chemistry revealed that a single nitroxide molecule could scavenge many free radicals and still remain active. A prominent nitroxide first utilized in experimental and then a clinical trial was Tempol. In vitro studies with radiation survival curves of cell lines demonstrated significant radiation protection by Tempol and the analog 4-Amino-Tempo [16, 17]. Studies in mice demonstrated that systemic toxicities of administration of Tempol limited its effectiveness as a radiation protector [69]. Primary toxicities involved renal failure, blood pressure abnormalities, and ineffectiveness below a significant level. One clinical trial utilized a topical cream containing Tempol placed on the scalp of patients receiving whole brain irradiation [70]. Radiation dermatitis was a common side effect associated with whole brain irradiation resulting from the tangential radiation beam crossing the scalp. Patients receiving Tempol cream in an application device holding the material on the scalp had significant reduction in radiation dermatitis. However, these studies pointed out the importance of relatively high doses of Tempol required for a therapeutic effect and suggested that organ specific application or systemic application would be difficult. Translating these technologies to oral cavity radioprotection was recently reported [71] in which a salve containing the equivalent of 150 mg/kg of Tempol resulted in significant reduction in radiation dermatitis in a mouse model [71]. While economically more feasible than production of MnSOD-plasmid for localized delivery, Tempol seemed unlikely to be of generalized usefulness, because of the requirement of a very high concentration.

The mechanism of irradiation-induced normal tissue damage has been demonstrated to be primarily one of prevention of apoptosis [12, 16, 17, 40, 45, 47, 49, 72]. While other forms of radiation killing of normal tissues were demonstrated including autophagy, necroptosis, intermitotic death, and bystander killing, the primary mechanism of normal tissue damage was demonstrated to be that of apoptosis. The mitochondrial mechanism of apoptosis was verified in vitro and in vivo and suggested that interrupting the mitochondrial step in the radiation-induced apoptotic pathway might be one strategy by which to increase the effectiveness of nitroxides, namely through mitochondrial targeting [49]. Two strategies of mitochondrial targeting of nitroxides were developed: Hemigramicidin linkage [16, 17, 67, 68] and triphenylphosphoniam targeting [21]. These two strategies served as controls for each other and documented the clear effectiveness of nitroxide dose reduction by mitochondrial targeting. A series of hemigramicidin linked nitroxides was recently developed and their effectiveness compared [73]. Depending on the structure of the hemigramicidin molecule, mitochondrial concentration ranged from 33 to 600 fold. These data demonstrated that equivalent radioprotection of cell lines in vitro could be achieved with significantly lower concentrations of Tempol when the hemigramacidin linker was used. Initial effectiveness studies were carried out with triphenylphosphonium linked nitroxide in vitro. The clinical translational capacity of the GS-nitroxides was preferable for several reasons.

The strategy of development of GS-nitroxides was that of exploiting the archaebacterial origin of mitochondria, through a process termed endosymbiosis. The antibiotic gramicidin developed for its bacterial targeting was, therefore, approached as a potential molecular construct by which to deliver nitroxide to the endosymbiotic structure in mammalian cells, mitochondria. Several categories of GS-nitroxides were compared and despite the difference in mitochondrial concentration ranging from 400 to 600 fold with the drug XJB-5-131 down to 33 fold with the smaller molecule JP4-039, the initial in vivo experiments in mice demonstrated minimal difference in the radioprotective capacity. Furthermore, in radiation mitigation experiments in which the drug was given either immediately after irradiation out to as much as 72 h after the LD 50/30 dose of total body irradiation, JP4-039 was demonstrated to be a highly effective protector [74, 75]. For fractionated irradiation in which administration of drug between fractions is in fact a protector for the next fraction, but a mitigator for the previous fraction, it was determined that the smaller molecule JP4-039 should be advanced in a preclinical model for organ specific radiation protection.

Application of JP4-039 (GS-Nitroxide) in Organ Specific Radioprotection

While seeking to achieve a different mission than that of clinical normal tissue radioprotection during fractionated radiotherapy, the Center for Medical Counter Measures Against Radiation (CMCR) Program of the National Institutes of Allergies and Infectious Diseases (NIAID) facilitated, through their research funding, the development of JP4-039 as a novel "dual" use drug [74]. The drug proved highly effective in radiation mitigation against total body irradiation in the mouse model and has been shown to be effective in C57BL/6NHsd as well as C3H/HeN mice administered drug I.V. or I.P. 24 h or later after total body irradiation [75, 76]. Furthermore, recent developments have demonstrated that topical administration of JP4-039 can ameliorate beta-irradiation (electron beam), skin burns, and also when delivered systemically through a biodegradable microneedle patch system, can achieve blood levels comparable to that of I.V. administration and is also effective in radiation mitigation when delivered 24 h after total body irradiation. Thus, the success of JP4-039 in the CMCR Program as a radiation mitigator boosts enthusiasm for potential use of this drug in clinical radiotherapy.

The model of radiation-induced esophagitis after single fraction or fractionated radiotherapy to the mouse thoracic cavity, was utilized to test the effectiveness of JP4-039 in vivo [23]. The formulation/emulsion system required for delivery of drug became an obvious focal point. The formulation appropriate for intravenous administration (F14) was developed to replace the more cumbersome and potentially toxic (for mice) cremphor el 10%-ethanol 10%-water 80%. The F14 emulsion facilitated delivery of comparable blood levels after intravenous injection compared to that of other administration vehicles without toxicity. However, F14 was successful largely because of the ease of transport of drug across cells and multiple layers of cells within tissue. This delivery system was not optimal for organ specific delivery where concentration of drug in the site of anticipated irradiation damage was desirable [77].

An organ specific radiation protection delivery system for JP4-039 was developed to take advantage of Tween's ability to hold drug at the site of administration [23]. First, studies were carried out with the F15 emulsion in the esophagus radiation protection model. Mice administered JP4-039 to the esophagus immediately prior to irradiation demonstrated significant radiation protection [23]. The effectiveness of radiation protection was comparable to that seen with MnSOD-PL.

GS-nitroxides were also shown to be effective in combined injury models, situations in which ionizing irradiation was sustained in combination with another mode of injury: burn, trauma, infection, or other associated tissue and organ damage inducing agents. GS-nitroxides were shown to be effective in ameliorating the effects of traumatic brain injury [78], and were effective in ameliorating the effects of aging in a mouse strain prone to accelerated aging [24].

GS-Nitroxides as Radioprotectors for the Oral Cavity/Oropharynx: Comparative Effectiveness to Other Agents in Pre-Clinical or Clinical Trials

To evaluate the effectiveness of JP4-039/F15 in radiation protection of the oral cavity, several potential outcomes were considered. The effectiveness of a small molecule radiation protector necessitates comparison with other utilized modalities. These are listed in Table 1 [30–34, 71, 79–81]. The mechanism of radiation-induced mucositis has been well established and involves both epithelial and vascular damage, and secondary opportunistic infections (Table 2). Therefore, evaluation of a potentially clinically relevant radioprotector requires comparative evaluation with these other modalities. Furthermore, the use of radioprotective small molecules in head and neck cancer patients, in the setting in which normal tissue radioprotection

Table 1 Agents considered for local administration to ameliorate radiation mucositis in head and neck cancer patients

Agent
KGF
G-CSF
Pilocarpine
GM-CSF
Amifostine
Tempol (4-Amino Tempo)

Table 2 Potential mechanism of radiation mucositis

Mechanisms
Epithelial Cell Damage
Endothelial Cell Damage
Opportunistic Infections
Inflammatory Cytokines
Induction of Stem Cell Senescence

would be required, necessitates demonstration that the radioprotector would not also protect tumor, particularly in a setting of microscopic tumor in the radiotherapy target volume.

Finally, the requirement for fractionated radiotherapy in the management of head and neck cancer patients necessitated the understanding of the pharmacokinetics and the pharmacodynamics of intraorally administered JP4-039/F15. Could the drug be given safely between multiple radiotherapy fractions? What was the time of administration prior to radiation fractions, how long did the drug remain active in tissues, and could there be complications of the administration of JP4-039/F15 in the setting of concomitant chemotherapy. Chemotherapy and radiotherapy in fact represent a form of "combined injury" in the cancer patient with respect to normal tissues.

Preclinical Evaluation of JP4-039/F15 as an Oral Cavity/Oropharyngeal Radioprotector: The Importance of Fanconi Anemia (FA) Mice

It is well known in clinical radiotherapy that there is significant heterogeneity in normal tissue response between patients. Radiation oncologists know that around 10 % of patients will suffer severe complications of fractionated radiotherapy while another 10 % will suffer nearly no detectable radiation side effects. The remaining 80 % demonstrate toxicities that follow a "bell-shaped" curve of varying degrees of normal tissue toxicity. Since there is currently no way to evaluate patients who might be in the "radiosensitive" 10 % of the dose response curve for 5 ½ −7 weeks of fractionated radiotherapy for head and neck cancer, and it was not known whether JP4-039/F15 would be potentially toxic to a subset of patients, a Fancd2-/- mouse model of a radiosensitive subset of patients was chosen [82, 83]. Two strains of mice (C57BL/6J and 129/SV) were independently modified through homologous recombination experiments to generate homozygous recombinant negative deletion Fancd2−/− mice. Both mouse strains were tested for radiosensitivity of the oral cavity/oropharynx and both were shown to be relatively radiosensitive compared to heterozygote Fanc2+/− and wild type Fancd2+/+ mice [82, 83].

Conclusions and Plans for Future Developments

Small molecule normal tissue radioprotectants will be of great value in the management of head and neck cancer patients over the next decades. Improvements in chemotherapy and more sophisticated radiotherapy techniques will allow for both enhanced combined modality therapy and dose escalation to increase radiocurability and local control. However, there will be an overlap between the efficacy of therapeutic modalities and the capacity of patients to sustain normal tissue toxicity. To address the needs emerging in enhanced therapeutic protocols, greater attention must be paid to protect the normal tissues in the radiotherapy treatment volume. Small molecule radioprotectors such as the GS-nitroxides (in particular JP4-039) will move into clinical trials where the focus will be on parameters common to past and current clinical trials, but also with concern for the future of the care of cancer patients, namely outcomes analysis. Therapeutic outcomes will focus not only on quality of life issues during and immediately after chemoradiotherapy, but on long-term outcomes including patient satisfaction, quality of life, and overall health care costs. Some issues of particular concern will be those related to long-term outcomes. A radiotherapeutic to limit normal tissue toxicity during radiotherapy will not be optimal if the increase in normal tissue tolerance facilitates higher irradiation doses that produce more significant late effects. Late effects in the oral cavity and oropharynx include radiation fibrosis, oropharyngeal muscle damage, osteo radionecrosis, and fistula and sinus formation.

As the number of long-term survivors (2 years and greater after completion of chemoradiotherapy) increases, there will be more available data for analysis. While the initial clinical trials on the use of small molecule radioprotectors in the oral cavity/oropharynx out of necessity focus on acute effects, the availability of long-term survivors in adequate numbers for appropriate evaluation will uncover potential late effects of chemoradiotherapy, which would not have been expected. The history of chemoradiotherapy has taught clinicians that “the good news always comes first” and nowhere has this been more dramatic than in pediatric radiotherapy. Initial successes, in using radiotherapy for treatment of pediatric tumors, led to successful improvement in survival and long-term followup studies, which unfortunately showed significant late side effects of radiotherapy. For this reason, radiotherapy is now eliminated or minimized in its use for many childhood cancers. The same principle has held true for the management of Hodgkin’s Disease in which previous protocols with total nodal irradiation have led to near uniform utilization of chemotherapy with radiotherapy only for sites of gross disease. These changes have come largely from an appreciation of significant late effects of radiotherapy. Management of patients with head and neck cancer, particularly those with post-operative management needs will likely follow the same sequence of events. Caution must be exercised in the implementation and utilization of small molecule radioprotectors with concern not only for minimizing acute radiation effects, but with concern for unexpected late effects.

References

1. Epperly, MW, Bray JA, Kraeger S, Zwacka R, Engelhardt J, Travis E, Greenberger JS. Prevention of late effects of irradiation lung damage by manganese superoxide dismutase gene therapy. Gene Ther. 1998;5:196–208.
2. Zwacka RM, Dudus L, Epperly MW, Greenberger JS, Engelhardt JF. Redox gene therapy protects human IB-3 lung epithelial cells against ionizing radiation-induced apoptosis. Human Gene Ther. 1998;9:1381–86.
3. Epperly MW, Bray JA, Krager S, Berry LA, Gooding W, Engelhardt JF, Zwacka R, Travis EL, Greenberger JS. Intratracheal injection of adenovirus containing the human MnSOD transgene protects athymic nude mice from irradiation-induced organizing alveolitis. Int J Radiat Oncol Phys. 1999;43(1):169–81.
4. Epperly MW, Sikora C, Defilippi S, Bray J, Koe G, Liggitt D, Luketich JD, Greenberger JS. Plasmid/liposome transfer of the human manganese superoxide dismutase (MnSOD) transgene prevents ionizing irradiation-induced apoptosis in human esophagus organ explant culture. Int J Cancer (Radiat Oncol Invest). 2000;90(3):128–37.
5. Epperly MW, Defilippi S, Sikora C, Gretton J, Kalend K, Greenberger JS. Intratracheal injection of manganese superoxide dismutase (MnSOD) plasmid/liposomes protects normal lung but not orthotopic tumors from irradiation. Gene Ther. 2000;7(12):1011–8.
6. Epperly MW, Epstein CJ, Travis EL, Greenberger JS. Decreased pulmonary radiation resistance of manganese superoxide dismutase (MnSOD)-deficient mice is corrected by human manganese Superoxide dismutase-plasmid/liposome (SOD2-PL) intratracheal gene therapy. Radiat Res. 2000;154(4):365–74.
7. Epperly MW, Gretton JA, DeFilippi SJ, Sikora CA, Liggitt D, Koe G, Greenberger JS. Modulation of radiation-induced cytokine elevation associated with esophagitis and esophageal stricture by manganese superoxide dismutase-plasmid/liposome (SOD-PL) gene therapy. Radiat Res. 2001;155:2–14.
8. Epperly MW, Travis EL, Whitsett JA, Epstein CJ, Greenberger JS. Overexpression of manganese superoxide dismutase (MnSOD) in whole lung or alveolar type II (AT-II) cells of MnSOD transgenic mice does not provideintrinsic lung irradiation protection. Radiat Oncol Invest. 2001;96:11–21.
9. Greenberger JS, Kagan VE, Pearce L, Boriseniao G, Tyurina Y, Epperly MW. Modulation of redox signal transduction pathways in the treatment of cancer. Antioxid Redox Signal. 2001;3(3):347–59.
10. Epperly MW, Kagan VE, Sikora CA, Gretton JE, Defilippi SJ, Bar-Sagi D, Greenberger JS. Manganese superoxide dismutase-plasmid/liposome (MnSOD-PL) administration protects mice from esophagitis associated with fractionated irradiation. Int J Cancer (Radiat Oncol Invest). 2001;96(4):221–33.
11. Epperly MW, Sikora CA, DeFilippi SJ, Gretton JE, Bar-Sagi D, Carlos T, Guo HL, Greenberger JS. Pulmonary irradiation-induced expression of VCAM-1 and ICAM-1 is decreased by MnSOD-PL gene therapy. Biol Blood Bone Marrow Transpl. 2002;8(4)175–87.
12. Epperly MW, Sikora C, Defilippi S, Gretton J, Zhan Q, Kufe DW, Greenberger JS. MnSOD inhibits irradiation-induced apoptosis by stabilization of the mitochondrial membrane against the effects of SAP kinases p38 and Jnk1 translocation. Radiat Res. 2002;157:568–77.
13. Kanai AJ, Zeidel ML, Lavelle JP, Greenberger JS, Birder LA, de Groat WC, Apodaca GL, Meyers SA, Ramage R, VanBibber MM, Epperly MW. Manganese superoxide dismutase gene therapy protects against irradiation-induced cystitis. Am J Physiol (Renal Physiol). 2002;44:1152–60.
14. Epperly MW, Defilippi S, Sikora C, Gretton J, Greenberger JS. Radioprotection of lung and esophagus by overexpression of the human manganese superoxide dismutase transgene. Mil Med. 2002;167(1):071.
15. Tarhini AA, Belani C, Luketich JD, Ramalingam SS, Argiris A, Gooding W, Petro D, Kane K, Liggitt D, Championsmith T, Zhang X, Epperly MW, Greenberger JS. A phase I study of concurrent chemotherapy (Paclitaxel and Carboplatin) and thoracic radiotherapy with

swallowed manganese superoxide dismutase (MnSOD) plasmid liposome (PL) protection in patients with locally advanced stage III non-small cell lung cancer. Hum Gene Ther. 2011;22(3):336–43.
16. Fink M, Macias CA, Xiao J, Tyurina YY, Delude RL, Greenberger JS, Kagan VE, Wipf P. Hemigramicidin-TEMPO conjugates: novel mitochondria-targeted antioxidants. Crit Care Med. 2007;35(9):5461–70.
17. Jiang J, Belikova NA, Xiao J, Zhao Q, Greenberger JS, Wipf P, Kagan VE. A mitochondria-targeted nitroxide/hemi-gramicidin S conjugate protects mouse embryonic cells against g –irradiation. IJROBP. 2008;70(3):816–25.
18. Kagan VE, Bayir A, Bayir H, Stoyanovsky D, Borisenko GG, Tyurina YY, Wipf P, Atkinson J, Greenberger JS, Chapkin RS, Belikova NA. Mitochondria-targeted disruptors and inhibitors of cytochrome c/cardiolipin peroxidase complexes: a new strategy in anti-apoptotic drug discovery. Mol Nutr Food Res. 2009;53:104–14.
19. Jiang J, Stoyanovsky D, Belikova NA, Tyurina YY, Zhao Q, Tungekar MA, Kapralova V, Huang Z, Mintz A, Greenberger JS, Kagan VE. A mitochondria-targeted triphenylphosphonium-conjugated nitroxide functions as a radioprotector/mitigator. Rad Res. 2009;172:706–14.
20. Rajagopalan MS, Gupta K, Epperly MW, Franicola D, Zhang X, Wang H, Zhao H, Tyurin VA, Kagan VE, Wipf P, Kanai A, Greenberger JS. The mitochondria-targeted nitroxide JP4-039 augments potentially lethal irradiation damage repair. In Vivo. 2009;23:717–26.
21. Belikova NA, Jiang J, Stoyanovsky DA, Greenberger JS, Kagan VE. Mitochondria-targeted (2-hydroxyamino-vinyl)-triphenyl-phosphonium releases NO and protects mouse embryonic cells against irradiation-induced apoptosis. FEBS Lett. 2009;583:1945–50.
22. Kagan VE, Wipf P, Stoyanovsky D, Greenberger JS, Borisenko G, Belikova NA, Yanamala N, Samhan AAK, Tungekar MA, Jiang J, Tyurina YY, Ji J, Klein-Seetharaman J, Pitt BR, Shvedovah AA, Bayir H. Mitochondrial targeting of electron scavenging antioxidants: regulation of selective oxidation vs random chain reactions. Adv Drug Del Rev. 2009;61(14):1375–85.
23. Kim H, Bernard ME, Epperly MW, Shen H, Amoscato A, Dixon TM, Doemling AS, Song L, Gao X, Wipf P, Wang H, Zhang X, Kagan VE, Greenberger JS. Amelioration of radiation esophagitis by orally administered p53/mdm2/mdm4 inhibitor (BEB55) or GS-Nitroxide. In Vivo. 2011;25(6):841–9.
24. Nasto L, Seo H, Tilstra J, Robinson A, Clauson C, Sowa G, Ngo K, Dong Q, Lee J, Robbins P, Niedernhofer L, Kang JD, Vo N. Inhibition of NF-kB activity ameliorates age-associated disc degeneration in a mouse model of accelerated aging. Pitt Orthop J. 2013;124:110.
25. Mehanna H, West CML, Nutting C, et al. Head and neck cancer-part 2: treatment and prognostic factors. Br Med J. 2010;41:721–5.
26. Nonzee NJ, Dandade NA, Markossian T, et al. Evaluating the supportive care costs of severe radiochemotherapy-induced mucositis and pharyngitis. Cancer. 2008;113:1446–52.
27. Murphy BA, Beaumont JL, Isitt J, et al. Mucositis-related morbidity and resource utilization in head and neck cancer patients receiving radiation therapy with or without chemotherapy. J Pain Symptom Manage. 2009;38:522–32.
28. Greenberger JS. Radioprotection. In Vivo. 2009;23:323–36.
29. Epperly MW, Lai SM, Mason N, Lopresi B, Dixon T, Franicola D, Niu Y, Wilson WR, Kanai AJ, Greenberger JS. Effectiveness of combined modality radiotherapy of orthotopic human squamous cell carcinomas in Nu/Nu mice using Cetuximab, Tirapazamine, and MnSOD-plasmid liposome gene therapy. In Vivo. 2010;24:1–8.
30. Sonis ST. Efficacy of palifermin (keratinocyte growth factor-1) in the amelioration of oral mucositis. Core Evid. 2009;4:199–205.
31. Vuyyuri SB, Hamstra DA, Khanna D, et al. Evaluation of D-methionine as a novel oral radiation protector for prevention of mucositis. Clin Cancer Res. 2008;14:2161–70.
32. Hamstra DA, Eisbruch A, Naidu MUR, et al. Pharmacokinetic analysis and phase I study of MRX-1024 in patients treated with radiation therapy with or without cisplatinum for head and neck cancer. Clin Cancer Res. 2010;16:2666–76.
33. Zheng C, Cotrim AP, Sunshine AN, et al. Prevention of radiation-induced oral mucositis after adenoviral vector-mediated transfer of the keratinocyte growth factor cDNA to mouse submandibular glands. Clin Cancer Res. 2009;15:4641–6.

34. Mitchell JB, DeGraff W, Kaufman D, et al. Inhibition of oxygen-dependent radiation-induced damage by the nitroxide mimic, tempol. Arch Biochem Biophys. 1991;289:62–70.
35. Guo HL, Seixas-Silva JA, Epperly MW, Gretton JE, Shin DM, Greenberger JS. Prevention of irradiation-induced oral cavity mucositis by plasmid/liposome delivery of the human manganese superoxide dismutase (MnSOD) transgene. Radiat Res. 2003;159:361–70.
36. Greenberger JS, Epperly MW. Radioprotective antioxidant gene therapy: potential mechanisms of action. Gene Ther Mol Biol (GTMB). 2004;8:31–44.
37. Greenberger JS, Epperly MW. Pleiotrophic stem cell and tissue effects of ionizing irradiation protection by MnSOD-plasmid liposome gene therapy. In Columbus F, editor. Progress in gene therapy. Nova Science Publications; 2005. pp. 110–8.
38. Epperly MW, Wegner R, Kanai AJ, Kagan V, Greenberger EE, Nie S, Greenberger JS. Irradiated murine oral cavity orthotopic tumor antioxidant pool destabilization by MnSOD-plasmid liposome gene therapy mediates tumor radiosensitization. Radiat Res. 2007;267:289–97.
39. Greenberger JS. Gene therapy approaches for stem cell protection. Gene Ther. 2008;15:100–8.
40. Epperly MW, Gretton JE, Bernarding M, Nie S, Rasul B, Greenberger JS. Mitochondrial localization of copper/zinc superoxide dismutase (Cu/ZnSOD) confers radioprotective functions in vitro and in vivo. Radiat Res. 2003;160:568–78.
41. Epperly MW, Goff J, Zhang X, Shields D, Hong WS, Hongmei, Franicola D, Bahnson A, Greenberger EE, Greenberger JS. Increased radioresistance, checkpoint inhibition and impaired migratory capacity of bone marrow stromal cell lines derived from SMAD3−/− mice. Rad Res. 2006;165:671–7.
42. Hongliang G, Epperly MW, Bernarding M, Nie S, Gretton J, Jefferson M, Greenberger JS. Manganese superoxide dismutase- plasmid/liposome (MnSOD-PL) intratracheal gene therapy reduction of irradiation- induced inflammatory cytokines does not protect orthotopic lewis lung carcinomas. In Vivo. 2003;17:13–22.
43. Epperly MW, Bernarding M, Gretton J, Jefferson M, Nie S, Greenberger JS. Overexpression of the transgene for manganese Superoxide dismutase (MnSOD) in 32D cl 3 cells prevents apoptosis induction by TNF-a, IL-3 withdrawal and ionizing irradiation. Exp Hematol. 2003;31(6):465–74.
44. Epperly MW, Osipov AN, Martin I, Kawai K, Borisenko GG, Jefferson M, Bernarding M, Greenberger JS, Kagan VE. Ascorbate as a "redox-sensor" and protector against irradiation-induced oxidative stress in 32D cl 3 hematopoietic cells and subclones overexpressing human manganese Superoxide Dismutase. IJROBP. 2004;58(3):851–61.
45. Kagan VE, Tyurina YY, Bayir H, Chu CT, Kapralov AA, Vlasova II, Belikova NA, Tyurin VA, Amoscato A, Epperly M, Greenberger J, DeKosky S, Shvedova AA, Jiang J. The "pro-apoptotic genies" get out of mitochondria: oxidative lipidomics and redox activity of cytochrome c/cardiolipin complexes. Chem-Biol Interact. 2006;163:15–28.
46. Belikova NA, Jiang J, Tyurina YY, Zhao Q, Epperly MW, Greenberger J, Kagan VE. Cardiolipin specific peroxidase reactions of cytochrome c in mitochondria during irradiation induced apoptosis. IJROBP. 2007;69(1):176–85.
47. Tyurin VA, Tyurina YY, Kochanek PM, Hamilton R, DeKosky ST, Greenberger JS, Bayir H, Kagan VE. Oxidative lipidomics of programmed cell death. Methods Enzymol 2008;19(442):375–93.
48. Kagan VE, Bayir HA, Belikova NA, Kapralov O, Tyurina YY, Tyurin VA, Jiang J, Stoyanovsky DA, Wipf P, Kochanek P, Greenberger JS, Pitt B, Shvedova AA, Borisenko G. Cytochrome c/cardiolipin relations in mitochondria: a kiss of death. Free Radic Biol Med. 2009;46:1439–53.
49. Epperly MW, Melendez A, Zhang X, Franicola D, Smith T, Greenberger BA, Komanduri P, Greenberger JS. Mitochondrial targeting of a catalase transgene product by plasmid liposomes increases radioresistance in vitro and in vivo. Radiat Res. 2009;171:588–95.
50. Epperly MW, Chaillet JR, Kalash R, Shaffer B, Goff J, Shields D, Dixon T, Wang H, Berhane H, Kim J-H, Greenberger JS. Conditional radioresistance of tet-inducible manganese superoxide dismutase bone marrow stromal cells. Radiat Res. 2013;180:189–204.
51. Stoyanovsky DA, Huang Z, Jiang J, Belikova NA, Tyurin V, Epperly MW, Greenberger JS, Bayir H, Kagan VE. A manganese-porphyrin complex decomposes hydrogen peroxide, com-

partmentalizes into mitochondria, inhibits apoptosis, and acts as a radiation mitigator in vivo. JACS Med Chem Lett. 2011;362:21–34.
52. Atkinson J, Kapralov AA, Yanamala N, Pearce L, Peterson J, Tyurina YY, Epperly MW, Huang Z, Jiang J, Maeda A, Feng W, Wasserloos K, Belikova NA, Tyurin VA, Fletcher J, Wang Y, Vlasova II, Klein-Seetharaman J, Stoyanovsky DA, Bayir H, Pitt BR, Greenberger JS, Kagan VE. A mitochondria-targeted inhibitor of cytochrome c peroxidase mitigates radiation induced death. Nature Commun. 2011;2:497.
53. Berhane H, Epperly MW, Goff J, Kalash R, Cao S, Franicola D, Zhang X, Shields D, Houghton F, Wang H, Sprachman M, Wipf P, Song L, Gao X, D'Andrea A, Guinan E, Parmar K, Greenberger JS. Radiobiologic differences between bone marrow stromal and hematopoietic progenitor cell lines form Fanconi Anemia (FancD2−/−) mice. Radiat Res. in press.
54. Epperly MW, Carpenter M, Agarwal A, Mitra P, Nie S, Greenberger JS. Intra-oral manganese superoxide dismutase plasmid liposome radioprotective gene therapy decreases ionizing irradiation-induced murine mucosal cell cycling and apoptosis. In Vivo. 2004;18:401–10.
55. Epperly MW, Guo HL, Jefferson M, Wong S, Gretton J, Bernarding M, Bar-Sagi D, Greenberger JS. Cell phenotype specific duration of expression of epitope-tagged HA- MnSOD in cells of the murine lung following intratracheal plasmid liposome gene therapy. Gen Ther. 2003;10:163–71.
56. Guo HL, Wolfe D, Epperly MW, Huang S, Liu K, Glorioso JC, Greenberger J, Blumberg D. Gene transfer of human manganese Superoxide dismutase protects small intestinal villi from radiation injury. J Gastrointest Surg. 2003;7:229–36.
57. Carpenter M, Epperly MW, Agarwal A, Nie S, Hricisak L, Niu Y, Greenberger JS. Inhalation delivery of manganese superoxide dismutase-plasmid/liposomes (MnSOD-PL) protects the murine lung from irradiation damage. Gene Therapy. 2005;12:685–90.
58. Epperly MW, Zhang X, Nie S, Cao S, Kagan V, Tyurin V, Greenberger JS. MnSOD-Plasmid Liposome gene therapy effects on ionizing irradiation induced lipid peroxidation of the Esophagus. In Vivo. 2005;19:997–1004.
59. Epperly MW, Shen H, Zhang X, Nie S, Cao S, Greenberger JS. Protection of esophageal stem cells from ionizing irradiation by MnSOD-Plasmid Liposome gene therapy. In Vivo. 2005;19:965–74.
60. Epperly MW, Smith T, Wang H, Schlesselman J, Franicola D, Greenberger JS. Modulation of total body irradiation induced life shortening by systemic intravenous MnSOD-plasmid liposome gene therapy. Rad Res. 2008;170(4):437–44.
61. Epperly MW, Wang H, Jones J, Dixon T, Montesinos C, Greenberger JS. Antioxidant-chemoprevention diet ameliorates late effects of total body irradiation and supplements radioprotection by MnSOD-plasmid liposome administration. Radiat Res. 2011;175:759–65.
62. Greenberger JS, Epperly MW, Gretton J, Jefferson M, Nie S, Bernarding M, Kagan V, Guo HL. Radioprotective gene therapy. Curr Gene Ther. 2003;3:183–95.
63. Niu Y, Epperly MW, Shen H, Smith T, Lewis D, Gollin S, Greenberger JS. Intraesophageal MnSOD-Plasmid Liposome administration enhances engraftment and self-renewal capacity of bone marrow derived progenitors of esophageal squamous epithelium. Gene Ther. 2008;15:347–56.
64. Niu Y, Wang H, Wiktor-Brown D, Rugo R, Shen H, Huq MS, Engelward B, Epperly M, Greenberger JS. Irradiated esophageal cells are protected from radiation-induced recombination by MnSOD gene therapy. Rad Res. 2010;173:453–61.
65. Rajagopalan MS, Stone, Rwigema J-C, Salimi U, Epperly MW, Goff J, Franicola D, Dixon T, Cao S, Zhang X, Buchholz BM, Bauer AJ, Choi S, Bakkenist C, Wang H, Greenberger JS. Intraesophageal manganese superoxide dismutase-plasmid liposomes ameliorates novel total body and thoracic irradiation sensitivity of homologous deletion recombinant negative nitric oxide synthase-1 (NOS1−/−) mice. Rad Res. 2010;174:297–312.
66. Epperly MW, Bahary N, Quader M, Dewald V, Greenberger JS. The Zebrafish—Danio rerio—is a useful model for measuring the effects of ionizing irradiation. In Vivo. 2012;26(6):889–7.
67. Jiang J, Kurnikov I, Belikova NA, Xiao J, Zhao Q, Vlasova IL, Amoscato AA, Braslau R, Studer A, Fink MP, Greenberger JS, Wipf P, Kagan VE. Structural requirements for optimized delivery, inhibition of oxidative stress and anti-apoptotic activity of targeted nitroxides. J Pharmacol, Exp Ther. 2007;320(5):1050–60.

68. Fink MP, Macias CA, Xiao J, Tyurina YY, Delude RL, Greenberger JS, Kagan VE, Wipf P. Hemigramicidin-TEMPO conjugates: novel mitochondria-targeted anti-oxidants. Biochem Pharmacol. 2007;74:801–9.
69. Hahn SM, Sullivan FJ, DeLuca AM, et al. Evaluation of tempol radioprotection in a murine tumor model. Fred Radic Biol Med. 1997;2:1211–6.
70. Metz JM, Smith D, Mick R, et al. A phase I study of topical tempol for the prevention of alopecia induced by whole brain radiotherapy. Clin Cancer Res. 2004;10:6411–7.
71. Cotrim AP, Yoshikawa M, Sunshine AN, Zheng C, Sowers AL, Zheng C, Sowers AL, Thetford AD, Cook JA, Mitchell JB, Baum BJ. Pharmacological protection from radiation + cisplatin-induced oral mucositis. Int J Radiat Oncol Biol Phys. 2012;83(4):1284–90.
72. Tyurina YY, Tyurin VA, Epperly MW, Greenberger JS, Kagan VE. Oxidative lipidomics of γ-irradiation induced intestinal injury. Free Radic Biol Med. 2008;44:299–314.
73. Frantz M-C, Skoda EM, Davoren J E, Wang Z, Epperly MW, Stripay JL, Tyurin VA, Fink B, Kapralov A, Greenberger JS, Bayir H, Robbins PD, Niedernhofer LJ, Kagan VE, Wipf P. Synthesis and biochemical analysis of mitochondria-targeted nitroxide conjugates based on gramicidin S. JACS (in press).
74. Rwigema J-CM, Beck B, Wang W, Doemling A, Epperly MW, Shields D, Franicola D, Dixon T, Frantz M-C, Wipf P, Tyurina Y, Kagan VE, Wang H, Greenberger JS. Two strategies for the development of mitochondrial-targeted small molecule radiation damage mitigators. Int J Radiat Oncol Biol Phys. 2011;80(3):860–8.
75. Greenberger JS, Kagan V, Bayir H, Lazo J, Wipf P, Song L, Gao X, Clump D, Epperly MW. Mitochondrial targeted small molecule radiation protectors and radiation mitigators. Front Radiat Oncol. 2012;1(59):1–12.
76. Goff JP, Epperly MW, Shields D, Wipf P, Dixon T, Greenberger JS. Radiobiologic effects of GS-nitroxide (JP4-039) in the hematopoietic syndrome. In Vivo. 2011;25:315–24.
77. Gao X, Huang Y, Makhov AM, Epperly M, Lu J, Grab S, Zhang P, Rohan L, Xie XQ, Wipf P, Greenberger J, Li S. Nanoassembly of surfactants with interfacial drug-interactive motifs as tailor-designed drug carriers. Mol Pharm. 2013 Jan 7;10(1):187–98. doi: 10.1021/mp300319m. Epub 2012 Dec 1.
78. Ji J, Kline AE, Amoscato A, Samhan-Arias AK, Sparvero LJ, Tyurin VA, Tyurina YY, Fink B, Manole MD, Puccio AM, Okonkwo DO, Cheng JP, Alexander H, Clark RS, Kochanek PM, Wipf P, Kagan VE, Bayir H. Lipidomics identifies cardiolipin oxidation as a mitochondrial target for redox therapy of brain injury. Nature Neuroscience. 2012 Oct;15(10):1407–13.
79. Cotrim AP, Sowers AL, Lodde BM, et al. Kinetics of tempol for prevention of xerostomia following head and neck irradiation in a mouse model. Clin Cancer Res. 2005;11:7564–8.
80. Spielberger R, Stiff P, Bensinger W, et al. Palifermin for oral mucositis after intensive therapy for hematologic cancers. N Engl J Med. 2004;351:2590–8.
81. Buentzel J, Micke O, Adamietz IA, et al. Intravenous amifostine during chemoradiotherapy for head-and-neck cancer: a randomized placebo-controlled phase III study. Int J Radiat Oncol Biol Phys. 2006;64:684–91.
82. Berhane H, Kalash R, Epperly MW, Goff J, Xu K, Franicola D, Zhang X, Dixon T, Shields D, Wang H, Wipf P, Song L, Gao X, Greenberger JS. Amelioration of irradiation induced oral cavity mucositis and distant bone marrow suppression in Fancd2−/− (FVB/N) mice by intraoral JP4-039/F15. Radiat Res. (submitted).
83. Berhane H, Goff J, Epperly MW, Xu K, Franicola D, Zhang X, Dixon T, Shields D, Wang H, Wipf P, Parmar K, Ferris R, Greenberger JS. Intraoral GS-Nitroxide JP4-039 protects oral mucosa of Fancd2−/− (C57BL/6) mice during irradiation of orthotopic tumors. Cancer Res. (submitted).
84. Epperly MW, Guo HL, Bernarding M, Gretton J, Jefferson M, Greenberger JS. Delayed intratracheal injection of manganese superoxide dismutase (MnSOD)-plasmid/liposomes provides suboptimal protection against irradiation-induced pulmonary injury compared to treatment before irradiation. Gene Ther Mol Biol. 2003;7:61–8.

Pharmacogenomics for Oral Disease

Kenneth S. Kornman, Gordon W. Duff, Kenneth Huttner, Peter J. Polverini and Jan Ching Chun Hu

Pharmacogenomics

Pharmacogenomics at its simplest is the selection of a therapeutic drug regime guided by genomic information. This can be the patient's germ-line genomic sequence, or somatically-mutated sequence data from tumors, or even genomic data from an infectious pathogen such as HIV [82].

Genomic information can be used to predict the pharmacokinetics of a drug (related to absorption, distribution, metabolism and elimination), and at present we understand a good deal about drug metabolism related to population genetic variants in the cytochrome P450 enzyme family. These were among the earliest human gene polymorphisms known to be relevant to drug metabolism and, after a long period of development are beginning to find applications in clinical medicine [136].

Increasingly, we are learning about genetic variations that affect the pharmacodynamics of a drug, how it interacts with a pharmacological target to achieve its therapeutic effect. In fact, we now develop drugs to be specific for mutations that may, for example, give a growth advantage to tumor cells. This area of precision medicine has achieved rapid and notable success in oncology where the prognosis in many cancers is improving continuously based on the use of highly targeted monoclonal antibodies and other inhibitors of cellular growth pathways.

K. S. Kornman (✉)
Interleukin Genetics, 135 Beaver Street, Waltham, MA 02452, USA
e-mail: kkornman@ilgenetics.com

G. W. Duff
University of Sheffield Medical School, Sheffield, UK

K. Huttner
Edimer Pharmaceuticals, Cambridge, MA, USA

P. J. Polverini · J. C. C. Hu
University of Michigan School of Dentistry, Ann Arbor, MI, USA

S. T. Sonis (ed.), *Genomics, Personalized Medicine and Oral Disease*,
DOI 10.1007/978-3-319-17942-1_17

These drug targets often relate to somatic mutations that are specific for the tumor cells as well as germ-line mutations that contribute to genomic instability and faulty DNA repair. The characterization of these genetic variants and their potential for drug targeting was, and continues to be, a scientific achievement of the first importance. However, it may be more challenging to define the variations in host defense systems that can pre-dispose to cancer. We can be optimistic that such advances will occur, for example our understanding of the mechanisms by which tumor cells evade immune-detection and destruction will open another rich area for molecular targeting within precision medicine [140].

Gene-directed therapy is already well established in clinical practice. As of May 2014, there were 161 FDA-licensed medicines with gene information in the label, and 45 medicines (with a preponderance of oncology drugs) have gene information in the indication.

(http://www.fda.gov/Drugs/ScienceResearch/ResearchAreas/Pharmacogenetics/ucm083378.htm)

These basic concepts can be broadened beyond drug treatment guided by genomics to include all modalities of therapeutic intervention guided by a range of biomarkers including, but not limited to, genomic data. This is the evolving notion of personalized medicine, stratified medicine or precision medicine, various terms that have been used to describe therapeutic interventions based on increasing cellular, molecular and biochemical characterization of pathology. Although there is the implication in the future of tailor-made drugs for a single individual, the practical application of pharmacogenomics in the near term involves stratification of individuals into biologic classes that are associated with differential therapeutic responses.

We can broaden the concept further to encompass population stratification across the entire healthcare spectrum, which would include predictive and preventive medicine at the individual, group and public health levels. This broad interpretation is consistent with the definition of personalized medicine as used by the U.S. President's Council on Advisors on Science and Technology, "'Personalized Medicine' refers to the tailoring of medical treatment to the individual characteristics of each patient…to classify individuals into subpopulations that differ in their susceptibility to a particular disease or their response to a specific treatment. Preventative or therapeutic interventions can then be concentrated on those who will benefit, sparing expense and side effects for those who will not" [73].

In an age of spiraling health costs, predictive medicine is an area of high interest both in terms of health improvement and health economics. The underlying principle is that an ability to predict future disease risk at an individual or group level would allow risk-mitigation measures to be targeted to where they would achieve most benefit.

It is relatively easy to design a therapeutic clinical trial where the population is stratified according to the presence or absence of a biomarker, and where the response to the drug can be compared in the pre-specified patient sub-groups. However it is much more challenging to design studies to test the predictive value of biomarkers associated with disease risk or that may, themselves, confer disease risk at some time in the future, perhaps decades away in diseases of ageing such as

Alzheimers. Knowledge of this kind will of course be forthcoming in the future as we accumulate complex, longitudinal health data from ever-larger populations, and develop competent information technology to link eHealth data with large biological datasets.

In the meantime this field is likely to develop through observational studies on such health databases as currently exist, and the retrospective recognition of unequal distribution of interventional events within the population. Such evaluation of retrospective interventions in existing datasets allows comparison of historic outcomes to inform predictive models of future outcomes. Later in this chapter an example of this approach will be considered in depth as a case study relevant to dental medicine [39].

In this chapter, we discuss both the specific use of genetic information to develop drugs for the management of rare genetic disorders that affect the oral cavity and to improve therapeutic outcomes with oral, head and neck cancers based on targeting drugs to specific somatic mutations that affect tumor growth. We also discuss personalized medicine approaches to stratify patients for more effective preventive or therapeutic interventions for the common oral diseases of periodontitis and caries.

Pharmacogenomic Approaches to Correct X-linked Hypohidrotic Ectodemal Dysplasia

The clinical presentation of abnormal tooth morphology and missing dentition, particularly in a pediatric patient, should raise the suspicion of syndromic oligodontia. A classic example is the severe oligodontia in patients with X-linked hypohidrotic ectodermal dysplasia (XLHED), a disorder described first in 19th century reports and now the target of a unique clinical trial for therapeutic correction of a developmental disorder involving dentition.

XLHED is the most common genetic disorder of ectoderm development, presenting with abnormalities of teeth, skin, hair and secretory glands [17]. XLHED is caused by mutations in the ectodysplasin gene (EDA) affecting synthesis, structure and function of the ectoderm signaling molecule EDA-A1 [59]. EDA-A1 is a potent molecular activator of ectoderm placode maturation, with a short window of exposure required to trigger maturation into permanent, fully functional epithelial appendages, e.g. glands, teeth and hair follicles [46].

As an X-linked disorder, affected males have a more consistently severe phenotype, including a risk for life-threatening hyperthermia secondary to an absence of sweating, and a predisposition to pneumonia. Affected (carrier) females have a more variable clinical presentation. Diagnosis in XLHED-affected males is made most commonly when infants present with delayed or absent primary dentition (Stanford, C, personal communication). The severity of dental abnormalities contributes to nutritional, psychosocial and financial issues for the patients and their families. Current dental management in XLHED involves dentures during childhood and subsequent dental implants [147].

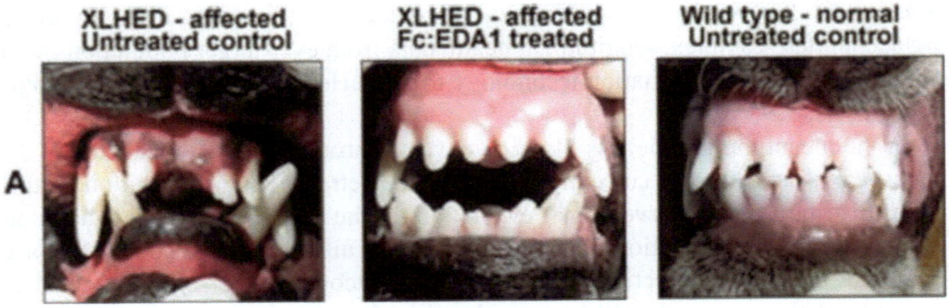

Fig. 1 Photographs of incisors and canines in *XLHED*-untreated, *XLHED*-treated and wild-type dogs (11, with permission)

Following the unraveling of XLHED genetics in 1996 [59], innovative scientists in Switzerland tested the hypothesis that EDA-A1 replacement, delivered at the right time to the right tissues at the right doses could substitute for the endogenous molecule and correct the XLHED phenotype including dentition. Important to their strategy was the presence of a well-established XLHED animal model, the tabby mouse. In a remarkable feat of translation medicine, they unraveled the key elements of the EDA-A1 molecule, rebuilt it as an IgG Fc:EDA1 recombinant fusion protein (Fc:EDA1), established conditions for its synthesis and purification, and demonstrated that injection either into the circulation of a pregnant tabby mouse (with transplacental fetal exposure) or directly into a newborn tabby pup normalized ectoderm development [37].

The tabby mouse results were validated in a second animal model, a dog strain derived from a German shepherd male with a spontaneous EDA mutation [11]. Here a short course of neonatal dosing with the Fc:EDA1 protein yielded a remarkable, sustained correction in the canine XLHED phenotype, most strikingly in restoring the adult dentition from 20 misshapen teeth to nearly normal adult dentition (canine adult teeth, $n=42$) (Fig. 1).

Translation of animal developmental timeframes to human is an uncertain exercise at best (Fig. 2), but preclinical data is an important element in designing an approach to human clinical trials. Dr. Casal's canine results with postnatal Fc:EDA1 dosing established the parameters of dosing, the timing required for optimal results, and an absence of off-target effects- all key issues for human studies.

What was lacking for initiation of a human intervention study was an appreciation of XLHED phenotypic variability in patients, critical data for clinical outcome measures. The initial report of the EDA gene sequence was accompanied by a short list of mutations detected [59], and prior to the availability of routine mutation testing, descriptions of XLHED phenotypic variability and genotype were inconsistent at best [109]. To address this dual gap in genotyping data and clinical description of XLHED, Edimer Phamaceuticals, Inc., a biotechnology company focusing on hereditary ectodermal dysplasia, has collaborated with investigators in the U.S. and Europe in a series of natural history studies with genotype/phenotype correlation (http://edimerpharma.com/publications-news/publications). In brief, the data from

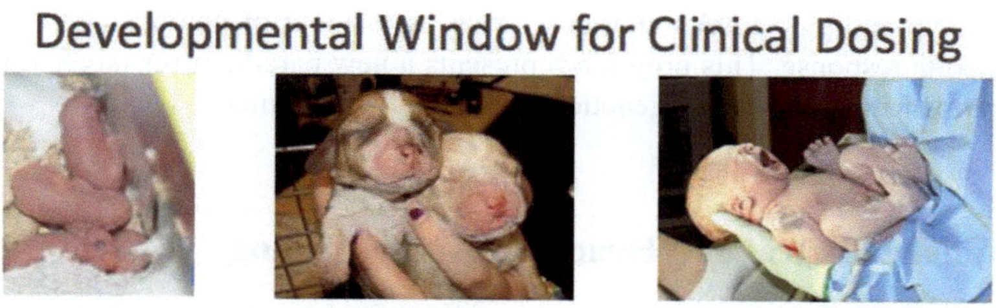

Species	Gestation	Relative Maturity at Birth	Time to Weaning	Primary Dentition	Permanent Dentition
Mouse	20 d	27 wks	3 wks	None	4 wks
Dog	60 d	36 wks	6 wks	12 wks	28 wks
Human	265 d	40 wks	2 yrs	3 yrs	12.5 yrs

Fig. 2 Relative developmental maturities for mice, dogs and humans including dentition [50]; with permission

these studies, merged with genotyping reports from across the globe, now present a much clearer picture of the XLHED phenotype and how it may predict pharmacogenomic response in a trial of EDA-A1 replacement.

Among XLHED-affected patients, 80–90 % have the classic phenotype associated with null mutations resulting in complete loss of EDA-A1 function. These patients have a consistently severe form of XLHED with anhydrosis, few to no adult teeth, sparse scalp hair that is lost early, and clinically significant pulmonary and ocular findings starting in childhood. The other 10–20 % have a distinct set of EDA "hypomorphic" mutations associated with a more variable phenotype including fewer facial features of XLHED, the presence of residual sweat glands that are partially responsive to stimulation, a greater tooth number and hair follicle count, as well as reduced severity of lung and eye involvement as assessed by standardized testing [10]. The location of EDA hypomorphic mutations suggests that some may affect EDA-A1 levels, while others alter monomeric structure or trimer interactions necessary for receptor activation (http://edimerpharma.com/website/wp-content/uploads/2012/06/2012-Nature-Miami-Genotype-Phenotype-Poster_020112.pdf). In either case, one could hypothesize (and test) that this subset of patients with under-developed rather than absent ectodermal-derived structures could be differentially responsive to pharmacologic levels of EDA-A1 replacement.

In summary, EDA-A1 replacement is well tolerated and biologically active in mouse and dog XLHED models of XLHED, with the canine results demonstrating that a single, short-course of replacement in the early neonatal period provides a significant and sustained health benefit including correction of adult dentition. Building on two decades of advances, Edimer Pharmaceuticals, Inc. has now completed the toxicology investigations required for clinical studies and has advanced an EDA-A1 replacement molecule (EDI200) into clinical trials in XLHED. An adult Phase 1 study (NCT01564225) has completed dosing and a Phase 2 neonate study (NCT01775462) is recruiting. The studies are enrolling patients with either null or hypomorphic EDA mutations for an assessment of pharmacogenomics in therapeutic response. This program represents a new paradigm for targeted rescue and permanent correction of genetic disorders of development.

Personalized Oral and Head and Neck Oncology

Personalized cancer therapy has been part of the standard armamentarium used by oncologists for more than a decade [16, 56, 76, 97, 108]. As advances in genomic technology and genetic profiling gain momentum the identification of gene expression patterns will reveal new phenotypic details that will facilitate earlier and more accurate diagnoses, enhance patient stratification and better match patient needs with precision-based therapies [76, 108]. Genetic profiling of patients with neoplastic disease have not only revealed a wide range of potential new targets for tumor therapy but have also provided greater insights into individual patient responsiveness to therapy. Some notable success stories include, the use of the tyrosine kinase Inhibitor, Imatinib (Gleevec), for the treatment of chronic myelogenous leukemia, Herceptin in breast cancer therapy and the B-RAF kinase inhibitor PLX4032 for the treatment of melanoma [25, 34, 126]. Although rapid advances in genomic medicine and have proven invaluable in linking targeted therapies to patient-specific needs, the application of this technology for the majority of tumors has been slow to develop. However, as new biomarkers are discovered and therapeutic targets are revealed the promise of personalized medicine will be more fully realized and become a routine strategy for treating a majority of cancers [55].

Oral and head and neck squamous carcinoma continue to be a worldwide public health problem [14, 121]. Approximately 500,000 new cases of oral and head and neck squamous carcinoma (OHNSCC) will be diagnosed this year with the majority of these patients presenting with advanced stage disease at the time of diagnosis [123, 121]. Despite advances in surgery, chemoradiation and improved diagnostic strategies, efforts to improve early diagnosis and minimize disease progression continue to be a challenge [9, 65, 86, 113, 142].

Molecularly Targeted Therapies Targeted therapies for the treatment of OHNSCC fall into several categories that encompass both unique and overlapping targets. These include oncogenes, biomarkers associated with metastasis, gene amplifications, mutations and translocations, epigenetic alterations including DNA methylation and alterations in histone proteins [40, 120, 144]. Some of the more promising targets include receptors for epidermal growth factor (EGFR), vascular endothelial growth factor (VEGFR) that disrupt or block signaling pathways and networks that govern cell growth, cell motility and survival; angiogenesis inhibitors that block tumor neovascularization and accelerate vascular normalization; activation of immune response to tumor associated antigens and unique biomarkers that identify less aggressive forms of OHNSCC, i.e., human papillomavirus-associated, and more recently, microRNAs [1, 8, 24, 30, 35, 54, 84, 87, 88, 92, 100, 102, 111,

115, 128, 134]. A number of these agents are now in clinical trials with other more promising targets soon to be forthcoming [8, 24, 30, 87, 88, 100, 128, 134].

One of the earliest targets identified for the treatment of OHNSCC was the tyrosine kinase receptor epidermal growth factor (EGFR) [30, 76, 105]. Because it's downstream effects on tumor growth, mobility, survival and therapeutic resistance EGFR was recognized early on as a potential therapeutic target [2, 30, 40, 105]. While monoclonal antibodies to EGFR, such as cetuximab, panitumumab, nimotuzumab, and zalutumumab have been shown interrupt down stream signaling they have limited clinical utility when used as single agents. However, when used in combination with chemoradiation significant improvement in loco-regional control has been reported [8, 100, 134]. Alternative strategies employing tyrosine kinase inhibitors that prevent EGFR phosphorylation have also proven to be better tolerated and associated with more prolong disease-free survival [127]. Another confounding factor is that despite the observation that EGFR is overexpressed in the majority of HNSCC, only a small subset of tumors have shown clinical responsiveness to EGFR therapy [30, 36, 87]. Since there are no biomarkers that reliably predict therapeutic responsiveness, the search for new targets that predict patient responsiveness with improved clinical efficacy continues.

The intercellular signaling pathway phosphatidylinositol-3′ kinase (PI3)/Akt/ mammalian target of rapamycin (mTOR) plays a central role in integrating both intracellular and extracellular signaling networks that regulate cell metabolism, growth, and survival [30, 75, 76, 88]. Activation of mTOR occurs in most human malignancies where it contributes to tumor initiation, progression and therapeutic resistance [24, 65, 88]. Preclinical studies suggest that the mTOR pathway is a potential therapeutic target for OHNSCC [75, 88]. Inhibition of mTOR by rapamycin and less toxic derivative inhibitors has antitumor activity when used alone or in combination with conventional chemotherapeutic agents [75, 88]. Despite the promise of mTOR as a therapeutic target, patient responsiveness to mTOR inhibitors has not been consistently observed in OHNSCC. With the identification of more reliable molecular targets treatment efficacy will certainly improve.

A number of target-specific therapeutic agents while having mixed results as single agents when used in combination with conventional chemotherapy and radiation therapy have been shown to increased their therapeutic efficacy while reducing toxic side effects [64, 76]. The combined use of chemoradiation with target-specific agents has proven to be successful for the treatment of human papillomavirus (HPV)-positive cancers of the head and neck [2, 86]. Due to the infectious nature of HPV and its association with oral and pharyngeal cancer, oral HPV infection may serve as a biomarker of this form of OHNSCC [102]. Although the causal relationship between HPV infection and its carcinogenic mechanism of action in the oral cavity and pharynx requires further validation a more thorough understanding of the role of this infectious agent in the etiology and progression of OHNSCC will likely guide new therapies in the foreseeable future.

With the discovery that tumors are angiogenesis dependent, angiogenesis inhibitors have long been proposed as an effective therapeutic target-specific agent for a wide variety of tumors [35, 84, 92]. The principal anti-angiogenic agents are vascular endothelial cell growth factor (VEGF) and a number of down stream components of the VEGF signaling pathway that regulate VEGF-mediated processes including angiogenesis, vascular normalization, cell survival and therapeutic resistance [35, 52, 138]. For example, Bevacizumab, a humanized VEGF monoclonal antibody, not only inhibits angiogenesis, but also facilitates the increased delivery of chemotherapeutic agents by decreasing microvascular permeability by stimulating vascular normalization and decreasing intratumoral pressure [41, 52, 92]. However, these antiangiogenic/antiVEGF agents when administered to patients with advanced stage disease have limited efficacy [72, 110, 145]. Recent studies have shown that as tumors adapt to chronic stress, activation of the unfolded protein response can induce the angiogenic switch and confer therapeutic resistance to tumors and tumor associated endothelial cells [138]. As with other target specific agents when used in combination with convention anticancer therapy, the results with antiangiogenic appear more promising.

Future Opportunities and Challenges As investigations into cancer genomics and the unique molecular architecture of cancers are unraveled, new therapeutic targets will no doubt be revealed. Biomarkers present in saliva have already revealed a number of new genetic and epigenetic targets [13, 50]. Alterations in patterns of DNA and histone methylations is a common feature of most cancers and can be passed on intergenerationally thus potentially serving as a predictive marker of cancer susceptibility [144].

New targeting agents such as broad spectrum kinase inhibitors show great promise in clinical trials. Other potential targeting agents include proteasome inhibitors, histone deacetylases, heat shock proteins and other molecular chaperones [29, 77]. Innovative approaches to therapy such as oncolytic viruses and systemic immunotherapy among others may prove valuable in the near future. While many challenges lie ahead new biomarkers will no doubt provide further insight into how to therapeutically navigate and target with precision the molecular networks and genetic mutations that drive neoplastic development and progression [85].

Tumor-associated antigens are specifically expressed by malignant cells and recognized by the host immune system [111]. As a result anticancer immune responses can be used to identify cancer-specific antigenic signatures and employed as a screening tool for early detection [111]. The routine screening of sera for the presence of cancer associated antibody signatures would facilitate early detection, assist in the identification of novel pathways involved in tumor initiation and progression, and aid in distinguishing between indolent and more aggressive tumors [111]. The identification of such aggressive tumors would allow for the selection of appropriate treatment strategies before there has been uncontrolled disease progression.

The clinical utility of microRNAs (miRNAs), small endogenous noncoding RNAs, has gained momentum as potential targets for cancer therapy [115]. They play critical roles in many important cellular processes including cell differentiation, proliferation and apoptosis [115]. As every tumor has a unique miRNA signature, it may be possible use these tumor-specific biomarkers to identify tumors during the initiation process thereby facilitating therapeutic intervention and altering the course of disease. In addition to their potential use as therapeutic targets, tumor-specific miRNAs also have potential clinical use as biomarkers for improved diagnosis, risk stratification and predicting prognosis [115]. As predictors of tumor progression miRNA-based therapies could be employed preemptively to block metastases thereby improving a patient's quality of life and overall survival [115].

In addition to identifying new biomarkers and molecular targets, greater attention will be paid to inhibitors that hit multiple targets and subvert the molecular cross talk that often undermines the success of current targeting strategies. A number of biologicals, i.e. protein kinase C beta and Aurora kinase when used in combination with chemotherapeutic agents cisplatinin and paclitaxel respectively, are being evaluated for management of OHNSCC and broad spectrum kinase inhibitors such Sorafenib hold promise as a chemopreventive agent [145]. Other potential targets include proteasome inhibitors, histone deacetylases, heat shock proteins and other molecular chaperones [65, 138]. The challenges oncologists face in overcoming biological redundancies underscore the need to better understand the molecular cross talk and communication within cell signaling networks, the universality of this approach in the face of tumor heterogeneity and the continued need to identify new biomarkers that may provide insight into how to therapeutically navigate these pathways.

The sequencing of the human genome has initiated a remarkable burst of scientific and clinical discovery [19, 18, 45]. With the unveiling of the "blueprint for genomics research" by the National Human Genome Research Institute, the "base pairs to bedside" strategy will rapidly redefine genomic medicine and stimulate the integration of genomic research, clinical research and patient care [45]. As progress in genomic medicine continues, and as a systems approach to disease diagnosis and treatment is deployed in the patient care environment this technology will no doubt lead to a deeper understanding of the biology and structure of genomes, accelerate an integrated response to the diagnosis and treatment of human disease, and advance the science of medicine and dentistry while improving the effectiveness of health care. We expect that patients with OHNSCC may be among the early beneficiaries of these advances.

Periodontitis

Chronic periodontitis (CP) is a bacterially induced inflammatory disease. Bacterial biofilms on teeth are essential for initiation and progression of chronic periodontitis, and individuals with severe generalized CP appear to have dysregulated immuno-inflammatory processes, different microbial ecosystems than seen in health or mild disease, and altered homeostasis of bone and connective tissues which directly result in destruction of tissues that support the teeth. Severe or progressive CP is associated with tooth loss, increased systemic inflammatory mediators [96] and increased risk for certain systemic diseases [23, 112].

Extensive clinical research beginning in 1965 established standard protocols focused on regular disruption of bacterial biofilms to predictably prevent and treat CP. These clinical approaches for the management of periodontal diseases have proven effective for the majority of patients but were often interpreted into overly simplified corollaries:

1. All individuals are equally susceptible to the bacterial challenge in terms of development of periodontitis.
2. Severity is a simple function of the magnitude of bacterial exposure and the duration. (NOTE: although specific bacteria are associated with periodontitis and have the molecular mechanisms to cause the disease, current concepts suggest that bacterial ecosystems develop and evolve together with the host responses, as opposed to a classical infection. Therefore, it is unlikely that the variance in severities of CP patients is explainable primarily due to initial differences in bacterial composition of the dental biofilm.)
3. Bacterial reduction and control, as described in longitudinal studies, will achieve predictably successful prevention and treatment of periodontitis in all patients.

Due to the expectation that all patients respond similarly to intervention protocols focused on bacterial reduction, one would not anticipate value from the inclusion of genomic information to guide prevention or therapy. Current evidence, however, indicates that subsets of patients are more likely to develop severe periodontitis and have less favorable responses to standard periodontitis prevention and treatment protocols [3, 4, 15, 27, 38, 39, 69, 79]. The distinctions between the majority of patients who respond predictably and those subsets that respond differently provide opportunities to manage the challenges of preventing and treating those patients most likely to have both oral and non-oral complications of severe periodontitis.

During validation of the key principles for prevention and treatment of CP, researchers and clinicians observed anomalies in the basic concepts. For example, a dog study that established the first evidence that bacterial accumulations over time lead predictably to periodontitis [70] actually reported that 2 of 10 dogs developed extensive plaque, calculus, and gingivitis but never developed periodontitis during the experimental period. Similar challenges to the key concepts were noted in studies of humans with essentially no oral hygiene. In Sri Lanka [71] and Tanzania [6] although most individuals had massive accumulations of plaque, calculus, and gingivitis for many years, a small percentage (8 % in Sri Lanka) developed severe generalized periodontitis. In addition studies of clinical outcomes following treatment and maintenance of moderate to severe periodontitis patients in clinical practices reported favorable long term clinical results for approximately 80 % of the cases, but a subset of patients continued to progress and lose teeth [47, 78]. A similar pattern was reported in a closely monitored patient population with excellent maintenance care [69] and in adults with no diagnosis of periodontitis who were monitored long-term while receiving routine preventive care by their dentists [39].

The generalizations that suggested the entire story was driven by bacterial exposure in a "one size fits all" manner were also challenged by evidence that emerged about the relationship of various risk factors to CP severity and progression. For

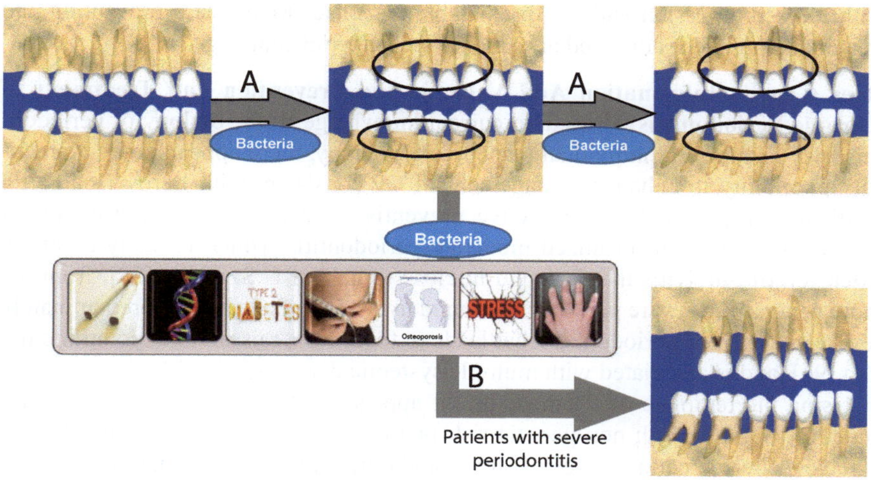

Fig. 3 Diagrammatic representation of epidemiologic and clinical classes of chronic periodontitis. Path A represents the majority of individuals who, with bacterial exposure, exhibit a predictable clinical presentation of mild to localized moderate periodontitis. Path B represents individuals who are exposed to a bacterial challenge but also have host-modifying risk factors, including smoking, uncontrolled diabetes, certain genetic variations, obesity, osteoporosis, certain types of stress, and other chronic inflammatory diseases. These Path B individuals are enriched for generalized moderate to severe periodontitis and also tend to respond differently to preventative and therapeutic care

example, reports in twins [81, 82] showed that 50% of the variance in severity of periodontitis in adults was attributable to genetics, and multiple reports that a small number of risk factors, such as smoking and diabetes, were strongly predictive of severe periodontitis that was less responsive to standard prevention and therapy [15, 38, 39, 66, 95].

All of the above observations define important modifications to the initial concepts of periodontitis pathogenesis and clinical management. These modified concepts speak to the evolving notion of personalized medicine. Current evidence indicates that most individuals (Fig. 3; Path A) respond predictably to the bacterial challenge of dental biofilms to develop mild to localized moderate CP. Disease in these individuals is predictably prevented by routine bacterial biofilm removal/disruption on a regular basis. If periodontitis develops, these patients can be predictably treated with minimal to no disease recurrence with diligent home care and regular professional maintenance care. However, individuals with one or more of a small set of risk factors appear to be on a different disease trajectory (Fig. 3; Path B). Individuals with these risk factors are enriched with severe generalized periodontitis. A small number of risk factors, specifically smoking, diabetes, and IL-1 gene variations, have been shown to precede periodontitis [89, 133] and alter expression of immune-inflammatory pathways that lead to destruction of bone and connective tissues and/or alter homeostasis of bone and connective tissues more directly (Fig. 3, Path B) [22, 38, 43, 57, 90, 95, 106]. These risk factors are associated with more progression and less predictable responses to standard prevention and treatment approaches directed at reducing the bacterial challenge.

Does Genetic Information Add Value in the Prevention and Treatment of Chronic Periodontitis? Based on current knowledge of the biology of periodontitis, risk factors, and individual responses to therapy, there appears to be a practical opportunity to enhance management of CP consistent with Leroy Hood's P4 medicine [48]:personalized, predictive, preventive, and participatory. It is important to distinguish mild to localized moderate periodontitis, which is likely a natural extension of gingivitis that is found in close to half of U.S. adults [28], from the more generalized severe periodontitis found in 8–15 % of adults. It is predominately severe generalized periodontitis that leads to tooth loss, raises systemic inflammation [96], and is associated with multiple systemic diseases [23, 112].

Using the template of P4 medicine, it appears that it may be possible to use genetics [63, 90, 148]; non-genetic risk factors including smoking, uncontrolled diabetes, obesity, certain types of stress, and other chronic inflammatory diseases [15, 38, 66, 95]; and other biomarkers [58, 91] which have been associated with periodontitis severity to more objectively *personalize* a patient's disease trajectory—i.e. identify whether they are more likely to be on a biologic path for long-term mild to localized moderate disease (Fig. 3; Path A) or a path to more progressive/severe periodontitis (Fig. 3; Path B). The key is to identify a patient's risk path prior to severe disease—i.e. *predictive*, and then to take action that modifies the risk for severe disease and associated complications and costs—i.e. *preventive*. If a patient already has moderate to severe generalized periodontitis the goal after active therapy is to prevent disease recurrence and associated complications of progressive disease.

Periodontal preventive care is effective, but limited evidence supports twice yearly prevention for all adults who do not already have periodontitis. One study (Michigan Personalized Prevention Study [39]) used a large dental claims database to determine if predefined risk factors could be used to stratify patients to differentiate long-term responses to preventive dental care. In 5,117 adults with no diagnosis of periodontitis patients were stratified, including genetics, to determine the influence of frequency of preventive dental care on tooth loss over 16 years [39]. The percentage of patients with none of three risk factors (smoking, diabetes, certain IL-1 gene variations) who lost teeth was not significantly lower in patients with two prophylaxes/year compared to those with one/year. Patients with any single risk factor benefited significantly from two prophylaxes/year compared to one/year, and patients with multiple risk factors appeared to require > 2 prophylaxes/year.

The findings of the Michigan Personalized Prevention Study were generally consistent with observations from a 10 year longitudinal study [4] in which a random sample of adults, all age 50, were monitored for periodontitis and tooth loss. Since they recruited a random sample by postal codes, the majority of patients had no to minimal periodontitis at the start of the study. Authors report that more than 95 % of the 320 participants were managed throughout the 10 years in a "needs-related" program [5], in which oral hygiene instructions and frequency of prophylaxes are adjusted based on clinical assessments of patient needs. In general, needs-related preventive care was very effective in preventing periodontitis in this adult population. Since the study does not have a control group for which a different prevention protocol was used, we cannot conclude that this level of preventive care is necessary to maintain periodontal health in this specific population. We can however compare how these patients responded over time based on smoking and IL-1 gene variations that were assessed at the conclusion of the study. Of the patients with both risk factors, almost four times as many patients (41 %) lost multiple teeth to periodontitis during the 10 year prevention and monitoring period, compared to patients with neither risk factor (11 %). Patients with one of the two risk factors were intermediate in tooth loss between the other two groups.

There is much less evidence to support added value from patient stratification to guide maintenance care of patients who have received active treatment for moderate to severe periodontitis. In 4 small studies (total $n = 689$) [27, 68, 79, 99] significantly more progression/tooth loss was reported in maintenance patients who smoked or carried IL-1 gene variations versus those without the risk factors, but 3 (total $n = 161$) did not show greater progression in those with the two risk factors [12, 26, 61].

It is expected that future studies will prospectively assess the value of this approach and determine clinical utility of other genetic and non-genetic risk factors in patient stratification for effective CP prevention/treatment.

Pharmacogenomics Opportunities Clinical studies have shown potential value for the use of certain host modifying pharmaceutical agents, such as bisphosphonates [67] statins [31, 101, 129], doxycycline [42], teriparatide [7], and non-steroidal anti-inflammatory drugs [53, 143], in the treatment of severe generalized periodontitis. One may postulate that patients with early signs of periodontitis plus risk factors that alter the biology to a pattern more conducive to disease progression may be appropriate patients for use of host modifying drugs to prevent severe disease and its complications.

Conclusions Given current evidence for the role of certain risk factors in chronic periodontitis, it is now possible to identify a substantial subset of patients at increased risk for severe/progressive periodontitis to more effectively manage prevention and treatment. One would expect [62, 91, 93] and recent evidence [58] demonstrates that biochemical and gene expression differences should differentiate patient populations that are on different disease trajectories relative to periodontitis severity/progression. The value of such approaches to stratifying patients above and beyond clinical information has not been directly evaluated but is implicit in studies that include long-term monitoring in clinical settings in which patient care could be modified based on clinical parameters [4, 39]. The use of genomic/genetic information to stratify periodontitis patients for more effective periodontitis management is at an early stage but certainly has made encouraging first steps to clinical application.

Dental Caries

Dental caries is a chronic, infectious disease that affects the general population at 5 times the frequency of asthma [130]. In developed countries, dental caries affects 60–90% of schoolchildren and the majority of adults and results in a lifelong affliction, and a global oral health burden [146]. At the turn of the Century, the water fluoridation and various fluoride application programs have effected a decrease in prevalence and severity of caries in permanent dentition. In recent years, however, disparities in caries prevalence, changing dietary habits, and limited access to dental care in many countries have produced a definitive increase in caries [60, 74, 114].

Caries results from sustained interactions among bacteria that produce acid, substrates that the bacteria metabolize, and host factors including tooth, saliva and hygiene. The non-shedding surfaces of teeth are conducive for microbial colonization the bacteria and their by-product produce a biofilm that decreases pH and causes tooth surface demineralization. Dental biofilm creates a microenvironment permissive for repetitive demineralization; this process if uninterrupted will lead to cavitation. Because of dental caries developes as the result of interactions among host, microbes and substrates, it is generally considered a preventable disease. The integrity of surface enamel, virulence of the microorganisms and availability of fermentable substrates collectively influence the initiation and progression of caries. Removing any one of those factors will hamper the decaying process. Many therapeutic agents have been tested to intervene or prevent dental caries but none besides fluoride has significantly impacted the caries burden [49]. Inhibiting demineralization and enhancing remineralization can effectively halt the caries progression. Prophylactically removing biofilm; antimicrobial agents targeting virulent strains; sugar substitutes reducing fermentable substrates; therapeutics promoting topical remineralization may have preventive effect but none of those measures are cost-effective for global implementation.

Caries Control by Recombinant DNA or Protein Technologies Infectious diseases may be most efficiently intercepted by activating host defense systems through immunization prior to the anticipated exposure. Such an approach has been practiced successfully to manage many childhood infectious diseases in a cost efficient and large-scale implementation manner. Not surprisingly, various immunization strategies have been pursued to control dental caries [123]. The concept that active host immune response can be triggered by utilizing recombinant DNA encoding virulent bacterial surface protein in conjunction with specific adjuvants to trigger protection via systemic IgG and/or mucosal IgA antibody production has been demonstrated to be effective in animal models [125]. Although the microbiological etiology of dental caries is complex, immunological prevention has targeted primarily *Streptococcus mutans*. Selected recombinant protein fragments and synthetic peptides of the surface antigens of mutans streptococci have been produced and tested for the purpose of developing caries vaccines [150]. Three specific protein molecules residing on the streptococcal surface have been studied extensively as candidate antigens, specifically, the fibrillar adhesins that control attachment of the bacteria to tooth surfaces [131], the glucosyltransferases that generate adhesive glucans from sucrose-containing substrates, and the cell-wall-associated glucan-binding receptors with the ability to facilitate glucan-mediated bacterial aggregation [132]. A complex DNA vaccine to boost host immunity and target bacterial antigens simultaneously involves robust and sophisticated designs that can be achieved with advanced cloning technologies. Its efficacy has been demonstrated in animal models but has not progressed into human clinical trials [44]. Passive immunity against dental carries has been explored as well. While dietary supplements of IgG or IgY antibody to glucosyltransferase, glucan binding receptors, or monoclonal antibody against *Streptococcus mutans* cell surface molecules have also been shown to reduce caries in experimental animals [124], sustaining the therapeutic concentration of such proteins in the model system can be unpredictable and costly.

Despite the abundant literature evidence supporting the efficacy of dental caries vaccines, all are limited to experimental investigations. Caries vaccines are not yet available for clinical applications. Unless an effective and cost-efficient caries vaccine can be unequivocally demonstrated, the strategy of using recombinant DNA or protein technologies for vaccine production to control caries remains an elusive goal.

Microbial and Host Genetics Influencing Caries Risk Determination of common risk factors is a rational approach to combat chronic diseases. Identifying and targeting the common risk factor(s) may be a sound approach to improve the management and prevention of dental caries among the high-risk populations.

Genome wide association studies (GWAS), linkage scans, gene expression analyses of animals and human have demonstrated a significant familial, racial and gender influence on dental caries. Shared genetic and environmental factors predict an individual's oral health [118]. Among the 640 Dunedin proband-parent groups evaluated, caries/tooth loss risk was significantly linked with familial history of caries/tooth loss. The gender differences in caries have been attributed to saliva composition and flow rate, hormone/pregnancy, dietary preference, genetic variations and social roles of the females [33]. Compounded by a slightly advanced dental age, the eruption of not yet fully matured dental enamel surface into the complex oral environment might have contributed to higher caries risk among females [104, 117]. Although the opposite was demonstrated in primary dentition among a group of 2 year old children, where males acquired a higher caries risk [80]. Genome-wide association scan of 1305 US children ages 3–12 yrs revealed several loci with plausible biological roles in dental caries [116]. Furthermore, the discoveries of caries risk associated genetic polymorphisms mapped by GWAS have been conducted in populations residing on various continents various continents. The estimated impact of host genetics on dental caries pattern and experience ranges from 30 to 65% [116].

Genetic variations of pathogens often render them advantageous in invasion [94]. Children with severe early childhood caries (SECC) harbored more genotypically diverse oral flora than caries-free children [103]. Despite the demonstration of a positive association between *Streptoccoccus sobrinus* clonal diversity and early childhood caries activity, little is known about how the genetic diversity of oral flora determines cariogenic virulence.

Based on the demonstrable familial, gender, and ethnic influences on caries susceptibility and the severity of early childhood caries likely influenced by genetically varied pathogenic strains, it is logical to hypothesize that genetic polymorphisms of the host and/or the pathogen predict caries risk or severity. Targeting host and/or microbial genetic factors therfore may be plausible caries control strategies.

Targeting Microbial or Host Genetic Factors Application of Next-Generation Sequencing techniques has confirmed the presence of over 800 distinct bacterial species in the oral cavity [122]. The concept of engineering genetically-altered bacterial surface polysaccharides to influence colonization or introducing genetically-modified bacteria to weaken biofilm integrity has been investigated [149, 151]. The application of such approaches for caries control *in vivo,* its impact on the oral microbiota and long-term prognosis have yet to be demonstrated.

Advances in genetics also made possible large population studies to discern linkage of complex disorders. Using high throughput screening to identify at-risk populations has been applied on independent cohorts with multifactorial traits. Caries, highly influenced by host and environmental factors is a multifactorial trait. To date, association studies, linkage analyses, gene expression analyses and critical reviews of literature have yielded ample evidence supporting the linkage of genetic polymorphisms and caries in humans. Genome wide association studies of cohorts from various ethnicities residing in different continents have identified caries susceptible and caries protective loci that not only influence caries risk but also affect variation in taste (6-n-propylthiouracil response), saliva compositions (AQP5), enamel proteins (KLK4), dentin structure (DSPP) and the oral milieu (TRAV4) [135, 137, 141].

However, the rapidly accruing susceptibility genes for caries are highly heterogeneous, and in many instances, independent cohorts do not share the same risk indicators [21, 98, 107, 137]. A systematic review of literature evidence summarized the finding of 1214 candidate caries genes; among them 53 are named dental caries genes. To determine the level of significance, caries candidate genes were prioritized by in silico analysis and genes with high degree relevance include specific interleukins (IL8 and IL1B), chemokine ligands (CCL2 and CCL7), matrix metalloproteinases (MMP3 and MMP9) and transforming growth factor-beta family members (TGFB1, TGFBR1 and TGFBR2) [139]. Not surprisingly, candidate genes for caries risk susceptibility are implicated in the process of host inflammatory response, tooth development and caries progression. Elucidation of the basis of the association between candidate genes and dental caries requires systematic biological validations the work that will provide critical evidences for developing pharmacogenomics tools for control of dental caries.

From Genomic Discoveries to Clinical Practice Public Health experts have proposed the use of community-based interventions, strengthening caries research, and advocating multidisciplinary collaborations for caries control [13]. Furthermore, with the identification of preventative dental caries genes being named, the potential of using molecular genetic technologies to control caries should be explored. The effort invested in identifying genes and genetic markers to improve diagnosis, prognosis and therapeutics in dental caries signifies an advance in the field of dentistry, which will need to be translated into affordable clinical practice.

The development of pharmacogenomic approaches for managing complex genetic disorders involves molecular genetic testing of causative genes allowing detection of precise mutations from the specific cohort of patients with the disorder. Such mutations are then categorized based on pathophysiologic defects, to provide essential information that can be used for the development of genotype-based therapeutic strategies. An emerging example is the disorder of cystic fibrosis (CF) [20]. The conventional supportive treatment has improved the mean life expectancy of CF patients from 15 years in the 1970s to about 35–40 years of age today. While the use of pharmacogenomic drugs to address the underlying cause of CF has promised significant improvement of quality of life as well as expectancy [119]. In order to apply novel therapeutic tools targeting cystic fibrosis transmembrane conductance regulator protein (CFTR), specific mutations responsible for its loss-of-function needed to be decoded [32]. The same applies to genetic disorders involving teeth, in order to utilize pharmacogenomic approaches to combat caries, genetic factors associated with susceptibility and resistance will need to be identified, validated experimentally, then screening strategies developed to accurately and efficiently detect such factors. Small molecules combined with high throughput testing can then be identified to target genetic polymorphisms to inhibit risk factors while polymorphisms that confer protection may be enhanced. Consequently, such screening strategies and pharmacogenomic tools will need to be made available and affordable globally (Fig. 4).

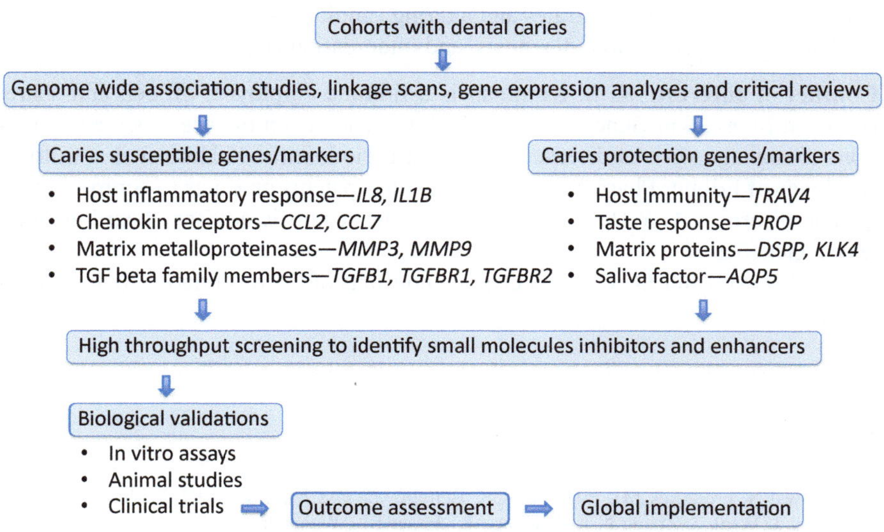

Fig. 4 A concept of translating genetic knowledge into pharmacogenomic tools for dental caries intervention. *IL8*: interleukin 8; *IL1B*: interleukin 1 beta; *CCL2*: chemokine ligand 2; *CCL7*: chemokine ligand 7; *MMP3*: matrix metalloproteinase 3; *MMP9*: matrix metalloproteinase 9; *TGFB1*: transforming growth factor beta 1; *TGFBR1*: TGFB receptor 1; *TGFBR2*: TGFB receptor 2; *TRAV4*: T cell receptor alpha chain variable gene cluster; *PROP*:6-n-propylthiouracil; DSPP: dentin sialophosphoprotein; *KLK4*: kallikrein 4; *AQP5*: aquaporin 5

Conclusion

Although dental caries is rarely a fatal disease, it is a multifactorial trait with painful and sometimes long lasting sequelae. It affects primarily young children and socially disadvantaged adults. Current practice of dental care is not likely to meet the global needs of caries management [114]. To apply the lessons learned from pharmacogenomic management of complex disorders, we will need to advance current understanding of the genetic compositions of dental caries.

Characterizing pathogen virulence factors, then utilizing the genetic sequences that encode such factors in conjunction with boosting host cytotoxic T-lymphocyte antigens to stimulate host immunity have been proven effective in controlling caries in rodents. Transition from animal models to human clinical trials to commercially manufacture the anti-caries DNA vaccines will require significant support and synchronized effort.

Assessing host genetic vulnerability, manipulating genetic codes of the virulent pathogens, and developing animal models that recapitulate human dental caries will allow systematic investigation of caries pathophysiologies for the purpose of devising and testing therapeutics that are effective for population-based application.

With gene-based therapy continuing to advance and broad-based clinical utilization of pharmacogenomics tools becoming a possibility, the prospect to apply effective interventions to complex dental diseases is unfolding. Specifically for dental caries, decoding the genetic composition of at risk populations, boosting host defenses, replacing the substrates and altering candidate pathogens will permit better dental caries prevention, diagnosis, and intervention (Fig.4).

Conflict of Interest: K. S. Kornman is a fulltime employee, officer, and shareholder of Interleukin Genetics, Inc., which has patents on the use of genetic variations to manage various diseases, including periodontitis. K Huttner is a fulltime employee of Edimer Pharmaceuticals, which has patents on drugs to control certain genetic disorders. GW Duff is a member of the scientific advisory board at Interleukin Genetics.

P.J. Polverini and J.C.C. Hu declare no conflicts of interest relative to this manuscript.

References

1. Agrawal N, Frederick MJ, Pickering CR, Bettegowda C, Chang K, Li RJ, Fakhry C, Xie TX, Zhang J, Wang J, Zhang N El-Naggar AK, Jasser SA, Weinstein JN, Trevino L, Drummond JA, Muzny DM, Wu Y, Wood LD, Hruban RH, Westra WH, Koch WM, Califano JA, Gibbs RA, Sidransky D, Vogelstein B, Velculescu VE, Papadopoulos N, Wheeler DA, Kinzler KW, Myers JN. Exome sequencing of head and neck squamous cell carcinoma reveals inactivating mutations in NOTCH1. Science. 2011;333(6046):1154–7.
2. Ang KK, Andratschke NH, Milas L. Epidermal growth factor receptor and response of head-and-neck carcinoma to therapy. Int J Radiat Oncol Biol Phys. 2004;58(3):959–65.

3. Axelsson P. Diagnosis and Risk Prediction of Periodontal Diseases. Carol Stream, IL, Quintessence. 2002a.
4. Axelsson P. Role of genetic and hereditary factors. Diagnosis and risk prediction of periodontal diseases. Carol Stream, Quintessence. 2002b;3:146–63.
5. Axelsson P, Nystrom B, Lindhe J. The long-term effect of a plaque control program on tooth mortality, caries and periodontal disease in adults. Results after 30 years of maintenance. J Clin Periodontol. 2004;31(9):749–57.
6. Baelum V, Fejerskov O, Karring T. Oral hygiene, gingivitis and periodontal breakdown in adult Tanzanians. J Periodontal Res. 1986;21(3):221–32.
7. Bashutski JD, Eber RM, Kinney JS, Benavides E, Maitra S, Braun TM, Giannobile WV, McCauley LK. Teriparatide and osseous regeneration in the oral cavity. N Engl J Med. 2010;363(25):2396–405.
8. Bernier J. Drug Insight: cetuximab in the treatment of recurrent and metastatic squamous cell carcinoma of the head and neck. Nat Clin Pract Oncol. 2008;5(12):705–13.
9. Bibault JE, Fumagalli I, Ferte C, Chargari C, Soria JC, Deutsch E. Personalized radiation therapy and biomarker-driven treatment strategies: a systematic review. Cancer Metastasis Rev. 2013;32(3–4):479–92.
10. Burger K, Schneider AT, Wohlfart S, Kiesewetter F, Huttner K, Johnson R, Schneider H. Genotype-phenotype correlation in boys with X-linked hypohidrotic ectodermal dysplasia. Am J Med Genet A. 2014.
11. Casal ML, Lewis JR, Mauldin EA, Tardivel A, Ingold K, Favre M, Paradies F, Demotz S, Gaide O, Schneider P. Significant correction of disease after postnatal administration of recombinant ectodysplasin A in canine X-linked ectodermal dysplasia. Am J Hum Genet. 2007;81(5):1050–6.
12. Cattabriga, M, Rotundo R, Muzzi L, Nieri M, Verrocchi G, Cairo F, Pini Prato G. Retrospective evaluation of the influence of the interleukin-1 genotype on radiographic bone levels in treated periodontal patients over 10 years. J Periodontol. 2001;72(6):767–73.
13. Chai RL, Grandis JR. Advances in molecular diagnostics and therapeutics in head and neck cancer. Curr Treat Options Oncol. 2006;7(1):3–11.
14. Chaturvedi AK, Anderson WF, Lortet-Tieulent J, Curado MP, Ferlay J, Franceschi S, Rosenberg PS, Bray F, Gillison ML. Worldwide trends in incidence rates for oral cavity and oropharyngeal cancers. J Clin Oncol. 2013;31(36):4550–9.
15. Chavarry NG, Vettore MV, Sansone C, Sheiham A. The relationship between diabetes mellitus and destructive periodontal disease: a meta-analysis. Oral Health Prev Dent. 2009;7(2):107–27.
16. Chin L, Andersen JN, Futreal PA. Cancer genomics: from discovery science to personalized medicine. Nat Med. 2011;17(3):297–303.
17. Clarke A, Phillips DI, Brown R, Harper PS. Clinical aspects of X-linked hypohidrotic ectodermal dysplasia. Arch Dis Child. 1987;62(10):989–96.
18. Collins F. Has the revolution arrived? Nature. 2010;464(7289):674–5.
19. Collins FS, Morgan M, Patrinos A. The Human Genome Project: lessons from large-scale biology. Science. 2003;300(5617):286–90.
20. Cystic Fibrosis. Chapter 14. In: European Lung White Book. Edited by G. John Gibson, Robert Loddenkemper, Yves Sibille and Bo Lundbäck. pp 160–75. Access: http://www.erswhitebook.org/chapters/cystic-fibrosis/.
21. Deeley K, Letra A, Rose EK, Brandon CA, Resick JM, Marazita ML, Vieira AR. Possible association of amelogenin to high caries experience in a Guatemalan-Mayan population. Caries Res. 2008;42(1):8–13.
22. Delima AJ, Karatzas S, Amar S, Graves DT. Inflammation and tissue loss caused by periodontal pathogens is reduced by interleukin-1 antagonists. J Infect Dis. 2002;186(4):511–6.
23. Dietrich T, Sharma P, Walter C, Weston P, Beck J. The epidemiological evidence behind the association between periodontitis and incident atherosclerotic cardiovascular disease. J Periodontol. 2013;84(4 Suppl):S70–84.
24. Dorsey K, Agulnik M. Promising new molecular targeted therapies in head and neck cancer. Drugs. 2013;73(4):315–25.

25. Druker BJ, Talpaz M, Resta DJ, Peng B, Buchdunger E, Ford JM, Lydon NB, Kantarjian H, Capdeville R, Ohno-Jones S, Sawyers CL. Efficacy and safety of a specific inhibitor of the BCR-ABL tyrosine kinase in chronic myeloid leukemia. N Engl J Med. 2001;344(14):1031–7.
26. Ehmke B, Kress W, Karch H, Grimm T, Klaiber B, Flemmig TF. Interleukin-1 haplotype and periodontal disease progression following therapy. J Clin Periodontol. 1999;26(12):810–3.
27. Eickholz P, Kaltschmitt J, Berbig J, Reitmeir P, Pretzl B. Tooth loss after active periodontal therapy. 1: patient-related factors for risk, prognosis, and quality of outcome. J Clin Periodontol. 2008;35(2):165–74.
28. Eke PI, Dye BA, Wei L, Thornton-Evans GO, Genco RJ, Gordon Douglass, Roy Page CDC Periodontal Disease Surveillance workgroup: James Beck. Prevalence of periodontitis in adults in the United States: 2009 and 2010. J Dent Res. 2012;91(10):914–20.
29. Elser C, Siu LL, Winquist E, Agulnik M, Pond GR, Chin SF, Francis P, Cheiken R, Elting J, McNabola A, Wilkie D, Petrenciuc O, Chen EX. Phase II trial of sorafenib in patients with recurrent or metastatic squamous cell carcinoma of the head and neck or nasopharyngeal carcinoma. J Clin Oncol. 2007;25(24):3766–73.
30. Erjala K, Sundvall M, Junttila TT, Zhang N, Savisalo M, Mali P, Kulmala J, Pulkkinen J, Grenman R, Elenius K. Signaling via ErbB2 and ErbB3 associates with resistance and epidermal growth factor receptor (EGFR) amplification with sensitivity to EGFR inhibitor gefitinib in head and neck squamous cell carcinoma cells. Clin Cancer Res. 2006;12(13):4103–11.
31. Fajardo ME, Rocha ML, Sanchez-Marin FJ, Espinosa-Chavez EJ. Effect of atorvastatin on chronic periodontitis: a randomized pilot study. J Clin Periodontol. 2010;37(11):1016–22.
32. Fanen P, Wohlhuter-Haddad A, Hinzpeter A. Genetics of cystic fibrosis: CFTR mutation classifications toward genotype-based CF therapies. Int J Biochem Cell Biol. 2014;52:94–102.
33. Ferraro M, Vieira AR. Explaining gender differences in caries: a multifactorial approach to a multifactorial disease. Int J Dent. 2010;2010:649643.
34. Flaherty KT, Puzanov I, Kim KB, Ribas A, McArthur GA, Sosman JA, O'Dwyer PJ, Lee RJ, Grippo JF, Nolop K, Chapman PB. Inhibition of mutated, activated BRAF in metastatic melanoma. N Engl J Med. 2010;363(9):809–19.
35. Folkman J. What is the evidence that tumors are angiogenesis dependent? J Natl Cancer Inst. 1990;82(1):4–6.
36. Frederick BA, Helfrich BA, Coldren CD, Zheng D, Chan D, Bunn PA Jr., Raben D. Epithelial to mesenchymal transition predicts gefitinib resistance in cell lines of head and neck squamous cell carcinoma and non-small cell lung carcinoma. Mol Cancer Ther. 2007;6(6):1683–91.
37. Gaide O, Schneider P. Permanent correction of an inherited ectodermal dysplasia with recombinant EDA. Nat Med. 2003;9(5):614–8.
38. Genco RJ, Borgnakke WS. Risk factors for periodontal disease. Periodontol 2000. 2013;62(1):59–94.
39. Giannobile WV, Braun TM, Caplis AK, Doucette-Stamm L, Duff GW, Kornman KS. Patient stratification for preventive care in dentistry. J Dent Res. 2013;92(8):694–701.
40. Glazer CA, Chang SS, Ha PK, Califano JA. Applying the molecular biology and epigenetics of head and neck cancer in everyday clinical practice. Oral Oncol. 2009;45(4–5):440–6.
41. Goel S, Wong AH, Jain RK. Vascular normalization as a therapeutic strategy for malignant and nonmalignant disease. Cold Spring Harb Perspect Med. 2012;2(3):a006486.
42. Golub LM, McNamara TF, Ryan ME, Kohut B, Blieden T, Payonk G, Sipos T, Baron HJ. Adjunctive treatment with subantimicrobial doses of doxycycline: effects on gingival fluid collagenase activity and attachment loss in adult periodontitis. J Clin Periodontol. 2001;28(2):146–56.
43. Graves DT, Oates T, Garlet GP. Review of osteoimmunology and the host response in endodontic and periodontal lesions. J Oral Microbiol. 2011;3.
44. Guo JH, Jia R, Fan MW, Bian Z, Chen Z, Peng B. Construction and immunogenic characterization of a fusion anti-caries DNA vaccine against PAc and glucosyltransferase I of Streptococcus mutans. J Dent Res. 2004;83(3):266–70.

45. Guttmacher AE, Collins FS. Genomic medicine–a primer. N Engl J Med. 2002;347(19):1512–20.
46. Headon DJ. Ectodysplasin signaling in cutaneous appendage development: dose, duration, and diversity. J Invest Dermatol. 2009;129(4):817–9.
47. Hirschfeld L, Wasserman B. A long-term survey of tooth loss in 600 treated periodontal patients. J Periodontol. 1978;49(5):225–37.
48. Hood L, Balling R, Auffray C. Revolutionizing medicine in the 21st century through systems approaches. Biotechnol J. 2012;7(8):992–1001.
49. Horst JA, Pieper U, Sali A, Zhan L, Chopra G, Samudrala R, Featherstone JD. Strategic protein target analysis for developing drugs to stop dental caries. Adv Dent Res. 2012;24(2):86–93.
50. Hu S, Arellano M, Boontheung P, Wang J, Zhou H, Jiang J, Elashoff D, Wei R, Loo JA, Wong DT. Salivary proteomics for oral cancer biomarker discovery. Clin Cancer Res. 2008;14(19):6246–52.
51. Huttner K. Future developments in XLHED treatment approaches. Am J Med Genet A. 2014;164A(10):2433–6.
52. Jain RK. Normalization of tumor vasculature: an emerging concept in antiangiogenic therapy. Science. 2005;307(5706):58–62.
53. Jeffcoat MK, Page R, Reddy M, Wannawisute A, Waite P, Palcanis K, Cogen R, Williams RC, Basch C. Use of digital radiography to demonstrate the potential of naproxen as an adjunct in the treatment of rapidly progressive periodontitis. J Periodontal Res. 1991;26(5):415–21.
54. Jimeno A, Kulesza P, Wheelhouse J, Chan A, Zhang X, Kincaid E, Chen R, Clark DP, Forastiere A, Hidalgo M. Dual EGFR and mTOR targeting in squamous cell carcinoma models, and development of early markers of efficacy. Br J Cancer. 2007;96(6):952–9.
55. Jorgensen JT. Companion diagnostics in oncology—current status and future aspects. Oncology. 2013;85(1):59–68.
56. Kalia M. Personalized oncology: recent advances and future challenges. Metabolism. 2013;62 Suppl 1:S11–4.
57. Karimbux NY, Saraiya VM, Elangovan S, Allareddy V, Kinnunen T, Kornman KS, Duff GW. Interleukin-1 gene polymorphisms and chronic periodontitis in adult whites: a systematic review and meta-analysis. J Periodontol. 2012;83(11):1407–19.
58. Kebschull M, Demmer RT, Grun B, Guarnieri P, Pavlidis P, Papapanou PN. Gingival tissue transcriptomes identify distinct periodontitis phenotypes. J Dent Res. 2014;93(5):459–68.
59. Kere J, Srivastava AK, Montonen O, Zonana J, Thomas N, Ferguson B, Munoz F, Morgan D, Clarke A, Baybayan P, Chen EY, Ezer S, Saarialho-Kere U, de la Chapelle A, Schlessinger D. X-linked anhidrotic (hypohidrotic) ectodermal dysplasia is caused by mutation in a novel transmembrane protein. Nat Genet. 1996;13(4):409–16.
60. Konig KG. Clinical manifestations and treatment of caries from 1953 to global changes in the 20th century. Caries Res. 2004;38(3):168–72.
61. Konig J, Ruhling A, Plagmann HC, Meisel P, Kocher T. Influence of interleukin (IL)-1 composite genotype on clinical variables in non-smoking, well-maintained compliant patients with chronic periodontitis. Swed Dent J. 2005;29(1):11–6.
62. Kornman KS. Mapping the pathogenesis of periodontitis: a new look. J Periodontol. 2008;79(8 Suppl):1560–8.
63. Kornman KS, Polverini PJ. Clinical Application of Genetics to Guide Prevention and Treatment of Oral Diseases. Clin Genet. 2014;86(1):44–9.
64. Kumar P, Benedict R, Urzua F, Fischbach C, Mooney D, Polverini P. Combination treatment significantly enhances the efficacy of antitumor therapy by preferentially targeting angiogenesis. Lab Invest. 2005;85(6):756–67.
65. Kundu SK, Nestor M. Targeted therapy in head and neck cancer. Tumour Biol. 2012;33(3):707–21.
66. Labriola A, Needleman I, Moles DR. Systematic review of the effect of smoking on nonsurgical periodontal therapy. Periodonto 2000. 2005;37:124–37.
67. Lane N, Armitage GC, Loomer P, Hsieh S, Majumdar S, Wang HY, Jeffcoat M, Munoz T. Bisphosphonate therapy improves the outcome of conventional periodontal treatment: results of a 12-month, randomized, placebo-controlled study. J Periodontol. 2005;76(7):1113–22.

68. Lang NP, Tonetti MS, Suter J, Sorrell J, Duff GW, Kornman KS. Effect of interleukin-1 gene polymorphisms on gingival inflammation assessed by bleeding on probing in a periodontal maintenance population. J Periodontal Res. 2000;35(2):102–107.
69. Lindhe J, Nyman S. Long-term maintenance of patients treated for advanced periodontal disease. J Clin Periodontol. 1984;11(8):504–14.
70. Lindhe J, Hamp SE, Loe H. Plaque induced periodontal disease in beagle dogs. A 4-year clinical, roentgenographical and histometrical study. J Periodontal Res. 1975;10(5):243–55.
71. Loe H, Anerud A, Boysen H, Morrison E. Natural history of periodontal disease in man. Rapid, moderate and no loss of attachment in Sri Lankan laborers 14 to 46 years of age. J Clin Periodontol. 1986;13(5):431–45.
72. Machiels JP, Henry S, Zanetta S, Kaminsky MC, Michoux N, Rommel D, Schmitz S, Bompas E, Dillies AF, Faivre S, Moxhon A, Duprez T, Guigay J. Phase II study of sunitinib in recurrent or metastatic squamous cell carcinoma of the head and neck: GORTEC 2006-01. J Clin Oncol. 2010;28(1):21–8.
73. Marburger J, Kvamme FE. Priorities for Personalized Medicine: Report of the President's Council of Advisors on Science and Technology. 2008.
74. Marthaler TM. Changes in dental caries 1953–2003. Caries Res. 2004;38(3):173–81.
75. Martins F, de Oliveira MA, Wang Q, Sonis S, Gallottini M, George S, Treister N. A review of oral toxicity associated with mTOR inhibitor therapy in cancer patients. Oral Oncol. 2013;49(4):293–8.
76. Matta A, Ralhan R. Overview of current and future biologically based targeted therapies in head and neck squamous cell carcinoma. Head Neck Oncol. 2009;1:1–6.
77. Mazumdar A, Henderson YC, El-Naggar AK, Sen S, Clayman GL. Aurora kinase A inhibition and paclitaxel as targeted combination therapy for head and neck squamous cell carcinoma. Head Neck. 2009;31(5):625–34.
78. McFall WT Jr. Tooth loss in 100 treated patients with periodontal disease. A long-term study. J Periodontol. 1982;53(9):539–49.
79. McGuire MK, Nunn ME. Prognosis versus actual outcome. IV. The effectiveness of clinical parameters and IL-1 genotype in accurately predicting prognoses and tooth survival. J Periodontol. 1999;70(1):49–56.
80. Menghini G, Steiner M, Thomet E, Roos M, Imfeld T. Caries prevalence in 2-year-old children in the city of Zurich. Community Dent Health. 2008;25(3):154–60.
81. Michalowicz BS, Aeppli D, Virag JG, Klump DG, Hinrichs JE, Segal NL, Bouchard TJ Jr., Pihlstrom BL. Periodontal findings in adult twins. J Periodontol. 1991;62(5):293–9.
82. Michalowicz BS, Diehl SR, Gunsolley JC, Sparks BS, Brooks CN, Koertge TE, Califano JV, Burmeister JA, Schenkein HA. Evidence of a substantial genetic basis for risk of adult periodontitis. J Periodontol. 2000;71(11):1699–707.
83. Michaud V, Bar-Magen T, Turgeon J, Flockhart D, Desta Z, Wainberg MA. The dual role of pharmacogenetics in HIV treatment: mutations and polymorphisms regulating antiretroviral drug resistance and disposition. Pharmacol Rev. 2012;64(3):803–33.
84. Mineta H, Miura K, Ogino T, Takebayashi S, Misawa K, Ueda Y, Suzuki I, Dictor M, Borg A, Wennerberg J. Prognostic value of vascular endothelial growth factor (VEGF) in head and neck squamous cell carcinomas. Br J Cancer. 2000;83(6):775–81.
85. Moon C, Oh Y, Roth JA. Current status of gene therapy for lung cancer and head and neck cancer. Clin Cancer Res. 2003;9(14):5055–67.
86. Morgan MA, Parsels LA, Maybaum J, Lawrence TS. Improving the efficacy of chemoradiation with targeted agents. Cancer Discov. 2014;4(3):280–91.
87. Morgillo F, Bareschino MA, Bianco R, Tortora G, Ciardiello F. Primary and acquired resistance to anti-EGFR targeted drugs in cancer therapy. Differentiation. 2007;75(9):788–99.
88. Nathan CO, Amirghahari N, Rong X, Giordano T, Sibley D, Nordberg M, Glass J, Agarwal A, Caldito G. Mammalian target of rapamycin inhibitors as possible adjuvant therapy for microscopic residual disease in head and neck squamous cell cancer. Cancer Res. 2007;67(5):2160–8.

89. Nelson RG, Shlossman M, Budding LM, Pettitt DJ, Saad MF, Genco RJ, Knowler WC. Periodontal disease and NIDDM in Pima Indians. Diabetes Care. 1990;13(8):836–40.
90. Nikolopoulos GK, Dimou NL, Hamodrakas SJ, Bagos PG. Cytokine gene polymorphisms in periodontal disease: a meta-analysis of 53 studies including 4178 cases and 4590 controls. J Clin Periodontol. 2008;35(9):754–67.
91. Offenbacher S, Barros SP, Singer RE, Moss K, Williams RC, Beck JD. Periodontal disease at the biofilm-gingival interface. J Periodontol. 2007;78(10):1911–25.
92. Olsson AK, Dimberg A, Kreuger J, Claesson-Welsh L. VEGF receptor signalling—in control of vascular function. Nat Rev Mol Cell Biol. 2006;7(5):359–71.
93. Page RC, Kornman KS. The pathogenesis of human periodontitis: an introduction. Periodontol 2000. 1997;14:9–11.
94. Palmer SR, Miller JH, Abranches J, Zeng L, Lefebure T, Richards VP, Lemos JA, Stanhope MJ, Burne RA. Phenotypic heterogeneity of genomically-diverse isolates of Streptococcus mutans. PLoS ONE. 2013;8(4):e61358.
95. Papapanou PN. Periodontal diseases: epidemiology. Ann Periodontol. 1996;1(1):1–36.
96. Paraskevas S, Huizinga JD, Loos BG. A systematic review and meta-analyses on C-reactive protein in relation to periodontitis. J Clin Periodontol. 2008;35(4):277–90.
97. Parkinson DR, Johnson BE, Sledge GW. Making personalized cancer medicine a reality: challenges and opportunities in the development of biomarkers and companion diagnostics. Clin Cancer Res. 2012;18(3):619–24.
98. Patir A, Seymen F, Yildirim M, Deeley K, Cooper ME, Marazita ML, Vieira AR. Enamel formation genes are associated with high caries experience in Turkish children. Caries Res. 2008;42(5):394–400.
99. Persson GR, Matuliene G, Ramseier CA, Persson RE, Tonetti MS, Lang NP. Influence of interleukin-1 gene polymorphism on the outcome of supportive periodontal therapy explored by a multi-factorial periodontal risk assessment model (PRA). Oral Health Prev Dent. 2003;1(1):17–27.
100. Pfister DG, Su YB, Kraus DH, Wolden SL, Lis E, Aliff TB, Zahalsky AJ, Lake S, Needle MN, Shaha AR, Shah JP, Zelefsky MJ. Concurrent cetuximab, cisplatin, and concomitant boost radiotherapy for locoregionally advanced, squamous cell head and neck cancer: a pilot phase II study of a new combined-modality paradigm. J Clin Oncol. 2006;24(7):1072–8.
101. Pradeep AR, Rao NS, Bajaj P, Kumari M. Efficacy of subgingivally delivered simvastatin in the treatment of patients with type 2 diabetes and chronic periodontitis: a randomized double-masked controlled clinical trial. J Periodontol. 2013;84(1):24–31.
102. Prince A, Aguirre-Ghizo J, Genden E, Posner M, Sikora A. Head and neck squamous cell carcinoma: new translational therapies. Mt Sinai J Med. 2010;77(6):684–99.
103. Qin XR, Zhou Q, Qin M. Genotypic diversity and virulence traits of streptococcus sobrinus isolated from caries-free children and children suffering severe early childhood caries. Chin J Dent Res. 2013;16(1):63–9.
104. Ramezani GH, Norozi A, Valael N. The prevalence of nursing caries in 18 to 60 months old children in Qazvin. J Indian Soc Pedod Prev Dent. 2003;21(1):19–26.
105. Rodemann HP, Dittmann K, Toulany M. Radiation-induced EGFR-signaling and control of DNA-damage repair. Int J Radiat Biol. 2007;83(11–12):781–91.
106. Rogus J, Beck JD, Offenbacher S, Huttner K, Iacoviello L, Latella MC, de Gaetano M, Wang HY, Kornman KS, Duff GW. IL1B gene promoter haplotype pairs predict clinical levels of interleukin-1beta and C-reactive protein. Hum Genet. 2008;123(4):387–98.
107. Rose EK, Vieira AR. Caries and periodontal disease: insights from two U.S. populations living a century apart. Oral Health Prev Dent. 2008;6(1):23–8.
108. Ross JS. Cancer biomarkers, companion diagnostics and personalized oncology. Biomark Med. 2011;5(3):277–9.
109. Rouse C, Siegfried E, Breer W, Nahass G. Hair and sweat glands in families with hypohidrotic ectodermal dysplasia: further characterization. Arch Dermatol. 2004;140(7):850–5.
110. Salama JK, Haraf DJ, Stenson KM, Blair EA, Witt ME, Williams R, Kunnavakkam R, Cohen EE, Seiwert T, Vokes EE. A randomized phase II study of 5-fluorouracil, hydroxyurea, and

twice-daily radiotherapy compared with bevacizumab plus 5-fluorouracil, hydroxyurea, and twice-daily radiotherapy for intermediate-stage and T4N0-1 head and neck cancers. Ann Oncol. 2011;22(10):2304–9.

111. Scanlon CS, D'Silva NJ. Personalized medicine for cancer therapy: Lessons learned from tumor-associated antigens. Oncoimmunology. 2013;2(4):e23433.
112. Scher JU, Ubeda C, Equinda M, Khanin R, Buischi Y, Viale A, Lipuma L, Attur M, Pillinger MH, Weissmann G, Littman DR, Pamer EG, Bretz WA, Abramson SB. Periodontal disease and the oral microbiota in new-onset rheumatoid arthritis. Arthritis Rheum. 2012;64(10):3083–94.
113. Scully C, Bagan JV. Recent advances in Oral Oncology 2007: imaging, treatment and treatment outcomes. Oral Oncol. 2008;44(3):211–5.
114. Selwitz RH, Ismail AI, Pitts NB. Dental caries. Lancet. 2007;369(9555):51–9.
115. Sethi S, Ali S, Sethi S, Sarkar FH. MicroRNAs in personalized cancer therapy. Clin Genet. 2014;86(1):68-73.
116. Shaffer JR, Wang X, Feingold E, Lee M, Begum F, Weeks DE, Cuenco KT, Barmada MM, Wendell SK, Crosslin DR, Laurie CC, Doheny KF, Pugh EW, Zhang Q, Feenstra B, Geller F, Boyd HA, Zhang H, Melbye M, Murray JC, Weyant RJ, Crout R, McNeil DW, Levy SM, Slayton RL, Willing MC, Broffitt B, Vieira AR, Marazita ML. Genome-wide association scan for childhood caries implicates novel genes. J Dent Res. 2011;90(12):1457–62.
117. Shaffer JR, Feingold E, Wang X, Tcuenco KT, Weeks DE, DeSensi RS, Polk DE, Wendell S, Weyant RJ, Crout R, McNeil DW, Marazita ML. Heritable patterns of tooth decay in the permanent dentition: principal components and factor analyses. BMC Oral Health. 2012;12:7.
118. Shearer DM, Thomson WM, Caspi A, Moffitt TE, Broadbent JM, Poulton R. Family history and oral health: findings from the Dunedin Study. Community Dent Oral Epidemiol. 2012;40(2):105–15.
119. Sheridan C. First cystic fibrosis drug advances towards approval. Nat Biotechnol. 2011;29(6):465–6.
120. Shirai K, O'Brien PE. Molecular targets in squamous cell carcinoma of the head and neck. Curr Treat Options Oncol. 2007;8(3):239–51.
121. Siegel R, Naishadham D, Jemal A. Cancer statistics, 2013. CA Cancer J Clin. 2013;63(1):11–30.
122. Siqueira JF, Jr., Rocas IN. The oral microbiota: general overview, taxonomy, and nucleic acid techniques. Methods Mol Biol. 2010;666:55–69.
123. Smith DJ. Dental caries vaccines: prospects and concerns. Expert Rev Vaccines. 2010;9(1):1–3.
124. Smith DJ. Prospects in caries vaccine development. J Dent Res. 2012;91(3):225–6.
125. Smith DJ, Mattos-Graner RO. Secretory immunity following mutans streptococcal infection or immunization. Curr Top Microbiol Immunol. 2008;319:131–56.
126. Sotiriou C, Pusztai L. Gene-expression signatures in breast cancer. N Engl J Med. 2009;360(8):790–800.
127. Soulieres D, Senzer NN, Vokes EE, Hidalgo M, Agarwala SS, Siu LL. Multicenter phase II study of erlotinib, an oral epidermal growth factor receptor tyrosine kinase inhibitor, in patients with recurrent or metastatic squamous cell cancer of the head and neck. J Clin Oncol. 2004;22(1):77–85.
128. Stransky N, Egloff AM, Tward AD, Kostic AD, Cibulskis K, Sivachenko A, Kryukov GV, Lawrence MS, Sougnez C, McKenna A, Shefler E, Ramos AH, Stojanov P, Carter SL, Voet D, Cortes ML, Auclair D, Berger MF, Saksena G, Guiducci C, Onofrio RC, Parkin M, Romkes M, Weissfeld JL, Seethala RR, Wang L, Rangel-Escareno C, Fernandez-Lopez JC, Hidalgo-Miranda A, Melendez-Zajgla J, Winckler W, Ardlie K, Gabriel SB, Meyerson M, Lander ES, Getz G, Golub TR, Garraway LA, Grandis JR. The mutational landscape of head and neck squamous cell carcinoma. Science. 2011;333(6046):1157–60.
129. Subramanian S, Emami H, Vucic E, Singh P, Vijayakumar J, Fifer KM, Alon A, Shankar SS, Farkouh M, Rudd JH, Fayad ZA, Van Dyke TE, Tawakol A. High-dose atorvastatin

reduces periodontal inflammation: a novel pleiotropic effect of statins. J Am Coll Cardiol. 2013;62(25):2382–91.
130. Surgeon General 2000. US Department of Health and Human Services. Oral Health in America: A Report of the Surgeon General. Rockville: National Institute of Dental and Craniofacial Research, National Institutes of Heath. 2000:308.
131. Takahashi I, Okahashi N, Matsushita K, Tokuda M, Kanamoto T, Munekata E, Russell MW, Koga T. Immunogenicity and protective effect against oral colonization by Streptococcus mutans of synthetic peptides of a streptococcal surface protein antigen. J Immunol. 1991;146(1):332–6.
132. Taubman MA, Holmberg CJ, Smith DJ. Immunization of rats with synthetic peptide constructs from the glucan-binding or catalytic region of mutans streptococcal glucosyltransferase protects against dental caries. Infect Immun. 1995;63(8):3088–93.
133. Thomson WM, Broadbent JM, Welch D, Beck JD, Poulton R. Cigarette smoking and periodontal disease among 32-year-olds: a prospective study of a representative birth cohort. J Clin Periodontol. 2007;34(10):828–34.
134. Vermorken JB, Mesia R, Rivera F, Remenar E, Kawecki A, Rottey S, Erfan J, Zabolotnyy D, Kienzer HR, Cupissol D, Peyrade F, Benasso M, Vynnychenko I, De Raucourt D, Bokemeyer C, Schueler A, Amellal N, Hitt R. Platinum-based chemotherapy plus cetuximab in head and neck cancer. N Engl J Med. 2008;359(11):1116–27.
135. Vieira AR, Marazita ML, Goldstein-McHenry T. Genome-wide scan finds suggestive caries loci. J Dent Res. 2008;87(5):435–9.
136. Walker L, Yip V, Pirmohamed M. Applications in Therapeutics: Adverse Drug Reactions. Handbook of Pharmacogenomics and Stratfield Medicine. 2014. pp. 405–26.
137. Wang X, Willing MC, Marazita ML, Wendell S, Warren JJ, Broffitt B, Smith B, Busch T, Lidral AC, Levy SM. Genetic and environmental factors associated with dental caries in children: the Iowa Fluoride Study. Caries Res. 2012a;46(3):177–84.
138. Wang Y, Alam GN, Ning Y, Visioli F, Dong Z, Nor JE, Polverini PJ. The unfolded protein response induces the angiogenic switch in human tumor cells through the PERK/ATF4 pathway. Cancer Res. 2012b;72(20):5396–406.
139. Wang Q, Jia P, Cuenco KT, Feingold E, Marazita ML, Wang L, Zhao Z. Multi-dimensional prioritization of dental caries candidate genes and its enriched dense network modules. PLoS ONE. 2013;8(10):e76666.
140. Weber J. Immune checkpoint proteins: a new therapeutic paradigm for cancer–preclinical background: CTLA-4 and PD-1 blockade. Semin Oncol. 2010;37(5):430–9.
141. Werneck RI, Mira MT, Trevilatto PC. A critical review: an overview of genetic influence on dental caries. Oral Dis. 2010;16(7):613–23.
142. Williams MD. Integration of biomarkers including molecular targeted therapies in head and neck cancer. Head Neck Pathol. 2010;4(1):62–9.
143. Williams RC, Jeffcoat MK, Howell TH, Rolla A, Stubbs D, Teoh KW, Reddy MS, Goldhaber P. Altering the progression of human alveolar bone loss with the non-steroidal anti-inflammatory drug flurbiprofen. J Periodontol. 1989;60(9):485–90.
144. Williams, S, Hughes T, Adler C, Brook A, Townsend G. Epigenetics: a new frontier in dentistry. Aust Dent J. 2014;59(Suppl 1):23–33.
145. Williamson SK, Moon J, Huang CH, Guaglianone PP, LeBlanc M, Wolf GT, Urba SG. Phase II evaluation of sorafenib in advanced and metastatic squamous cell carcinoma of the head and neck: Southwest Oncology Group Study S0420. J Clin Oncol. 2010;28(20):3330–5.
146. World Health Organization. What is the burden of ral disease? 2000. https://www.who.int/oral_health/disease_burden/global/en#.
147. Wright JT, Grange DK, Richter MK. Hypohidrotic ectodermal dysplasia. GeneReviews(R). Pagon RA, Adam MP, Ardinger HH et al. Seattle (WA). 1993.
148. Wu X, Offenbacher S, Lomicronpez NJ, Chen D, Wang HY, Rogus J, Zhou J, Beck J, Jiang S, Bao X, Wilkins L, Doucette-Stamm L, Kornman K. Association of interleukin-1 gene variations with moderate to severe chronic periodontitis in multiple ethnicities. J Periodontal Res. 2014;50(1):52–61.

149. Xue X, Sztajer H, Buddruhs N, Petersen J, Rohde M, Talay SR, Wagner-Dobler I. Lack of the delta subunit of RNA polymerase increases virulence related traits of *Streptococcus mutans*. PLoS ONE. 2011;6(5):e20075.
150. Yan H. Salivary IgA enhancement strategy for development of a nasal-spray anti-caries mucosal vaccine. Sci China Life Sci. 2013;56(5):406–13.
151. Yoshida Y, Ganguly S, Bush CA, Cisar JO. Carbohydrate engineering of the recognition motifs in streptococcal co-aggregation receptor polysaccharides. Mol Microbiol. 2005;58(1):244–56.

Personalized Medicine, Genomics and Oral Diseases: The Future

Stephen T. Sonis

Introduction

As has been noted throughout this book, there are at least three key applications for genetics and genomics that ultimately impact clinical care [1]. First, genomics can be used to predict disease risk. Second, genomics can help define the pathogenesis and pathobiology of an illness. And third, in the context of pharmacogenomics, drug targets can be identified, responder populations to specific agents can be determined, and toxicity risk established. Given that the success of a drug is defined by its efficacy and safety, one would conclude that these latter two elements would provide a powerful incentive to include pharmacogenomics as a routine component in the clinical development schema. However, as of 2012 only 0.19 % of trials reported on the ClinTrials.gov database had pharmacogenomics outcomes [2]. Clearly we have a ways to go. This chapter focuses on both the potential and the challenges associated with the translation of genomics to the clinic.

Pharmacogenomics

The promise of genomics-based personalized medicine has already started to be realized relative to the diagnosis, management and monitoring of oral disease. Commercially available genomic tests are available to assess the risk of periodontal disease and HPV-related oral cancer. And, as described abundantly in the proceeding chapters, genetic associations have been identified for a broad range of oral diseases. While the genetic aspects of disease etiology can be leveraged into risk assessment strategies, they only reach their true clinical value when they result in an actionable event that prevents the actual development of a condition or interferes

S. T. Sonis (✉)
Brigham and Women's Hospital, Boston, MA, USA
e-mail: ssonis@partners.org; sonis@biomodels.com

S. T. Sonis (ed.), *Genomics, Personalized Medicine and Oral Disease*,
DOI 10.1007/978-3-319-17942-1_18

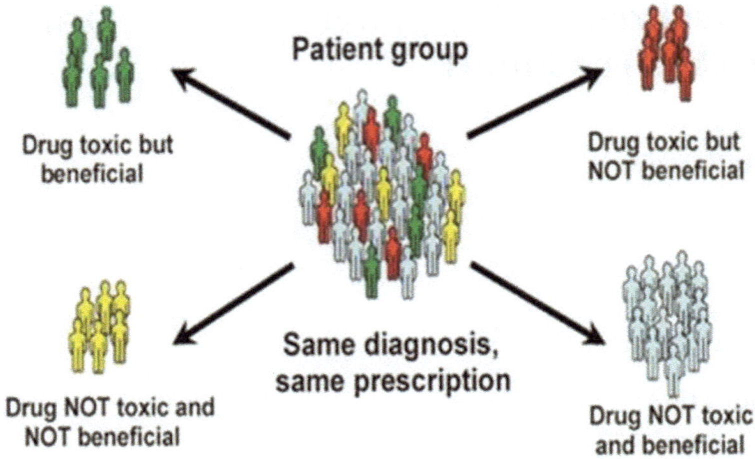

Fig. 1 Optimal personalized treatment depends on being able to successfully differentiate between those patients most likely to respond to therapy with the lowest risk of toxicity and those who are unlikely to derive any benefit from the intervention and those who are at most risk of adverse treatment-related events. (From Chakma J. JYI 2009;16)

with its progression. An obvious example might be individualizing the aggressiveness of a treatment plan for a patient with a high risk of advanced periodontal disease. In such a case the clinician might opt for extraction and implant placement rather than a more conservative and protracted treatment approach. But what if the disease necessitates a pharmacological intervention? Genomics has the potential to provide two pieces of critical information: what is the probability that an individual patient will respond to a given drug, and equally important, what is the likelihood that the same patient will have an adverse event associated with the drug.

Ultimately, the big picture objective is to prospectively identify patients at risk for a particular disease and then develop a hierarchical option list for interventions based on most to least efficacious and least to most toxic (for the patient). Minding the current reimbursement environment, the cost of treatment could also be figured in. The ideal drug is one which is consistently efficacious, has minimal toxicity and is of low cost (Fig. 1). In general, we are willing to bear a toxicity risk and financial cost if a drug is consistently efficacious. And our tolerance for these costs increases with the severity of the disease being treated. If the disease has only modest morbidity, our appetite for expensive or potentially toxic agents is limited. These facts have been consistently noted by the pharmaceutical industry and, no doubt, impact their enthusiasm for certain indications. Probably the most consistent examples of this phenomenon are cancer drugs, where even very modest efficacy is viewed positively despite extraordinary toxicity risks and very high costs. Interestingly, the application of these parameters to "quality of life" drugs has introduced the additional parameter of patient preference into the equation. No doubt, you're aware of the vast direct-to-patient marketing that routinely presents itself on TV. Acne, hair loss and erectile dysfunction are examples of what might be considered to be quality of life indications for which the major use motivation is patient-driven. The cost of prescriptions for these indications is not trivial: hair loss, $ 3.5 billion and $ 5 billion for OTC and prescription treatments for acne, and $ 5 billion for ED. And patient desires might supersede other considerations. Nonetheless, patient decisions, especially if substantive co-pays exist, could be driven by predictive knowledge of efficacy and risk.

While the conceptual application of individualizing treatment is not new among providers, it has largely been based on the clinician's judgment and trial and error. Patients and providers have titrated doses of drugs, balancing efficacy and side effects in an observational way. The reported finding that only 30–50 % of patients taking a particular medicine respond to the recommended dose as predicted is alarming [3]. And those percentages are worse for cancer drugs. Clearly the term "standard" dose is a misnomer. Pilocarpine provides a relevant example. Pilocarpine, a parasympathomimetic alkaloid stimulates salivary flow in patients with disease- or radiation-induced xerostomia by non-specifically stimulating the parasympathetic nervous supply to the salivary gland [4]. Since its mechanism of action is non-specific, other parasympathetically innervated organs and tissues are effected resulting in an adverse event portfolio that includes sweating, bradycardia, bronchospasm and diarrhea. The recommended dosing schedule is relatively broad—5 to 10 mg three or four times daily. Typically patients are started on a mid-level dose, say 5 mg three times per day and then asked to report how they do. The dose is then titrated up or down to find the therapeutic 'sweet spot', i.e. the dose and schedule with the best saliva impact and least collateral damage. Since the side effects of pilocarpine are generally mild and reversible, the cost of such an approach is mostly time and sometimes money.

But what happens when the toxicity is severe and the therapeutic index is minimal. In those cases, there is a significant risk that NOT individualizing dosing will have a dramatically negative effect on outcome.

More broadly, true implementation of genomics to patient management is drug-related, pharmacogenomics. From the clinical perspective, pharmacogenomics impacts patients at multiple levels and as drug development becomes more complex and expensive and as third-party payers grapple with any opportunity to control healthcare costs, the ability to prospectively understand how drugs impact patients attains added value. Drug development from both efficacy and regulatory standpoints is currently dependent on a 'one-size fits all' algorithm. Success of clinical trials is based on demonstrating statistical significance between a control population and the study cohort. One of the key components in study design is the consideration of inclusion and exclusion criteria. There is often an inherent conflict between the biological and genomic inclusion/exclusion qualifiers which select a specific patient population and commercial objectives for a new drug which argue that the bigger the market (the most inclusive) the better. Of course, an alternative commercial strategy is to focus an indication on diseases or patient cohorts that are so extensively defined and limited that they meet the FDA's criteria for Orphan Drug Status [5]. Not only are there financial incentives from the FDA, but pricing for such agents is typically extraordinary. For example, in 2012 Vertex received approval for a new drug which was specifically developed to treat 4% of patients with cystic fibrosis who demonstrated a unique genomic marker. For this small number of patients (about 1200), the drug is a godsend. But the annual cost of $ 294,000 is dramatic.

Pharmacogenomics is generally described at two levels: pharmacokinetics and pharmacodynamics. Genes associated with pharmacokinetics are probably the best known and studied and are related to the control the enzymes associated with drug metabolism.

The impact of genetics on pharmacokinetics was first reported years ago and refers to its impact on drug metabolism and deposition. Of about 170 genes noted to have these activities [6], variations in cytochrome p450 have probably been best described and impact a wide range of commonly used medications [7] including omeprazole, phenytoin and celecoxib. Other genes affect the metabolism of anti-cancer agents including 5-fluorouracil and methotrexate. The impact of these genes varies, but can be categorized functionally in three ways. First, if a patient is programmed with enzymes that are 'hyper-efficient' in metabolizing a drug, the agent could be chewed up so quickly as to be ineffective. Even if the drug was potentially efficacious, the standard dose would be insufficient to keep up with the rate of metabolism. Conversely, if a patient was deficient in the gene controlling the production of the metabolizing enzyme, the drug would not be processed resulting in a buildup which could result in toxic levels. This observation has been reported in cancer patients being treated with standard doses of common regimens including, for example, 5-FU who then go on to develop severe toxicities [8]. The third option is the presumably the most common—the patient is programmed to produce sufficient enzyme to metabolize the drug efficiently, but not so much as to abrogate its therapeutic effect. Of course there are likely to be many patients who fall somewhere in between such definitive categories.

Because metabolizing enzymes are specific, easily described and impact both clinical outcomes and the therapeutic index their underlying genomics have been popular and fruitful areas of study. Results of studies have been effectively leveraged into genetic tests. While these tests have been successful in providing insight into the potential benefit of pharmacogenomics in terms of individualizing treatment, they have also demonstrated just how complicated and challenging actually applying pharmacogenomics testing to clinical practice can be.

A Pharmacogenomics' Story—Pharmacokinetics

Did you ever wonder why there are so many dosing options for Coumadin? (Fig. 2)

While Coumadin is not a drug that is typically used for the treatment of oral diseases, it does provide excellent insight into the complexity and challenges of applying pharmacogenomics clinically.

Warfarin (aka Coumadin) was approved as an anticoagulant in the mid-1950s and remains the most widely used drug for that indication being taken by patients at risk

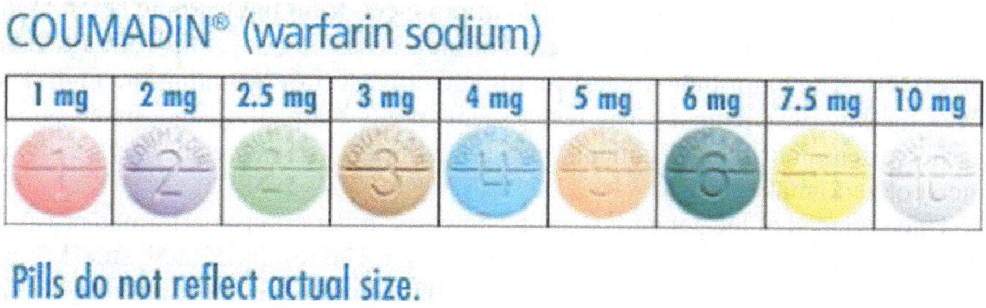

Fig. 2 Nine different doses are available for Coumadin

for pulmonary embolism, deep vein thrombosis and atrial fibrillation. While it is extremely effective, it has a narrow therapeutic index—too little drug and potential lethal clot formation might occur, too much and the patient is at risk of spontaneous hemorrhage. It is also clear that dosing requirements vary widely from patient to patient and that genotype impacts dose requirements [9]. Historically the optimum dose for an individual patient is determined by assessing a proscribed starting dose and then measuring its effectiveness using a standard clinical blood test called the international normalized ratio or INR. Patients who's INR falls below the targeted level have dose adjustments which to increase their Coumadin and those above have their Coumadin dose reduced. Given the clinical importance of establishing and maintaining therapeutic and safe INR levels, an alternative approach to dosing seemed desirable.

A series of genomic studies demonstrated that, in addition to clinical and demographic differences, two genes associated with Coumadin metabolism, cytochrome P450, family 2, subfamily C, polypeptide 9 (CYP2C9) and vitamin K epoxide reductase complex, subunit 1 (VKORC1) [10] were found to be associated with Coumadin activity. The data were so compelling that the FDA made note of a genetic association and Coumadin activity in 2007 and updated the label in 2010 noting that genetic information should be considered in determining dose. Since the early clinical studies showed a seemingly consistent and predictive association between these genotypes and Coumadin dosing, one would assume that clinical adaptation would be a no-brainer. And yet is not. While pharmacogenetic testing kits are readily available and not exceptionally expensive ($ 250–400 per test) given the potential savings in costs associated with management of Coumadin complications, insurance companies nonetheless consider testing experimental and do not pay.

Why is this the case? The data supporting the use of genetic data as a means to individualize Coumadin dosing has been inconsistent. In 2009, the International Warfarin Pharmacogenetics Consortium reported the results of a large (over 5000 patient) study in which they developed a pharmacogenetic algorithm to guide Coumadin dosing. The study design was consistent with many similar trials—it consisted of two study cohorts: one of 4043 subjects who were used to create the algorithm (I'll call them the learning set) and the other of 1009 subjects to which the algorithm was applied to test its validity (the validation set). When the validation subjects were tested against the algorithm produced from the learning set, it could identify patients who needed low doses or high doses of Coumadin with a p value of <0.001. The algorithm was especially effective at differentiating the high dose from the low dose patients. The result seems pretty compelling and, based on our earlier discussion of how patients metabolize drugs (slow metabolizers and fast metabolizers) makes sense.

Unconvinced of the genomics' utility, however, other groups performed prospective, randomized trials in larger groups of patients. The results of these studies have been inconsistent, and provide insight into the complexities of translating seemingly clear laboratory results into a clinically actionable tool.

The inconsistencies identified are illustrative of barriers to the adaptation of other pharmacogenetic testing. The issue is capsulized by a series of papers that described clinical trials assessing the predictive value of genomic testing in Coumadin dosing which were published in the December 12, 2013 issue of the New England Journal of Medicine. One study [11] concluded that:

> Genotype-guided dosing of warfarin did not improve anticoagulation control during the first 4 weeks of therapy.

In contrast, but more consistent with the Wafarin Consortium's conclusions, another study, reporting the results of a randomized trial, Pirmohamed et al [12] noted:

> Pharmacogenetic-based dosing was associated with a higher percentage of time in the therapeutic INR range than was standard dosing during the initiation of warfarin therapy.

Commenting on the papers, Zineh et al [13] remarked that there were a number of possible reasons, some procedural, some based on end points, and others to account for this lack of consensus. And the recent report by Drozda et al (Pharmacogenetics and Genomics 2014) provides some additional insight into the multifactorial nature of genomic influences on drug metabolism.

A comparison of subject demographics between the Kimmel and Pirmohamed studies reveals an interesting contrast. Whereas less than 2% of participants in the Pirmohamed study were Black, they constituted 27% of subjects in the Kimmel study. Drozda et al, noting that African Americans have worse outcomes with genotype-guided dosing, speculated that there were ethnic differences in genotype frequencies. Clearly such a conclusion has enormous repercussions, not only in clinical applications of therapy, but also in drug development, clinical trial design and guidelines for drug use. The question seems so obvious, but has been largely ignored.

To test their hypothesis, Drozda et studied 274 African Americans [9]. In addition to studying the three genetic variants associated with Coumadin dosing identified in Europeans, they also studied four alleles unique to individuals of African descent. They found that marked errors in dosing occurred when these alleles were not included, but were improved when they were added.

So while genomic applications applied to pharmacokinetics have traditional viewed as low hanging fruit and seemingly straight-forward, it is clear that the multi-factorial nature of genomics relative to pharmacokinetics requires a comprehensive approach in its translation to assessing toxicity risk.

Genomics and Pharmacodynamics

In contrast to pharmacokinetics, genomics influences pharmacodynamics by impacting a drug's biological target, i.e. its mechanism of action. For example, for patients with a number of inflammatory diseases like ulcerative colitis, one class of agents targets tumor necrosis factor-alpha (TNF-α), a known mediator of the condition. Biological agents directed against this pro-inflammatory cytokine such as infliximab (Remicade) and adalimumab (Humira) are extremely popular with total annual sales of over $ 12,000,000,000. While these therapies work extremely well for some patients, their ability to mitigate acute symptoms and/or provide disease remission is inconsistent, ranging from about 20 to 40 % [14].

Genomics is a significant driver of the variable response to anti-TNF therapy. The pharmacodynamics of TNF production is likely dependent on a series of biological events or pathways, all under potential genetic influence. The gene coding for TNF is dependent on the transcription factor nuclear factor kappaB and functional polymorphisms of NF-κB pathway have been found to predict patients' response to anti-TNF therapy [15]. Alternatively, although the impact on anti-TNF response seems modest, polymorphisms of the gene coding for TNF production also are associated with treatment effect. Largely through studies of patients with arthritis, another inflammatory disease treated with anti-TNFs, it is becoming clear that the TNF-signaling does not act in isolation, other canonical pathways are impacted and, no surprise, these too are genetically controlled [16]. Finally, it is also possible that there may be heterogeneity in the pathogenesis of a particular disease phenotype [17]. While all patients with ulcerative colitis share a final common clinical phenotype that tags them with the disease, there may be more than one way to trigger the key symptom drivers. While one pathway may be genomically associated with the TNF route noted above (and so susceptible to treatment), another may not and be coupled with another mediator and a different set of genes. There is little doubt that the biological signals trigger the development of a specific clinical sign can be diverse. However, it also seems likely that there is probably a final common pathway that leads to a specific disease phenotype. For example, diarrhea can be caused viral or bacterial infection, auto-immune disease, etc. But the signal at the end organ (the gut) is probably the consequence of one of two possibilities: hypermobility of the GI tract or necrosis or destruction of intestinal cells. From the standpoint of genomics and pharmacodynamics, this observation suggests that identification of genes at this biological control point might provide a very robust target for new therapies.

A Pharmacogenomics' Story—Pharmacodynamics

Oral mucositis is a common, devastating side effect of many forms of cancer therapy [18]. It can be induced by radiation therapy for head and neck cancers or by chemotherapy used to treat a broad range of solid or hematological malignancies. It results in devastating ulcers throughout the mouth and oropharynx that cause severe pain, limit patients' ability to eat, increase the risk of local and systemic infection and contribute dramatically to the use of medical resources resulting in higher health care costs. Indeed, the incremental cost of mucositis in patients receiving treatment for head and neck cancer is over $ 17,000 [19].

The use of genomics to study mucositis is illustrative of the translational power of the field. Genomics has contributed to the understanding of the pathogenesis of the condition, identified drugable targets, differentiated responders and non-responder characteristics to treatment and identified patients at risk for developing the mucositis.

The application of genomics to patients at risk for cancer treatment-related toxicities presents a unique opportunity. Unlike genomic studies in patients with chronic diseases which are initiated when the individual already manifests the condition, cancer patients are treatment naïve before they start therapy and so baseline gene studies can be obtained which can then be used to compare changes in gene expression as toxicities (mucositis, fatigue, nausea, vomiting, cognitive dysfunction, etc.) develop. Thus these patients provide a human model in which differential expression of specific genes can be temporally tracked to phenotype development. Furthermore, since much of the underlying biology associated with cancer treatment-related toxicities parallels chronic diseases, the information gained from this population may be transferable non-oncology populations.

Mucositis is representative of many of the side effects associated with cancer therapy and is among the most studied, both clinically and biologically [20]. Once thought to be singularly a consequence of indiscriminant cell injury to the rapidly dividing cells of the oral mucosa—it was assumed that radiation or chemotherapy couldn't distinguish between rapidly dividing cancer cells and rapidly dividing normal cells—the price of tumor kill was collateral damage of the mucosa. While close examination of cell kinetics did not support that hypothesis, it wasn't until genomic studies were applied that a more realistic understanding of mucositis pathogenesis emerged.

A look at changes in gene expression from both radiated mucosal tissue and from peripheral blood in animals and humans who developed mucositis demonstrated remarkable consistencies [21, 22]. It became clear that radiation and chemotherapy induced a biological cascade that ultimately culminated in tissue injury [23]. Initially elements of the innate immune response, the inflammasome, and oxidative stress were detectable. Subsequently, transcription factors, including NF-κB and their associated genes were activated (at least 14 canonical pathways could be identified) resulting in a burst of pro-inflammatory cytokine production, activation of the ceramide pathway and destruction of connective tissue elements. Importantly, it became clear that, like most biological processes, it was not a single gene that was driving the process, but instead networks of genes that together defined ontological pathways.

While individual genes contribute to the pathogenesis of mucositis and other diseases, they don't do so in an equivalent way. Some genes are more significant and controlling than others, but it is the 'team' of genes which is responsible for the outcomes. The ability to create a statistical hierarchy of gene contribution and then arrange the genes into networks in which the connections are defined probabilistically provides a roadmap for drug development.

How does this work? Suppose you decide to leave science or clinical practice to work for a new airline as its advertising manager. Your boss tells you that your priority is to build the business so your company can effectively compete with United—a monumental task and especially challenging since you have a limited budget. You have to decide where to advertise to achieve the biggest impact. What's the most efficient way to identify your potentially best markets? Well, you could look at a list of cities served by United, about 242. But your budget would be destroyed or you wouldn't have enough to spend at any one place to achieve a meaningful return. So what do you do? You go to the United website and pull-up their route map, a representation of the United Airlines' extensive network and, by the way, very similar in appearance to the kind of gene networks that are built from gene expression data. It becomes clear that the most efficient sites to place your advertising are the key airports which constitute the network hubs or nodes. You now go from 242 airports marketing targets to a key 9.

Well the same approach could work for drug discovery. If you can develop a gene network associated with the condition of interest, say mucositis, you can identify the biological hubs, i.e. targets [22]. Figure 3 shows part of a network that was identified from genes expressed in patients who received concomitant chemora-

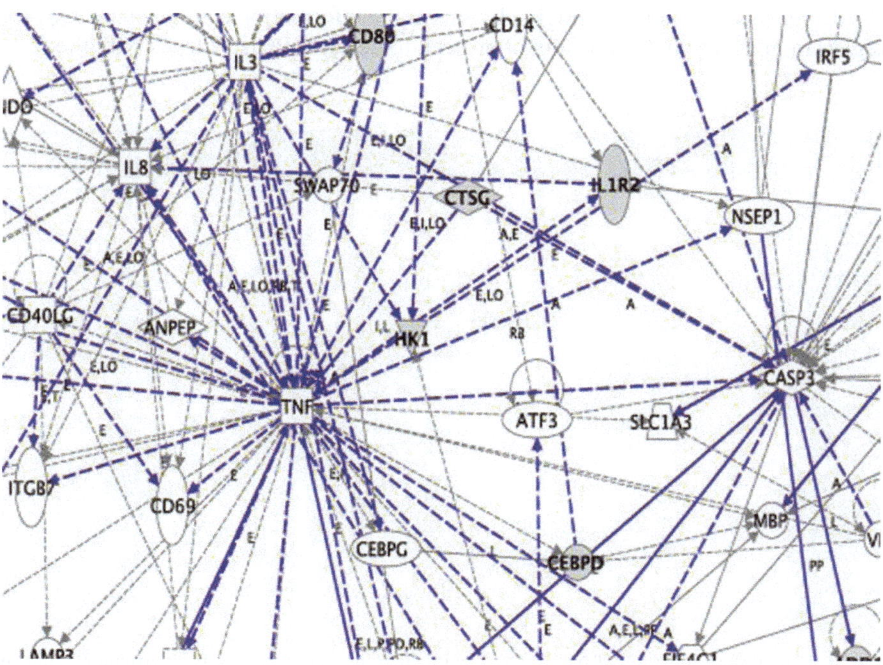

Fig. 3 Portion of gene network expressed by patients with head and neck cancer being treated with concomitant chemoradiation who then went on to develop severe oral mucositis. (From [23])

diation for head and neck cancer and then went on to develop severe mucositis in which TNF is a hub. This is informs us not only a TNF target, but also targets associated with pathways associated with TNF production. And so it is not surprising that one of the 14 canonical pathways identified by ontological analysis of genes expressed in animals and patients who develop mucositis is the NF-kappaB pathway. So if you're looking for a drug equivalent of your airline advertising strategy you now have two potential targets, NF-kappaB or an anti-TNF. As noted in the ulcerative colitis examples above, this information leads to at least three intervention approaches: shutting down TNF production, mitigating its levels or blocking its target.

A second clinical application of genomics having implications in drug development and personalized care has been its ability to differentiate and define responder and non-responder populations using prospective genomic data [2, 24–26]. The potential advantages of such a scheme to both clinical trial design and patient treatment are profound, but not totally free from some disadvantages (see Challenges' section below). From the clinical trial standpoint, using genomic criteria to enrich a study population has obvious benefits (Fig. 4). First, it means that only subjects most likely to respond to the test agent would be included, limiting exposure of the study drug and potential toxicities from individuals unlikely to benefit. Second, since the proportion of subjects likely to respond is high, the number of individuals included in a study is reduced which favorably affects the time and cost of the trial. And third, clarity of an efficacy signal should be more easily obtained. The ideal time to include responder/non-responder studies in clinical trial design is unclear. While phase 2 trials are seemingly obvious as the information could then be used for registration trials, study sponsors often seek to perform Phase 2 trials with the few-

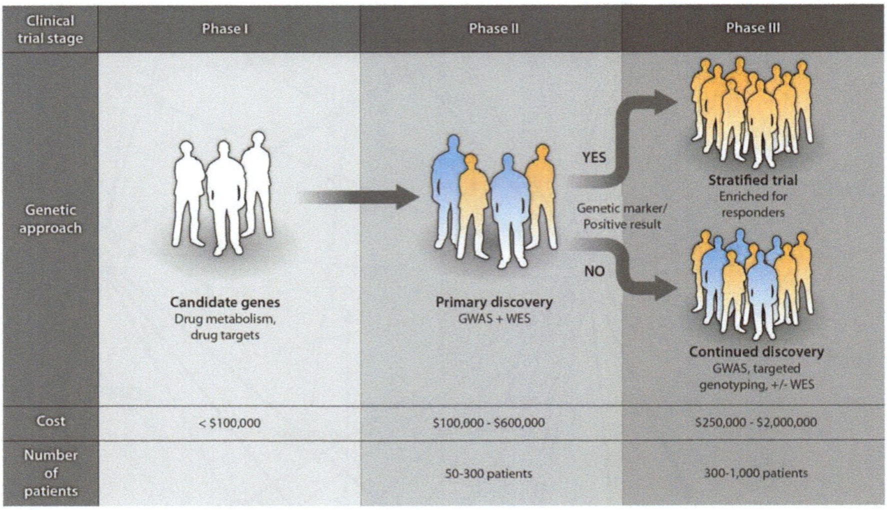

Fig. 4 Potential application of genomics to clinical trials demonstrates the particular value of including genomics in trial design, especially to stratify patients in the transition from Phase 2 to Phase 3. (From [26])

est possible number of subjects. This decision has the potential of compromising the strength of the genomic analysis (although one might argue that some information is better than no information). It could be argued that genomic studies should optimally be included in all phases of clinical trial development and sequentially and cumulatively assessed to contribute to genetically defining response/non-response and toxicity risk. Even post-approval studies would provide valuable information that could be leveraged to optimize treatment populations while at the same time minimizing complication risks.

A decision to use any drug is based on the balance between its benefits and risks (and recently its cost). This is especially true of anti-cancer agents in which a side effect profile that is often severe is tolerated because of the potential life-saving benefit of treatment. As noted above, oral mucositis is among the most common and onerous toxicities of many forms of cancer therapy. For patients undergoing hematopoietic stem cell transplants (HSCT), mucositis is often associated with the rigorous conditioning regimens that are used. Typically these patients receive aggressive combinations of chemotherapy that, in the absence of transplant rescue, would be lethal. The rationale is that the ferocity of the chemotherapy which may also be accompanied by total body irradiation, destroys residual cancer cells. For patients treated in such a way, palifermin (Kepivance) has been shown to be effective in largely attenuating the incidence and duration of severe mucositis [27]. However, to be effective palifermin has to be given prophylactically, prior to chemotherapy infusion and so days before there is any expectation of mucositis development. Further complicating its use are the facts that it must be given intravenously for three consecutive days prior to chemotherapy infusion, that it costs about $ 10,000 per course (6 doses), and that it is administered intravenously. The timing and route of administration are significant because it means that patients have to come to the hospital or the infusion center for three additional days so that the drug can be delivered. Both of these factors increase the overall cost of palifermin use and are of some inconvenience for patients. On the other hand, transplant patients who develop mucositis generally have longer hospital stays and require more resources during their stay resulting in an incremental cost of mucositis that far exceeds the cost of palifermin [28]. So you think, based on its efficacy and ultimate cost savings the use of palifermin in HSCT patients makes sense. But there's a problem. Of patients receiving the most common conditioning regimens in preparation of HSCT, only 40 % (I use the term 'only' in a guarded way) will develop severe mucositis and so would benefit from palifermin. Thus, in order for palifermin to be cost effective and clinically meaningful, understanding which patients are at risk for mucositis is critical.

Since risk of a disease or toxicity may be indicated by SNPs or genes, there are alternative approaches to adapting clinically. SNPs are sourced from DNA which provides logistical advantages of being much more stable than RNA (needed for gene-based studies) and is easily sourced from a saliva or buccal swab sample. Conveniently DNA can be stored easily at room temperature (think of DNA recovered from dinosaur bones as an example of its stability) [29]. Simply be collecting a DNA sample, differences in SNP networks can be determined for patients who develop mucositis and those who do not. This approach has been shown to be effective in a pilot study of patients undergoing conditioning regimens in preparation for autologous stem cell transplants and provides a good example of genomics' comprehensive clinical utility.

Challenges of Applying Genomics to Individualized Medicine

The history of medical science is rich with oversimplification of concepts that have hindered their clinical adaptation. Thinking that the application of genomics to solve all issues involved in individualizing medical care is naïve. Certainly there will not be a single magic bullet. The discussion that follows is not meant to burst the genomics/personalized medicine balloon. Rather, it is included to stimulate an approach that recognizes that the diagnosis and management of diseases is complex, not only for the patient and the provider, but also because it must be put in a societal context that includes economics.

It is clear that non-genetic factors can influence the risk of disease and response to drugs. For example, chemical or physical challenges can be so overwhelming that their impact supersedes genetic modification. While much has been published about genomics and cancer development, inherited mutations are only associated with 5–10 % of cancers which limits the utility of genetic testing to determine cancer risk for most malignancies. A gene-based etiology for the majority of cancers has been further cast into doubt by the recent report of Tomasetti and Vogelstein suggesting that much of cancer risk is random [30]. Likewise, factors completely independent of genetic control may impact drug efficacy and toxicities.

It is often the case that patients take more than one drug—not only is the performance of drugs independently affected by pharmacogenomic factors, but it seems likely that one drug could impact the pharmacogenomics (kinetics of dynamics) of another concomitantly administered agent and thereby potentially negating the genomic signal generated by a single drug. Studies of polypharma effects will be complex and potentially difficult to sort out.

And even where genes impact drug response and toxicity risk, other factors that impact outcomes cannot be ignored. First, as was discussed above in the Coumadin story, there is population variance in genomic regulation of both efficacy response and toxicity. To completely define risk or drug response, the study population must be large enough to assure that contribution of demographic factors can be robustly evaluated. Clearly, there is a serious risk in presuming that efficacy signals found in one population are readily applicable to another. As a consequence, this creates a challenge for drug developers in that the size and cost of studies escalates dramatically as accrual and diversity requirements are expanded. As noted above, an alternative approach is to use genetics to enrich study populations by defining those individuals with the highest likelihood of responding to a treatment and with the lowest risk of toxicity. As discussed above, such an approach might markedly increase the efficiency of clinical trials, provide results more quickly and result in drugs being most appropriately given to the 'right' patients. However, there are also potential disadvantages to this approach which are not trivial [31, 32]. The study population could be so exclusive as to mislead conclusions about the generalizability of efficacy or toxicity risk to broad populations. Additionally, the true toxicity of a drug might be clouded when the study population is not demographically or genetically inclusive. And finally, it is possible that broad genetic inclusions would be obscured.

Assuming that a single gene regulates either pharmacokinetic or pharmacodynamics processes is probably naïve. Generally, drugs are metabolized by more than one enzyme and their mechanism of action and/or the determinants of response/non-response are attributable to a group of synergistically functioning genes. Thus candidate gene approaches seem unlikely to provide a comprehensive assessment of drivers of response or toxicity risk. GWAS studies with large numbers of subjects or analytical approaches using inferred outcomes seem more likely to provide more meaningful endpoints.

Conclusions

Despite all that genomics promises, the path to its complete and comprehensive integration of into clinical medicine has not been easy. Aside from all the scientific challenges addressed throughout this book and elsewhere, the burgeoning number of stakeholders in drug discovery, regulation, delivery and payment results in groups with a wide range of potentially competing agendas [33]. The resistance of some insurance companies to pay for disease-specific genetic testing for risk determination illustrates the need to include substantive and positive cost/benefit analyses as a component of genomics' development. The need for studies which include large and diverse populations speaks to the desirability of shared databases which contain standardized demographic fields, comprehensive health and pharmacological inputs, and quality genetic outputs. The recent initiative for pharmaceutical companies to share data, and particularly data from placebo- or standard of care study arms has the potential to provide a treasure trove of information, especially if genetic information is included.

The study of genetics and genomics has come a long way since Mendel started planting peas. The potential for genomics to favorably impact health has still not been realized. But it is clear that clinical applications which include disease risk prediction, optimization of treatment, and minimization of toxicity are already having an influence how patients receive care. But the best is yet to come.

References

1. Festen EA, Weersma RK. How will insights from genetics translate to clinical practice in inflammatory bowel disease? Best Pract Res Clin Gastroenterol. 2014;28:387–97.
2. Burt T, Dhillon S. Pharmacogenomics in early-phase clinical development. Pharmacogenomics. 2013;14:1085–97.
3. Wang L, McLeod HL, Weinshilboum MD. Genomics and drug response. N Engl J Med. 2011;364:1144–53.
4. Berk L. Systemic pilocarpine for treatment of xerostomia. Expert Opin Drug Metab Toxicol. 2008;4:1333–40.
5. Hall AK, Carlson MR. The current status of orphan drug development in Europe and the US. Intractable Rare Dis Res. 2014;3:1–7.
6. Katz DA, Murray B, Bhathena A, Sahelijo L. Defining drug disposition determinants: a pharmacogenetic-pharmokinetic strategy. Nat Rev Drug Discov. 2008;7:293–305.
7. Bhathena A, Spear BB. Pharmacogenetics: improving drug and dose selection. Curr Opin Pharmacol. 2008;8:639–46.
8. McLeod HL, Sargent DJ, Marsh S, Green EM, King CR, Fuchs CS, Ramanathan RK, Williamson SK, Findlay BP, Thibodeau SN, Grothey A, Morton RF, Goldberg RM. Pharamcogenetic predictors of adverse events and response to chemotherapy in metastatic colorectal cancer: results from North American Gastrointestinal Intergroup Trial N9741. J Clin Oncol. 2010;28:3227–33.
9. Drozda K, Wong S, Patel SR, Bress AP, Nutescu EA, Kittles RA, Cavallari LH. Poor warfarin dose prediction with pharmacogenomics algorithms that exclude genotypes important to African Americans. Pharmacogenet Genomics. 2015;25:73–81.
10. International Warfarin Pharmacogenetics Consortium. Estimation of the warfarin dose with clinical and pharmacogenetic data. N Engl J Med. 2009;360:753–64.
11. Kimmel SE, French B, Kasner SE, Johnson JA, Anderson JL, Gage BF, et al. A pharmacogenetic versus a clinical algorithm for warfarin dosing. N Engl J Med. 2013;369:2283–93.
12. Pirohamed M, Burnside G, Eriksson N, Jorgensen AL, Toh CH, Nicholson T, et al. A randomized trial of genotype-guided dosing of warfarin. N Engl J Med. 2013;369:2294–303.
13. Zineh I, Pacanowski M, Woodcock J. Pharmacogenetics and coumarin dosing—recalibrating expectations. N Engl J Med. 2013;369:2273–5.
14. Zampeli E, Gizis M, Siakavellas SI, Bamias G. Predictors of response to anti-tumor necrosis factor therapy in ulcerative colitis. World J Gastrointest Pathphysiol. 2014;5:292–303.
15. Bank S, Anderson PS, Burisch J, Pedersen N, Roug S, Galsgaard J, et al. Associations between functional polymorphisms in the NFκB signaling pathway and response to anti-TNF treatment in Danish patients with inflammatory bowel disease. Pharmacogenomics. 2014;14:526–34.
16. Prajapati R, Plant D, Barton A. Genetic and genomic predictors of anti-TNF response. Pharmacogenomics. 2011;12:1571–85.
17. Emery P, Dorner T. Optimising treatment in rheumatoid arthritis: a review of potential biological markers of response. Ann Rheum Dis. 2011;70:2063–70.
18. Sonis ST. Mucositis: the impact, biology and therapeutic opportunities of oral mucositis. Oral Oncol. 2009;45:1015–20.
19. Nonzee NJ, Dandade NA, Patel U, Markossian T, Agulnik M, Argiris A, Patel JD, et al. Evaluating the supportive care costs of severe radiochemotherapy-induced mucositis and pharyngitis: results from a Northwestern University Costs of Cancer Program pilot study with head and neck and nonsmall cell lung cancer patients who received care at a county hospital, a Veterans Administration hospital, or a comprehensive cancer center. Cancer. 2008;113:1446–52.
20. Russi EG, Raber-Durlacher JE, Sonis ST. Local and systemic pathogenesis and consequences of regimen-induced inflammatory responses in patients with head and neck cancer receiving chemoradiation. Mediators Inflamm. 2014;2014:518261.

21. Al-Dasooqi N, Sonis ST, Bowen JM, Bateman E, Blijlevens N, Gibson RJ, et al. Emerging evidence on the pathobiology of mucositis. Support Care Cancer. 2013;21:3233–41.
22. Sonis ST, Scherer J, Phelan S, Lucey CA, Barron JE, O'Donnell KE, et al. The gene expression sequence of radiated mucosa in an animal mucositis model. Cell Prolif. 2002;35(Suppl 1):93–102.
23. Sonis S, Haddad R, Posner M, Watkins B, Fey E, Morgan TV, et al. Gene expression changes in peripheral blood cells provide insight into the biological mechanisms associated with regimen-related toxicities in patients treated with head and neck cancers. Oral Oncol. 2007;43:289–300.
24. Roses AD. Pharmacogenetics and drug development: the path to safer and more effective drugs. Nat Rev Genet. 2004;5:645–56.
25. Roses AD. Pharmacogenetics in drug discovery and development: a translatable perspective. Nat Rev Drug Dis. 2008;7:807–17.
26. Urban TJ, Goldstein DB. Pharmacogenetics at 50: genomic personalization comes of age. Sci Translat Med. 2014;6:1–9.
27. Spielberger R, Stiff P, Bensinger W, Gentile T, Weisdorf D, Kewalramani T, et al. Palifermin for oral mucositis after invasive therapy for hematologic cancers. N Engl J Med. 2004;351:2590–8.
28. Elting LS, Shih YC, Bensinger W, Canter SB, Cooksley C, Spielberger R, et al. Economic impact of palifermin on costs of hospitalization for autologous hematopoietic stem-cell transplant: analysis of phase 3 trial results. Biol Blood Marrow Transplant. 2007;13:806–13.
29. Sonis S, Antin J, Tedaldi M, Alterovitz G. SNP-based Bayesian networks can predict oral mucositis in autologous stem cell recipients. Oral Dis. 2013;19:721–7.
30. Tomasetti C, Vogelstein B. Cancer etiology. Variation in cancer risk among tissues can be explained by the number of stem cell divisions. Science. 2015;347:78–81.
31. Shah RR. Pharmacogenetics in drug regulation: promise, potential and pitfalls. Phil Trans R Soc Lond B Biol Sci. 2005;360:1617–38.
32. Shah RR, Shah DR. Personalized medicine: is it a pharmacogenetic mirage? Br J Clin Phamacol. 2012;74:698–721.
33. Weinshilboum RM, Wang L. Pharmacogenetics and pharmacogenomics: development, science and translation. Ann Rev Genomics Hum Genet. 2006;7:223–45.

Index

S. T. Sonis (ed.), *Genomics, Personalized Medicine and Oral Disease*,
DOI 10.1007/978-3-319-17942-1

MIX
Papier aus verantwortungsvollen Quellen
Paper from responsible sources
FSC® C105338

If you have any concerns about our products,
you can contact us on
ProductSafety@springernature.com

In case Publisher is established outside the EU,
the EU authorized representative is:
Springer Nature Customer Service Center GmbH
Europaplatz 3, 69115 Heidelberg, Germany

Printed by Libri Plureos GmbH
in Hamburg, Germany